Joe Webel
Agriculture Education Field
1401 South Maryland Drive
Urbana, IL 61801

Y0-ADZ-408

in Our Lives

An Introduction to Agricultural Science, Business, and Natural Resources

AgriScience and Technology Series
Series Editor
Jasper S. Lee, Ph.D.

INTERSTATE PUBLISHERS, INC.
Danville, IL

AGRISCIENCE IN OUR LIVES, Sixth Edition. COPYRIGHT © 1994 BY INTERSTATE PUBLISHERS, INC. All rights reserved. Prior editions, 1956, 1964, 1973, 1978, 1984. Prior editions published under the title **Agriculture in Our Lives**. Printed in the United States of America.

Cover Photo Credits:

Upper left, lower left and right courtesy of Office of Agricultural Communications, Mississippi State University

Upper right courtesy of Danville Area Community College

Order from
Interstate Publishers, Inc.
510 North Vermilion Street
P.O. Box 50
Danville, IL 61834-0050
Phone: (800) 843-4774
FAX: (217) 446-9706

Library of Congress Catalog Card No. 93-78224

ISBN 0-8134-2965-X

1 2 3
4 5 6
7 8 9

Foreword

The food and fiber industry of the United States is the envy of the world. Each farmer in the U.S. produces enough food for 120 other people. This production capability did not come about by accident, however. Agricultural production has increased because of farmers' use of sound scientific principles—developed and improved throughout years of research and practice.

Government-funded agricultural education and research led directly to the current success enjoyed by America's agricultural industry. In the late 1800's and early 1900's, political leaders with vision provided the United States with a system of agricultural education and research that laid the groundwork for later success. The Morrill Land Grant Act of 1862 provided for the development of colleges of agriculture. The Hatch Act of 1887 resulted in experiment station research to accompany the colleges of agriculture. The Morrill Land Grant Act of 1890 led to more colleges of agriculture, specifically to provide the same opportunities for minorities. The Smith-Lever Act of 1914 provided for the development of the cooperative extension service, designed to disseminate information to the masses. The Smith-Hughes Act of 1917 provided for formal education in agriculture and other vocational areas at the secondary level.

This public investment in agriculture provided a system that accounts for a large portion of the nation's economic well-being. Agricultural products account for a larger percentage of exports than any other item. On average, Americans spend only 13% of their disposable income for food, by far the lowest of any country on earth. This frees up money to invest in other sectors of the national economy and provides for a high standard of living.

In recent years, the need for training of students to work on a farm, although still very important, has been decreasing. Technological advances have allowed farm size to increase and the need for physical labor to decrease. As a result, the opportunities for employment in agriculture have shifted from production to other areas. These areas include management, communications, research, marketing, processing, and many others. The focus of many secondary agricultural education programs has changed to meet the needs of students in these areas.

AgriScience in Our Lives is designed as an introductory text in agriculture. It focuses on the principles of science, management, and production that make agriculture such an important part of American industry. It gives students an awareness of agriculture that will help them develop into intelligent, responsible citizens.

To the Teacher

AgriScience in Our Lives provides its readers with a broad, general introduction to the scientific principles which undergird agricultural production, research, marketing, and management. As such, it can be useful in a number of different settings. Some of these are discussed below.

The major use of *AgriScience in Our Lives* is as a student text in an introductory agriscience or exploratory agriculture class at the seventh, eighth, or ninth grade level. Students can gain a broad orientation and understanding of agriculture that will prepare them for further study in agriculture or just make them more agriculturally literate, no matter what field they decide to enter.

Another use of *AgriScience in Our Lives* is as a general reference for background information in advanced agriculture courses. Reading an appropriate section from this book may help students understand the importance of a particular topic they encounter in a more advanced class. For example, a student in an advanced class on plant genetics could benefit from reading Chapter 15, "Improving Farm Crops—Biotechnology." Another example might involve an upper level student who enrolls in an agriscience course for the first time as a junior or senior. A reading assignment from *AgriScience in Our Lives* can provide the student with vital background information needed to do well in the subject.

Other teachers may use *AgriScience in Our Lives* as a reference book for specific applications to their subjects. Biology, chemistry, and physics teachers can use specific examples and activities from agriculture to make their subject matter more interesting and useful to their students. The book can also provide them with the general background information they need to understand how some of their scientific principles relate to agriculture and real-life situations.

As a general text, *AgriScience in Our Lives* provides an introduction to many areas of agriculture and agriscience. A list of references is provided at the end of each chapter to allow students to study these areas in greater depth.

The photographs, charts, graphs, and other visuals are designed to highlight and clarify concepts and statistics related to agriculture. They supplement the text to assist in interpreting the information and provide a different approach for learners who are visually oriented.

Table of Contents

Foreword v

To the Teacher vi

1. The Industry of Agriculture 1
2. Rural Life in the United States 37
3. Youth and Rural Living 75
4. Agricultural Careers 101
5. Landscaping the Home Grounds 139
6. Growing Food for the Family 167
7. Conservation—The Wise Use of Natural Resources 201
8. The Mechanization of Farming 235
9. Characteristics of Farm Animals 271
10. Improving Herds and Flocks—Biotechnology 319
11. Feeding Livestock 351
12. Caring for Livestock and Their Products . . 375
13. Keeping Animals Healthy 401
14. Characteristics of Farm Crops 427

15 Improving Farm Crops—Biotechnology **461**

16 Providing Fertile Soils **489**

17 Planting and Cultivating Farm Crops **523**

18 Marketing Farm Products **551**

19 Managing the Farm Business **585**

20 Cooperatives in Agriculture **613**

21 Keeping Our Environment Clean **637**

22 Starting a Small-Scale Agricultural
   Business—Entrepreneurship **655**

23 Frontiers of Agriculture **669**

Index **689**

Chapter 1

The Industry of Agriculture

Most people think of agriculture as it was defined for many years—*the science and art of cultivating the soil, including harvesting crops and raising livestock*. This was also the definition of farming, and it served as a good description of agriculture (farming) when people obtained almost all their food and clothing directly from farmers or produced their own. Today, however, everyone must depend on a great many people and businesses for food, clothing, the materials with which to build homes, and the many other items used in daily living. Many of the activities that farmers carried out are now done off the farm by other people. And so, the definition of agriculture has changed. Agriculture is now often considered to be *all activities that involve the production, processing, distribution, and marketing of farm products*. Thus, anyone who provides services to farms is also included in this broad definition of agriculture.

The industry touches the life of every person in many ways. Every day, whether individuals are awake or asleep, at work or at play, they are surrounded by and depend upon the products of the agricultural industry.

OBJECTIVES

1. Define the industry of agriculture.
2. Describe the changes occurring in agriculture.
3. Identify the principal types of farms in the United States.
4. Discuss the importance of agriculture and rural life.

Agriculture Is the Nation's Biggest Industry

The agricultural industry is the nation's biggest industry, with assets of over $1,092 billion. It is also the nation's largest employer, with a payroll that includes almost one of every five jobs in private enterprise.

Nearly 23 million people work in some part of the agricultural industry. Farming itself uses 3.4 million workers, as many workers as there are in the transportation, steel, and automobile industries combined. In a recent year, over 2.4 million workers did some farm work for hire.

Agriculture needs the services of about 20 million workers to store, transport, process, and merchandise the output of the nation's farms. The following examples of where these workers are help to describe the industry.

> 350,000 in the meat and poultry industry, including meat-packing, prepared meats, and poultry dressing plants.
>
> 175,000 in the dairy industry, including the manufacturing of products such as fluid milk, concentrated and dried milk, cheese, butter, and ice cream.
>
> 225,000 in the baking industry, including plants for making bread, biscuits, and crackers.
>
> 240,000 in canned, cured, and frozen food plants.
>
> 175,000 in cotton mills and finishing plants.

Another 5 million people provide the seeds, fertilizers, and other supplies farmers use for production and family living.

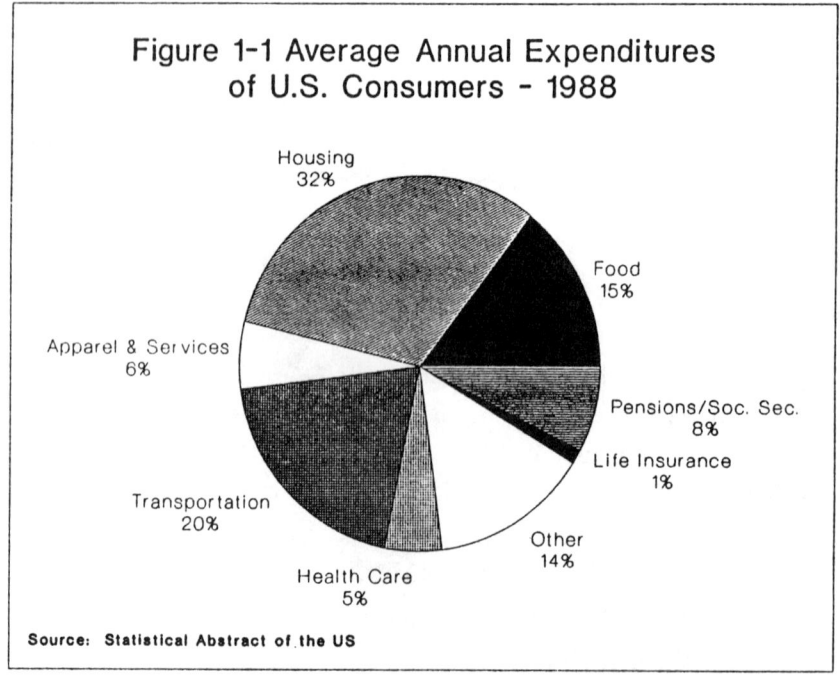

Fig. 1-1. Consumer expenditures for foods are a necessary part of total living costs. (Courtesy, U.S. Department of Agriculture)

Consumers spent over $285 billion for U.S. farm foods, about 16.2 per cent of their disposable personal income.

The foundation for the agricultural industry is the business of commercial farming. Thousands of businesses depend on the products of the farm for their operations. In addition, thousands of other businesses thrive on the purchases made by farmers for farming operations. Billions of dollars are used for farm operating expenses, such as:

$3.9 billion for seed.
$18.4 billion for feed.
$10.5 billion for livestock.
$8.8 billion for fertilizer and lime.
$10.3 billion for hired labor.
$21.7 billion for depreciation and other consumption of farm capital.
$16.3 billion for repairs and operation of capital items.
$7.3 billion for interest.
$4.0 billion for property taxes.

Cash expenses for farms in 1992 totaled over $125 billion.

Agriculture Is a Continuously Changing Industry

The agricultural industry is also a continuously changing industry. Many states have established planning and development units within their departments of agriculture. These units deal with long-range issues such as the building needs of rural areas and the effectiveness of rural area levels of government. Government regulation in agriculture also continues to increase with regard to product quality, accurate labeling, food purity, fertilizers, pesticides, feed, and seed. The future welfare of the country will become increasingly dependent on future growth and efficiency in the production, assembly, processing, and distribution of agricultural renewable resources such as food, fiber, forest, and seafood products. New occupations continue to develop as the tasks performed on the farm become sufficiently specialized to lead to the creation of off-farm agricultural businesses. The agricultural industry is truly one of the most dynamic industries, as well as being the oldest industry.

What Changes Are Taking Place in Farming and Rural Life?

Farming is the oldest occupation. The Bible includes many passages about sowers of grain, keepers of vineyards, shepherds and their flocks, and tillers of the soil. The story of farming through the ages shows many changes. The central theme of it all has been the struggle of people to secure food. Hendrik Van Loon, a famous writer, once said, "The history of man is a story of a hungry animal in search of food." What does this statement mean to you?

Until about 10,000 B.C., most people probably were mere gatherers of food. That is, they used the plants that grew around them in a natural state. Following that period, they started to engage in the organized hunting of animals, which frequently meant that they needed to be on the move and to live as nomads. Later came the stages of herding animals and domesticating them. Tilling the soil for producing food crops probably was started after the hunting stage. Thus, as people became more civilized, they lived a more settled mode of life and tried to provide a steady food supply.

Farming and farm life are still undergoing many changes. These changes have been especially great in recent years and are continuing at a rapid pace. Someone has estimated that more changes have taken place in farming in the last 75 years than in the previous 75 centuries. In order to gain some understanding of farming in the modern age, we must become familiar with the changes that are taking place. Some of these are mentioned in the portions of this chapter that follow and, in greater detail, in other chapters in this book.

Changes in farm homes. The homes on the farms of today are quite different from the log cabins, sod shanties, and adobe huts of the pioneers. Pioneer homes with their fireplaces for heating and cooking and the other crude equipment of that time are in marked contrast with modern homes and their conveniences now found on many farms.

Changes in transportation and communication. Transportation for farmers has improved greatly. The changes from horseback to oxcart to horsedrawn vehicles, and finally to automobiles and airplanes, constitute a chain of events that have aided the farmer in overcoming the isolation of the past. Rural mail delivery, the telephone, improved roads, radio, and television are also important developments in country life that have helped to eliminate the isolation that formerly existed. Two-way radio communication for use by operators of large farms and ranches as they supervise their far-flung operations is becoming more common.

The development of computer programs to provide farmers with technical information for making farm operation decisions, for analyzing farm records, for selling crops and livestock, for helping diagnose plant and animal disease and insect problems, and for conducting financial transactions will bring computers into nearly every farm operator's home.

Changes in mechanization. One of the greatest changes during the past century has been the development of improved machinery and power equipment for farming. Tractors and tractor-drawn equipment have revolutionized the methods of rasing crops. There is now, on the average, one tractor per 78 acres of cultivated land in the United States. Electricity for power and light has been made available for nearly all farms. These developments have increased the output per worker in other industries. The average farm worker in the United States produces enough food and fiber for about 115 persons. The mechanization of farming in this country is in marked contrast with the primitive methods of farming that still prevail in many foreign countries.

Changes in methods of producing crops and livestock. In addition to the mechanization of farming, science has brought many changes in the methods of producing crops and livestock. New varieties of crops, effective methods of fertilizing the soil, bettter methods of improving and feeding livestock. Of special significance have been the improved methods of controlling diseases and insects of crops and livestock. The rapid development of chemical pesticides, in particular, has made possible an abundance and certainty of food and fiber almost undreamed of before.

While we have not learned to control the weather, farmers are able to

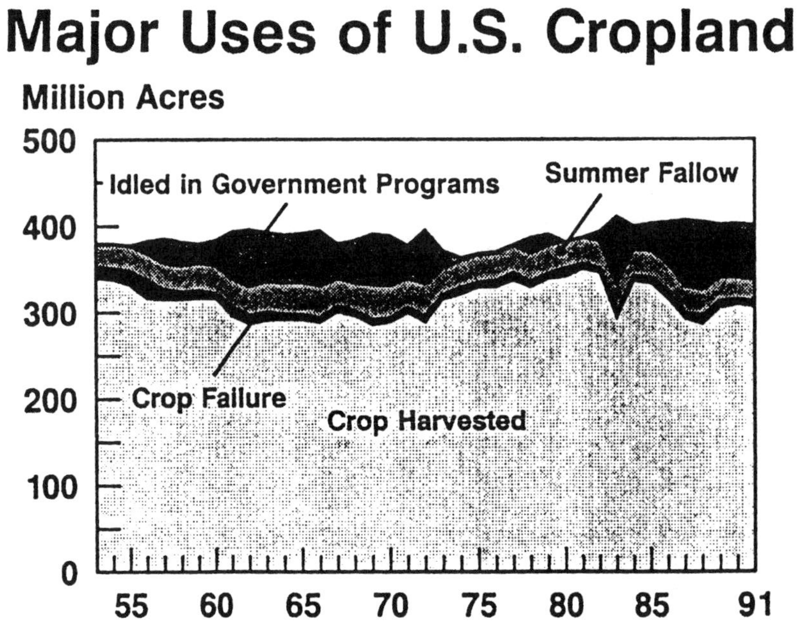

Fig. 1-2. Improvements in farming and increased acres planted to crops continue to push farm production higher. (Courtesy, U.S. Department of Agriculture)

reduce some of its harmful effects. Hardy varieties of crops, sturdy breeds of livestock, drainage and irrigation to provide favorable moisture conditions for crops, techniques in farming to reduce the loss of moisture from the soil, and equipment to lessen frost injury to crops are examples. Weather forecasts made by trained government employees enable farmers to learn about weather conditions a few hours or even a few days in advance. Cloud seeding with chemicals to cause rain is still in the experimental stages.

Changes in size and number of farms. Farms are increasing in size and decreasing in number (see Table 1-1). The average farm today is over 100 acres larger than in 1960, increasing from an average size of 297 acres in 1960 to 468 acres in 1992. More than half the farmland is now in farms of 500 acres or more. The trend toward larger farms is due, to a large degree, to mechanization, which makes it possible for a farm family to farm a large acreage. One result is that a large percentage of the farm products that go to market are raised on a relatively few farms. Of the total value of farm products sold in a recent year, about 76 per cent were raised on 15.5 per cent of the farms and the other 24 per cent on the remaining farms.

Table 1-1. Number, land in farms, and average size of U.S. farms, 1985-1992

Year	Number of farms	Land in farms	Average size of farm
		(1,000 acres)	(acres)
1985	2,292,530	1,012,078	441
1986	2,249,820	1,005,883	447
1987	2,212,960	998,923	451
1988	2,197,140	994,543	458
1989	2,170,520	991,158	457
1990	2,140,420	987,420	461
1991	2,105,060	982,766	467
1992	2,095,740	980,063	468

Source: U.S. Department of Agriculture.

While there is an increasing number of large farms, it is interesting to note that there is also a fairly stable number of small farms and small layouts too small to be classed as farms. This trend has come about because of improved transportation that makes it possible for persons living in the country to hold jobs in nearby towns or cities. Decreased working hours in industry and the help of the family make it possible to carry on these operations.

Changes in farm population. Due largely to increased mechanization and other improved methods of farming, the percentage of people engaged in farming in the United States is decreasing. For example, in 1991, only about 1.9 per cent of the population lived on farms. A century earlier, about

The Industry of Agriculture 7

85 percent lived on farms. The number of people on farms continues to decrease, though the total population is increasing rapidly. In 1991, about 4.6 million people lived on farms, as compared to about 14.8 million in 1961 and 30 million in 1940. The total population of the United States exceeded 187 million in 1961 and reached over 250 million in 1991.

The shift from farm to city has made it possible to release many people to engage in occupations that provide goods and services for all the people of the United States. Thus, a high standard of living has been brought about

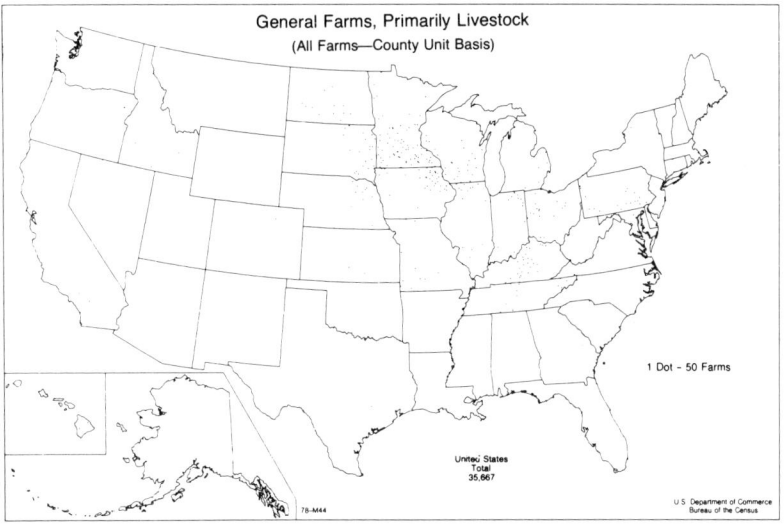

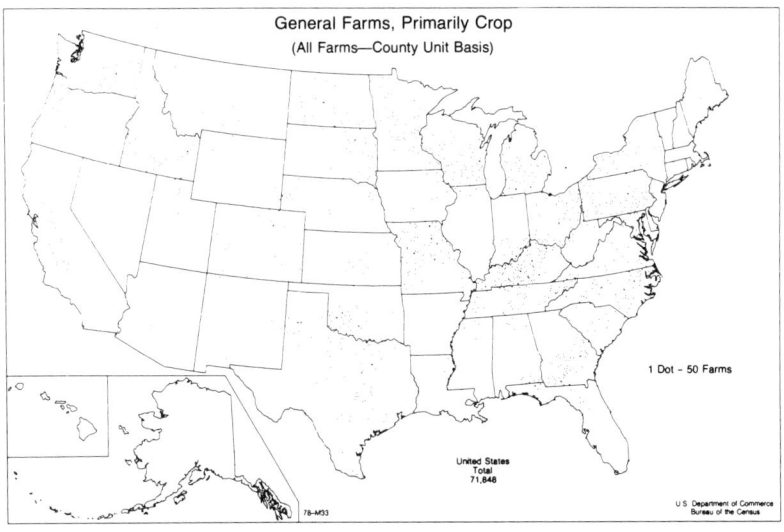

Fig. 1-3. The number of general farms is decreasing with increased specialization. (Courtesy, Bureau of the Census, U.S. Department of Commerce)

in this country. In marked contrast, in some of the countries of the world more than four-fifths of the people are still engaged in farming, and the standards of living are extremely low.

Changes in uses of farm products. The development of new uses for agricultural products is another way in which science is aiding farmers and other people. The science of finding new industrial uses for agricultural products is called *chemurgy*. Through it, many uses have been found for corn, peanuts, soybeans, potatoes, and other farm products raised in abundance. Most materials that formerly went to waste in the slaughter of animals at the central packing plants are now converted into useful products of many kinds.

Changes in relationships on local, national, and world levels. Rural people and others in the United States are realizing more than ever before that they have important relations with all groups in the United States and with the world at large. Modern transportation and communication have eliminated the isolation of small groups within this country and the isolation of this nation from the rest of the world. Farmers must buy large amounts of products and equipment; therefore there is a growing interdependence among agricultural, labor, industrial, and other groups that comprise the public.

Peace and security for all peace-loving people are dependent on our keeping our nation strong and all groups within it prosperous. Our responsibilities to other nations include helping them to improve their living conditions by showing them how to farm better so they can provide themselves with more food and more of the other products that come from the soil. Economic relations among nations should be improved, so that it will be possible for the United States to trade increased amounts of agricultural and industrial products for needed products from other nations. As never before, the world of today requires the application of the Golden Rule among persons, groups, and nations.

Changes in economic conditions. Over the years, rural people have been affected by changes in prices and in the purchasing power of the dollar. During the serious depression of the 1930's, farm prices fell more than the prices of items the farmers bought, with the result that farmers were in a difficult situation. Urban people got food and clothing at lowered prices, but farmers were not able to purchase many of the products of industry. The prices of the products sold by farmers tend to change more rapidly than the prices of the items they buy. Consequently, the purchasing power of farmers fluctuates widely over a period of years.

Even with the rather wide fluctuations in price levels, the farmer's share of each dollar spent for food by the consumer has seldom been more than 50 cents. In 1992, the farmer's share of the food dollar was about 30 cents. This

The Industry of Agriculture

means that 70 cents of every dollar the consumer spends for food products, on the average, goes to pay the costs of transportation, processing, other marketing services, and profits after the products leave the hands of the farmer.

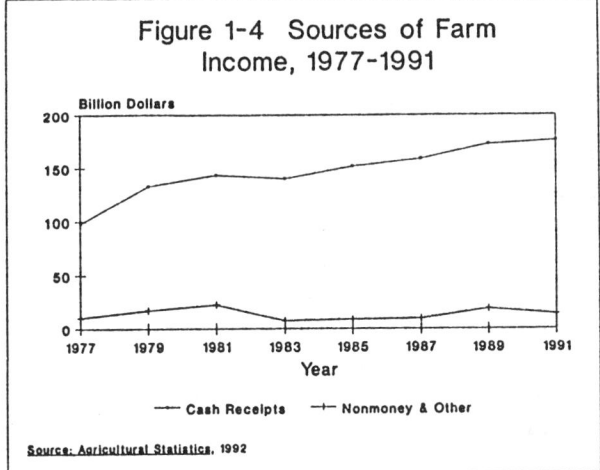

Fig. 1-4. Cash receipts from marketing and non money income are the main sources of farm income. Net farm income has not kept pace with rising costs. (Courtesy, U.S. Department of Agriculture)

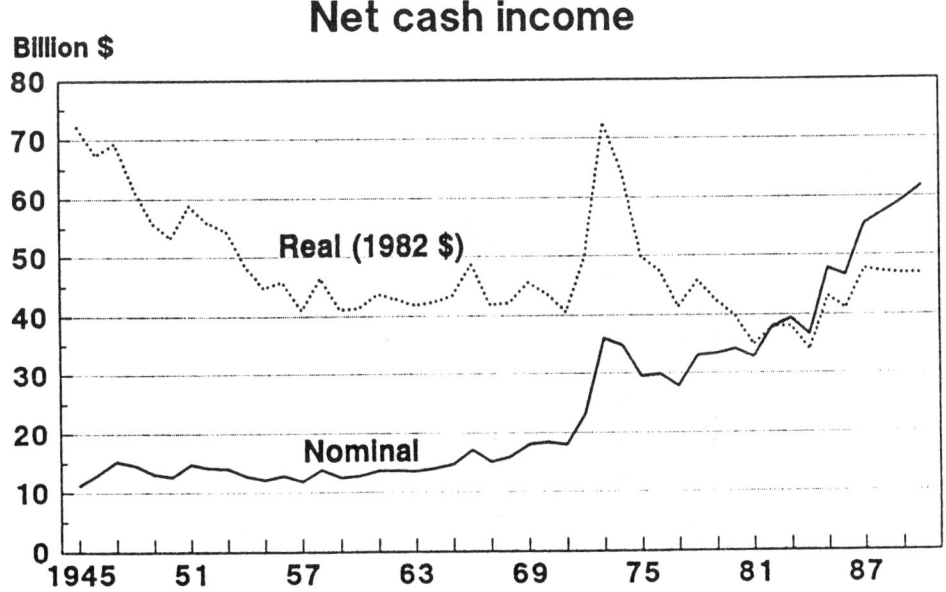

Fig. 1-5. More than one-half of the total farm family income comes from off-farm sources. (Courtesy, U.S. Department of Agriculture)

Changes in cropland. U.S. farmers have abandoned millions of acres of cropland to other uses since World War II. However, they have brought into production 26.7 million acres of cropland. The actual acres of cropland harvested decreased from 256 million acres in 1978 to 234 million in 1992.

The new cropland is concentrated geographically in very definable areas. In southern Florida, irrigation and drainage combined to create new vegetable acreage and orchards. Farther west, clearing and drainage benefited the Mississippi Delta area, creating new land, mostly for soybeans and cotton.

Texas has two crop areas that developed as a result of irrigation. One area is at the southern tip, near the Gulf of Mexico, and the other is in the Texas High Plains. Irrigation also opened acreage in Arizona, California, Colorado, Idaho, Kansas, and Washington.

Dry land farming improvements led to many new winter wheat acres in northern Montana. Drainage, clearing, contouring, and leveling were used to expand cropland acres in the Corn Belt.

The main stretch of abandoned acreage runs through much of the eastern United States from Mississippi on the south to Maine on the north. Another large area of retired acres stretches from northeastern Texas, through Oklahoma, into western Arkansas and Missouri.

Much of the land was taken out of production due to urbanization, low

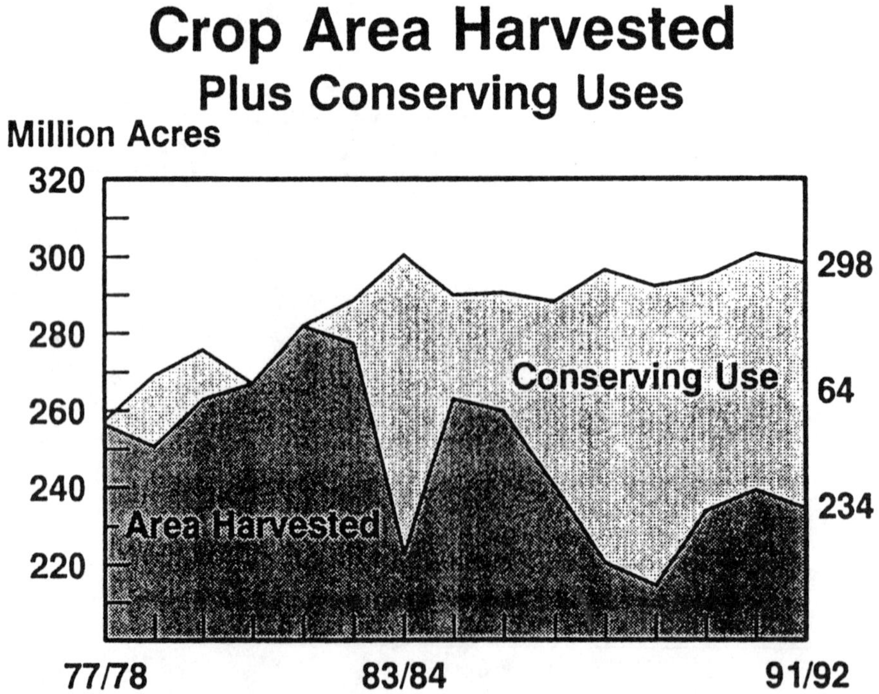

Crop Area Harvested
Plus Conserving Uses

Fig. 1-6. Land area harvested has decreased since 1978. (Courtesy, U S. Department of Agriculture)

fertility, or poor adaptability to modern machinery. Rising labor costs were also a factor.

About a third of the crop acreage in the United States is planted to the four feed grains—corn, grain sorghum, oats, and barley.

Changes in migratory labor force. Although the plight of the migrant farm worker is one of the most discussed problems of agriculture, the number of migrants is only a small segment of the farm labor force. The number of migrant farm workers decreased from 466,000 in 1965 to 217,000 in 1979. Recently, however, the number has remained at about 200,000.

Regional changes in the use of hired farm labor during the past 15 years reflect the continuing mechanization of agriculture leading to the need for long-term hired labor.

Changes in census of agriculture.[1] The very first census of agriculture was taken in 1840. It dealt mainly with an inventory of the chief classes of livestock, the production of wool, the value of poultry, the value of dairy products, and the production of the main crops. The first five-year census of agriculture was conducted in 1925.

In 1840, the United States had 6 million milk cows; 24 million pigs; and 19 million sheep. The first wheat census was taken in 1839. Production was 85 million bushels. The first census of the number of farms was in 1850, when there were 1.5 million. The peak year for the number of farms was 1935, when 6.8 million were recorded.

In 1945, the first detailed breakdown of farms by sales classes was reported. In 1964, computers were used extensively for the first time to record and analyze census data. The 1969 census was the first in which data were collected on veterinarian operations, crop sprayers, fruit packers, cotton ginners, corn shellers, hatcheries, and horticulture consultants.

The 1974 census was the first to be conducted by mail rather than by census takers going from door to door. As more information is obtained, the census can be expected to tell us more about changes in our agricultural history.

What Is the Nature of Farming in the United States?

Farming is a big industry in itself. The 2.37 million farms in 1987 operated 980 million acres of land. The states with the largest acreage of farm land were:

    Texas 138,500,000
    Montana 62,100,000

---

[1]*Farm Index*, U.S. Department of Agriculture, December, 1976.

Kansas 48,500,000
Nebraska 47,700,000
New Mexico 47,400,000
South Dakota 44,700,000
North Dakota 41,700,000

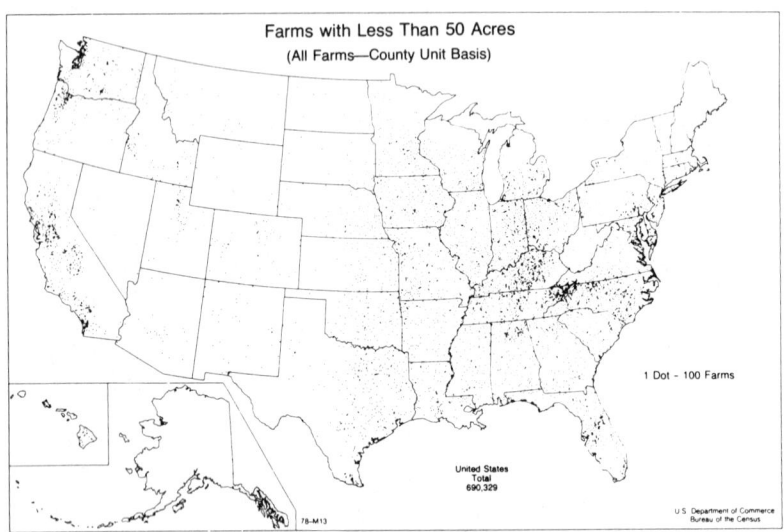

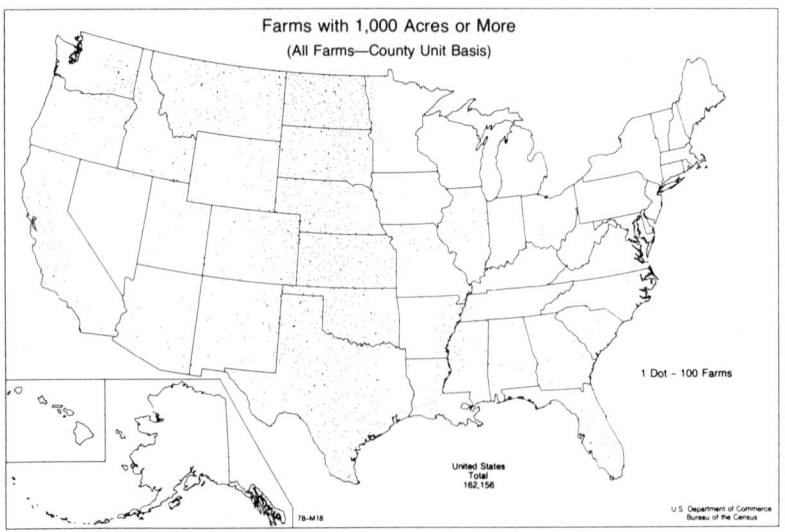

Fig. 1-7. Farms vary widely in size of operation. (Courtesy, Bureau of the Census, U.S. Department of Commerce)

The Industry of Agriculture

In a recent year, farm size varied from an average of about 5,558 acres in one western state to about 83 acres in one South Atlantic state.

The 50 states have a total area of about 2.3 billion acres. Of the total, farms take up just over 50 per cent; forests take 25 per cent; parks and wastelands, such as the western desert, take 16.6 per cent; and cities, towns, roads, and airports take up the rest.

A total of 750 million acres, or about one-third of the country, is owned by the federal government.

Farming is a business

If the average person were asked to name some forms of business, he/she would probably mention grocery stores, clothing stores, hardware stores, manufacturing firms, banks, and many others before farming.

Is farming a business? The average farmer has an investment of thousands of dollars. Balance sheet assets have risen rapidly over the years (see Table 1-2). A farmer must make decisions, develop plans, and solve problems; all of these require "head-work." A typical farm has high operating costs for interest, taxes, repair and replacement of equipment, fertilizers, wages, fuel, electricity, and many other items.

Table 1-2. Balance sheet assets on U.S. farms

Year	Current$	Deflated$
	(Billions)	(1982)
1987	772.5	658.0
1988	805.1	663.7
1989	819.7	649.0
1990	834.6	634.7
1991	845.0	617.2
1992	850-860	600-610

Source: U.S. Department of Agriculture (1992 estimated).

The cost of producing corn has risen dramatically, and the amount continues to increase. In one state, it is over $400 per acre. In another state, approximate costs for the various items needed to produce an acre of corn are estimated to be as follows:

Seed	$ 8.00 to	$ 12.00
Hours per acre (3.0 – 4.5)	—	—
Machinery costs	50.00 to	55.00
Insecticides/herbicides	8.00 to	12.00
Fertilizers	30.00 to	40.00
Building storage, etc.	8.00 to	12.00
Land—taxes and interest	50.00 to	110.00
Total	$154.00 to	$241.00

Farmers need to obtain yields of 100 bushels or more per acre in order to earn a reasonable return on their labor costs.

The private enterprise system is the backbone of the U.S. way of life. Farmers operate the greatest number of private enterprises in the country. They have much in common with businesspersons and others in towns and cities. Of course, there are many farms that are not operated in a businesslike manner, but the same may be said of many other kinds of business establishments.

Farming is a complicated occupation

In order to be successful, farmers must have a wide knowledge of many things. They must have some of the background of a scientist, mechanic, salesperson, accountant, financier, and manager. They must be able to do many tasks, including numerous kinds of tasks connected with raising livestock and crops, operating complicated equipment, constructing and repairing buildings and equipment, controlling weeds, draining land, improving the soil, and controlling insects. They must keep informed about new scientific developments and be able to use them in their own situations. They must work and plan with their families and other people in order to improve farming and farm living conditions. They must know how to carry on the business aspects of farming.

Even the farmers who specialize in one type of livestock or one kind of crop have to perform nearly all of the tasks listed and many more. The operation and management of a farm is truly one of the most complex and demanding occupations.

Farming is a way of life

A farm is a place to live as well as a place to make a living. Thus, in farming, there is a close relationship between home life and the occupation that provides the income. The entire family has day-to-day contacts with farming, and most members of the family take an active part in operating the farm.

Farming varies widely in the United States

There are many kinds of farming in the United States, depending on conditions such as the nature of the soil, the topography, the climate, the rainfall, and the markets. The principal types of farms are described briefly in the following paragraphs.

Cash grain farms obtain 50 per cent of their total sales income from wheat, rice, corn, sorghums, soybeans, other small grains, cowpeas, dry field and seed beans and peas, popcorn, lentils, and mustard and safflower seeds. In a recent year, cash grain farms comprised 24 per cent of all farms and averaged 439 acres in size and $41,419 in value of products sold. The greatest concentration of cash grain farms is in the Corn Belt, where corn and soybeans are the principal crops sold.

1. *Wheat farms* are cash grain farms that obtain 50 per cent or more of their total sales income from wheat. They account for 12.7 per cent of all cash grain farms and average 985 acres in size and $33,555 in value of farm products sold. The largest concentrations of wheat farms are in North Dakota, South Dakota, Nebraska, Kansas, Oklahoma, Montana, Washington, and Oregon.
2. *Rice farms* are cash grain farms that receive 50 per cent or more of their sales income from rice. They are located mostly in Arkansas, Mississippi, Louisiana, and California and on the Gulf Coast of Texas.
3. *Corn farms*, which obtain 50 per cent or more of their total sales from corn, comprise 26.4 per cent of all cash grain farms. These farms are concentrated in the East North Central states.
4. *Soybean farms* receive 50 per cent or more of their sales income from soybeans, account for 32.8 per cent of all cash grain farms and 7.9 per cent of all farms. They are located primarily in the North Central states, east of the Mississippi River, and along the Mississippi River in Missouri, Arkansas, Tennessee, Mississippi, and Louisiana.
5. *Cotton farms* account for about 1.3 per cent of all farms. They average 719 acres in size and $94,672 in value of farm products sold, and they are located almost entirely in the South and in irrigated areas of the Southwest.
6. *Tobacco farms* comprise 5.8 per cent of all farms. They are the smallest in terms of size of farm (99 acres) and in terms of value of products sold ($16,679). Nearly all tobacco farms are in six southern states: North Carolina, South Carolina, Virginia, Kentucky, Tennessee, and Maryland.
7. *Sugar crop farms* get 50 per cent or more of their total sales income

from sugar beets or sugarcane. These farms are located in the irrigated areas of the western states (sugar beets) and in Florida, Louisiana, and Hawaii (sugarcane).

8. *White potato farms* receive 50 per cent or more of their sales from white potatoes. Most of these farms are located in Maine, Idaho, and North Dakota.
9. *Vegetable and melon farms* obtain at least 50 per cent of their sales income from vegetables and melons. The farms are widely scattered, with concentrations in California, Michigan, Wisconsin, New Jersey, New York, Florida, and Texas.
10. *Fruit and tree nut farms* receive at least 50 per cent of their sales income from berry crops, grapes, tree nuts, citrus fruits, and deciduous tree fruits (apples, peaches, pears, etc.).
11. *Horticultural specialty farms* tend to be located close to large urban areas. They obtain at least 50 per cent of their sales income from ornamental plants, nursery products, and food crops grown under cover.
12. *General farms, primarily crop,* obtain at least 50 per cent of their sales income from crops, but less than 50 per cent from any one crop group.
13. *Beef cattle feedlots* receive 50 per cent or more of their sales income from beef cattle fattened for a period of at least 30 days. Beef cattle fed on a fee or contract basis are included.
14. *Beef cattle farms, except feedlots,* derive 50 per cent or more of their sales income from nonfattened beef cattle.
15. *Hog farms* obtain 50 per cent or more of their sales income from hogs and pigs. These farms are concentrated in the Corn Belt states.
16. *Sheep and goat farms* receive at least 50 per cent of their sales income from sheep, goats, goats' milk, wool, and mohair. Sheep and goat farms are located mostly in Texas, in California, and in other western states.
17. *Dairy farms* receive 50 per cent or more of their sales income from milk and cream. These farms average 294 acres in size and $74,492 in value of products sold. Dairy farms are concentrated in the North Central and Middle Atlantic states. Wisconsin, Minnesota, New York, Pennsylvania, Ohio, and Michigan account for 62 per cent of all dairy farms. Because of its large dairy farms, California ranks second in value of dairy products sold.
18. *Poultry and egg farms* receive 50 per cent or more of their sales income from chickens, chicken eggs, broilers, turkeys, turkey eggs, hatched chicks, and other poultry products. The greatest concentrations of poultry and egg farms are in the broiler-producing areas of Georgia, Alabama, Mississippi, Arkansas, Delaware, and Maryland.
19. *Animal specialty farms* derive 50 per cent or more of their total sales

The Industry of Agriculture

income from fur-bearing animals, such as rabbits, horses, mules, donkeys, and ponies, as well as from honey, fish, worms, and other animal specialties.

20. *General farms, primarily livestock,* obtain at least 50 per cent of their sales income from livestock and livestock products, but less than 50 per cent from any single livestock or livestock product group. Animal specialties are included.

The following is a list of selected farm enterprises and the leading states in the production of each (the top-producing state is listed first).

Barley: Idaho, North Dakota, California, Montana, Minnesota, Washington, Colorado, South Dakota

Beans, dry, edible: Michigan, California, Idaho, North Dakota, Nebraska, Colorado, Washington

Corn, grain: Iowa, Illinois, Minnesota, Nebraska, Indiana, Ohio, Wisconsin, Michigan

Cotton: Texas, California, Arizona, Mississippi, Louisiana, Arkansas, Alabama, Oklahoma

Hay: Wisconsin, Iowa, California, Minnesota, Nebraska, New York, South Dakota, Texas

Honey: California, Florida, North Dakota, Minnesota, Montana, South Dakota, Texas, Missouri

Maple syrup: Vermont, New York, Wisconsin, Ohio, Michigan, Pennsylvania, New Hampshire, Massachusetts, Maine

Mint oil: Washington, Oregon, Idaho, Indiana, Wisconsin, Michigan

Oats: Minnesota, South Dakota, Iowa, Wisconsin, Ohio, Pennsylvania, New York, Nebraska

Peanuts: Georgia, Texas, North Carolina, Alabama, Florida, Oklahoma, Virginia, New Mexico

Popcorn: Nebraska, Indiana, Iowa, Ohio, Illinois, Kentucky, Michigan, Missouri

Rice: Arkansas, California, Texas, Louisiana, Mississippi, Missouri

Sorghum: Texas, Kansas, Nebraska, Missouri, Oklahoma, Colorado, California, South Dakota

Soybeans: Iowa, Illinois, Indiana, Minnesota, Missouri, Ohio, Louisiana, Arkansas

Sugar beets: California, Minnesota, Idaho, North Dakota, Michigan, Nebraska, Colorado, Wyoming

Sugarcane: Florida, Louisiana, Hawaii, Texas

Sunflowers: North Dakota, Minnesota, South Dakota, Texas

Tobacco: North Carolina, Kentucky, South Carolina, Tennessee, Georgia, Virginia, Wisconsin, Pennsylvania

Wheat: Kansas, Oklahoma, North Dakota, Washington, Texas, Montana, Nebraska, Colorado

Vegetables (all): California, Florida, Texas, New York, Oregon, Minnesota, Washington, Michigan, Arizona, Ohio, New Jersey, Colorado

Asparagus: California, Washington, Michigan, Illinois, New Jersey

Beans, green lima: California, Delaware, Wisconsin, Maryland

Beans, snap: Wisconsin, Oregon, New York, Michigan, Illinois, New Jersey, Florida, Tennessee

Beets: Wisconsin, New York, Oregon

Broccoli: California, Texas, Oregon, Arizona

Cabbage: New York, Wisconsin, Ohio, Indiana, Colorado

Carrots: California, Texas, Washington, Michigan, Wisconsin, Oregon, Minnesota, New York

Potatoes, white: Idaho, Washington, Maine, Oregon, California, Wisconsin, North Dakota, Colorado, Minnesota, New York

Potatoes, sweet: North Carolina, Louisiana, California, Texas, Georgia, Alabama, Mississippi

Meat, dairy products, poultry

Broilers: Arkansas, Georgia, Alabama, North Carolina, Mississippi, Maryland, Texas, Delaware, California, Virginia

Cattle and calves: Texas, Iowa, Nebraska, Kansas, Missouri, Oklahoma, California, Wisconsin

Chickens: California, Georgia, Arkansas, Indiana, Pennsylvania, Alabama, North Carolina, Texas, Florida

Eggs: California, Georgia, Pennsylvania, Arkansas, Indiana, North Carolina, Texas, Florida

Hogs and pigs: Iowa, Illinois, Minnesota, Indiana, Missouri, Nebraska, North Carolina, Georgia, Ohio

Milk: Wisconsin, California, New York, Minnesota, Pennsylvania, Michigan, Ohio, Iowa

Sheep and lambs: Texas, California, Wyoming, South Dakota, Utah, New Mexico, Montana, Colorado

Wool: Texas, California, Wyoming, Colorado, South Dakota, Utah, Montana, New Mexico

Turkeys: Minnesota, North Carolina, California, Arkansas, Missouri, Virginia, Texas, Iowa, Indiana

Fruits and nuts

Apples: Washington, New York, Michigan, Pennsylvania, California, Virginia, Oregon

Apricots: California, Washington, Utah

Blueberries: Michigan, New Jersey, Maine, North Carolina, Washington, Oregon

Cantaloupes: California, Texas, Arizona, Indiana, Michigan, Georgia, Colorado, South Carolina

Cherries, sweet: Washington, California, Oregon, Michigan, New York, Utah, Idaho

Cherries, tart: Michigan, New York, Utah, Wisconsin

Cranberries: Massachusetts, Wisconsin, New Jersey, Washington, Oregon

Grapefruit: Florida, Texas, California, Arizona

Grapes: California, New York, Washington, Pennsylvania, Michigan, Arizona, Ohio, Arkansas

Lemons: California, Arizona

Limes, tangelos, temples: Florida

Oranges: Florida, California, Texas, Arizona

Papayas: Hawaii

Peaches: California, South Carolina, Georgia, New Jersey, Pennsylvania, North Carolina, Michigan, Virginia

Pears: California, Washington, Oregon, New York, Michigan, Colorado, Pennsylvania, Utah, Connecticut

Pecans: Georgia, Alabama, New Mexico, Louisiana, South Carolina, Florida, Mississippi, Oklahoma

Persimmons: California

Pomegranates: California

Prunes and plums: California, Oregon, Washington, Michigan, Idaho

Strawberries: California, Florida, Oregon, Michigan, Washington, New York, Ohio, Pennsylvania

Tangerines: Florida, California, Arizona

Tomatoes: Florida, California, South Carolina, New Jersey, Texas, Pennsylvania, New York, Alabama

Watermelons: Florida, Texas, Georgia, South Carolina, Indiana, Mississippi, North Carolina, Alabama

The animal enterprises with the greatest value of production for the United States are cattle and calves, dairy cattle, hogs, and poultry. The leading crops in value of production in the United States are corn for grain, soybeans, wheat, and cotton. Among the fruits, the leaders in product value are grapes, oranges, and apples; almonds are on top in the nut category.

The leading states in value of cash receipts for all commodities are California, Iowa, Texas, Illinois, Minnesota, Nebraska, Kansas, Wisconsin, Indiana, and North Carolina.

Farming Regions of the United States

Because of climatic and other geographic conditions, it is possible to divide the country into 10 agricultural regions. Together, these 10 regions provide a vivid picture of the nation's agriculture.

The *Pacific Region* consists of the three mainland states on the Pacific Ocean (California, Oregon, and Washington) plus Alaska and Hawaii. In the northern part of the region, wheat, fruits, and potatoes are most important; in the southern part, vegetables, fruits, and cotton are most important. Cattle are raised throughout the entire region. Sugarcane and pineapples are the major crops in Hawaii.

The *Mountain States* from Idaho and Montana on the north to New Mexico and Arizona on the south comprise another quite distinctive region. Vast areas of this region are best suited to raising cattle and sheep. Irrigation provides the valleys with water for hay; sugar beets, potatoes and other vegetables; and fruits. Wheat is grown in the northern part.

The *Northern Plains* and the *Southern Plains* extend from Canada on the north to Mexico on the south and from the Mountain States on the west to the Corn Belt on the east. The western part of these regions is hampered by low rainfall and the northern part by cold winters and short growing seasons. About three-fifths of all winter and spring wheat is grown in these regions. Other small grains, grain sorghum, hay, forage crops, and pastures are grown for cattle. Cotton is produced in the southern part.

The *Corn Belt*, consisting of parts of nine states—Ohio, Indiana, Illinois, Iowa, Missouri, Kansas, Nebraska, South Dakota, and Minnesota, has rich soil, a good climate, and sufficient rain for excellent farming conditions. Corn, beef cattle, hogs, and dairy products are the major farm outputs. Also of great importance are other feed grains, soybeans, and wheat.

The *Delta States* of Mississippi, Louisiana, and Arkansas produce soybeans, cotton, rice, and sugarcane. Livestock production is gaining in importance with the gradual improvement in pastures of the region. This is also a major broiler-producing region.

The *Southeast Region*, which includes Alabama, Florida, Georgia, and South Carolina, is important for cattle, broilers, fruits, vegetables, and peanuts. Florida is a major producer of citrus fruits and winter vegetables.

The *Appalachian Region* includes Virginia, West Virginia, North Carolina, Kentucky, and Tennessee. This is the major tobacco-producing area of the country. Peanuts, cattle, and dairy products are also important.

The *Northeastern States* from Maine to Maryland, and the *Lake States* of Michigan, Wisconsin, and Minnesota, make up the principal milk-producing areas. The climate and soil in these states are suited to raising feed grains and forage for cattle and for providing pastures for grazing. Broiler production is important in Maine, Delaware, and Maryland; fruits and vegetables are also important in these regions.

Farms vary widely in size of business

The farms and ranches throughout the United States vary widely in the size of their business operations, even within the same region or community, just as do other kinds of businesses.

Since 1975, the definition of a farm has been any place from which $1,000 or more of agricultural products are sold or normally would be sold during the year the census is taken. Previously, a place with less than 10 acres was called a farm if $250 or more of agricultural products were sold or normally would have been sold during the census year. A place with 10 acres or more was classified as a farm if $50 or more in agricultural products were sold or normally would have been sold during the census year.

About 90 per cent of all farm products going to market are produced on farms with gross sales of $20,000 or more per year. Doing most of the buying and selling, it is this large group of farms that has created the agricultural industry.

The number of large farms, with sales of $100,000 or more in a year, is rapidly increasing. From 1970 to 1992, these farms increased in number to 324,840 and now make up 15.5 per cent of the total number of farms. Large farms produce most of the cotton, potatoes, orchard crops, eggs, broilers, and turkeys each year. in 1992, they took in 76 per cent of all cash receipts in farming.

Farms with annual sales of between $20,000 and $40,000 are decreasing in number. In today's economic climate, these farms seem to be too large for part-time farming and too small for full-time farming. The actual investment required for a full-time commercial farm is now estimated to be more than $500,000. Thus, it is becoming more difficult for young people to get started.

Farm families are becoming less dependent on farming for their total income. By 1980, nonfarm sources provided 62 per cent of the average farm

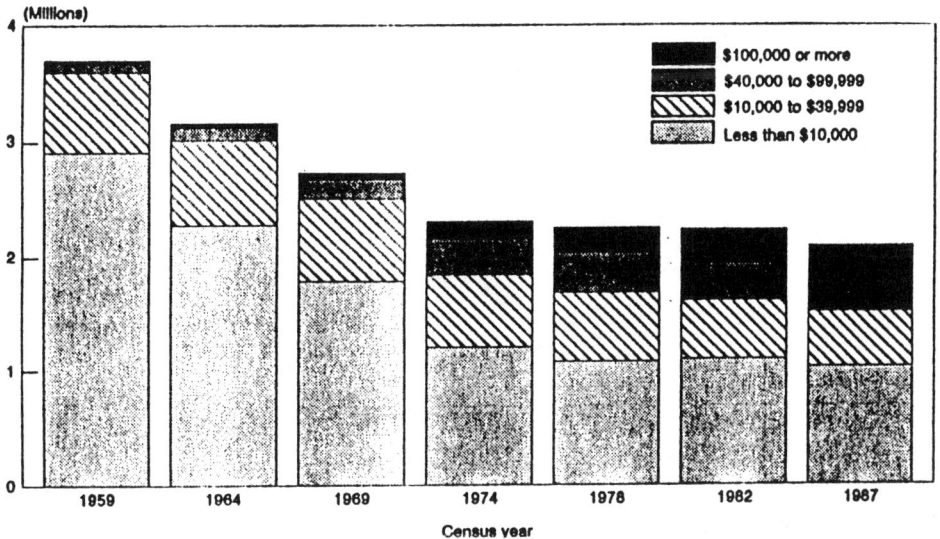

Fig. 1-8. In 1992, farms with sales of over $100,000 made up 15.5 per cent of the number of farms and had 76 per cent of the cash receipts. (Courtesy, U.S. Department of Agriculture)

family's income. The percentage was higher for families on smaller farms and lower for those on larger farms.

What Is Necessary for Success in Farming?

In 1869, a farm woman in Wisconsin wrote to her husband, who was then traveling in a covered wagon in Iowa and looking for a farm on which to settle. In her letter, the woman included the following: "I hope that wherever you may conclude to pitch our tent it may be healthful and have a fair prospect at least of possessing good social and religious advantages. This is the most I care for."

All individuals, whether or not they live on farms, have their own definition of what success is. "What do I really want most in life?" is a prevailing question. How would you answer this question?

In part, (1) securing a satisfactory income, (2) developing good homes and communities, and (3) carrying out responsibilities as citizens are factors influencing successful living on a farm or elsewhere. Thus, because everyone is influenced by the success and welfare of farmers, the ways that these three factors may be applied to farming are considered in the following paragraphs.

Securing a satisfactory income in farming

The prosperity of farmers is closely related to the prosperity of the entire nation. For farmers to be prosperous, the entire nation must be prosperous. If payrolls of industry increase, farm incomes usually increase. If a recession or a depression occurs, farm incomes fall. In this situation, farmers buy less machinery, fewer automobiles, and smaller amounts of other products.

Farmers can do a great deal to help themselves by farming efficiently. In any year, there is a wide spread between the profits of the most efficient farmers and the least efficient farmers who are farming under similar conditions. The most efficient farmers usually give attention to many of the following factors:

1. *Controlling soil erosion and improving soil fertility.* Farmers use good cropping programs on tillable land, grow legumes for feed and for soil improvements, use practices that check soil losses, return manure and plant residues to the soil, and add commercial fertilizers, if they are found to be profitable.
2. *Growing crops high in value for feed or sale.* Certain crops, such as alfalfa and corn, produce a large amount of feed per acre. Some crops, such as soybeans, wheat, potatoes, sugar beets, tobacco, and cotton, may be grown for sale as cash crops.

The Industry of Agriculture

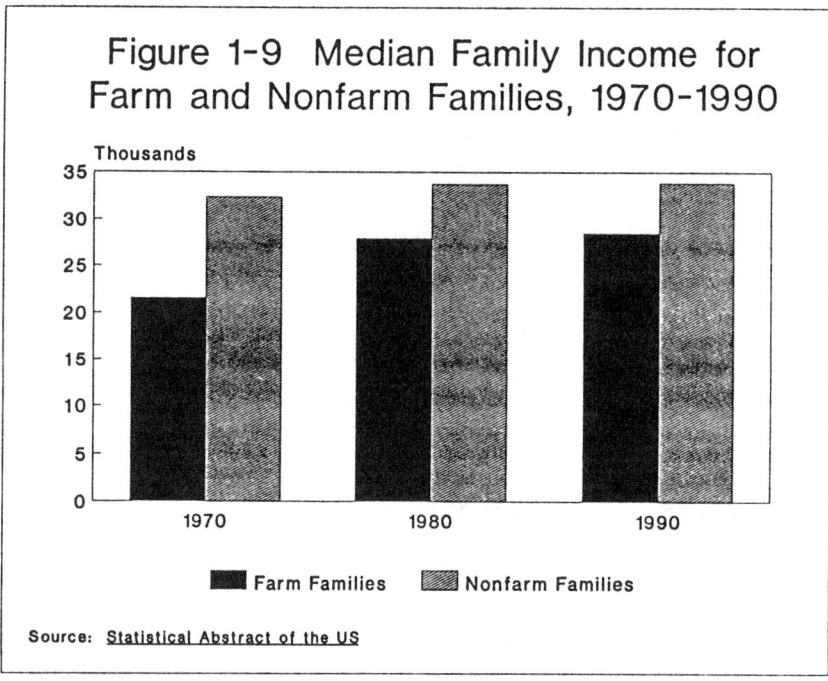

Fig. 1-9. Disposable income for farm families continues to be less than that for nonfarm families. (Courtesy, U.S. Department of Agriculture)

3. *Obtaining high yields of crops.* This is done through improved soil fertility, good seed, effective cultural methods, and better control of plant enemies, such as harmful insects, diseases, and weeds.
4. *Raising suitable kinds and numbers of livestock.* Livestock of kinds suitable for the farm, the farm family, and the market are raised for profit on many farms and ranches.
5. *Securing high production from the livestock raised.* High production is obtained by using good-quality animals, maintaining animal health, feeding balanced rations, and providing proper care.
6. *Obtaining the best possible prices for the products sold.* This is done by producing good-quality products for sale at times of the year when prices are likely to be highest for the products raised. In some cases, farmers secure better returns by marketing their products cooperatively.
7. *Using labor efficiently.* This is done by using labor-saving equipment and planning the work so that farm operators and their employees are kept busy at productive tasks.
8. *Spending money wisely.* The expenditures for buildings, equipment, labor, and services are made carefully so as to provide good returns. They buy what is needed for the comfort and well-being of their families, but they watch their chances to purchase products of good qual-

ity at favorable prices. Many farmers save money by buying some goods and services cooperatively.
9. *Producing considerable food for their families.* Many thrifty farm families are able to produce a large portion of the food needed for the family table. With careful processing and storage, this supply may be utilized throughout the year.
10. *Developing a business large enough to provide a satisfactory living for their families.* This may be accomplished by increasing the size of their farms, or farming intensively, or both.

Developing good homes and communities

While it is important to give attention to the production side of farming and to the various other aspects of the farm business that affect income, the improvement of rural living should also be considered. In the final analysis, most of us want to make money so we can live satisfactorily and have some of the comforts of life. By making money, we are also able to contribute to the welfare of others by paying taxes, contributing to charities and social agencies, and helping our young people get started in life.

Good homes, churches, and schools should be a top priority. Some communities need to develop better opportunities for recreation and social activities. Youth need to have guidance in selecting their life work. Hospitals and medical facilities need to be developed. Library services need to be provided. Soil, other natural resources, and the natural beauty of the countryside need to be considered.

Carrying out responsibilities as citizens

All of us have obligations as citizens in a democratic society. In meeting these responsibilities, we need to give attention to ways of improving the welfare of all groups in our nation. Farm people have an important and essential part in building and maintaining a strong nation, and they can secure much satisfaction from the contributions they make.

U.S. farm people are citizens of a great nation, and increasingly they are taking part in national affairs. Farmers have representation in state and national legislative bodies in greater proportions than their numbers in the total population. Hence, they have an unusual responsibility for influencing the welfare of state and nation.

Surplus farm products of many kinds produced in the United States are needed in other parts of the world, and farm people and others should help bring about relationships that will make these products available through world trade. The United States is contributing to the welfare of some of the developing nations by helping them to use improved methods of farming. Many of the methods of farming found profitable in this country are being adopted elsewhere in the world. Some U.S. farm people and other farming experts are teaching rural people in other nations to develop better farming methods for their conditions. For example, hybrid corn is widely used in this country and higher yields have resulted from its use. Some years ago, an agency of the United Nations provided $40,000 to finance the development of hybrid corn for European countries. Six years after it was provided, the corn yields had increased so much that $24 million was added to the value of the corn crop in one year.

Why Should Individuals Become Better Acquainted with Agriculture and Rural Life?

Becoming better aquainted with agriculture and rural life will help us to better understand the importance of agriculture in our lives and to learn to use this knowledge in many ways.

The following are some reasons why the study of agriculture is important to everyone.

1. *We can become better citizens of our democratic society if we understand agriculture and rural life and some of the problems of rural people.* To be intelligent citizens, we must be familiar with both the agricultural and urban elements of our national life. Farming is the nation's *largest* and *most important* industry, employing nearly 2,848,000 persons. In addition there are nearly 20,000,000 persons employed in other agricultural occupations. The story of the history of our country is, to a large extent, the story of the development of the agricultural industry, especially farming.
2. *We can become better world citizens if we understand the important part that agricultural products play in maintaining the peace and in improving the living conditions of humankind.* More than half of the people in the world do not get enough to eat and many of these are living on a starvation diet. The population of the world is increasing more rapidly than the supply of food. Much of the unrest in the world today stems from many people being poorly fed, poorly clothed, and poorly housed. For our national security and for humanitarian reasons, we must utilize the resources of agriculture for the improvement of all people. No other country has so much surplus food available for export.
3. *We all have a "stake" in farming because we eat and wear agricultural products.* Everyone expects to get three square meals a day and to have ample clothing. Each person eats nearly three-quarters of a ton of food each year, but yet only 10 per cent of our food is imported. However, many people know very little about how food

Fig. 1-10. Agriculture provides food, clothing, shelter, and other necessities for daily living.

and fiber are produced and what happens to them from the time they leave the farms until the products are bought at stores. Studying agriculture will help us understand the price structure in agriculture so that we can become intelligent consumers of agricultural products.

4. *The study of plants and animals raised and produced on farms is an interesting and desirable part of our education.* Whether a person lives in a city or in the country, part of a complete education is becoming familiar with farm plants and animals from which come so much of our food and clothing.

5. *We can enjoy our contacts with the rural environment more fully if we become familiar with farming and rural life.* Some of us live on farms, some of us visit farms, and many of us see the country as we ride on highways and railroads or travel by plane. Through a study of agriculture, we are better able to enjoy these contacts with the rural portions of our nation.

6. *We can develop interesting hobbies and spare-time activities of an agricultural nature.* A knowledge of their possibilities and how to carry them out will help us to select agricultural activities and to succeed with them. For example, some of us may learn how to raise flowers, vegetables, or animals for pleasure and even for profit.

7. *We can be more useful to our families if we learn about agriculture.* Some of us can help to produce a part of the family food supply by raising vegetables, fruits, and chickens or other livestock. We can help become better buyers of foods if we know how to judge their

The Industry of Agriculture

Fig. 1-11. A man, a horse, and a rope are still needed for working with beef cattle. (Courtesy, *Agricultural Research*, U.S. Department of Agriculture)

quality. We can help develop good lawns and plantings around our homes.

8. *We can choose our occupations more wisely* if we study a wide range of possibilities, including occupations connected with agriculture. There are many agricultural occupations, in addition to farming, that we should consider in choosing our life work.

9. *We can increase our appreciation and understanding of certain books, poems, songs, and art works* if we become familiar with agriculture and rural life. Many notable writings and pieces of art are related to rural life and agriculture.

10. *We can better understand the importance of wisely using the soil and other agricultural resources.* All of us are dependent on these resources. Soil is our most important natural resource and its conservation is important to urban people as well as to rural people. Soil is the source of food and clothing and lumber. If we fail to conserve our soil, water, and forests, we will all be affected.

Everyone Can Learn About Agriculture

Many people have had interesting and valuable experiences on farms. These contacts with farming may have been gained by living on farms or by visiting the farms of relatives or friends. Some people may not have been this fortunate. Whether or not we have lived on farms or in other ways have had experiences in farming, there is much of interest that we may learn about agriculture and how it affects our lives.

Because many people know little about farming, opportunities are

Fig. 1-12. It's lunchtime for this lamb at an exhibit in the Chicago Museum of Science and Industry.

Fig. 1-13. Baby chicks hatch every day in full view of many fascinated people. This exhibit is in the Chicago Museum of Science and Industry.

The Industry of Agriculture

being provided to help them learn about it. As part of the Bronx Zoo, which is located in New York City, a "Farm in the Zoo" is operated to show what farm animals look like and how they behave. This portion of the zoo includes dairy cows, chickens, sheep, goats, hogs, and horses. Each year, hundreds of thousands of people watch the cows being milked, pet the animals, and observe the caretakers as they provide feed and daily care.

Museums, such as the Museum of Science and Industry in Chicago, maintain exhibits of modern farming. Each year, millions of people view these exhibits.

The *National Herb Garden*, established at the National Arboretum in Washington, D.C., is really three gardens in one, with each garden providing an opportunity for people to see the plants that are grown for herbs and to learn how the herbs are used. The National Herb Garden should be of special interest to those individuals who want to grow their own herbs.

Fig. 1-14. A lecture tour for the public in the National Herb Garden in Washington, D.C. (Courtesy, *Agricultural Research*, U.S. Department of Agriculture)

Agriculture Can Be Studied in School

The elementary schools in the Los Angeles Public School System have special trucks equipped as Mobile Farm Units. There are four major units: a "Mobile Dairy Unit"; a "Small Stock and Poultry Unit"; an "Agriculture Science and Conservation Unit"; and a "California Wildlife Unit." These units are taken from school to school where they can be viewed and studied by the students. In addition, thousands of city youth in the public schools are enrolled in various kinds of courses in agriculture in which they gain experiences in doing gardening, producing fruit, and raising livestock. Some of these young people are preparing to engage in various types of farming, but most of them are learning about agriculture as a part of their general education, regardless of the occupation they may select for their life work.

Discovering some of the realities of farm life firsthand is a regular part of the curriculum for third graders in Crowley, Louisiana.[2] The program is sponsored by the National Future Farmers of America Foundation as part of its "Food for America" project. Its aim is to show children where their food comes from. The organizers in Crowley are FFA chapter members at Crowley Senior High School. The demonstration farm is divided into two parts. Two acres, behind the high school, are used for a vegetable garden and greenhouse. Nearby, there are 10 acres for sheep, beef and dairy cows, geese, turkeys, and rabbits, and there is a pond for fish and ducks. Mrs. Grace Brown, the teacher of the children, said, "Most people assume that children come in contact with farm animals and food crops. But more than half don't. It is impossible to overestimate the value of their direct, sensory experience. A child who pokes through soil to find the roots of a vegetable, confronts a cow face-to-face, smells a barnyard, and sits at the edge of a pond to watch ducks is not likely to forget the experience."

On the 4,700 acres of the Smokey House Project in Vermont, local high school students are developing job skills in forestry, agriculture, carpentry, and auto mechanics. The Soil Conservation Service has assisted with conservation planning for the project.

Smokey House youth help to manage about 2,500 acres of forestland, a sheep farm, a solar greenhouse, and a garden where they grow vegetables to sell. Three dairy farms are managed through a lease arrangement. While they experiment with different careers, they receive the minimum wage for their work and earn academic credit. The Smokey House Project, for students who have had difficulty in regular schools, teaches youth specific job skills along with management of resources and respect for the environment.

Many other schools throughout the United States are providing oppor-

---

[2]From article, "First Impressions of a Farm," by Gene Warren, Public Information Officer, Soil Conservation Service, U.S. Department of Agriculture, published in *Soil Conservation*, February, 1977.

Fig. 1-15. Agriculture courses are a basic part of the Los Angeles Public School programs. (*Top*) Lath house practice is part of the gardening experiences of every student (John Muir Junior High School, K. Holland, teacher). (*Bottom*) Bob Randolph *(left)* and the nursery project at Van Nuys High School. The net profit of $3,000 (over a three-year period) was used for college. (Courtesy, Los Angeles Public Schools)

tunities for young people to become familiar with agriculture and rural life. Courses in general or nonvocational agriculture are provided by some of these schools. Through these courses, young people are becoming acquainted with agriculture. In about 9,000 high schools, vocational agriculture is provided for students who are interested in farming and farm-related occupations.

What Can the Study of Agriculture Do for Me?

Your purposes for studying agriculture may differ in part from those of other persons in your group, but many of these purposes are probably similar. Consider carefully the question of what a study of agriculture will do for you. Some of the reasons for studying agriculture are listed earlier in this chapter. These, and other information in this chapter, may suggest some of the reasons which you feel are important to you. What are some of your reasons or purposes for studying agriculture?

Fig. 1-16. Johnny Webb, FFA member; Mrs. Grace Brown, teacher, Crowley, Louisiana; and her third grade class are pulling radishes from the FFA farm garden. (Courtesy, Soil Conservation Service, U.S. Department of Agriculture)

The Industry of Agriculture

Fig. 1-17. Tommy Faulk, FFA member, is showing Mrs. Grace Brown's third grade class the FFA greenhouse in Crowley, Louisiana. (Courtesy, Soil Conservation Service, U.S. Department of Agriculture)

SUGGESTED ACTIVITIES

1. Describe to your class your most interesting experiences on a farm or your experiences with plants and animals elsewhere. In what ways were these experiences of value to you? Describe the most unusual farm experience you have had.

2. Describe the principal kinds of farming in some region of the United States in which you have travelled. What factors seem to account for the kinds of farming found there?

3. Write a description of a pioneer home and one of a modern farm home. Compare the conveniences and comforts. What changes are most recent?

4. Interview a farmer who has farmed for many years. Ask the farmer to describe some of the changes that have taken place during his/her lifetime.

5. With others in your class, make a trip to the farm of a successful farmer. Ask what the farmer does to make his/her business successful. Note evidences of a fertile soil, good crops, healthy livestock, and good living conditions.

6. Visit some farm that has been in the same family for several generations. What are some of the prized possessions that have been handed down from generation to generation? How was the land first obtained by this family? What price did they pay? What improvements have been made? If possible, find other interesting incidents and then make a report to your class.

7. Start a library of your own of books and bulletins on various phases of agriculture and rural life in which you are especially interested. (Many bulletins can be obtained from the agricultural college in your state and from the U.S. Department of Agriculture. Bulletins from the latter source can best be procured through a representative or a senator.)
8. As you read the daily newspaper, note the kinds of news items and other information related to agriculture and farming. What items are of interest to both rural and urban people? To farmers only?
9. Start a notebook in which you write interesting ideas and a scrapbook in which you place pictures about farming and rural life.
10. Interview the owner-manager or some of the workers in a non-farm agricultural business. Find out how the business developed and how it serves people.
11. Concepts for discussion:
    a. The agricultural industry is big business and includes much more than farming.
    b. Agriculture is a continuously changing industry.
    c. About half the land area of the states is taken up by farms.
    d. There are many kinds of farming in the United States.
    e. Farms vary in size.
    f. Farming requires a large investment.
    g. We are all dependent on agriculture for food, clothing, some parts of our shelter, and other needs in daily living.
    h. The food we eat may have been produced in the corner of the nation farthest from our home.
    i. Millions of jobs are provided in the agricultural industry.
    j. Agriculture can be studied in the public schools.
    k. Workers for agriculture must be highly trained.
    l. A knowledge of agriculture will help us to live better and to be more informed citizens.

REFERENCES

Lee, J. S., and D. L. Turner. *Introduction to World AgriScience and Technology.* Interstate Publishers, Inc., Danville, IL 61832.

National Plant Food Institute. *Our Land and Its Care.* National Plant Food Institute, 1700 K Street, N.W., Washington, DC 20006.

U.S. Department of Agriculture. *Agricultural Situation.* U.S. Government Printing Office, Washington, DC 20402.

U.S. Department of Agriculture. *Americans in Agriculture: Portraits of Diversity.* Yearbook of Agriculture: 1990, U.S. Government Printing Office, Washington, DC 20402.

U.S. Department of Agriculture. *Our American Land.* Yearbook of Agriculture: 1987, U.S. Government Printing Office, Washington, DC 20402

U.S. Department of Agriculture. *Food—From Farm to Table.* Yearbook of Agriculture: 1982, U.S. Government Printing Office, Washington, DC 20402.

U.S. Department of Agriculture. *1983 Fact Book of U.S. Agriculture.* Misc. Pub. No. 1063, U.S. Government Printing Office, Washington, DC 20402.

U.S. Department of Agriculture. *Agriculture Outlook '92: New Opportunities for Agriculture.* Washington, DC 20402.

U.S. Department of Agriculture. *Agricultural Statistics, 1992*. U.S. Government Printing Office, Washington, DC 20402.

Chapter 2

Rural Life in the United States

The great strength of the U.S. agricultural industry is in those people who work in it. Many of the people in agriculture have chosen to live and to work on the land; other people also have chosen to live in the country. About 4.8 million persons, or about 1.9 percent of the nation's total population, lived on farms in 1989. The proportion of the population living in rural areas ranged from 13 percent in the Pacific states to 44 percent in the East South Central states, and this proportion is continuing to increase.

Country living has had a romance to it that belongs to no other way of life. Country life is thought of as being free from noise, crowds, and smog. Farm people are looked on as the last truly independent segment of the population. Rural families are viewed as closely knit groups, working and playing together a great deal.

Although conditions are gradually changing, much that has been valued and cherished of country life still exists. The countryside, still scenic and picturesque, provides the "playground" for much of the leisure time of the entire population. Many people still are looking forward to "life in the country" on their own "small piece of land."

OBJECTIVES

1. Identify those who are included in the rural population.
2. Discuss how rural homes and family life can be improved.
3. Discuss basic food groups and proper nutrition.
4. Identify organizations that serve rural people.
5. Discuss how better rural communities can be developed.

Who Lives in Rural USA ?

Before talking about rural life in the United States, we need to consider

Farm Population

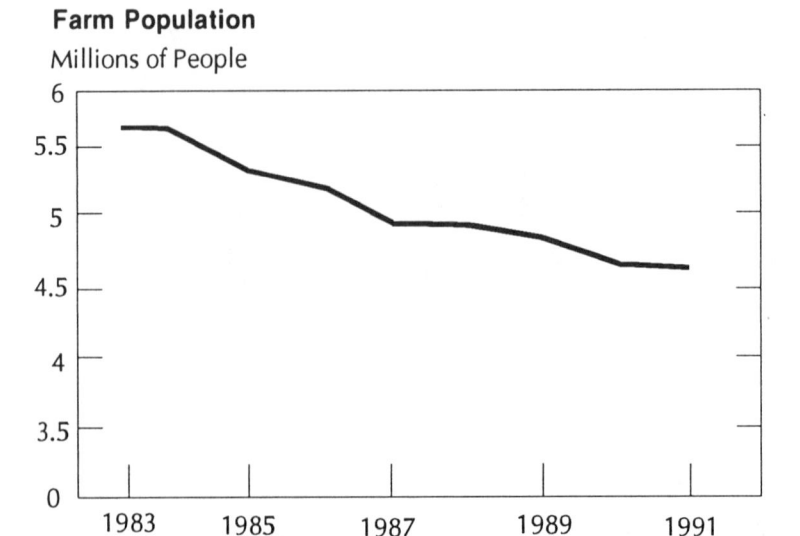

Millions of People

Fig. 2-1. Only about 1.8 per cent (new definition) of the nation's population, or 4.6 million people, lived on farms in 1991. (Courtesy, U. S. Department of Agriculture, *Agricultural Statistics*)

Urban and Rural Population

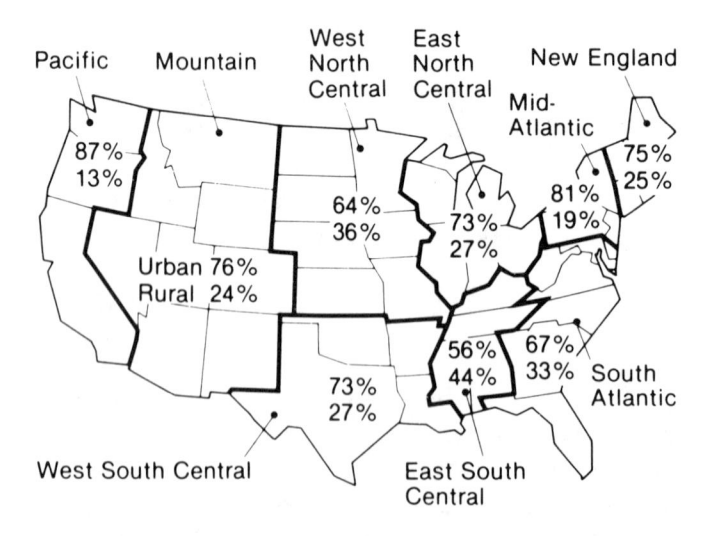

Pacific 87% / 13%
Mountain — Urban 76% / Rural 24%
West North Central 64% / 36%
East North Central 73% / 27%
Mid-Atlantic 81% / 19%
New England 75% / 25%
West South Central 73% / 27%
East South Central 56% / 44%
South Atlantic 67% / 33%

1980 data. Pacific includes Alaska and Hawaii.

Fig. 2-2. Most people in the United States live in urban communities. (Courtesy, Bureau of the Census, U.S. Department of Commerce)

Rural Life in the United States

who is included in the rural population. As classified by the U.S. census, the people who live in rural areas include those who live on farms, those who live in the country but do not live on farms, and those who live in villages and towns of fewer than 2,500 people. The people who live in densely settled areas around large cities are classed as urban dwellers, together with all persons who live in towns and cities of 2,500 people or more.

The number of people who actually live on farms is slowly decreasing. However, the rural nonfarm population is increasing at a rate fast enough to more than make up for the decrease in the farm population. The flow of people into rural areas is about equally divided between those from central cities and those from suburban areas. While part of the growth in rural population is due to factors such as the decrease in the movement of farmers and their families off the farm, the improvement of highways, and the development of retirement settlements, much of the growth appears to be because people in rural areas are more satisfied with where they live than are their town and city "cousins."

Contrary to the popular belief, many farm residents don't depend on agriculture for a living. About 2.4 million persons did some hired farm work during 1981. About 4 out of 10 of these farm workers did some nonfarm work during the year, the largest numbers being in the wholesale and retail trade (13 per cent) and in manufacturing (19 per cent). Nearly 74 per cent of these workers were white, while blacks and others accounted for 13 per cent and Hispanics accounted for 13 per cent. Over one-third of all hired farm worker families had incomes below $10,000, while about 18 per cent had incomes of $25,000 or more.

In 1980, over half of the farm women did the bookkeeping for their farms and raised food for their families. Almost half of the farm women ran

Net nonmetro migration to and from metro areas, 1988-89

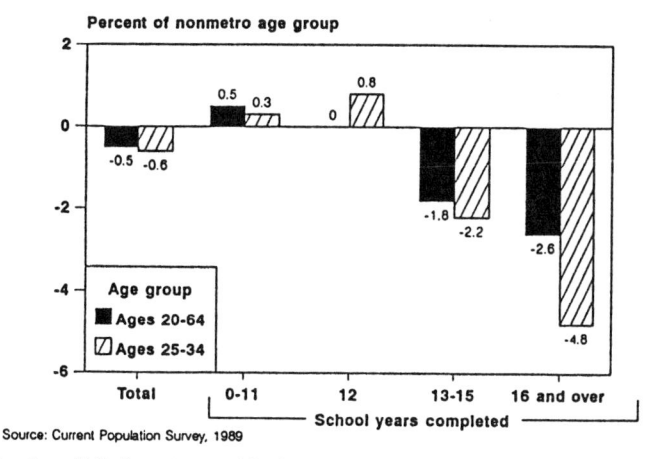

Fig. 2-3. (Courtesy, U.S. Department of Agriculture)

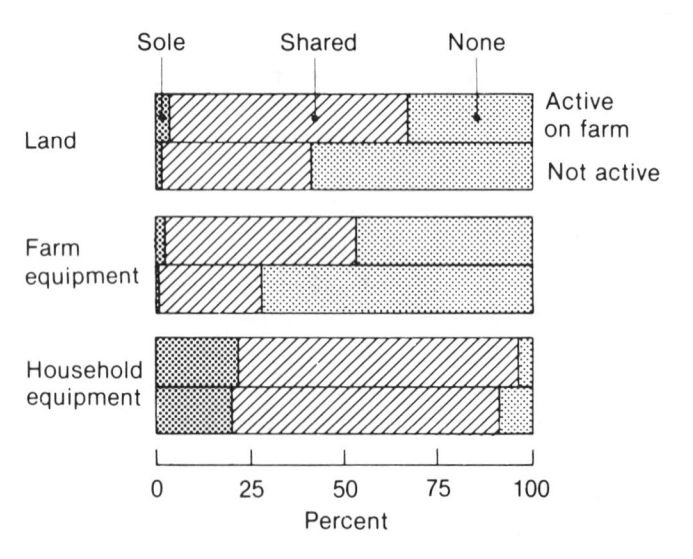

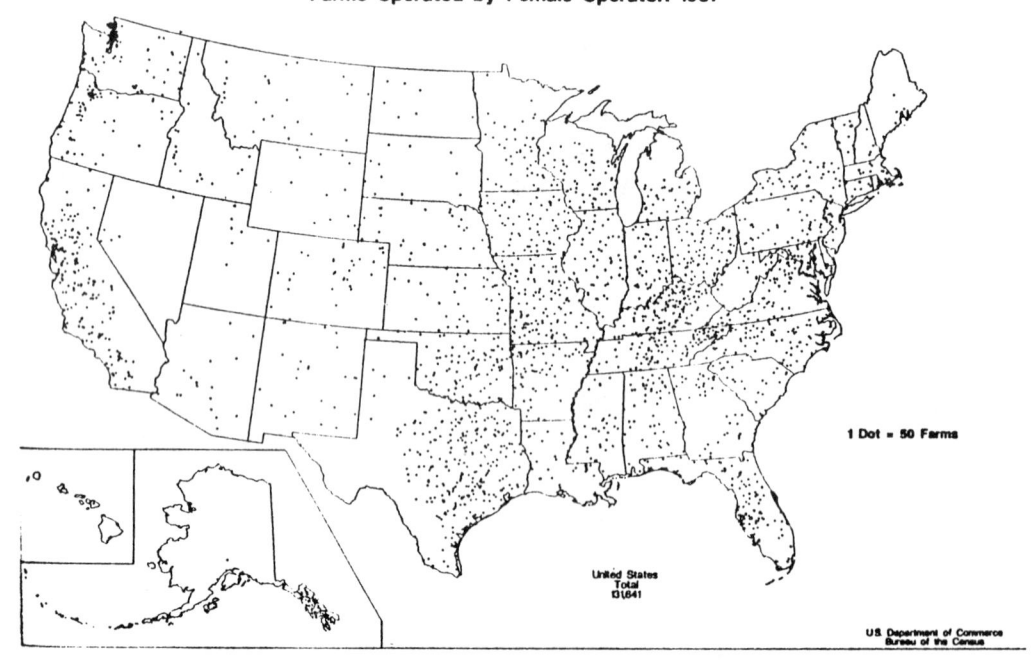

Fig. 2-4. Women play an important part in the operation of the nation's farms. (Courtesy, U.S. Department of Agriculture)

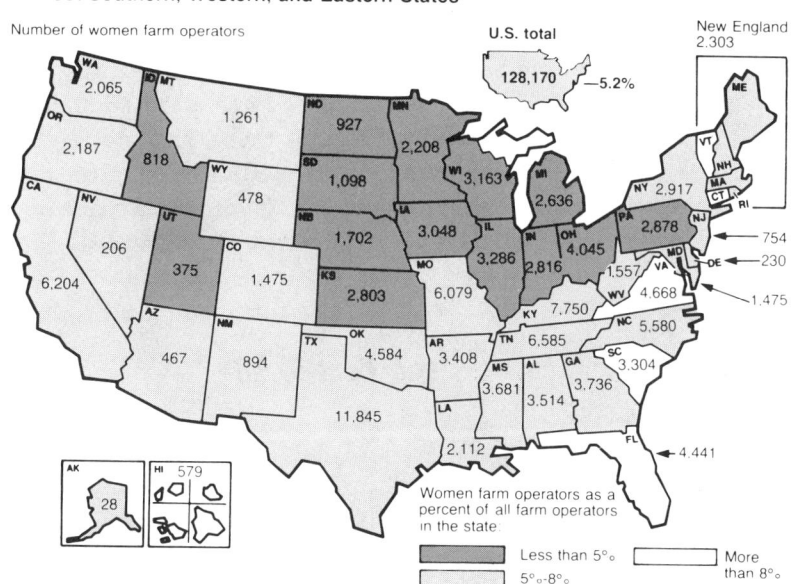

Rural Life in the United States

Women Account for at least 5% of All Farm Operators In Most Southern, Western, and Eastern States

Number of women farm operators — U.S. total 128,170 — 5.2% — New England 2,303

WA 2,065; OR 2,187; ID 818; MT 1,261; ND 927; MN 2,208; WI 3,163; MI 2,636; NY 2,917; ME (New England 2,303); VT/NH/MA/CT 754; RI 230; PA 2,878; NJ 1,475; SD 1,098; WY 478; CA 6,204; NV 206; UT 375; CO 1,475; NE 1,702; IA 3,048; IL 3,286; IN 2,816; OH 4,045; MD 1,557; WV 4,668; KS 2,803; MO 6,079; KY 7,750; TN 6,585; NC 5,580; AZ 467; NM 894; TX 11,845; OK 4,584; AR 3,408; MS 3,681; AL 3,514; GA 3,736; SC 3,304; LA 2,112; FL 4,441; AK 28; HI 579

Women farm operators as a percent of all farm operators in the state:
- Less than 5%
- 5%–8%
- More than 8%

Fig. 2-5. The number of women farm operators is slowly increasing. (Courtesy, U.S. Department of Agriculture)

errands and did other farm work, such as harvesting crops and caring for animals. Women are also taking over the management responsibilities on many farms. At the present time, over 5 per cent of all farms are operated by women.

About 22 per cent of farm women had nonfarm jobs in 1980. Most of the women gave financial reasons for seeking outside employment; about 25 per cent said they worked to provide money for the farm. Family income, measured in constant dollars, increased only slightly during the 1970's, although dollar income doubled.

Rural Homes and Family Life

Although there are differences of opinion with regard to what a "good" life is, most people would probably include (1) homes with favorable living conditions and a wholesome family life; (2) communities with good schools and churches, adequate health facilities, modern means of communication, and an efficient government; and (3) amenable relations between rural and urban people so that all work together in developing a democratic society with all sharing in its benefits and responsibilities.

If rural USA were to be studied, some needed improvements would be apparent. In some respects, as a whole, people in urban areas have progressed more than people in rural areas. For example, the farther farm families live from cities, the lower their level of living is likely to be. In counties with the highest percentages of population on farms are found the smallest percentages of the people supplied with doctors and hospitals and other health facilities. The remoteness of many accidents and the absence of any witnesses to help speed medical attention for the victims help explain why farming is still ranked as one of the most dangerous of industries. In recent years, only construction and mining have been considered more dangerous. Educational levels are also lower, and modern conveniences are fewer in the homes.

How Can Rural Homes and Family Life Be Improved?

The business side of farming is important—farming must be done efficiently if a good farm income is to be obtained. However, a farm is more than a business. It is a business and a home combined. Most people would agree that one of the main reasons for earning money in farming, or in any

Fig. 2-6. In some areas, installing gas lines is part of improving rural living. (Courtesy, M. J. Clark)

Rural Life in the United States

other line of work, is that this money generally makes possible a better home and better living conditions for the family.

Improving homes

Comfortable homes on farms, as elsewhere, contribute to the enjoyment of living. Various conveniences are desirable, such as electric lights, running water, complete bathrooms, and modern kitchens. These save time and eliminate much drudgery, thus giving individuals more time for family life and for other activities that contribute to good living.

The slow growth in total housing available has stalled progress in improving the quality of rural housing. Although most farm homes are provided with electricity, many are not supplied with running water—either hot or cold—and modern bathroom facilities. Many homes have no central heating system and are without some of the other modern conveniences found more frequently in town and city homes. This is particularly true of homes on rented farms and of homes for farm laborers.

Most rural families still prefer single-family homes. In recent years, much has been done to design homes appropriate for rural areas. Some of the features included in these designs, in addition to those mentioned above, are a well-equipped laundry, a pleasant living room, a recreation room, and enough bedrooms for the size of the family.

Inadequate Nonmetro Housing

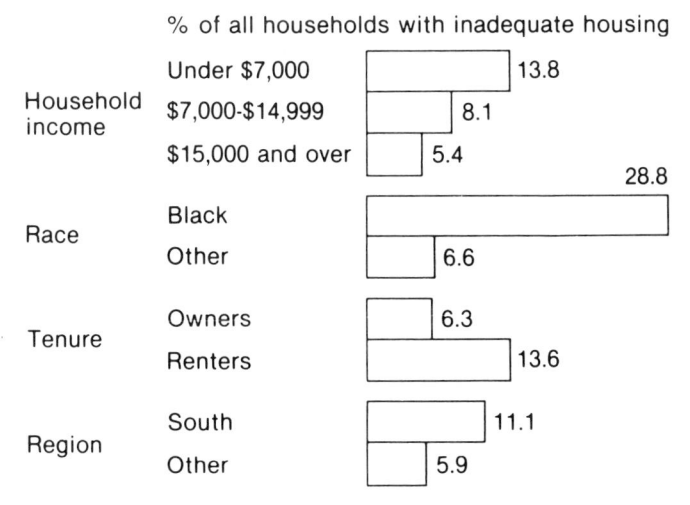

% of all households with inadequate housing

Category	Subcategory	%
Household income	Under $7,000	13.8
	$7,000-$14,999	8.1
	$15,000 and over	5.4
Race	Black	28.8
	Other	6.6
Tenure	Owners	6.3
	Renters	13.6
Region	South	11.1
	Other	5.9

Inadequate housing units lack complete plumbing and/or have more than one person per room. Complete plumbing includes hot and cold piped water, bath or shower, and flush toilet.

Fig. 2-7. Much rural housing still needs to be improved. (Courtesy, Bureau of the Census, U.S. Department of Commerce)

Type of Housing Units

Percent

Single family: 63.5 / 79.4

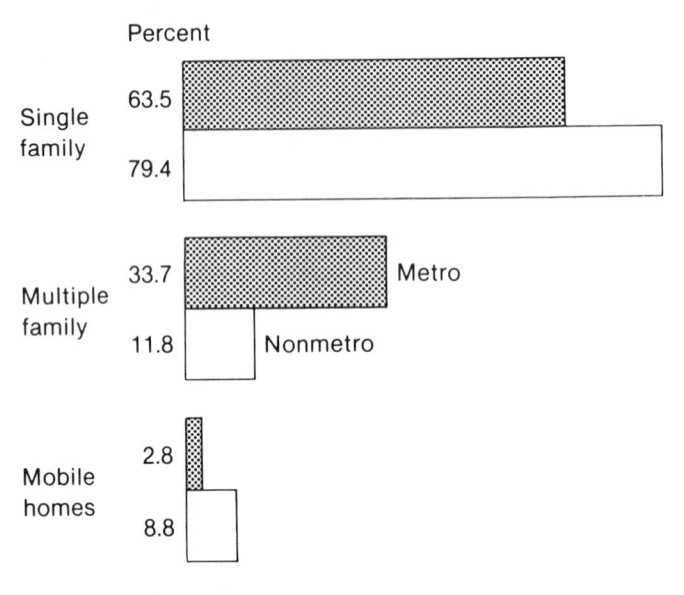

Multiple family: 33.7 (Metro) / 11.8 (Nonmetro)

Mobile homes: 2.8 / 8.8

Fig. 2-8. Rural families prefer single-family homes. (Courtesy, Bureau of the Census, U.S. Department of Commerce)

Providing attractive surroundings

Much enjoyment from rural life comes from having attractive surroundings. As individuals drive through the country, they may be attracted by farmsteads on which the buildings are in good repair and neatly painted, the fences are in good condition, and the trees and other plantings around the home and farmstead are arranged attractively.

With some effort and a little expense, much can be done to improve the surroundings of rural homes. Young people can assist by helping care for the lawn, flowers, and shrubs. Some of the same types of improvements may be made in the surroundings of urban homes. Many suggestions for improving the surroundings of the home are given in Chapter 5.

Improving family life

To have a good home, a family needs more than modern conveniences and attractive surroundings. Edgar Guest, a poet, has said, "It takes a heap of livin' to make a house a home."

Congenial life within the family circle and the cooperation of the various members in performing the work of the farm are important. Every young person can be of help to the family in various ways. A friendly spirit in the family can be developed if all the members plan certain phases of the work

Rural Life in the United States 45

together. Recreational activities, such as picnics and games, should include the entire family.

Within some farm families who own farms, there is a strong desire to keep each farm in the same family generation after generation. Under those conditions, much attention is often given to improving the farm and conserving the soil so that each generation can build "on the shoulders" of the preceding generation. Keeping the home farm in the family may help to develop strong family ties.

Eating proper foods

To be strong and healthy, a nation must be well fed. Even in the United States, a recent survey showed that not more than half of the families had nutritious diets, and at least one-third of the families had diets that were definitely poor. The encouraging thing for persons living in the country is that their families, in general, have somewhat better diets than village and city families, although improvements are needed in many cases.

The liberal use of milk, eggs, green and yellow leafy vegetables, fruits, whole-grain cereals, and meats is needed. Otherwise, the diets are likely to be low in proteins, minerals, and vitamins. The Basic Seven food guide indi-

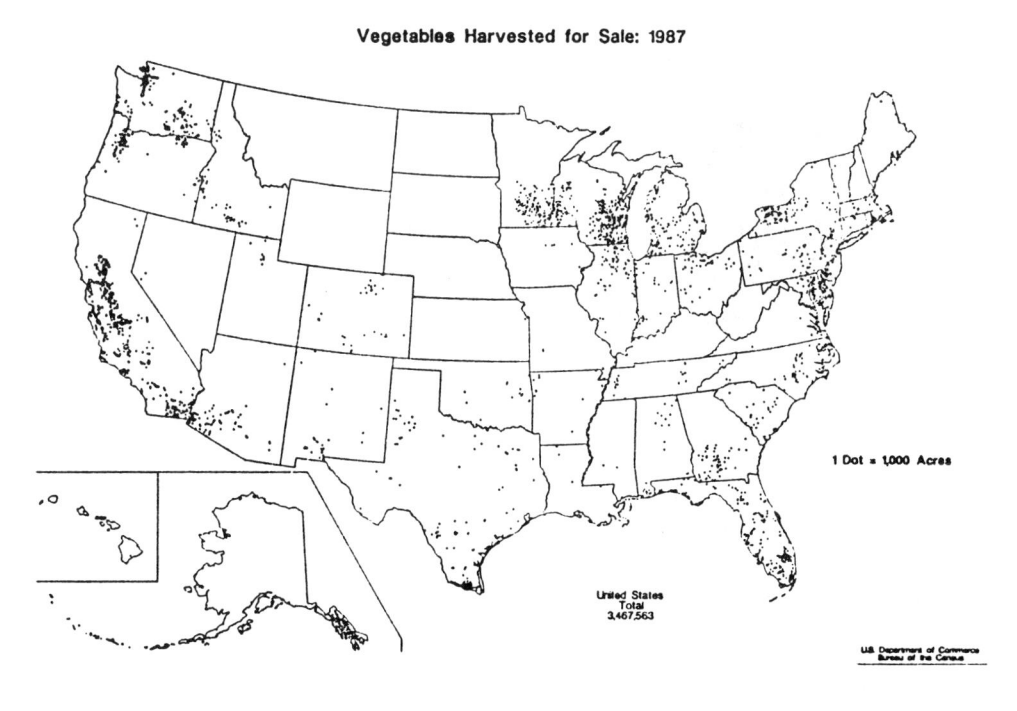

Fig. 2-9. The consumption of fresh fruits has increased, while the consumption of fresh vegetables and processed fruits and vegetables has decreased. (Courtesy, U.S. Department of Agriculture)

Fig. 2-10. The Basic Seven guide helps us to select foods for a well-balanced diet. Some of these foods may be homegrown. (Courtesy, U.S. Department of Agriculture)

cates the foods that should be included daily in the diet (see Fig. 2-10). In many schools, warm lunches are served to children from farms and to others who do not go home for lunch. These lunches are usually provided at a moderate cost and include milk and other wholesome foods, some of which are surplus food given to the schools by the federal government.

The "Basic Four" foods. Very similar to the Basic Seven food guide is the Basic Four. Both were prepared by the U.S. Department of Agriculture. The food groups in the Basic Four are meat, fruits and vegetables, dairy products, and bread and cereals. Value, alternates, and daily servings of these food groups are indicated as follows.

MEAT GROUP

Beef, veal, pork, lamb, variety meats, poultry, fish and shellfish, eggs.

Value: High-quality protein and, in varying degrees, iron, niacin, riboflavin, thiamine (especially in pork).

Alternates: Dry beans, dry peas, lentils, nuts, peanut butter.

Daily servings: Two or more.

FRUIT AND VEGETABLE GROUP

Value: Citrus fruits or other fruits and vegetables, such as broccoli, cantaloupes, peppers, fresh strawberries, for vitamin C (ascorbic acid). Dark green or deep yellow vegetables for vitamin A.

Alternates: All other fruits and vegetables, including potatoes.

Daily servings: Four or more.

DAIRY PRODUCTS GROUP

Milk in any form, cheese, ice cream.

Value: High-quality protein, riboflavin, other vitamins and minerals. Whole milk and some fortified milk also offer vitamin A. Most homogenized and evaporated milks also carry vitamin D.

Daily servings: Three or four cups of milk for children; four or more for teenagers; two or more for adults. When cheese and ice cream replace milk, more servings are needed to provide the calcium equivalent of milk.

BREAD AND CEREALS GROUP

Bread, cereals (ready-to-eat or cooked), cornmeal, grits, crackers, macaroni, rice, spaghetti, flour, quick breads, other baked goods.

Value: Protein, iron, several of the B vitamins, food energy (calories).

Daily servings: Four.

Using income wisely

The wise use of income is important in family living. Food, clothing, and shelter are considered basic. To the extent that earnings permit, most people like to secure some of the luxuries in life. Careful planning of personal and family finances makes it more likely that they will be able to provide the necessities of life and, in addition, some of the things that make life pleasant.

The amount of spendable income from farming varies widely over a period of years and thus complicates the effective planning of expenditures. Careful planning makes it possible to divide expenditures most effectively among various items, such as food, household operation, car and other transportation, clothing, health care, recreation, church and charities, gifts, and savings. Farmers must make expenditures for the business phases of farming. These include taxes, interest on money borrowed, payments on loans, fuel for tractors and trucks, repairs and replacement of buildings and equipment, hired labor, etc.

Fig. 2-11. Federal price support loans and crop insurance are important considerations in the financial planning of farmers.

Providing security for retirement

Most farm people, as well as others, look forward to the time when they can retire with enough income to provide a comfortable living. This calls for careful planning and the wise choice of ways to invest money. Insurance of various kinds, land, bonds, stocks, and savings accounts in banks represent some of the types of investments that can be made.

Social Security laws cover farm workers and farm operators. Under these laws, employers deduct a specified percentage from the wages of employees and add an equal amount, all of which is sent at specified intervals to the Internal Revenue Office in their district. These funds are used to provide income upon retirement. It should always be kept in mind, however, that Social Security is intended only to make sure that a retired worker has *some* income. Social Security income will not provide enough money for many of the luxuries in life. In order to have many of the extras they may want during retirement, people should save some portion of their total earnings each year and invest them in secure ways until the money is needed.

Participating in wholesome recreation and hobbies

In many rural homes, all members of the family engage in wholesome types of recreation and develop hobbies. Young people are encouraged to bring their friends into their homes for various types of recreational activities.

Hobbies of various kinds are of interest to both the young and the old. Some persons become interested in growing flowers, painting, raising pets, playing a musical instrument, or engaging in other activities. As an example, one farmer in Indiana, who is a very successful corn grower, became interested in raising gladioli as a hobby to give him a change from farm work. Over a period of several years, he expanded his hobby to 100 different varieties, which he exhibited at local and state fairs. He also sold gladioli to local florists and others.

Almost everyone has heard about Grandma Moses who became interested in painting in the later years of her life. Since she had lived on a farm

Fig. 2-12. The number of horses is increasing, in large part for their recreational use. (Courtesy, Appaloosa Horse Club, Inc.)

all her life, it was only natural that she should paint rural scenes. Some of these were so well done that she became famous. In several states, artists are employed at the state universities to encourage rural people and others to develop their talents along these lines.

Many people obtain enjoyment from reading and from studying various paintings and other works of art. Much excellent literature and many works of art are based on rural life and the rural environment. Some of these are listed at the end of this chapter.

In some places, rural community clubs provide entertainment and recreational activities for the young and old through meetings at various farm homes. Organizations such as the Farm Bureau, the Grange, and the Farmers' Union have regular meetings for the discussion of farm problems and for recreation.

Ruritan, a civic service organization, provides support for recreational activities for rural youth.

The Little Country Theatre movement has been organized by people in some communities. As a result, rural people are participating in plays. At intervals, they present plays and other productions for the entertainment of themselves and the people who attend.

In some states, attention is given to organizing community choruses, bands, orchestras, folk games, and old-time square dances. One farmer expressed his enjoyment in participating in such activities by saying, "You work enough better and feel enough better the next day to make up for lost sleep. When you sing good music, it doesn't all go out of you."

Preventing accidents

Farming is more dangerous than most other occupations. While farmers in general live longer than people in most other occupations, they face certain hazards that may result in minor injuries and the risk of some serious accidents. With the increased use of tractors, corn pickers, combine harvesters, electrical gadgets, and other mechanical equipment, accidents from these sources have increased. A study at the University of Missouri showed that the corn picker was by far the most dangerous farm machine, and the tractor the least dangerous. However, it has been estimated that tractors kill 1,000 persons each year.

The chance of a farm worker being killed is nearly three times greater than for a worker in a manufacturing industry. In a recent year, over 6,000 farm residents were killed in accidents. In addition, about 500,000 received disabling injuries in accidents. All farm workers, young and old, should take every precaution possible to insure safety for themselves and others.

Farm life is unique in its combination of work and recreation activities. About 60 per cent of all farm fatalities are work related. Drowning causes nearly 15 per cent of the farm deaths; machinery-related accidents, 24 per cent; and deaths caused by individuals' being crushed or struck by objects, 18 per cent. And about 77 per cent of these accidents are from April to October of each year. While older person make up only one-fourth of the work force, they are involved in one-third of the fatal accidents. Older people are the chief victims in machinery accidents. Younger people are the most frequent victims of drownings and suffocations. Other causes of death on the farm are falls, shootings, electric shock, burns, and poisons.

Most farm accidents are the result of farmers' being careless—hurrying, working when tired, or working when ill or emotionally disturbed. By using safety precautions in operating farm machinery, in caring for livestock, and in doing all the various daily tasks on the farm and in the farm home, farmers can avoid many accidents.

Highway accidents are one of the chief sources of injuries and deaths among rural people as well as among urban people. Careful driving would prevent most of these injuries and deaths. Studies have shown that teen-age drivers of cars have the worst accident record of any age group. Instruction in safe driving by some high schools has done a great deal to reduce these accidents among those taking the driver-training courses. Young people can be the best drivers.

Preventing fires

Fires in rural areas are of serious concern to farmers and to other people who like living in the country. Some rural areas do not have firefighting facilities, while weak wiring and construction standards and poor heating equipment in the homes tend to increase the chances of fires.

According to the National Fire Protection Association, at least 50 per cent of rural fire losses are due to faulty construction of buildings, and carelessness of various kinds is the cause of most others. Fire and lightning strike 2 out of every 100 farms each year. In a recent year, there were over 400,000 fires in rural structures and on wild land. The volunteer fire departments in rural areas are not equipped or trained to handle that many fires, many of which are miles away from the fire stations. As a result of the number of fires and the difficulty in dealing with them, fire insurance rates in rural areas are higher than in urban areas.

Fire losses in rural areas total over $1 billion each year, and that amount does not include the cost to the communities for emergency medical aid, food, and shelter for the victims.

During an average year in the United States, forest fires start in 125,000 places, burn 5,000,000 acres, consume 113,000,000 tons of wood, produce 165 cubic miles of smoke, emit 364,000 tons of hydrocarbons, burn 25 persons to death, and injure 1,350 other people.

Some rural communities have developed fire protection districts and have provided firefighting equipment with either volunteer or paid firefighters. These services are usually financed by special taxes, frequently in cooperation with towns and cities located in their areas.

Providing rural firefighting facilities is largely the responsibility of the local communities. Over 500 new fire departments have been organized and thousands of firefighters have been trained through programs set up by state foresters with money provided by the Cooperative Forestry Assistance Act of 1978. This act authorizes the U.S. Department of Agriculture to assist state foresters in organizing, training, and equipping local rural fire protection forces. Two other federal sources are available to assist rural fire departments: the Rural Community Fire Protection Program, which provides some grant money for planning, training, and purchasing equipment; and the Farmers Home Administration, which makes loans to support a wide variety of community facilities, including fire and rescue services. In some states, training is provided for firefighters at the state land-grant universities. Community leaders in other areas should contact their state fire marshall to see if similar services are available. Each country home should have a good source of water, such as a cistern, pond, or stream for use in fighting fires. Chemical fire extinguishers and other firefighting equipment should be kept in convenient places on every farm. Of course, in every home, fire-preventing precautions should be followed.

What Organizations Serve Rural People?

Many kinds of organizations serve rural people. Some organizations have been started at the grass roots by the farmers themselves. Some of these are national in scope, some are more or less local in their influence. Some are primarily recreational in nature, such as community clubs of various kinds. Some are religious, such as various organizations sponsored by churches. Some are educational, such as parent-teacher associations. Cooperatives of various kinds aid farmers in buying and selling.

Three of the largest farm organizations are the Grange, the Farm Bureau, and the Farmers' Union.

The Grange

The Grange is the oldest national farm organization. The full name is the National Grange of the Order of Patrons of Husbandry, but it is usually called the Grange. The organization was started in 1867. In a recent year there were about 450,000 members. Frequently, two or more members of a family are included as one membership.

The Grange places emphasis upon improved family life, and entire families participate as members. The Grange also seeks to develop cooperation and leadership among farm people and to improve community life. It encourages cooperation with schools and churches. The organization takes an active part in promoting legislation favorable to farm people, in providing certain types of insurance, and in helping its members understand national and world problems. Songs and other forms of music have a prominent place on the programs of the Grange. The basic unit of organization is the community Grange. Considerable attention is given to the interests of rural youth.

The Farm Bureau

The American Farm Bureau Federation was established as a national organization in 1921. It has been active in securing national legislation related to farm prices, credit, health, soil, conservation, roads, schools, and rural youth. The Farm Bureau is the largest general farm organization, with more than 3.2 million member families in over 2,800 county farm bureaus. The Farm Bureau is a voluntary organization of farm and ranch families. It conducts programs and activities at the local, state, national, and international levels.

The highest priority goal of the Farm Bureau is to bring about changes in federal programs to make life better for farmers and farm families. A professional staff assists in the preparation of materials on land use, water rights, conservation programs, pollution control, wildlife management, farm chemicals, and energy conservation. The commodity staff has the responsibility for supervising the American Agricultural Marketing Association (AAMA) fowl sales and its fruit and vegetable programs and for coordinating state association marketing programs.

The Farm Bureau Young Farmers and Ranchers program provides opportunities for leadership development and encourages young adults to become participating members in the Farm Bureau.

The Farmers' Union

The Farmers' Educational and Cooperative Union has been in existence as a national organization since 1902. In 1983, there were 270,000 paid memberships, based on one membership per family. Its main purpose is to improve the conditions of farmers through cooperation, education, and legislation. Youth activities are sponsored.

The National Farmers' Union also sponsors Green Thumb, a program funded by the U.S. Department of Labor and by contributions from local and state agencies. The program offers employment opportunities in local community projects to rural low-income senior citizens age 55 and over. Some of the types of projects involved are outdoor beautification and recreation, transportation, home repair, nutrition, historical site preservation, outreach, day care, fire prevention, crime control, food production, and land stewardship.

Other organizations

There are many other organizations that serve rural people in many ways. Some of these organizations serve both rural and urban dwellers. In many communities there are community clubs, cooperatives of various kinds, parent-teacher associations, and others.

National Farmers Organization. The National Farmers Organization (NFO) was formed to help farmers obtain a higher price for their products. The organization was formed by farmers, with farmers being the only persons eligible for membership. The functions and activities of the organization are limited strictly to cooperative marketing of the farm produce of its members. The members of the organization are attempting to obtain control of enough of the total production of each of the various kinds of farm products to be able to make it attractive and necessary for agricultural businesses dealing in raw farm products to contract with the National Farmers Organization for buying these products from its members. This is basically an effort of farmers to solve the problems of low farmer income, without the help of the federal government, through one kind of cooperative marketing activity. Cooperatives are discussed further in Chapter 23.

Ruritan. Ruritan is one of the leading civic service organizations. Club members include farmers, business and professional men, and other concerned citizens of local communities.

The slogan of Ruritan is Community Service, Fellowship, and Goodwill. The organization strives to create a better understanding between rural and urban peoples and, through volunteer community service, to make rural areas better places in which to live and work.

What Is the Place of Part-Time, Small-Scale Farming in Rural Life?

In recent years, there has been a back-to-the-land trend in the form of part-time, small-scale farming. The number of small farms under 3 acres in size and the number of places too small to be classed as farms have increased considerably in the last few years.

The trend toward living on a little land is especially noticeable in the areas surrounding many cities and along the improved highways leading from towns and cities. Most of these families receive their major income from jobs in factories, mines, building trades, stores, or other lines of work in nearby towns or cities. They live in the country and own a few acreas of land for a variety of reasons. Some of the reasons for owning the land are:

1. To use the land only for extra living space.
2. To enjoy hobbies and recreational activities.
3. To reduce living costs by growing food for the family.
4. To increase the family income by growing food for sale or by starting some other small business.
5. To satisfy a personal desire to own land.
6. To provide space for children to play and to have pets.

Unfortunately, living on a few acres is not without its problems. A primary difficulty stems from romantic notions of rural life. Many people know nothing about or ignore animal and plant diseases, insect pests, and other production and marketing problems they have to solve. As a result, they often find themselves in serious trouble. The success of even a small farm operation depends on how well the family members are able to cope with potential disadvantages and problems and how strongly they feel the advantages of rural living outweigh the disadvantages.

In addition to people who establish their homes on small pieces of land in the country, some people on farms take jobs in towns or cities for at least

Housing Costs

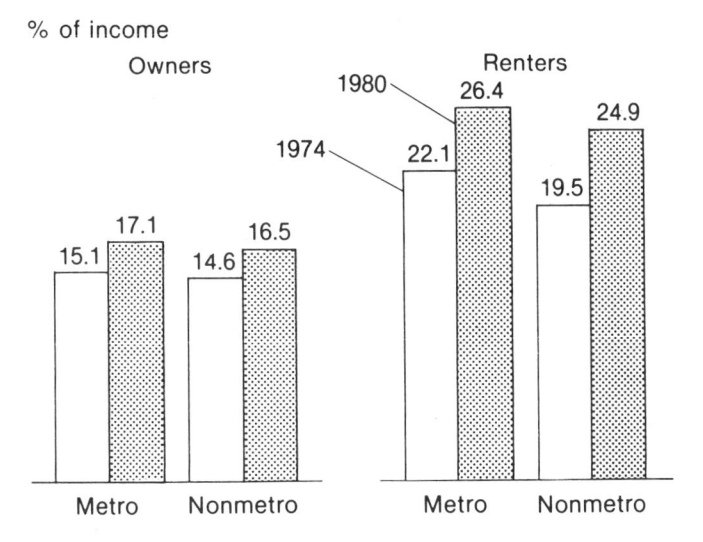

Includes cost of mortgage, taxes, insurance, and utilities for owners; gross rent for renters.

Fig. 2-13. The proportion of total income spent for housing is increasing. (Courtesy, Bureau of the Census, U.S. Department of Commerce)

a part of the year. These part-time farmers are increasing in number in sections of the country where the land is not highly productive and where city jobs are attractive.

The trends toward part-time, small-scale farming in many places are changing the old type of community life. In some cases, the newcomers to rural life do not have much in common with the farm families who have lived in the same community for many years. Gradually, the areas near the cities build up with homes and become a part of the cities.

Fig. 2-14. David Wood's headquarters for freighting line near Ridgway, Colorado. (Courtesy, Library, State Historical Society of Colorado)

What Is the Relation of Government to Agriculture?

Farmers have many problems, as has been noted. Some of these concern the individual farmers, as they strive to do a good job of farming. In operating their farms, they are interested in learning how to use the latest scientific methods. They may learn about some of these methods by reading farm journals, books, and other publications. They may also obtain assistance from teachers of vocational agriculture, their county agricultural agent, their state college of agriculture, and the U.S. Department of Agriculture.

Kinds of governmental services and agencies

Various governmental agencies and services are provided to aid farmers in solving some of their farm problems. Some people believe that governmental assistance to farmers is a recent development. Actually, various forms of assistance have been provided for many years. In the early days, farmers were aided in securing land by the Homestead Acts of Congress. Land-grant colleges were made possible through an act passed by Congress and signed by President Abraham Lincoln in 1862. In 1890, another act was passed to provide support for land-grant colleges for blacks.

Fig. 2-15. Instruction in agriculture did much to accelerate the change from picking corn by hand to using mechanical harvesters.

It should be emphasized that governmental assistance also has been provided for groups other than farmers. For example, tariffs were set up to protect young and growing industries. Down through the years, tariffs have benefited industries and labor groups much more than farmers.

State colleges of agriculture are made possible through appropriations provided by federal and state governments. These institutions provide education for students who attend them. They also conduct research related to new methods of farming and improved farm living. Extension specialists and publications are provided to carry information to farmers.

The U.S. Department of Agriculture, a division of the federal government, was created in 1862 by an act of Congress. It sponsors research and the preparation of various publications for farmers. It cooperates with state colleges of agriculture in some types of research and service to farmers. It also provides funds channelled through state colleges of agriculture, together with state and local funds, for employing county agricultural agents who assist farmers in every county in the United States.

The Forest Service of the U.S. Department of Agriculture manages 190 million acres in the national forest system. This includes 18 per cent of the nation's commercial forestland. In addition to trees, the national forests and grasslands provide a home for over 4 million big game animals and 79 species of threatened or endangered wildlife. In these and many other ways, the U.S. Department of Agriculture provides services that benefit farmers and the general public as well.

The Farm Credit Administration is an independent government agency, operating under a 13-member part-time Federal Farm Credit Board that provides loans to farmers and cooperatives through the Cooperative Farm Credit System. The system is actually owned by the farmers and cooperatives using it.

The USDA Farmers Home Administration (FmHA) also provides loans for farmers, making higher risk loans than other lending agencies are able to do. In these high risk loans, and in housing loans, the borrower is expected to refinance the loan with a private lender as soon as possible.

Another agency of the U.S. Department of Agriculture, the Rural Electrification Administration (REA), makes loans to rural electric systems, mostly nonprofit cooperatives. It also makes loans to improve and extend telephone services in rural areas.

Using crops in storage as security, farmers can also secure loans from the Commodity Credit Corporation (CCC). If the prices rise above the loan rate, the farmers can pay off their loans, with interest, and sell their crop on the market.

The Federal-State Marketing Improvement Program, under the USDA Agricultural Marketing Service, studies problems at the local and state levels. Projects include improving marketability of crops, finding new markets, improving marketing efficiency, and improving marketing information.

Federal services to agriculture include various marketing laws and orders. Some of these are the following:

1. The Perishable Agricultural Commodities Act—to encourage fair trade practices.
2. The Federal Seed Act—to require truthful labeling.
3. The Plant Variety Protection Act—to extend patent-type protection to developers of plants.
4. The Agricultural Fair Practices Act—to prevent discrimination against producers.
5. The U.S. Warehouse Act—to help prevent loss of stored products.
6. Federal marketing orders—to apply to fruits, vegetables, specialty crops, dairy products, and cotton.
7. The Agriculture and Consumer Protection Act of 1973—to help producers find and develop new markets.

8. The Packers and Stockyards Act—to maintain fair and open competition in marketing livestock, poultry, and meat.
9. The U.S. Grains Standards Act of 1976—to inspect all grain destined for export as it is loaded aboard ships. This inspection is done by the Federal Grain Inspection Service, which also supervises the inspection by state departments of agriculture and private firms of domestic grain marketed at inland locations.

The USDA Foreign Agricultural Service is responsible for expanding foreign markets for agricultural products.

The Soil Conservation Service gives technical assistance on erosion control and land and water use to farmers, ranchers, and all governmental units at any level. It works through 2,950 local conservation districts.

The Agricultural Trade Development and Assistance Act helps farmers by financing the sale of agricultural products to developing countries.

The Food Purchase Programs—In a recent year, over 1.8 billion pounds of food was donated by the U.S. Department of Agriculture through various acts and programs to schools, low-income families, public institutions, and disaster victims. This activity helps to provide a market for agricultural products.

Other USDA agricultural services, benefiting both consumer and producer, include USDA grading programs for beef, eggs, fresh and frozen fruits and vegetables, butter, canned fruits and vegetables, turkeys, chickens, other poultry, cotton, and tobacco. Standards exist for about 400 food and farm products.

The Food and Nutrition Service provides food assistance to those in need, inspects food to assure safety and quality, and deals with consumer and policy issues.

The U.S. Department of Agriculture assists the states in controlling insects, weeds, and plant diseases when these are a serious threat to the nation.

The Animal and Plant Health Inspection Service's programs are carried out in cooperation with the states. The programs are designed to keep foreign diseases from entering this country, to eradicate diseases that do get into the country, to fight major domestic animal disease outbreaks, and to provide for humane treatment of animals.

A Commodity Futures Trading Commission has also been established to regulate all futures trading.

The Department of Transportation concentrates on solving some of the major transportation problems facing rural areas, particularly those related to the deterioration of rural roads and bridges.

The Agricultural Stabilization and Conservation Service administers various commodity and land use programs to support farm prices, to adjust farm production, and to protect natural resources.

The Federal Crop Insurance Corporation administers an all-risk crop insurance program made available to farmers through private insurance companies.

The Extension Service conducts educational programs to help farmers, processors, handlers, farm families, communities, and consumers. It also provides leadership for the 4-H program for youth.

The federal government provides funds to states and local school districts for use in financing instruction in agriculture/agriscience education and certain other vocational subjects. These funds are administered through the U.S. Department of Education, but the states and school districts administer the program and provide additional funds. Agriculture teachers provide instruction for farmers of all ages, as well as for school students.

The National Agricultural Library provides comprehensive information services for the food and agricultural sciences. The library has 1.6 million volumes on agricultural subjects.

The Agricultural Research Service conducts research in the fields of livestock, plants, soil, water, air quality, energy, food safety and quality, nutrition, food processing and marketing, nonfood agricultural products, and international development. It also provides leadership for educational and training programs in the food and agricultural sciences.

The Cooperative State Research Service administers federal funds for research in many agricultural areas, such as photosynthesis, nitrogen fixation, genetic engineering, biological stress, and human nutrition.

The Economic Research Service analyzes topics related to agriculture and to rural communities.

The Statistical Reporting Service compiles the information about agriculture that we read about daily in the newspapers.

One of the problems in the development of rural areas is that of making effective and appropriate use of the land that is available. In a few states, county zoning laws have made it possible to set aside land not suitable for farming to be used for forests and recreation. Many states have passed various laws that deal with zoning. Some of these laws make it possible for rural governments to prevent the establishment of undesirable commercial activities that have an unfavorable influence on community life. In some cases, separate zones have been set aside for residential, commercial, and industrial uses of the land. These do not interfere with the use of land for most farming purposes. By planning and zoning, rural communities can prevent the formation of rural slums and encourage the development of attractive and wholesome communities.

Other federal and state agencies aid agriculture in various ways. A state department of agriculture is found in each state. This state agency gathers and distributes information about agriculture in the state, inspects food processing and distributing firms, gives technical information to farmers, and provides other services.

How Can Better Rural Communities Be Developed?

One of the pleasant features of rural life since pioneer days has been the willingness of neighbors to help each other. They trade work and turn out in groups to do the farm work for a sick neighbor. Neighborhood groups do various things together, including participating in recreational activities of various kinds.

In addition to activities by small groups of families within the same neighborhood, rural people over larger areas have various activities and interests in common. These areas are called communities. Within each community, we frequently find a town or a city in which churches, hospitals, and recreational facilities are available. Many people use facilities located in two or more communities. Some of these services are provided in the open country in addition to those available in towns and cities.

While there is much in rural life that is wholesome and desirable, there is also considerable room for improvement. The rural communities themselves hold the key to much of this improvement. Some communities have "withered on the vine," while some have gone forward in providing conditions for improved living for both rural and urban people. Communities must give attention to their schools, libraries, health facilities, churches, recreational activities, roads, and other means of communication. Good local governments, friendly relations among rural people and urban people, and effective leadership are important. These are discussed in the paragraphs that follow.

Improving schools

The public school is one of the basic institutions in a community. The district school serving a small neighborhood of a few square miles was common in former years and is still found in some areas. However, in most states, larger units are being developed for both elementary and secondary schools. A fairly common system in some states is to develop enlarged districts, each of which includes a town or small city, together with the farms and small villages in the surrounding area. These districts often cover as much as 100 square miles and some of them include much larger areas. The high schools in these districts frequently become "community schools" by providing a center for many types of educational and other activities for an entire community. The elementary schools are often used as neighborhood centers.

During recent years, many improvements have been made in schools that serve rural communities. The schools in enlarged districts have usually improved most, because they are able to provide better facilities at a reason-

Fig. 2-16. A modern rural school can provide excellent educational opportunities for youth.

able cost per student. New school buildings are being built; old buildings are being renovated. Much remains to be done, however, to provide adequate school buildings for rural youth throughout the United States.

In many schools, improvements have been made in the instructional programs. Guidance is provided. Vocational courses are offered in agriculture, home economics, business, trades and industries, and others. Other courses emphasize home and family life, citizenship, health, leisure, and other aspects of rural and urban living. Teachers who are familiar with rural life as well as urban life are employed. In many schools, hot lunches are served, and buses are used to transport rural students to and from school. Health checkups and assistance are given when necessary to provide medical and dental care. Some school systems employ nurses, doctors, and dentists for these services. Some systems provide classes for adults who wish to learn more about farming, homemaking, other occupations, or various aspects of improved living.

Although many improvements have been made, rural children in general are attending schools fewer days per year and for fewer years than children in cities. In some rural communities, a large percentage of the farm youngsters drop out of school at the end of the eighth grade or before they finish high school.

Providing library service

Many people like to read. Abraham Lincoln liked to read so well that he walked miles to borrow books. Library facilities are being provided for rural people in an increasing number of communities in the United States. Books are made available in various ways. Some public libraries in towns or cities provide borrowing privileges to rural people without charge or for a small fee. A method used in some states is to locate branch libraries or book stations in schools, stores, or other places to which books are brought from a central library operated by a city or a state. Some libraries are supported by taxes paid by both rural and urban people in the communities being served, but the costs are small.

In every state capital or statehouse, there is a state library. In most states, books may be borrowed from the state library by anyone in the state. Usually the only cost is the postage both ways for the books borrowed. In these cases, the people are as close to their state libraries as they are to their own mailboxes.

Many states provide state funds for aid to new or existing libraries in counties or regions of these states. Some state, regional, and county libraries provide libraries on wheels, called bookmobiles, which carry books to local communities. These bookmobiles make scheduled stops at stations, where the books on the shelves of the truck may be inspected and those of interest may be borrowed.

There are still many millions of rural people and some urban people who do not live near public libraries, where they may choose and borrow

Fig. 2-17. Many rural areas are served by regional libraries.

books readily. Each community needs to study its situation and make the improvements necessary to provide all its people with reading materials that may be obtained easily.

Providing health facilities

Many people think the open country provides the healthiest place in which to live. Fresh air, sunshine, outdoor exercise, and good foods are conditions of rural living favorable to good health. Although these factors may help, health progress has been most rapid in cities.

During World War II, a higher percentage of farm men were rejected in the armed forces because of physical defects than were men from other major occupations. The death rate of infants is higher in rural than in urban areas.

Fewer doctors in proportion to population are serving rural communities than are serving cities. Some rural communities are without doctors or hospital facilities, except at great distances. Some communities are organizing to improve these facilities through the construction of community hospitals that are as modern as the best found in cities. In these and other ways, doctors are being encouraged to choose rural areas. Medical services and hospital facilities in some of these cases are made available to each person or family at a nominal fee paid annually. Some rural families carry health insurance policies, which provide money when needed for hospital bills and certain kinds of medical services. Cooperative organizations for these services are being formed at a few places.

Many counties have health departments with regularly employed nurses who work in schools and aid out-of-school people in protecting health and controlling contagious diseases.

Providing churches

Rural churches have a prominent part in the life of rural people in some communities. The churches that are of greatest service employ ministers who understand the problems of rural people. Some of the most capable of these leaders are able to inspire their congregations to appreciate the beauties of their surroundings and to recognize the satisfaction of being close to the soil.

One minister of a rural church preached forcefully that farmers are stewards of the soil, and that to be faithful to this trust, they must maintain and improve the soil. One farmer said, "I'll never see another gully without thinking someone has committed a great sin." A priest in a rural parish helped save his community during a severe flood and later worked actively in developing a system of flood control to prevent future calamities. In these

Fig. 2-18. A new church building indicates the vitality of a small rural community in New York State.

and other ways, rural-minded clergy tie their sermons and their services to the problems of farming and farm living in the communities that they serve.

Providing good roads and other means of communication

One of the developments in the United States that has aided rural and urban people is the construction of a vast system of *improved highways*. In some states, local farm-to-market roads of an "all-weather" type reach most of the rural people. These need to be extended in some states, and many main highways need to be repaired so they will be more suitable for heavy traffic and increased speed. All people, whether rural or urban, benefit from good roads and highways. Why is this so?

Financing highway construction and repair is a problem in most states. People who use highways should expect to help pay for them through gasoline taxes and license fees for cars and trucks.

Telephones have become commonplace in rural areas in recent years. The combination of an increasing number of telephones and a decreasing number of farms has resulted in the percentage of farms with telephones being at an all-time high.

Radio has become a much used type of communication and entertainment for rural people. *Television* is available to most rural homes. Both radio and television have many good features. Questionable kinds of advertis-

Fig. 2-19. There are still many rural areas served only by dirt or gravel roads.

ing and some programs of low standard are still quite common and need correcting. Some state colleges and universities are sponsoring programs over the air, which, in general, are high in educational and recreational values. Many people believe that the *public* television stations offer the best kinds of programs for family viewing.

Although not as recent as some other forms of communication, *railroads* have long served rural communities in many ways. Increasingly, airplanes are aiding rural people as well as urban people through commercial air lines. Private planes owned by "flying farmers" are fairly common in some states.

Delivery of mail to rural areas was started by the federal government more than 65 years ago. This helped establish contact between rural people and the outside world.

Improving local governments

Local governments in villages, towns, cities, townships, and counties have contributed much to the improvement of rural life. While the original role of local government was a limited one, each local governmental unit is now being called on to increase both the level of its services and the types of services it provides.

Today, if a local government were to stop functioning, the residents of that community would have no police and no fire protection. The schools would close. The health department, which inspects the restaurants where they eat, the milk they drink, and the food they buy, would close. No one would be able to check a book out of the library or get a marriage license. The roads and bridges would become impassable, and there would be no courts or jails to deal with the criminals. In some communities, electrical service would no longer be available. Of course, local governments are not going to close down. Too often, however, people give little thought to the public services that they use and rely upon every day.

In general, rural people are getting valuable services in return for their tax dollars. However, in some cases, tax money has not been used as efficiently as might be desired. As a result, some of these services are less effective than they should be. For example, crimes and rackets are often thought to be confined to cities. This is not entirely true, as petty crimes, and even "organized" forms of crime, have come to rural communities. One of the alarming features is the number of rural young people involved in these undesirable activities.

Each form of local government continually needs to study the needs of the local community and then to make the necessary changes that will meet these needs.

To improve rural conditions, all citizens must do their part in selecting capable officials and insisting that they fulfill their responsibilities. Furthermore, every citizen should work toward having good homes, schools, and churches in the community and providing various recreational opportunities for young people so that they will be encouraged to engage in wholesome activities.

All residents should keep informed of what their local problems are and how the local governmental officials are dealing with them; they should join with others in community organizations that carry out projects to improve the way of life in the local area; they should become involved in local political organizations and assist in the election process; they should talk with local officials about their views on local issues; and they should vote whenever they have an opportunity to do so.

The federal government and rural community improvement

One of the major factors affecting the improvement of rural communities and rural areas generally has been federal assistance to rural area development programs.

The Food and Agriculture Act of 1962 gave the Secretary of Agriculture the authority to develop Resource Conservation and Development Areas. This is a USDA program to help people care for and use their natural resources to improve their community's economy, environment, and living standards. The Soil Conservation Service (SCS) oversees this program. It will fund only soil and water conservation measures, such as critical erosion control, flood prevention, farm irrigation, fish and wildlife protection, water-based recreation, and agricultural pollutants control. The coordinator for the program will help the local community get funds for other activities from local, state, and federal agencies.

The Soil Conservation Service has developed an Agricultural Land and Site Availability Assessment System to help state and local governments identify and preserve important farmland. The system gives states and local governments criteria for identifying farmland that is of state or local importance and for computing preferential tax assessments to preserve farmland. In this way, the rural character of a community can also be preserved.

Under the Rural Development Act of 1972, the U.S. Department of Agriculture is responsible for a national rural area development program. The program is designed to help rural areas secure (1) electricity; (2) better telephone service, health services, housing, streets and highways, and schools; (3) more protection for natural resources, family farm ownership, and investments in businesses and farms to create jobs; and (4) improvements in local government.

Growth and expansion in rural areas are the goals, with the state and federal governments *helping* local people and local organizations rather than *doing for them*. Some of these actions and programs have been effective. In recent years, jobs have been created faster in rural areas than in urban areas. Rural areas in the South, particularly, have attracted many new industries. Further decentralization of food processing, packaging, and distribution as well as the growing demand for recreational facilities will add to rural area economic growth.

Fig. 2-20. Modern developments have become a part of country life.

Fig. 2-21. Some rural homes still lack an adequate water source. (Courtesy, U.S. Department of Agriculture)

How Can Relations Between Rural and Urban People Be Improved?

Often, rural people do not understand urban people, and urban people are not familiar with the problems and conditions of rural people. It is to the advantage of both groups to understand each other and to work together for the benefit of all.

Improving rural and urban relationships

Some people in towns and cities are not fully aware of how much of their trade comes from rural people. They see the effects of paydays in factories in large amounts of spending and the payment of bills during certain periods each month. Farm people spend money more uniformly throughout each month. Actually, in many towns, well over half of the money spent in stores and for various services comes from farmers. Some communities are aware of these facts.

Various agencies are helping to develop better relations between rural people and urban people. The local newspaper may do a great deal to help build these relationships. Rural and urban organizations of various kinds also are helpful along these lines.

In rural commmunities, many people are needed as leaders who understand the problems of farmers and other rural people. In some cases, community councils and other local planning groups are determining needed improvements and are helpful in getting desired changes made. An example is a community in Michigan, consisting of a small town and a large agricultural area with below average soil conditions. Through encouragement by the school superintendent and others, a group of people met regularly to study the situation and to develop a program for community improvement. Efforts were united, and improved schools, farming, and community life resulted. Representatives of various organizations helped in these improvements. These included the county agricultural agent, persons in the Soil Conservation Service, the high school vocational agriculture teacher, the school superintendent, and the board of education. An association was developed to promote the gladiolus industry. A cooperative nursery was started to produce trees for reforestation. Winter sports, hunting, and fishing were promoted as recreational activities for local people and as attractions for tourists. Library facilities and services were improved. A community health center was developed. Adult education in several forms was provided by the school. The educational program for school-age students was changed to meet developing community needs and student interests.

Providing taxes to support public services

Since all people are faced with increased taxes, they should learn how their local tax dollars, as well as their state and federal taxes, are spent. They should elect responsible public officials who will promote economical government spending. In return for their tax dollars, they should receive schools, highways, health facilities, police protection, libraries, military protection, etc. What other services locally and nationally are provided by various kinds of taxes?

Improving nationwide relationships

As indicated earlier, farm people may at times be handicapped by wide fluctuations in farm prices. Periods of decreasing farm prices occur, and this may happen when prices for products that farmers buy are on the increase. Increasingly, people are recognizing that a prosperous agriculture is needed for a prosperous and strong nation. This is possible only as all people, rural and urban, understand the problems of agriculture and work together to solve these problems in ways that benefit all groups. It is likewise important that rural people understand the problems of urban people and help them to find satisfactory solutions to these problems.

A part of the responsibilities of all citizens is to encourage good international relationships. If more people would learn about the importance of good foods and other necessities to the peace of the world, they would be more sympathetic toward helping other nations. This help may be through increased trade and educational programs that assist them to farm better. Actually, the United States has gained much from foreign countries because almost all the animals on farms and many of the crops were originally imported. Even today, the United States has representatives looking for new plants of promise in foreign countries.

A changing relationship

The force of rural community improvement programs is slowly but surely changing the character and meaning of "life in the country." Rural areas have an abundant work force and plenty of space for living, working, and playing. To take advantage of these resources, rural area and community improvement programs are gradually bringing industry to the resources. Although some small towns are disappearing, other small towns are growing into large towns, and rural residential areas are developing. A new relationship between farmer, rural dweller, and urban dweller is in the making.

SUGGESTED ACTIVITIES

1. Hold a class discussion on the advantages and disadvantages of country life. What are some advantages that urban people have over rural people?
2. How would you answer the question "What is a good life for rural people?" How is this similar to and different from a good life for urban people?
3. Inspect your home and its surroundings. Note some features that should be improved. With other members of your family, discuss possible ways of making these changes. What are some improvements that you yourself can make?
4. List the foods you have eaten the past week. Which of the Basic Seven or Basic Four were included daily? Which ones were omitted? What foods should you eat more of or add to your diet to improve it?
5. What injuries have occurred recently on farms and elsewhere in your community? What injuries have occurred to members of your family? How were these accidents caused? How could they have been prevented?
6. What fire losses have occurred in your community the past year? What were the causes? How might they have been prevented?
7. With others in your class, start a safety and fire prevention campaign for your homes. Make a list of the hazards that might cause fires or accidents. Indicate the necessary steps to correct these situations.

8. What are some of the features of your high school that you feel are especially desirable? Why? What suggestions do you have for improving your school?
9. What library facilities are provided in your community? What provisions are made for rural people to borrow books? What suggestions do you have for improving these facilities?
10. What health facilities are available to your community? What improvements do you suggest?
11. What recreational activities are available in your commmunity? Which ones interest you? What activities might be improved or added?
12. What are some of the rural organizations in your community or in some other community with which you are familiar? What are some of their activities? Which ones have activities for youth?
13. With the help of your teacher, secure a copy of one of the following paintings or of some other famous painting related to rural life.[1]

Title	*Artist*
End of the Day	Adan
The Horse Fair	Bonheur
Oxen Plowing	Bonheur
Shepherd and His Flock	Bonheur
The Harvest	Breton
Song of the Lark	Breton
The Cornfield	Constable
One of the Family	Cotman
A Cornfield	DeWint
The Old Mill	Innes
Peace and Plenty	Innes
Shoeing the Horse	Landseer
Sheep, Spring	Mauve
Angelus	Millet
End of the Day	Millet
Feeding Her Birds	Millet
The Gleaners	Millet
The Man with the Hoe	Millet
The Sower	Millet
Interior of a Stable	Morland
A Boy with a Rabbit	Raeburn
Landscape with Mill	Rhysdael
Return to the Farm	Tryon

In an encyclopedia or elsewhere, study the life of one (or more) of these artists. Be prepared to describe the interesting events of the artist's life and the interesting features of the picture to your class. How did the person prepare himself/herself to be a great artist?

14. Read a book centered around rural life. Give a report on this book to your class, with special emphasis on the feature that interested you most. The following list will help you in selecting a book.

---

[1]Reproductions of many paintings may be secured at a low cost from the Perry Pictures Company, Malden, Massachusetts.

Rural Life in the United States

Title	Author
A Lantern in Her Hand	Bess Streeter Aldrich
Song of the Years	Bess Streeter Aldrich
The Farm	Louis Bromfield
Pleasant Valley	Louis Bromfield
Red Rust	Cornelia Cannon
Look to the Mountain	Le Grand Cannon, Jr.
As the Earth Turns	Gladys Hasty Carroll
Moon Valley	John Case
Peace Valley Warrior	John Case
Tom of Peace Valley	John Case
My Antonia	Willa Cather
A Goodly Heritage	Mary Ellen Chase
Windswept	Mary Ellen Chase
Hostages to Fortune	E. R. Eastman
The Hoosier Schoolmaster	Edward Eggleston
Always the Land	Paul H. Engle
Cimarron	Edna Ferber
So Big	Edna Ferber
Boy Life on the Prairie	Hamlin Garland
A Son of the Middle Border	Hamlin Garland
The Young Man in Farming	A. K. Getman
Barren Ground	Ellen Glasgow
Adventures in Contentment	David Grayson
The Friendly Road	David Grayson
Maria Chapdelaine	Louis Hemon
The Mississippi Bubble	Emerson Hough
Lone Cowboy	Will James
Smokey, the Cow Horse	Will James
But Look, the Morn	McKinlay Kantor
Free Land	Rose Lane
How Green Was My Valley	Richard Llewellen
American Ballads and Folksongs	John and Alan Lomax
Spoon River Anthology	Edgar Lee Masters
The Country Kitchen	Della T. Lutes
Country Schoolma'm	Della T. Lutes
My Friend Flicka	Mary O'Hara
The Brown Mouse	Herbert Quick
Vandermark's Folly	Herbert Quick
The Yearling	Marjorie K. Rawlings
Out to Old Aunt Mary's	James Whitcomb Riley
The Great Meadow	Elizabeth Madox Roberts
Giants in the Earth	O. E. Rolvaag
Peder Victorious	O. E. Rolvaag
Always the Young Strangers	Carl Sandburg
New Land	Sarah Lindsay Schmidt
Wagons Westward	Armstrong Sperry
Man with a Bull-Tongue Plow	Jesse Stuart
The Thread That Runs So True	Jesse Stuart
Country People	Ruth Suckow
The Earth Is Ours	Marion Teal
Fireweed	Mildred Walker
The Virginian	Owen Wister

15. Start your own library of books and bulletins on various phases of agriculture and rural life in which you are especially interested. (Many bulletins can be obtained from the agricultural college in

your state and from the U.S. Department of Agriculture. Bulletins from the latter source can best be procured through a representative or senator.)

16. Concepts for discussion:
    a. Life in the country is still very different from life in the city.
    b. Life in the country can be pleasant.
    c. Health and educational facilities have not developed as well in some rural areas as they have in some cities.
    d. A farm is still a business and a home combined.
    e. People living in rural areas tend to have better diets than do people in the cities.
    f. Farm families must plan for purchases carefully because of variations in the times of greatest income and greatest need for expenditures.
    g. Farm families work and play together more than do nonfarm families.
    h. Farming is a hazardous occupation.
    i. There are many organizations for rural people.
    j. The federal government affects the agricultural industry in many ways.
    k. There are many ways in which rural living can be improved.
    l. Often, rural people and urban people do not understand each other or the problems each has.
    m. Much music, literature, and art deals with rural living.
    n. Improved roads and means of communication have made rural living more enjoyable than it used to be.

REFERENCES

U.S. Department of Agriculture. *The Face of Rural America*. Yearbook of Agriculture: 1976, U.S. Government Printing Office, Washington, DC 20402.

U.S. Department of Agriculture. *Living on a Few Acres*. Yearbook of

Agriculture: 1978, U.S. Government Printing Office, Washington, DC 20402.

U.S. Department of Agriculture. *Our American Land*. Yearbook of Agriculture: 1987, U.S. Government Printing Office, Washington, DC 20402

U.S. Department of Agriculture. *Agriculture Outlook '92: New Opportunities for Agriculture*. Washington, DC 20402.

U.S. Bureau of the Census. Statistical Abstract of the United States: 1991, U.S. Government Printing Office, Washington, DC 20402

Chapter 3

Youth and Rural Living

In recent years, there have been many changes in how people live and what people value. Many family traditions and customs have disappeared. As a result, many of today's youth do not know what is expected of them. The rules of behavior are not clear, and adult guidance is not as firm as it should be. Adults have a great responsibility for helping youth to solve their problems and to become useful members of society. Likewise, youth have a challenge to help themselves and society. This is as true for rural areas as it is for urban areas.

OBJECTIVES

1. Discuss the needs and problems of young people.
2. Identify and describe organizations that have activities for young people.
3. Describe considerations that help make a youth organization successful.

What Are the Needs and Problems of Rural Youth?

Rural youth have many needs and problems similar to urban youth. They all need to learn how to be good citizens, to earn a living, to establish good homes, to engage in wholesome leisure-time activities, to maintain good health, and to do the other things expected of people in a democratic society. As rural youth "grow up," many of them secure jobs and establish homes in cities, while few urban youth become commercial farmers.

Some of the problems of rural and urban youth are not easy to solve to the satisfaction of all concerned. If these problems are to be solved in the best interests of youth and by democratic procedures, youth themselves must become sufficiently interested to study these problems and to assume part of the responsibility for solving them. Many adults are interested in "helping youth to help themselves" through advice and encouragement. Most youth appreciate help from adults who are sympathetic to their interests and viewpoints. This has become more noticeable because of a change in attitude on the part of youth towards adults. Youth want to be part of their families and to belong to groups. They seem to be more interested in traditional family values, patriotism, conservation, and the out-of-doors. They want adults to help them learn right from wrong, but as one youth put it, "We don't want adults to do things *for* us; we want them to do things *with* us."

Deciding how to earn a living

Many young people in school and many out of school think seriously about their life work. Frequently, they get helpful guidance from teachers and other adults who understand youth and can assist them with their problems. In some rural and urban communities, few youth get assistance from adults. However, large percentages who have received such assistance report that it has been helpful.

Some years ago, "Down on the Farm" was a popular song. In it was the refrain "How are you going to keep 'em down on the farm?" As a matter of fact, the plain answer is "You can't." It is true that many people can and should find places for themselves in farming as their life work. However, less than half of the young people reared on farms can find places in farming for earning a satisfactory living. The remainder must find jobs elsewhere. In brief, fewer and fewer people are needed to produce sufficient food and other agriculture products for the nation. Increased mechanization on farms has decreased the human labor needed for farming and increased the average size of farms, with a corresponding decrease in the total number of farms and people who live on farms. This has led to decreased opportunities for individuals to engage in farming. At the same time, there has been a great increase in the number of nonfarm agricultural occupations that offer job opportunities for rural youth. This situation is discussed in greater detail in Chapter 4.

Staying in school

"Should I stay in school or should I go to work?" This is a question many young people must ask. Some farm youth and others must face that they are needed at home or that their parents cannot afford to keep them in school.

A higher percentage of youth in urban centers than in rural communities graduate from high school. Rural youth who seek employment in cities

Fig. 3-1. An outdoor conservation class at Squillchuck State Park in Washington. (Courtesy, Soil Conservation Service, U.S. Department of Agriculture)

are in competition with urban youth, many of whom are better prepared. Furthermore, many who remain in rural areas are not as well prepared for employment as they should be. But, there are also many city youth of school age who do not attend schools of any kind.

Why should young people attend school? Whether or not they are now attending school, both rural and urban youth might well ask themselves this question. Attendance at school can help youth to become better citizens; to prepare for a vocation; to become better members of their families; to use leisure time more advantageously; to appreciate music and art; to stay healthy; and to read, write, and have basic math skills. A great many persons who are unemployed today are unable to perform even simple tasks requiring the use of reading, writing, and arithmetic.

If rural people hope to continue to improve themselves, they must be as well educated as the rank and file of urban people. Rural people should strive to become leaders in organizations and in governmental positions. They must select people for governmental positions who understand the problems of rural areas and who are broad-minded enough to seek solutions for these problems with due regard for the best interests of society as a whole. In developing these qualities, education has an important place.

Youth should realize the advantages of getting an education while they are still young. Rarely, if ever, do adults complain that they had too much schooling; but frequently older persons say that they regret that they did not attend school longer or "get more out of it" when they did attend. Young

Fig. 3-2. Students in schools offering agriscience programs learn job skills needed for a career in agriculture.

Fig. 3-3. A student learning to use power equipment in agricultural mechanics. Note the safety goggles and the push stick.

people can be seriously handicapped in life unless they get the best education possible. This is true whether they become farmers, homemakers, or mechanics, or whether they secure jobs along other lines.

Of special importance to youth is learning job skills while they are in high school. Whether they plan to go to college or to get a job after finishing high school, having some job skills can be very helpful. Youth who plan to go to college may find job skills useful in obtaining part-time work to help earn money to meet their expenses. Youth who plan to get a job after finishing high school may find that job skills may convince employers to hire them rather than someone else. In most communities, the youth who study vocational education subjects in high school are able to get jobs much more quickly than those who do not take vocational subjects. In some states, students are required to have a saleable skill or adequate preparation for college before they can receive their high school diplomas.

Earning pocket money

Many young people find it necessary to earn money during the years they are attending school. Some of this money may be needed to help pay their school expenses and to purchase clothing or other essentials. Young people also like to have some money to pay for social activities and other expenses involved in having a good time.

Certain hobbies and other activities may lead to opportunities for earning money. For example, youth on farms and small areas of land may raise livestock and crops for profit. In most cases, this may be done without seriously reducing the income for the rest of the family on a given farm. Popcorn, strawberries, raspberries, and vegetables are profitable crops in some sections. In some cases, youth may construct and sell certain articles, such as lawn equipment, rustic signs, and bird houses, in which they use native materials. They may collect and prepare for sale natural products such as berries, nuts, maple syrup and maple sugar, herbs, winter holly, and zoological and botanical specimens. They may sell materials such as cornstalks, ears of corn with husks, pumpkins, oak leaves, pine cones, and gourds for use as fall decorations in homes and stores.

Often, young people can earn money by establishing attractive roadside stands and selling products to tourists and others. For some periods of the year, they may find special employment in factories, forests, and resorts. Providing accommodations for tourists and guide service for hunters may be possibilities also. It is getting more and more difficult for both rural and urban youth to find summer jobs. Every opportunity needs to be seriously considered.

Rural youth should not overlook the opportunities for partnership with their parents in certain farming ventures. For example, they might assume much of the responsibility for the home dairy herd. By keeping records and

Fig. 3-4. In addition to earning money, this young person is securing valuable work experience. (Courtesy, U.S. Department of Agriculture)

using certain other practices of value, they may increase the profits from the herd, with a share going to them for their efforts. They may care for the home flock of poultry with similar arrangements and improve the income from it through better feeding, culling, improved marketing, and other practices.

Rural youth may also be able to work for local farmers. In one small town in Illinois, a local poultry farmer employed more than 20 boys, ranging in age from 11 years to 14 years, as egg sellers. Each boy had an egg route, much as paper carriers have paper routes. The farmer delivered the eggs to the boys' homes and the boys took over from there. The boys were paid a certain price per dozen for all eggs delivered, plus a bonus for each new customer obtained.

Some urban youth seek work on farms during the summer months. They feel that even though they do not expect to farm, work experiences of this kind are valuable and interesting. Such persons can expect a modest wage for their work while they are learning how to do the various jobs on farms. In some cities, the state employment service and the public schools help youth to secure jobs on suitable farms.

Young people should use their money wisely. In buying clothing or other articles, they should learn how to get the most for their money. They may put some money into savings accounts or government bonds. Youth on farms may decide to buy livestock with some of their earnings so that they can increase their income and in some cases aid themselves in getting started in farming.

Understanding the drug problem

One of the major problems youth face today is that of using drugs that have not been prescribed by a doctor. Young people and adults hear arguments that using marijuana and some other drugs is no more harmful than drinking alcohol and smoking tobacco—and we know how harmful the use of them can be. The terms *amphetamine, peyote, LSD, speed, marijuana, DMT, mescaline, cocaine, joint, acid head, trip, roach, angel dust, pipe, fix, shooting,* and *junkie* form a bewildering vocabulary for persons involved with drug abuse.

As in making other decisions, youth must decide for themselves if they want to try using drugs. Some factors for them to consider are the following:

1. The use or possession of marijuana, cocaine, heroin, and other drugs is illegal. Arrest and conviction can ruin a person's career.
2. Many hard drug users start out on marijuana—but that does not mean that using marijuana leads automatically to using hard drugs.
3. A drug habit can be very costly and can lead many youth to crime. Heroin addicts, alone, steal over a billion dollars' worth of merchandise each year.
4. Some persons have used marijuana for years with no outward ill effects—but this does not mean that using marijuana is not harmful.
5. Research has shown that using drugs can be very harmful. PCP and heroin can damage humans in many ways—physically, mentally, and emotionally. Genetic damage from drug use may extend to grandchildren.
6. Many drugs used for medical purposes (various kinds of barbiturates, for example) can be harmful when they are overused—thus the term *drug abuse.*
7. At first, drugs can give individuals the sensation of being completely free; later, these individuals may feel that they have become "slaves" to these same drugs.
8. Over 90 per cent of heroin addicts who undergo treatment for their addiction return to their habit. For many addicts, only death ends their addiction.

9. Youth who use drugs, both those who use them casually, such as the marijuana smoker, and those who use them heavily, such as the heroin addict, are found in all socio-economic classes.

Youth give many and varied reasons for using drugs. Some are just curious; others are rebelling against authority or trying to escape from their problems. All drug users seem to seek the exaggerated sense of well-being that comes from using drugs. Some youth feel they should be permitted to use marijuana; after all, their parents and other adults use and misuse alcohol and tranquilizers, both drugs. Other youth point out that a mistake in judgment by an adult is not a good reason for a youth to make a mistake in judgment.

Whatever their reasons may be for starting to use drugs, youth need first to take a long look at what happens to drug users and then determine whether or not they really want to pay the same price.

Solving personal problems and developing self-confidence

Like adults, young people have personal problems and worries. Some of these may cause youth to lose confidence in themselves. These problems, which youth frequently indicate if given the opportunity, include how to feel at ease with persons of their own age and how to plan for the future in times of uncertainty. Many believe they should be given more help in understanding themselves and in learning about sex and reproduction.

Many youth worry about whether or not others of their own age like them. In many cases, rural youth do not feel at ease in social groups. This is, in part, understandable since many rural youths work alone or with their own families and do not have many occasions to mingle in large crowds or groups.

A major factor in young people's lack of positive feelings about themselves may be the lack of opportunities for working and earning money for themselves. Studying about and learning job skills in the school vocational programs can help young people obtain part-time jobs, can help prepare them for starting small, part-time businesses of their own, and can help prepare them for participating in a family business enterprise. Knowing that he/she has the skill to perform work to earn money and to contribute to the family helps a person feel good about himself/herself.

Youth also express concern about other problems discussed in other sections of this chapter, such as earning money, choosing an occupation, continuing in school, having a good time, and participating in various organizations.

Youth themselves can do a great deal to solve their personal problems and thus build greater assurance in themselves. By paying attention to their dress and grooming and by learning the proper ways to meet people, they can overcome much of their shyness in mingling with groups. Some schools provide special training in dressing attractively, meeting people, and making friends.

Youth should take advantage of opportunities in school to participate actively in class discussions and in other school activities. Usually, teachers take an interest in the personal problems of their students and in helping them to understand themselves. Some schools have counselors. Instruction in science, social studies, family living, and other classes can help youth with their personal problems.

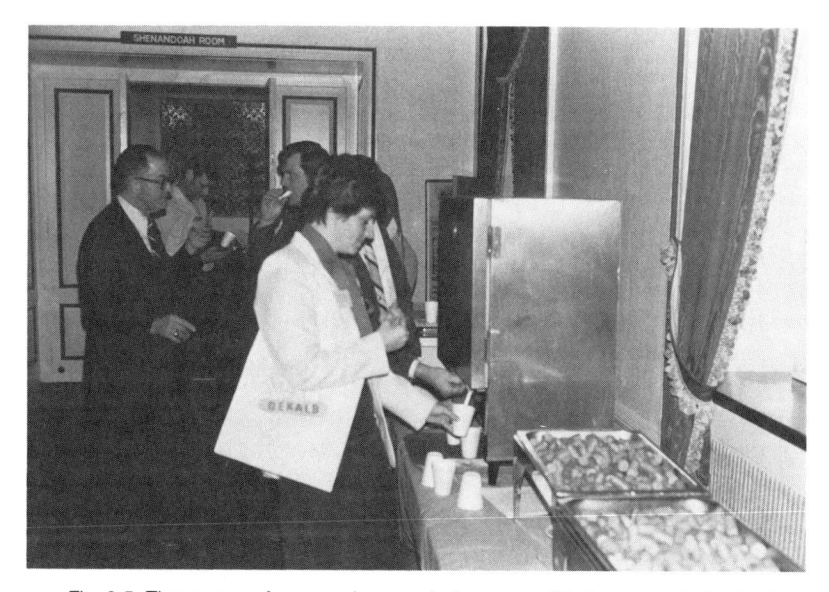

Fig. 3-5. These young farm people are enjoying some of their own products at a dairy bar. (Courtesy, Young Farmers of Virginia)

By taking part in school activities, such as music, hobby clubs, athletics, and speaking and debating, young people can build confidence in themselves. One high school student expressed this viewpoint in these words: "By joining school clubs, participating in school activities, and learning to do things well, you gain poise and confidence. You get so interested you don't think about yourself. Once you get the courage to break the ice, the rest comes easier."

As discussed elsewhere in this chapter, various organizations are available to youth. By becoming members of such organizations, young people have an opportunity to mingle together and to share in group activities of value to them.

In mingling with others of the same age, young people like to be part of the crowd; thus, they frequently do whatever the group suggests. In most cases, this may be desirable, but there are times when the group may engage in activities that are undesirable. Young people should not be afraid to live up to their convictions—they may have to show courage by suggesting a change in group plans or, failing in this, by refusing to go along with the group.

Some young people tend to be upset because they cannot solve their problems quickly. It is quite natural that they should worry about these situations. However, most young people adjust quite readily, especially when they have the help and counsel of adults, many of whom have had similar difficulties. These kinds of problems are not much different from those of young people of other generations. In the past, people usually learned to cope with the situations that confronted them, and there is every reason to feel confident that the present generation of youth will do the same.

Having a good time

"Happy is the person who has something to do." This applies to youth as well as to adults. Rural youth, like urban youth, face the problem of how to use their free time. They frequently indicate that they wish there were desirable places for them to go and organizations of interest to which they might belong. In fact, when rural youth are asked to state their problems, they usually emphasize the lack of social and recreational opportunities.

Most youth, if given the chance, choose wholesome types of activities. One of the problems facing rural people, formerly thought to be chiefly a problem of the city, is the temptation for some young people to engage in petty crimes or other undesirable types of activities.

Many young people in rural communities and elsewhere have developed hobbies of special interest to themselves. However, fewer rural youth report hobbies than city youth. How do you account for this? What hobbies have you had at previous times in your life?

Many persons who select hobbies may become so interested that they become experts or specialists in the activities chosen. A hobby may serve as a means for making money, and in some cases, it may develop into a vocation. The following list includes many hobbies and other activities appropriate for youth, especially rural youth.

AGRICULTURAL ACTIVITIES

Beekeeping (apiculture)
Landscaping the home grounds
Raising chickens
Raising Christmas trees
Raising farm animals
Raising farm crops
Raising flowers
Raising fur animals
Raising herbs
Raising plants without soil
Raising rabbits
Raising small fruits
Raising vegetables
Reforesting
Rock gardening
Tree grafting and budding

Youth and Rural Living

Fig. 3-6. Winter sports in the rural United States. (Courtesy, U.S. Department of Agriculture)

Fig. 3-7. Bicycle trail riding has become popular as recreation for youth. (Courtesy, Soil Conservation Service, U.S. Department of Agriculture)

COMMUNITY SERVICE

Developing a youth recreational center
Helping in a community cleanup campaign
Helping to raise funds for the Red Cross or United Way
Participating in collection of scrap and paper
Serving as playground assistants

CRAFTS

Basketry
Bird house construction
Blockprinting
Carpentry
Cooking
Furniture refinishing
Game and puzzle construction
Home decoration
Home workshop development
Leather crafts
Model airplane construction
Painting
Photography
Printing
Rug making
Sewing
Sketching
Soap carving
Taxidermy
Weaving
Wood carving

DRAMATICS AND SPEAKING

Acting in plays
Debating
Developing marionette and hand puppet shows
Organizing amateur and stunt shows
Organizing minstrel shows
Participating in discussion groups and forums
Speaking before groups
Writing plays

MUSIC

Participating in a family musical group
Playing a special instrument
Playing in a band or an orchestra
Singing folk songs
Singing in a glee club, a chorus, a choir, or an operetta

SOCIAL ACTIVITIES

Arranging tours to historic and scenic places
Dancing (modern, folk, and square dancing)
Developing quiz programs
Holding banquets for youth
Holding parties
Organizing spelling bees
Participating in harvest festivals and fairs

SPECIAL OUTDOOR ACTIVITIES

Camping
Collecting nature specimens (insects, fossils, rocks)
Fishing
Hiking
Raising wild fowls for release
Studying astronomy
Studying birds
Studying trees
Studying weeds
Studying wild flowers
Studying wildlife

SPORTS

Archery
Badminton
Baseball
Basketball
Biking
Bowling
Boxing
Croquet
Golf
Hockey
Horseshoes
Shuffleboard
Skating
Swimming
Table tennis
Tennis
Tobogganing
Tumbling
Volleyball
Wrestling

MISCELLANEOUS ACTIVITIES

Collecting coins
Collecting stamps
Organizing a plowing match
Participating in scouting
Performing tricks of magic

Taking part in group activities

Normally, young people desire to join groups and thus to develop a feeling of "belonging." Some organizations for rural youth are controlled primarily by adults, so youth do not have a chance to plan activities of interest to them. Frequently, youth organizations are not available, or youth do not know about them.

Young people on farms may feel they have few opportunities to participate in group activities with other young people. (Various group recreational and other social activities have been suggested in the preceding portions of this chapter.) Organizations of several kinds for youth are found in most communities. Some of these are discussed in this chapter.

In addition to activities of a purely social or recreational type, youth need to consider seriously their responsibilities as citizens in the local community, in the nation, and even in the world at large. Schools are helping a great deal along these lines, and various community organizations for youth also give attention to developing leadership and citizenship.

Developing youth centers

Youth like to have their own meeting places where they may play games, sit and chat, dance, and eat snacks. In many communities, no such places are available, so youth find undesirable places to go to. In some communities, youth centers have been developed through the active efforts of youth and interested adults. Youth are usually given a great deal of responsibility in running these centers, with supervision by responsible adults.

Living a good home life

Most young people are loyal to their own families and have a great deal of pride in their homes. Some, however, feel that their parents have no understanding of the problems of today's youth. Some young people feel that they do not share sufficiently in making decisions that affect themselves as members of the family group, such as establishing curfew hours, using the family car, planning work which they are expected to do, and improving the surroundings of the home.

As young people mature, they become concerned about finding suitable mates and establishing homes of their own. A situation of special concern to those who want to farm is that many potential mates now on farms would prefer to establish homes in cities rather than on farms.

Fig. 3-8. The Sierra National Forest provides a beautiful setting for horseback riding. (Courtesy, U.S. Department of Agriculture)

Most rural youth have unusual opportunities for a good home and family life. Many activities of farming involve the joint efforts of parents and their children. Young people may work with their parents to improve the surroundings of the home. They have the opportunity to become self-reliant by raising livestock and crops of their own and, at the same time, to earn money for themselves.

Farm families as a whole can be together much more than is usually the case with city families where several members of the family may work at occupations away from home. Farm families as well as city families may join together in various leisure-time activities, such as picnics, neighborhood meetings, or home family activities.

Modern conveniences now found in many rural homes make life pleasant for all members of the family. Much of the drudgery has been eliminated by modern kitchens, home laundries, and house-cleaning equipment. In such homes, young people like to bring their friends for snacks and good times. Most parents encourage them to feel free to use their homes in this manner.

Some rural youth, like urban youth, feel their parents are too strict. However, in most cases, this is in the best interests of youth. Through family discussions of matters involving the use of the car, time to get home at night, and similar "controls," parents and their children are usually able to arrive at decisions that are fair for all concerned. If young people show that they are responsible and use good judgment, their parents are more likely to allow them greater freedom in making their own decisions.

Some rural youth may feel that they have to work long hours. Instead of feeling sorry for themselves, they should realize that a reasonable amount of work on a regular schedule is excellent training for later life.

Organizations for Rural Youth

There are several organizations that have activities of interest to rural youth. Among the state and national organizations that include activities for young people, as well as for adults, are the Farm Bureau, the Grange, and the Farmers' Union. Organizations primarily for rural youth are 4-H Clubs and FFA. Organizations for rural and urban youth are Future Homemakers of America, Boy Scouts, Girl Scouts, and Camp Fire, Inc. In many communities, there are organizations that are purely local in nature for both youth and adults.

Young people can benefit from joining with some group or groups of persons their own age. If certain organizations are already in existence for which they are eligible, they may wish to become members of one or more. If none is available, young people may wish to take the necessary steps to organize one.

4-H Clubs

In a recent year, nearly 5 million young people between the ages of 10 and 21 were enrolled in 4-H Clubs in the United States. Their motto is "To Make the Best Better." While this organization was started for farm youth, only about 20 per cent of the membership is now from farms, while 24 per cent is in cities and suburbs with populations of 50,000 or more. The rest is in the smaller cities and nonfarm rural areas.

The meaning of "4-H" is explained in the 4-H pledge. Note the four key words that start with the letter h.

> I pledge
> My head to clearer thinking,
> My heart to greater loyalty,
> My hands to larger service,
> and My health to better living.

All young people who enroll in 4-H Clubs choose individual projects. Among the most popular projects have been raising corn, potatoes, beef calves, dairy calves, pigs, sheep, and poultry; canning fruits and vegetables; sewing clothing; and creating home furnishings.

In the programs of 4-H Clubs, attention is given to health, personal appearance, good manners, safety, music, dramatics, conservation, and handicrafts. Demonstration teams appear on local programs and at fairs. Exhibits of projects are given at fairs and special shows. In developing their projects, young people obtain valuable experiences along various lines. Annual achievement days are held in most counties at which awards are made to various members who merit them.

Activities of 4-H Clubs are especially popular for youth under 15 years of age. (The average age of all members is about 13 years.) In some states, those age 16 or over are given special considerations in Senior 4-H Clubs or Older Youth Clubs. The central agency that sponsors these clubs is the Office of Cooperative Extension Work, U.S. Department of Agriculture, Washington, D.C. The extension division of the college of agriculture in each state develops these activities through county agricultural agents, county 4-H Club leaders, and local leaders in the various communities. If you are now or have been a 4-H Club member, describe your experiences.

FFA

The motto of FFA is as follows:
> Learning to do,
> Doing to learn;
> Earning to live,
> Living to serve.

Fig. 3-9. Youth organizations promote learning by doing. *(Top)* A Virginia FFA member developing his leadership ability by participating in a public speaking contest. *(Bottom)* Junior 4-H leader teaching younger 4-H members bicycle care and safety, 1 of 50 "learn-by-doing" 4-H projects.

This organization recently had an active membership of over 500,000 young people enrolled in agriculture/agriscience in about 8,000 high schools that have local chapters. The FFA was founded as a national organization in 1928. Agriculture teachers are local advisors for FFA chapters.

The primary aim of the FFA is the development of agricultural leadership, cooperation, and citizenship. The specific purposes for which this organization was formed are as follows:

1. To develop competent and aggressive agricultural leadership.
2. To create and nurture in its members a love of agricultural life.
3. To strengthen the confidence of students of agriculture/agriscience in themselves and their work.
4. To create within members more interest in the intelligent choice of agricultural occupations.
5. To encourage members in the development of individual occupational experience programs in agriculture and in the establishment of agricultural careers.
6. To encourage members to improve their homes and surroundings.
7. To encourage members to participate in worthy undertakings to improve the agricultural industry.
8. To develop character within members, to train members to be useful citizens, and to foster patriotism within members.
9. To encourage members to participate in cooperative efforts.
10. To encourage members to be frugal.
11. To encourage members to improve their scholarship.

Fig. 3-10. Back in the saddle again! This young man established in ranching was chosen Star Farmer of America in the FFA. (Courtesy, Rod Turnbull, formerly of the weekly *Kansas City Star*)

12. To provide organized recreational activities and to encourage the development of others.

Active members in the FFA, through their achievements, become eligible for certain "degrees." They start as Green Hands and may earn chapter degrees after one year of successful work in agriculture/agriscience. Each year, some FFA members in each state are selected for state recognition on the basis of outstanding achievements in leadership, work experience programs, investments, and community service. Outstanding members from each state are selected each year for American FFA degrees. Recognition for progress toward establishment in agriculture is given through the Star Green Hand, Star Chapter Farmer, Star Chapter Agribusiness man, Star State Farmer, Star State Agribusiness man, Star Farmer of America, and Star Agribusiness man of America awards.

A program of activities is planned annually by each local chapter. This program includes activities related to agricultural experience programs, cooperation, community service, leadership, thrift, scholarship, and recreation.

Many changes have been made in the FFA to make it an organization that will serve equally well both youth interested in production agriculture and youth interested in agribusiness.

The formation of the FFA Alumni Association was an effort to provide stronger support for the FFA through the participation of former members.

Future Homemakers of America

The Future Homemakers of America (FHA) is an organization for high school youth enrolled in home economics or homemaking courses. In a recent year, there were 12,477 chapters with a total membership of 380,791. Teachers of homemaking in high schools are advisers of local chapters of this organization. The motto of this organization is "Toward New Horizons." The main goals of the organization are: (1) developing greater understanding among homes of the world, (2) training its members to be more democratic in all phases of life, (3) helping members to realize and accept their responsibilities in their homes, and (4) helping members to understand what homemaking can contribute to their future at home or in business.

Members of the FHA work together on many activities that contribute to better family living and community life. These include nutrition, home improvement, safety, child care, and recreation.

Boy Scouts

Many rural and urban boys belong to the Boy Scouts of America. Rural scouts provide activities of special interest to rural boys, including those of

Lone Scouts on isolated farms. Special interests may be selected in scouting, many of which deal with farming, country life, nature study, health, and safety. In recent years, money management, computers, cooking, traffic hazards, household emergencies, and crime reporting have become popular interests.

Merit badges are awarded for achievements along these and other lines. Scouts stress training for citizenship, as well as the development of special interests along many lines. Scout organizations in rural communities encourage joint activities with organizations such as the FFA and 4-H Clubs. Recently, membership has grown to over 4½ million young people.

Girl Scouts and Camp Fire, Inc.

These two separate organizations are, in many respects, similar to the Boy Scouts. These organizations are available in some rural communities. In a recent year, membership in the Girl Scouts was about 2¼ million. The Camp Fire organization, formerly for girls only, now admits boys, who now make up 12 per cent or more of its membership.

Fig. 3-11. Girl Scouts of the U.S.A. Brownies in the city learn about agriculture by visiting a supermarket. Brownies are also introduced to consumer education. (Courtesy, Girl Scouts of the U.S.A.)

Churches and other religious organizations

Churches in rural and urban communities sponsor activities of interest to youth. Some of these are especially active and interested in rural youth. The YMCA and YWCA are also available to rural youth in many communities. These organizations promote mutual sharing activities and wholesome recreation, as well as provide religious training.

Ruritan

Ruritan is a civic service organization whose purpose is to create better understanding between rural and urban people. Nearly all Ruritan clubs work with the local FFA and 4-H Clubs, and a third of the Ruritan clubs sponsor Boy Scouts. Through the Ruritan National Foundation and the Rising Senior (high school) Program, Ruritan gives awards and scholarship grants to youth. Many clubs provide and supervise community recreational centers and sponsor Little League and other athletic programs.

Organizations for older youth

Various kinds of organizations are being developed for young people in their late teens and early twenties. This age group, in many communities, has not been served very well, either by youth organizations or by adult organizations.

Fig. 3-12. Dave Kearney, outstanding student graduate of the Spring Creek High School, North Carolina, receives a plaque and Savings Bond from the Spring Bank Ruritan Club. (Courtesy, *The Ruritan*)

Fig. 3-13. Young people attend Young Farmers Association meetings to improve their farming skills. (Courtesy, Young Farmers of Virginia)

Fig. 3-14. Youth in the Young Adult Conservation Corps help keep the National Herb Garden ready for public viewing. These young people are watering rosemary and clipping basil in the Culinary Specialty Garden. (Courtesy, *Agricultural Research*, U.S. Department of Agriculture)

The agricultural extension service of colleges of agriculture through county agricultural agents in some states is sponsoring county and local organizations for older rural youth. In some states, the Farm Bureau is taking an active part in developing older rural youth activities.

In some states, out-of-school young farmers who attend classes provided for them in departments of vocational agriculture in high schools organize local groups known as Young Farmers Clubs or Young Farmers Associations. A few states have state associations of these local groups. The FFA Alumni Association, previously mentioned, may help to serve older youth.

The Rural Youth of the USA is another organization for young people with rural interests. A national conference is held annually. It acts as a coordinating organization for various activities and interests of rural youth.

International Farm Youth Exchange Program

Some organizations in the United States are cooperating with other countries in a program for exchanging visits of selected farm youth. Through this means, young people from the United States are sent to foreign countries to spend several months living with farm families in those countries. These persons are sponsored by organizations such as the National 4-H Club Foundation, the National Grange, and the FFA Foundation. Some foreign countries are likewise providing funds for sending selected farm youth to the United States. Under these arrangements in a recent year, 280 U.S. youth went to foreign countries, while 212 youth came to the United States where they lived with farm families in 35 states. These carefully selected youth are ambassadors of good will who help to promote international understanding and cooperation.

Fig. 3-15. Four Virginia 4-H members receiving orientation from Dr. C. D. Allen of the state 4-H staff prior to departure to other countries.

What's Ahead for Rural and Urban Youth?

"We don't want things handed to us. All we ask is a fair chance to get ahead." These statements express the viewpoints of many youth. Certainly it is to their credit that many youth are willing and eager to improve themselves. While some of their problems are not easy to solve, to a considerable extent, youth can help improve their own lot. Adults can provide valuable advice and cooperation.

In some rural and urban communities, community councils have been organized to help bring about improvements for young people and adults. Members of these councils represent many groups in their communities. Youth are usually represented and encouraged to participate. Does your community have such a council? If not, would it be desirable to have one?

Establishing new organizations

Youth in most communities have several organizations from which to choose. If there are none, youth should take the initiative to form organizations of interest and value to themselves.

An organization does not run itself. It must have good leadership and members who are loyal and willing to help. Successful organizations for rural youth have found the following suggestions to be of value.

1. Reach as many youth as possible.
2. Plan well-balanced programs with attention to recreation, education, vocations, community service, and other activities that appeal to young people.
3. Give the young people the major responsibility in planning the program, with counsel and advice as needed.
4. Set definite goals to be reached during the year.
5. Select capable leaders and members of committees.
6. Give all members a chance to take part.
7. Arrange for adult counselors who understand the problems of youth and who are willing to assist in various ways.
8. Keep good records and accounts.
9. Assist other community organizations and agencies.
10. Discuss the progress and accomplishments and note the places needing improvement at the end of the year.

As suggested in this chapter, youth can do much to help themselves, provide wholesome recreation, and improve their homes and communities. While they engage in these activities they are learning to become better citizens.

Youth and Rural Living 99

The future has many uncertainties, but youth can do a great deal by analyzing the situation objectively, securing the counsel of adults, and making the best plans possible. One sociologist has said, "If young people are to be successful, they need first to establish worthy goals; second, to organize their efforts to reach these goals; and third, to have the stick-to-it-iveness to accomplish their goals."

Youth must expect changes that will require adjustments in plans for the future. By preparing themselves and keeping alert, youth will be able to meet these challenges and look forward to the future with confidence.

SUGGESTED ACTIVITIES

1. Have members of your class indicate their problems. One way to do this is to have each person list his/her problems on a piece of paper and place it unsigned in a "problem box." What problems are listed most frequently? Which problems are similar to those discussed in this chapter? Discuss some of these problems in your class.

2. Specify some of the things you wish to learn in high school. What subjects and activities are aiding you in accomplishing these things? What additional subjects and activities do you need?

3. In your community, what are some of the reasons for young people quitting high school before graduation? What changes in the school might improve this situation? What other suggestions do you have that might encourage increased numbers to continue in school and benefit from it?

4. With others in your class, list the leisure-time activities of this group. What activities are similar for rural and urban youth? What ones are different for these two groups? What improvements in leisure-time activities do you suggest?

5. With your classmates, make a list of hobbies in which you and they participate. Plan a hobby show and have each person describe his/her hobby.

6. Hold a class discussion on the money-making activities in which the class members engage. What ones seem most appropriate for rural youth? Urban youth? What do they do with the money earned?

7. What are some of the things you could do at home to make it a better place for you and your family? What things could you do with the help of your family?

8. Describe to your class some youth organization of which you are a member or with which you are otherwise familiar. How might you get the most benefit from this or some other organization?

9. Secure information on some youth organization that interests you. Write to the headquarters for literature.

10. With other members in some organization to which you belong, plan a program of suitable activities that it will undertake. What activities might you include for recreation? Individual development? Group discussion? Community improvement?

11. Concepts for discussion:
    a. The needs of youth are similar regardless of where youth happen to live.
    b. Youth must assume part of the responsibility for solving their problems.
    c. Youth need to recognize the values of education.
    d. There are many ways in which youth can earn pocket money.
    e. Work experience is valuable for reasons other than earning money.
    f. Youth need to learn to use wisely the money they earn.
    g. Youth need to learn where to seek help for solving personal problems.
    h. Youth need to learn how to feel at ease in a social group.
    i. Youth help themselves by taking part in school activities.
    j. Youth need to have the courage to live up to high standards of conduct.
    k. There are many wholesome leisure-time activities in which youth may participate.
    l. Working, playing, and discussing in a family unit help youth develop into happy, responsible citizens.
    m. Rural youth can help improve relationships with other countries.
    n. Youth can do much to make their homes and communities better places to live.

REFERENCES

Binkley, H. R., and C. W. Byers. *Handbook for Student Organizations in Vocational Education.* Interstate Publishers, Inc., Danville, IL 61832.

Blue Cross and Blue Shield. *Drug Abuse: The Chemical Cop-Out.* Blue Cross and Blue Shield, Roanoke, VA 24000.

Russell, K. L. *The "How" in Parliamentary Procedure.* Interstate Publishers, Inc., Danville, IL 61832.

U.S. Department of Agriculture. *Americans in Agriculture: Portraits of Diversity.* Yearbook of Agriculture: 1990, U.S. Government Printing Office, Washington, DC 20402.

Chapter 4

Agricultural Careers

A century ago, there were only a few hundred different kinds of occupations in the United States. Today, there are many thousands of different ways to earn a living. Over 20,000 separate job titles are listed in the Dictionary of Occupational Titles.[1]

OBJECTIVES

1. Identify employment opportunities in the industry of agriculture.
2. Discuss agricultural industry career opportunities.
3. Discuss agricultural service career opportunities.
4. Identify professions in agriculture.
5. Identify and describe factors to be considered when choosing an occupation.

In 1988, about 118 million people were employed in the United States. By 2000, the number of people projected to be employed is over 136 million. Of the 118 million in 1988, about 21 million worked in agricultural occupations of various kinds. These consisted of over 3.5 million workers who labored on farms, 5 million workers who produced materials for farmers and provided special services for them, and 12 million workers who processed and distributed farm products. In addition, about 250,000 scientists served agriculture directly. There are more than 500 distinct occupations in agriculture, some of which require a college education, many of which do not.

There is no field of work that offers to young people a greater variety of opportunities for jobs stemming from the occupations themselves than do farming and ranching. Everyone must eat, and those tasks that individuals perform themselves to provide their own food are tasks farmers perform on a full-time basis. For example, raising food in a garden is something like farming, only on a very small scale.

---

[1] Available from the Superintendent of Public Documents, U.S. Government Printing Office, Washington, D.C. 20402

Farming involves constructing and maintaining buildings for livestock, for tools and equipment, and for crops. Thus, farmers need skills in carpentry, painting, electricity, plumbing, roofing, masonry, metal work, designing, and glazing. They must operate and maintain all kinds of tools and equipment. The skills needed by some farmers and ranchers include welding, metal lathe work, metal tempering, sheet metal work, tool conditioning, engine repair, forge work, drilling, and lubrication. Farmers treat plants and animals for diseases and insects, as well as for injuries of other kinds, thus engaging in some of the activities of the veterinarian, the exterminator, the plant disease specialist, and the insect control specialist. In harvesting and marketing produce, they undertake the processing, preserving, handling, and transporting. Farmers also carry out many tasks that are part of the work of the forester, the wildlife specialist, the conservationist, and the engineer. They often serve as weather forecasters. They must be at home in the business world since they keep records, manage finances, buy materials, sell produce, analyze business practices, employ workers, prepare tax returns, contract for special jobs, and deal with legal problems. In fact, farmers, even when their farms are specialized, must be able to perform a broad variety of tasks that other persons work at full time. In addition, they must know enough about many other tasks to be able to secure the proper help in dealing with problems that arise.

Farming is, indeed, an occupational field rich in both opportunities for employment within itself and opportunities for full-time employment in a specialty, in which one of the hundreds of skills and abilities the farmer has had to acquire is used. It is the specialties that make up so much of the non-farm agricultural employment opportunity and that also lead to full-time employment in parallel non-agricultural jobs. In addition to providing career opportunities, the skills and abilities learned for farming are also of use, almost daily, in maintaining and caring for the home.

No occupational field is more varied and complex in opportunity than is the field of agriculture. It offers opportunities for every kind of work from that of day laborer to that of teacher and scientist. This great variety of opportunities contributes to the problems young people have in planning their futures. Because of the number of occupations available today, young people find that making the choice of a life work is a more complicated problem than it ever was before.

On the following pages, materials that should help young people on farms and in cities decide on the type of life work in which they will engage are presented. These materials are centered primarily on agricultural occupations. Of course, many young people on farms and elsewhere will consider occupations in fields other than agriculture, but it is to their advantage to learn about opportunities in various agricultural occupations before they make their choices.

What Are Some of the Agricultural Occupations in Which People Engage?

There is an old saying that the grass on the other side of the fence always looks greener. Many young people who have been reared on farms or in small towns have the idea that they can get ahead only by going to the city. While it is true that many kinds of jobs are available in cities, it should not be assumed that opportunities are lacking elsewhere.

Some young people who have grown up on farms fail to realize that there is a wide range of agricultural occupations in which their own farm experience may be valuable. In other words, it is to their advantage to consider carefully various agricultural occupations, including farming, in which there are challenging opportunities. Also, many urban youth, even though they have had no background in farming, should consider agricultural occupations in which there are opportunities for careers. For all of these, a study of the agricultural materials in this book will be helpful. A study of agriculture is also an asset to people who become merchants, doctors, lawyers, ministers, or teachers in areas where they deal with rural people. Why is this true?

The number of people employed in farming is on the decrease. However, considering the millions engaged in farming, we can expect that many opportunities will continue to be available for those working in various farming occupations. In addition, the number of persons in agricultural occupations other than farming is steadily increasing, and there are many opportunities for qualified people to secure employment.

The various agricultural occupations can be classified in the following major groups:

1. *Occupations in production agriculture.* These include persons who operate farms as owners or renters and persons who operate farms as paid managers. Also included in these are hired workers and supervisors.
2. *Occupations in business and industry of an agricultural nature.* These include persons engaged in the processing and distribution of farm products and in the manufacture and distribution of products used by farmers.
3. *Occupations in agricultural services.* These include persons employed in providing various agricultural services to farmers and other people.
4. *Professions in agriculture.* These include persons in positions that require technical and professional preparation, usually received from an approved agricultural college or university.

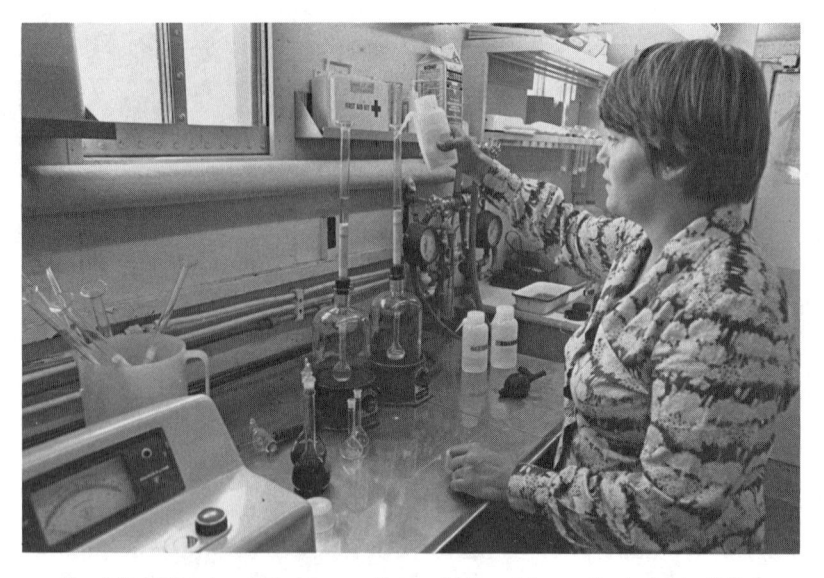

Fig. 4-1. A laboratory technician analyzes a high-protein, xanthophyll-rich alfalfa concentrate for poultry and swine to determine correct protein content. (Courtesy, *Agricultural Research*, U.S. Department of Agriculture)

The following lists of agricultural occupations are suggestive of the many kinds under each of these four groups.

OCCUPATIONS IN PRODUCTION AGRICULTURE

Beekeeper (apiarist)

Crop farmer:
1. Cash grain
2. Cotton
3. Fruit
4. Tobacco
5. Vegetable and truck crops
6. Miscellaneous crop specialties

Farm and ranch worker, supervisor, ranch hand

Farm hand—grain crop worker, ranch hand, laborer, herder, stable worker, dairy herd worker, and many others

Farm homemaker

Farm manager

Florist—grower, supervisor, worker

General farmer

Greenhouse operator, worker

Agricultural Occupations

Livestock farmer or rancher:
1. Beef cattle
2. Dairy cattle
3. Poultry
4. Sheep
5. Swine
6. Miscellaneous animal specialties

Nursery operator, specialist, supervisor, worker

Seed producer

OCCUPATIONS IN BUSINESS AND INDUSTRY OF AN AGRICULTURAL NATURE

Chick hatchery—operator, employee

Cooperatives—manager, employee

Farm appraisal, insurance, loans—agent, adjustor, agency operator

Farm supplies (feeds, fertilizers, seeds, etc.)—business operator, employee

Grain elevator work—operator, employee

Lumbering—logger, scaler, sawmill operator, faller

Manufacturing, distributing, and servicing agricultural equipment and products:
1. Electrical equipment for farms, farm machinery, tractors, and other equipment—dealer, demonstrator, manufacturer, mechanic, salesperson, service and maintenance specialist, etc.
2. Feeds—dealer, manufacturer, salesperson
3. Fertilizers—dealer, distributor, salesperson
4. Fungicides, herbicides, insecticides, etc.—dealer, manufacturer, salesperson
5. Seeds—dealer, distributor, salesperson

Pig hatchery—operator, employee

Processing and distributing farm products:
1. Cotton—gin operator, buyer, warehouse operator, grader
2. Dairy products—butter maker, cheese maker, distributor, ice cream maker, inspector, processing plant operator, produce dealer, worker, and many others
3. Fruits and vegetables—grader, packer, cannery operator, worker, and many others
4. Grains—commissioner, grader, elevator operator, grain exchange operator
5. Livestock—commissioner, buyer, trucker, meat inspector, meat salesperson, auction market operator, stockyards operator, packing-house employee
6. Poultry products—produce plant operator, buyer, egg grader, poultry dresser, packer
7. Tobacco—auctioneer, grader, auction market operator

Fig. 4-2. The county extension agent in the center uses farm records to counsel a young farmer family *(top)*. The home extension agent at the right provides nutrition counseling for a migrant worker *(bottom)*. (Courtesy, New York State College of Agriculture and Life Sciences, Cornell University)

OCCUPATIONS IN AGRICULTURAL SERVICES

Agricultural engineering—drainage specialist,[2] rural electrification specialist,[2] irrigation engineer[2]

Airplane dusting of crops—manager, pilot, helper

Artificial breeding—technician, manager

Auctioneer—livestock auctioneer, tobacco auctioneer, etc.

Country butcher

Country carpenter

Crop spraying service—manager, operator

Custom machine work—operator and owner of corn husker, corn sheller, grain combine, hay baler, etc., to perform work for farmers

Dairy Herd Improvement Association—field representative

Electrical service—electrician

Farm management service and general farm service—farm management specialist,[2] various specialists

Farm organizations—education specialists,[2] research specialist,[2] public relations specialist,[2] etc.

Farm record service—field representative

Feed grinding and mixing—operator of portable outfit or equipment in fixed location

Fertilizer service—mixer, operator of equipment for spreading fertilizer

Forestry—fire guard, fire ranger, forest ranger,[2] forest technician,[2] lookout, forest nursery supervisor,[2] scaler, timber cruiser

Fruit caretaker service (tillage, pruning, spraying,[2] picking, packing,[2] marketing)—manager, worker

Fruit spraying—spray outfit operator

Landscape gardening—gardener, helper

Livestock—trader and buyer

Mobile blacksmith shop—blacksmith

Mobile repair shop—operator

Poultry—blood tester, caponizer, chicken sexer, culler

Rural recreation—camping expert, guide, owner of hunting and fishing equipment

Farm supplies sales (feed, fertilizer, seeds, insecticides, fungicides, herbicides, etc.)

Seed cleaning and seed treating service—operator

Sheep shearing

Soil testing service—soil technician,[2] soil tester

Weed spraying—operator

Well drilling

---

[2] May also be classified as an agricultural profession.

108   AGRISCIENCE IN OUR LIVES

Fig. 4-3. A two-man crew is running a cruise of an area to gather data necessary for making a timber sale. (Courtesy, U.S. Forest Service)

Fig. 4-4. An auctioneer and ring worker selling beef cattle to the highest bidder. (Courtesy, U.S. Department of Agriculture)

Agricultural Occupations 109

Fig. 4-5. A technician places soil in a calcium chloride solution and heats it for 16 hours at 121°C under 15 pounds pressure per square inch. The extract is concentrated by centrifuging and is then tested. (Courtesy, U.S. Department of Agriculture)

PROFESSIONS IN AGRICULTURE[3]

Agricultural experiment station—research specialist in any one of many lines (usually in connection with an agricultural college or the U.S. Department of Agriculture)

Agricultural extension—county agricultural agent, 4-H club agent, home demonstration agent, specialist in any one of many lines

Civil service and non–civil service—agricultural chemist, agricultural economist, agricultural engineer, agricultural statistician, agronomist, animal husbandry specialist, bacteriologist, botanist, conservationist, entomologist, farm credit specialist, farm management specialist, food technologist, forester, geneticist, horticulturist, market specialist, rural sociologist, plant pathologist, geologist, seed analyst, seed certification inspector, soil scientist, and many others.

---

[3]Most of these professions require specialized education in a college of agriculture. Some require graduate study beyond a bachelor's degree.

Journalism and other communications—advertising specialist, agricultural editor, market reporter, photographer, radio and television broadcaster for farm audiences, agricultural writer and reporter

Landscape architecture

Teaching—high school agriculture teacher, college or university teacher in any one of many special fields

Veterinary medicine—food inspector, private practitioner, technician in commercial production of livestock medical supplies.

The chart entitled "Areas of Career Opportunities in Agriculture" is another way to visualize the extent of occupational opportunities in the field of agriculture. In every community, some persons can be found who are employed in the areas listed.

Careers in ecology

Special interest is being focused on environmental protection as a developing field for career opportunities. In this field there are many familiar occupational titles, such as geologist, meteorologist, oceanographer, forester, range manager, soil and water conservationist, wildlife conservationist or biologist, fisheries worker, recreational and parks employees, and technicians related to those occupations named.

Fig. 4-6. A Natural History Survey photographer taking pictures of thin sections of the internal organs of birds and mammals. (Courtesy, Illinois Natural History Survey)

Agricultural Occupations

Some other career titles have developed, such as that of ecologist and those of various environmental workers. All the occupational titles named deal with pollution prevention and control or environmental improvement for the health and enjoyment of all.

Fig. 4-7. Being a forest ranger is only one of many agricultural occupations in the area of wildlife management. (Courtesy, U.S. Forest Service)

Some of the familiar workers, such as farmer and commercial pesticide applicators, will need to learn new tasks and meet new regulations. Many widely used pesticides have been listed by the U.S. Environmental Protection Agency for restricted use. Farmers and commercial applicators need to be trained, licensed, and certified to buy and use these restricted pesticides.

Some occupations in architecture, engineering, and landscaping also deal with some aspects of environmental improvement.

AREAS OF CAREER OPPORTUNITIES

SERVING AGRICULTURAL PRODUCERS

Research

Agricultural Chemicals
Animal Science
Economics
Farm Engineering
Plant Science
Soil Science

Communications

Advertising
Exhibiting
Magazine Publishing
Market Reporting
Motion Pictures
News Reporting
Photography
Radio
Television

Education

Business Programs
College Instruction
Extension Programs
Government Programs
High School Teaching

FARMING AND OTHER

Business & Industry

Agricultural Chemicals
Banking and Credit
Cooperative Management
Custom Services
Farm Buildings and Utilities
Farm Equipment
Farm Management
Farm Supplies
Feeds
Fertilizers and Plant Foods
Forestry
Insurance
Land Appraisal
Pharmaceuticals
Sales and Management
Seeds and Grain
Transportation

Services

Agricultural Consulting
Agricultural Law
Agricultural Statistics
Agricultural Technology
Inspection and Regulation of
 1. Feed
 2. Seed
 3. Chemicals
Plant and Animal Quarantine
Quality Control and Grading
U.S. Department of Agriculture
Veterinary Medicine

Conservation & Recreation

Forest
Pollution Control
Range
Soil
Water
Wildlife

International Technical Aid Programs

Specialists in All Areas of Agriculture

Agricultural Occupations

IN AGRICULTURE

SERVING CONSUMERS

Research	*Communications*	*Education*
By-products	Advertising	Business Programs
Economics	Magazine Publishing	College Instruction
Food Distribution	Market Reporting	Extension Programs
Food Packaging	News Reporting	Government Programs
Food Processing	Photography	High School Teaching
New Products	Radio	
New Uses	Television	

AGRICULTURAL OCCUPATIONS

Business & Industry	*Services*	*Conservation & Recreation*
Dairy Marketing	Agricultural Consulting	Forest
Dairy Processing	Agricultural Statistics	Pollution Control
Fats and Oils Processing	Agricultural Technology	Range
Food Distribution and Marketing	Inspection and Regulation of	Soil
Food Packaging	1. Food	Water
Food Processing	2. Chemicals	Wildlife
Grain Marketing	3. Packaging	
Grain Storing	Meat Inspecting	
Meat Packing	Plant and Animal Quarantine	
Transporting	Quality Control and Grading	
	U.S. Department of Agriculture	
	Veterinary Medicine	

Adaptation of chart by W. Wessels, College of Agriculture, University of Illinois.

Careers in agriculture involving computer technology

In recent years, computer technology has been applied in many ways to the industry of agriculture. Thus, numerous new opportunities in agriculture have opened up for those persons who are interested in working with computers.

Many of the new opportunities involve the same jobs in agriculture, but in addition to having agricultural knowledge and skills, individuals must be able to use a computer to do the work.

Some examples of the kinds of jobs in agriculture that now require the ability to use the computer to some extent are as follows:

> Vocational agriculture teachers—for instructing students and working with farmers and other agricultural businesspersons who have computers.
>
> Agricultural extension workers—for operating computer management information programs for farmers and for working with farmers.
>
> Managers of computerized sales operations—for buying and selling livestock to persons and businesses in all parts of the country.
>
> Accountants—for keeping and analyzing agricultural business records and preparing financial reports.
>
> Personnel in livestock improvement associations and in cattle breeding associations—for keeping records and analyzing breeding programs.
>
> Secretaries—for preparing reports and sending correspondence.
>
> Agricultural researchers—for collecting data, analyzing data, and preparing research reports.
>
> Business managers—for studying business trends, keeping inventories, deciding what to buy and when to buy it, deciding when to sell, and preparing financial reports.
>
> Soil experts—for preparing maps.
>
> Weather experts—for making forecasts and preparing weather maps.

The list of jobs requiring the ability to use computers will continue to grow. It will not be too many years before nearly every home and every business will have a computer or a computer terminal.

What Should Be Considered in Choosing an Occupation?

The matter of choosing an occupation is a challenging task. To do it wisely, each person must be willing to spend considerable time and study. One of the first steps is to become familiar with a large number of occupations. A start has been made in this direction on the preceding pages. In choosing an occupation, each person should consider several factors. Some

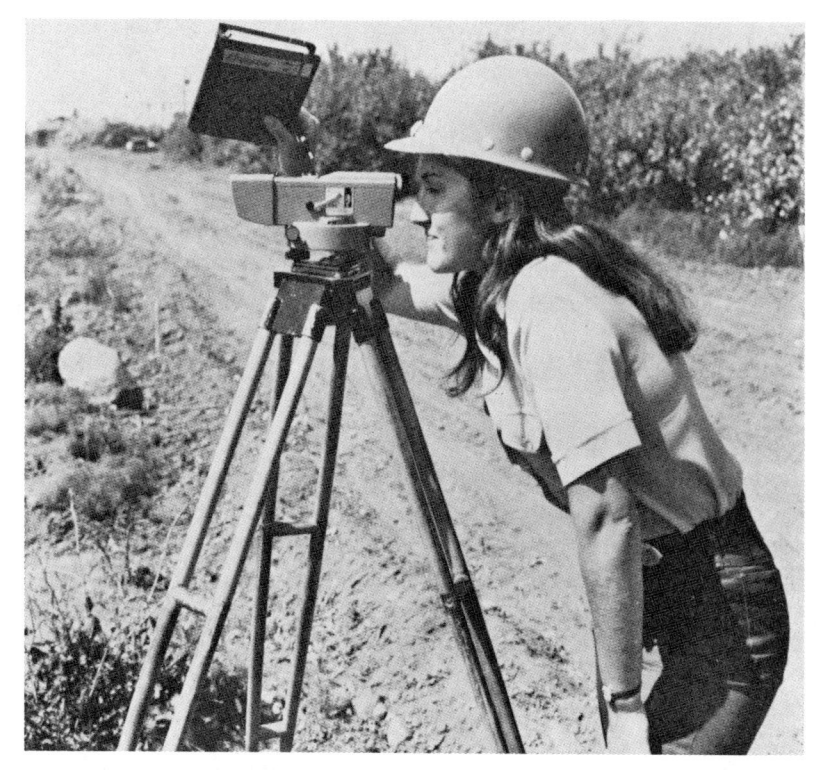

Fig. 4-8. A district conservationist surveying for a project in Colusa, California. (Courtesy, Soil Conservation Service, U.S. Department of Agriculture)

of the most important are included in the following questions and discussed in the following paragraphs.

1. What is the nature of the occupation?
2. What are the non-financial rewards and satisfactions?
3. What personal qualities, interests, and aptitudes are important for success?
4. What are the opportunities for securing employment?
5. What education and special training are required?
6. What are the financial rewards?
7. What are the advantages and disadvantages?

Nature of the occupation

Each person should become familiar with the nature of each occupation being considered. Individuals should consider the kind of work involved, the opportunities for service to society, the kind of people with whom they will be associated, the hours they will be expected to work, and various other features of the working conditions. (See the Job Questionnaire, a sample form that may be used for systematically obtaining and recording job information.)

In securing information about various agricultural jobs and other types of work, students should talk to their parents, teachers, school counselors, and other individuals who work in the fields under consideration. They should study information in books and bulletins that are available in schools and public libraries. They should secure publications from various other sources, such as federal agencies and universities. (Some of these materials are listed at the end of this chapter.)

Non-financial rewards and satisfactions

In addition to the money, there are various rewards and satisfactions to be gained from many jobs in agriculture and other fields. For example, some occupations in agriculture provide opportunities for service to people and to society. Some of the greatest satisfactions in life come from helping persons to improve themselves and their living conditions and to become responsible citizens. These features should be given serious consideration by individuals in choosing their life work.

Fig. 4-9. USDA plant quarantine inspectors examining cactus plants shipped into this country from Mexico for propagation. The inspectors are looking for pests and diseases. (Courtesy, U.S. Department of Agriculture)

Agricultural Occupations

JOB QUESTIONNAIRE

Date _____
      (Month)   (Year)

Name or title of occupation or job being studied _____

Name of firm or business where interview is being conducted _____

_____

Name of person interviewed:

    Employer or personnel manager _____

    Person employed in job being studied _____

Information to be obtained from employer about job being studied

I. Qualifications needed to work at job (in business where interview is being conducted):

    A. Education: Grade school ____ High school ____ College ____

        Vocational school ____ Other _____

    B. Special training and abilities _____

    C. Previous work experience _____

    D. Physical abilities _____

    E. Leadership and supervisory abilities _____

    F. Other qualifications _____

II. Job specifics (in business where interview is being conducted):

    A. Opportunity for placement:

        1. Number of workers at same job _____

        2. Turnover rate _____ Reasons _____

        _____

        3. Future employment increase or decrease _____

        4. Number of new workers needed next year _____

        Explain _____

B. Opportunities for advancement:
      1. Within firm _____
      2. Related occupations *(specify)* _____
      _____

   C. Salary and wage:
      1. Beginning wage _____ 2. Maximum wage _____

   D. Work benefits *(yes or no):*
      1. Group insurance _____ 4. Company investment program _____
      2. Retirement _____ 5. Other benefits _____
      3. Job security _____   _____

   E. Initial costs (uniform, bond, equipment, etc.) _____

   F. How to apply for this job _____

III. Areas of knowledge or subjects with which worker must be familiar (examples: soils, water, beef cattle, mechanics, accounting, etc.)
_____
_____
_____

Information to be obtained from person employed in job being studied

   I. Activities performed and/or duties of job _____
   _____
   _____

   II. Tools and equipment used _____
   _____

   III. Working conditions *(circle):* hot, cold; dry, wet; inside, outside; dusty; toxic; air-conditioned; other
   _____

> IV. Personal and social factors of job:
>
>   A. Size of community _____
>
>   B. Size of community in which this kind of job is usually found
>
>   _____
>
>   C. Number of years in job _____
>
>   D. Occupations of four or five friends: 1. _____
>
>   2. _____ 3. _____
>
>   4. _____ 5. _____
>
>   E. Recreation or hobby interests *(personal)* _____
>   _____
>
>   F. Organizations of which a member _____
>   _____
>
>   G. Non-financial rewards and satisfactions _____
>   _____

Personal qualities, interests, and aptitudes

Many agricultural jobs require relationships with other people. Hence, it is important for a person to have a pleasing personality and be able to get along with people. More failures in jobs are due to this factor than to lack of technical training.

Early in life, most individuals discover that they are more interested in certain types of activities than others. They may find that they like working with livestock, plants, or machinery, or that they enjoy several lines of work. Frequently, the lines in which they are interested are those in which they have developed some knowledge and ability, although this may not necessarily be the case. Sometimes young people become interested in occupations because those occupations have some special glamour or because they admire some individuals engaged in them. Later, they may find that even if they are interested in those occupations, they may not necessarily be cut out for them. In other words, they may not have the aptitude for performing the kind of work involved. Young people can gain some idea of their interests and aptitudes by determining what school subjects they like and do well in. Their parents, teachers, and school counselors can help them in this analysis. Most people, provided they are willing to put forth the necessary effort, will probably be successful in any one of several occupations.

Occupational opportunities

All individuals, before deciding definitely on their occupations, should consider whether or not there are opportunities for getting started in that line of work. A study of occupations in the home community should be helpful. If farming is the desired occupation, the individuals should consider opportunities for getting started on their home farms or elsewhere. If they are interested in some other agricultural occupations, they should secure information on employment opportunities in those occupations.

Certain trends are opening up new fields in agriculture or in other ways expanding opportunities in agricultural occupations. For example, there is a tendency at present for farmers with fairly large farms to increase the size of their farms. This is accompanied by increased use of tractors and large-scale machinery in general. More and more, farm workers who understand machinery and can operate tractors efficiently are being hired. On the other hand, there is a movement at the other end of the scale for an increased number of small farms. Some persons with a farm background and suitable training may wish to undertake small-scale farming as a sideline to their employment in nearby industries.

Some of the fields of agricultural occupations likely to continue in importance are (1) various types of soil specialties, (2) farm management, (3) farm equipment maintenance, (4) electrical service, and (5) many other service occupations previously listed. Furthermore, opportunities will continue for many occupations in agriculture for which professional preparation is required, as discussed under the next topic.

It is of course important for young people to understand that no guarantee can be made that jobs will be waiting for them in any field, or that success is assured once they have obtained a job. The supply of jobs is influenced by changing economic conditions. Individuals who wish to secure certain types of jobs must expect that they will have to move to areas where there are employment opportunities in those fields. Unless they are willing to do this, they will have to select from the kinds of jobs available within commuting distance of where they especially want to live.

Education and special training required

Most occupations require professional preparation, special training, or both. For example, to become teachers of vocational agriculture, individuals must graduate from an approved agricultural college and also have special

Agricultural Occupations

training for teaching. They must also have agricultural work experience. Certain personal qualities are necessary as well. They must be of good character and have a pleasing personality. They must be able to work successfully with others, including high school youth, young farmers, and adults.

Young people who have a desire to attend college and who are likely to succeed in college should consider the many opportunities in agricultural occupations open to persons with specialized college education and other qualifications. The fields include (1) agricultural research, (2) agricultural business and industry, (3) agricultural education, (4) agricultural communications, (5) agricultural conservation, (6) agricultural services, and (7) others. Helpful information may be secured from the state colleges of agriculture.

For most kinds of agricultural jobs, a high school education is a minimum requirement. Young people may gain some specialized training and experience while they are in high school and secure further training on the job.

It is also important to think in terms of the occupational flexibility provided by the particular educational program followed. Some programs may prepare persons for only one kind of work, especially educational programs beyond high school. According to some authorities, most people will have to prepare themselves for three occupations during their lifetime. With this in mind, it may be wise to think in terms of a particular group of occupations and plan an educational program for the occupational group rather than for

Fig. 4-10. Mite specialists douse a sample of bees and then examine the drain water to determine if the bee colony is infested with mites. (Courtesy, *Agricultural Research*, U.S. Department of Agriculture)

Fig. 4-11. A wildlife biologist removing tick from dead rabbit. (Courtesy, Illinois Natural History Survey)

Fig. 4-12. The marketing of flowers provides many jobs. This wholesale flower firm in Los Angeles, California, specializes in commission buying for retailers. (Courtesy, U.S. Department of Agriculture)

one specific occupation. This would make a change of occupational direction easier if a change became desirable or necessary.

The following groups of agricultural occupation titles illustrate occupations with enough common elements that young people can follow similar educational programs in high school to prepare for them.

GROUP I

Agricultural engineer
Agricultural lawyer
Conservationist
County agent or farm advisor
Farmer
Farm management specialist
Field representative, cannery
Field representative, Dairy Herd Improvement Association
Herder
Nursery operator
Vocational agriculture teacher, high school
Vocational agriculture teacher, college

GROUP II

Animal scientist
Geneticist
Horticulturist
Laboratory technician
Plant pathologist
Soil scientist
Soil technician
Veterinarian
Vocational agriculture teacher, high school

GROUP III

Farm equipment salesperson or dealer
Farm insurance salesperson
Farm supplies dealer
Fertilizer, seed corn, or hatchery business salesperson
Grain elevator operator
Livestock buyer
Livestock commissioner
Seed plant manager

The lists just given are not intended to be complete. The readers can add to these lists or develop their own group of occupational titles.

One other factor should be noted about the occupational flexibility desirable in occupational and educational planning. Individuals should also examine their agricultural occupation choices for flexibility in being able to change to non-agricultural fields. For example, the following occupations all have their counterparts in non-agricultural work: high school vocational agriculture teacher, geneticist, college vocational agriculture teacher, agricultural chemist, farm insurance salesperson, agricultural businessperson, laboratory technician, agricultural statistician, agricultural engineer, and many others. Youth can have confidence that a decision to study in the broad field of agriculture will yield a great variety of desirable occupational choices.

Financial rewards

Most people want to earn enough to provide a good living for themselves and their families. Beyond this, the desires of individuals vary. In some occupations, favorable retirement systems provide an income in the later years of life.

In addition to considering the starting salary or income in given occupations, individuals should consider the opportunities for increases in income as they gain experience and assume additional responsibility.

Advantages and disadvantages

Before selecting an occupation, a person should study it from the standpoint of its advantages and disadvantages. Nearly every job has some disadvantages, but usually the advantages far outweigh the disadvantages. What are considered disadvantages by some people may not be considered as such by others. The opportunities in agricultural occupations are so great that nearly everyone interested in agriculture and rural life can find opportunities according to his/her likes. The same may be said for many other kinds of occupations.

Looking for a job

Young people can obtain information about possible jobs from many sources. Some of these sources are (1) placement offices, (2) advertisements, (3) notices posted in public buildings, (4) friends and relatives, (5) teachers and counselors, and (6) companies. Job seekers should keep in mind that they do not have to locate actual job openings before they approach companies and individuals about working for them. If they will complete applications for the companies that do have the kinds of jobs they wish to secure, then the companies will be able to contact them about job interviews when openings do occur.

Once individuals have located job openings or companies for which they wish to work, they should contact these companies about filing a job application and possibly having a job interview. When looking for a job, young people should present themselves in the best possible way to their prospective

Agricultural Occupations

employers. Some of the questions potential employers will probably ask young job seekers are as follows:

1. What are your plans for the future?
2. What subjects did you study in school? What subjects did you like best? Which did you like least?
3. What extra-curricular activities did you enjoy most in school?
4. Do you have any hobbies that would help you in this job?
5. Which jobs are you interested in? Why?
6. Why do you want to work for this company?
7. Have you worked before? If so, where? How did you like the job? What military service experience have you had?
8. Why do you think you can be successful in the kind of work you wish to do?
9. In what part of the country do you want to live and work?
10. What salary or wage would meet your needs? How much do you hope to earn 10 years from now?
11. Have you had any experience in managing money?
12. Are you willing to travel on the job or to go where a job is located?
13. How would this job help meet your personal ambitions and lifetime goals?
14. What responsibility have you had for the work or activities of other people?
15. How would you describe a good boss and a good company to work for?

Fig. 4-13. Signs such as these indicate opportunities for jobs in agriculture.

Some of the questions are asked just to give a job seeker something to talk about so the interviewer can learn how able the person is at self-expression and what kind of person he/she is. Other questions are asked to help the interviewer find out whether or not the job seeker has the right preparation and experience for the job. In either case, it is important that the job seeker be able to speak easily and well in response to all questions asked. Some of the specifics about the job seeker that the interviewer will be looking for are as follows:

1. Neat personal appearance.
2. Ability to plan for the future.
3. Ability to work with others.
4. Willingness to work for what the company can pay.
5. Positive attitude toward any previous employers.
6. Positive attitude toward school.
7. Ability to deal with a difficult situation involving the job or other people.
8. Promptness.
9. Ability to make decisions.
10. Ability to accept responsibility for own behavior.
11. Poise and vitality.
12. Clear self-expression.
13. Enthusiasm and ambition.
14. Sense of humor.
15. Tolerance of others; courtesy.
16. Good moral character.
17. Willingness to do things for the general good.
18. Appreciation for the time and help of others.
19. Knowledge of and skills relating to the job desired.
20. Broad range of interests.
21. Maturity.
22. Ability to manage finances.
23. Appreciation for the value of experience and education.

How Can People Become Established in Agricultural Occupations?

People can become established in agricultural occupations in many

Agricultural Occupations

ways. For some, specialized education is needed before they can secure a position. This is true of the agricultural professions, most of which require education from a college of agriculture. For some jobs, particularly in business and industry, considerable training is given on the job. For some, partial preparation is secured through technical institutes or short courses in colleges of agriculture.

Certain agricultural jobs can be secured only after extensive work experience in other kinds of agricultural jobs. For nearly all agricultural occupations, a knowledge of agriculture, as acquired through high school courses in general agriculture and vocational agriculture, and through farming, is an asset.

Some agricultural occupations require knowledge and skill in various phases of mechanics. These include rural electricians, repair workers, and service technician for farm equipment and machinery, and many others. Some of these may be entered through apprenticeships and some by varied experiences that lead to the development of the necessary skills. Some high schools provide specialized courses for developing mechanical skills.

Young people use many routes to get established in farming on a full-time basis. Because of the limited opportunities to farm, the broad training needed, and the amount of capital required, many young people who become farmers are children of farmers. In many cases, in the early stages of establishment, family partnerships are developed. In a majority of the cases,

Fig. 4-14. An engineering technician checks equipment that records water run-off and soil loss from clean, cultivated corn. (Courtesy, Soil Conservation Service, U.S. Department of Agriculture)

Fig. 4-15. A science instructor points out features of good wetland habitat to a small group of Minnesota students. (Courtesy, Soil Conservation Service, U.S. Department of Agriculture)

young people who get established in farming receive help of some kind from either their parents or other relatives.

Those who engage in part-time farming frequently do so by purchasing small acreages and establishing homes on the land after they have become heads of families. These individuals secure most of the family income from jobs in industry and other lines of work.

Regardless of the occupation entered, success usually comes only with effort and planning. In most occupations, conditions are changing rapidly and require continuous study and thought. Farming and most other agricultural occupations are no exceptions. Thomas Edison once made the following statement: "There are three things which insure success, to-wit: ambition, imagination, and the will to work. Of these, the will to work accomplishes the most. Education of the right sort will short-circuit the process and get quicker results."

Who Are Some Individuals Who Have Had Successful Careers in Agricultural Occupations?

A study of people who have been successful in agricultural occupations

Fig. 4-16. A plant breeder working with sorghum to develop drought-resistant varieties. (Courtesy, U.S. Department of Agriculture)

should be interesting and helpful. Individuals in these fields have possibilities for achieving distinction and satisfaction, as is true also for people in other occupations. Following are several brief sketches of successful careers of men and women engaged in agricultural occupations. No doubt, you can give similar examples from your own observations and readings.

A farmer who became a noted corn breeder

Lester Pfister, a corn breeder, has won national fame for himself in the development of superior varieties of hybrid corn. The story of his determination, in the face of apparent failure, and finally of his success reads like a novel. He gained his inspiration many years ago when talking with Henry A. Wallace, former Secretary of Agriculture and former Vice-President of the United States. Wallace got him interested in corn breeding along lines that were relatively new at that time. Pfister worked for years with many strains of corn in an attempt to produce superior seed. He did a great deal of his work in a field enclosed by a hedge, because some of his neighbors made fun of his "freak" ideas. When it seemed he was on the point of developing some corn of special merit, there was a severe drought, which threatened to destroy the corn in his breeding plots and thus destroy years of work. His neighbors advised him to haul water to save the seed plots. His reply was "Let the weaklings die." The weaklings did die, but some strains did not. The strains that survived were used for further experimentation. Finally, with his farm heavily in debt, and after years of painstaking work, he developed some hybrid corn that proved superior. Only then did he feel he was in a position to produce seed he could recommend to sell to others. At present, the annual seed sales of his firm run into many thousands of dollars. However, his greatest satisfaction comes from being able to achieve success in what he started out to do—namely, develop superior varieties of corn. It was a true mark of progress when the one-millionth bushel of Pfister hybrid seed corn was sold. At a special banquet to celebrate the occasion, Pfister said, "This millionth bushel means more to me than the production and sale of the vast amount of seed. It means that into the pockets of farmers like yourself have gone many millions of dollars of added profits." To date, many million bushels of hybrid seed corn have been produced under the direction of Pfister.

A city man who became a successful country editor

James M. Savell moved from the city to the country and made good. He became editor and publisher of a weekly country newspaper and publishing firm in a town in Missouri. Although he had previous experience in printing, he did not enter the newspaper field until he was 50 years old, at the time he moved from New York City to take over a country newspaper. In addition to using the newspaper to report local news and the usual type of items found in such papers, he started to work for a better community. He wrote editorials on what improvements the community needed, and he helped the town get better streets and other improvements. He also helped to organize a "Good Government League" to encourage qualified people to run for office. This man liked country life and gave his best efforts to achieve success.

A remarkable farm woman and leader

Mrs. Raymond Sayre of Iowa has been called the best-known farm woman in the world. She was, first of all, a successful homemaker and housekeeper on a large farm. Even while rearing a family of four children, she found time to participate actively in community affairs. She became state president and then national president of the Associated Women of the American Farm Bureau Federation. Later, she was elected president of the Associated Country Women of the World, a worldwide organization. With other farm women from this country, she went to Europe to visit farm homes in foreign countries and thus learn about their problems first hand. In 1950, she was selected as Iowa Mother of the Year. She made the statement that the sign of a good homemaker is a healthy, happy family. She believes that the greatest task of women everywhere is to help build a world in which families can live in peace. Although she did not move to a farm until she married, Mrs. Sayre soon saw the need for better family and social life in rural areas and proceeded to do something about it. Though she had many responsibilities away from home, she continued her responsibilities as a homemaker. For her outstanding contributions to society, she was awarded the Doctor of Laws degree by Iowa State College.

A famous writer and plant specialist

Liberty Hyde Bailey, one of the grand old men of agriculture, grew up on a farm where he learned early in life how to work hard. Serving for many years as professor of horticulture and as dean at Cornell University, he became the world's leading authority on garden plants, palm trees, and blackberries. Early in this century, President Theodore Roosevelt appointed him as chairman of the Country Life Commission to study the problems of rural life. During his lifetime, he travelled to nearly every corner of the world to study and collect plants—he celebrated his ninetieth birthday while on a scientific hunt for plants in the West Indies. He expressed his philosophy of life in these words: "The measure of life is the living of it. The earth is good, and it is a privilege to live thereon." In 1950, at age 92, he was awarded the first Distinguished Service Award of the American Agricultural Editors Association. His illustrious career ended with his death at nearly 97 years of age, but his contributions to society, including the 156 books he wrote and/or edited, will continue to live on.

An efficient manager of a farm cooperative organization

Andrew G. Lohman was called in as an auditor to straighten out some tangled accounts in a struggling cooperative. The cooperative then gave him a part-time job at $50 per month as a bookkeeper, and persuaded him to take the job as manager when it was about to fold up. Through his efforts, the sales of the firm grew to more than $3 million per year. His assets were an eighth grade education, some training in a business school, a farm background, and a willingness to work and tackle new projects. During the many years that he was manager, the firm as a whole and every department in it made money every year. Through constant expansion and efficient management, the business added new services, which paid off in every case. Mr. Lohman had a capable staff of workers, who at scheduled times got together with him to discuss problems and to receive instruction. He was continually trying to find new ways of doing things. A wide variety of products and services was provided by this firm. Mixing and selling fertilizer and feed, marketing eggs, selling petroleum products, providing trucking service for certain products, manufacturing crates for vegetable crops, and providing garage services were some of the activities in which the firm engaged. When asked what he felt attributed to his success, he replied, "I believe in running the business of a cooperative as well as I would my own business."

A first-rank chemurgist

Dr. George Washington Carver was a black scientist at Tuskegee Institute in Alabama for more than 50 years. Born a son of slave parents and once traded for a racehorse, he became one of the world's famous research specialists in agriculture and a first-rank chemurgist. He worked his way through high school and through Iowa State College. During his years of study and experimentation at Tuskegee Institute, he developed more than 300 uses for the peanut and 118 uses for the sweet potato, as well as making many other contributions of value to agriculture. For many of these experiments, he constructed much of his own equipment from materials costing little or nothing. Once he was invited to present some of his findings to a committee in the U.S. Congress. Scheduled to speak only a few minutes, he so fascinated his listeners that they kept him nearly two hours. In addition to

Fig. 4-17. George Washington Carver, famous black scientist, made many scientific discoveries, including the development of numerous uses for peanuts and sweet potatoes. (Courtesy, P. H. Polk, Tuskegee Institute)

being a great scientist, he was a man of many talents and interests. He was an excellent musician, a thorough student of the Bible, an artist, an expert cook, and a skilled sewer and weaver. Dr. Carver lived very modestly and deprived himself of many comforts he might have had. He was imbued with a desire to serve society and to improve the lot of his own people. He donated his entire life savings for use in furthering research. A national monument and a museum have been established at his birthplace near Diamond, Missouri. In 1948, a special issue of a U.S. postage stamp was designed to commemorate his contributions to humankind.

A successful farmer and community leader

Leslie Heiser, a farmer in Fisher, Illinois, has been very successful as a farmer and an agricultural leader. With only eight years of formal schooling, he took advantage of opportunities to improve himself through reading, attending adult-farmer classes in the local school, and using services provided by the college of agriculture and other agencies. He was selected as a member of an advisory council for the department of vocational agriculture in the high school and worked enthusiastically for adult education and community improvement. He helped organize a Swine Herd Improvement Association among the local farmers, was its first executive secretary, and later

Fig. 4-18. Leslie Heiser *(left)* became a successful farmer, community leader, and state official in the Illinois Department of Agriculture. He is shown with the agriculture teacher who was his instructor in adult-farmer courses.

became the first president of a statewide organization for this work. He became a teacher of adult-farmer classes in the local high school and has been an active member of his church. Through his activities, he helped promote soil conservation on the farms in his community. He operated his 120-acre farm so successfully that the income equaled that of a typical farm twice as large. With the help of his wife, he modernized the house and improved the farmstead. He was selected as executive secretary of the Association of Illinois Soil Conservation Districts. For several years he was in charge of soil conservation in the Illinois Department of Agriculture and manager of the farms connected with state institutions.

When asked how he could spend so much time on activities away from his farm business, he replied, "Is it more important for me to try to accumulate as much money as possible for my sons when I am gone, or is it better to spend as much time as possible in making this a better community in which they will grow up?"

An exceptional farmer and a national farm leader

Allan B. Kline started farming in Benton County, Iowa, with very little money and on a farm with run-down soil. He developed a farming program of corn, legumes, and hogs. He managed to survive a serious depression and to acquire more land. By 1943, he had a 440-acre farm, debt-free, and was raising hogs by the carload. Even during the low-income years, he and Mrs. Kline were able to equip their home with many modern conveniences. Among the things that made farm life attractive for the family and friends were a swimming pool and a tennis court. For many years, he was active in community life and farm organizations but still had time to participate in family life. He earned degrees from two colleges. He was especially interested in debating, singing, and studying economics and philosophy. He received a Master Farmer Award in 1937. He made several trips to Europe to study and advise European farmers. Over a period of about 20 years he progressed from president of the county Farm Bureau to vice-president and then to president of the Iowa Farm Bureau Federation. After his presidency there, he became vice-president and then president of the American Farm Bureau Federation.

A distinguished gladioli grower

Mrs. Mary Kinyon of Michigan became famous as a gladioli grower. She started growing gladioli many years ago when a friend gave her some bulbs. This became a hobby that she shared with her husband. She became interested in the delicate process of hand pollination and through this method developed many new varieties of outstanding beauty. She received many awards for outstanding varieties she produced. When asked for her recipe for a happy life, she said, "I guess the secret is to be interested in one thing especially. Then you never grow old."

Outstanding farm homemakers

In some states, outstanding farm homemakers are selected each year. These are frequently designated as master farm homemakers. The individuals selected have done an unusually good job of homemaking and rearing a family, and they have spent considerable time in community affairs. They have been leaders in schools, churches, and other community organizations. Their accomplishments are a challenge to others and illustrate that homemakers on farms have much to contribute.

The dean of the college of agriculture in one of the states, in presenting the award to one of the homemakers, said, "She has done well as a homemaker; has sensed that the field of homemaking includes not only the house but the community and the public at large; has by her leadership, though quiet and unassuming, been effective; and has helped to make rural living more enjoyable and fruitful."

SUGGESTED ACTIVITIES

1. Make a list of the occupations in which you are most interested. Why are you interested in these occupations? What additional possibilities have you considered as a result of studying this chapter?

2. With persons in your class, make a survey of the agricultural occupations in your county, including various kinds of farming and other agricultural occupations. Compile a list of these occupations. Which ones from this group are of special interest to you? Why?

3. Of the agricultural occupations you have listed, which ones may be entered with no college training? Which require college training? What other special requirements must be met for each occupation?

4. Make a "Career Book" for yourself in which you assemble information for several occupations in which you are interested. For each, gather information and pictures, and discuss the advantages and disadvantages, the opportunities for employment, the requirements, and other factors that you should consider in making a choice. Use this information in helping you decide on occupations for which you might prepare yourself.

5. Interview someone who has been successful in an agricultural occupation in which you are interested. What advantages and disadvantages does this person give? What other items of interest relating to the job does he/she indicate? Add this information to your Career Book.

6. From an encyclopedia or other references, secure information about the life and accomplishments of some person who has made an outstanding contribution as a leader in agriculture. In what way did this person render services of value to farming? To society in

general? (Selections may be made from Stephen M. Babcock, William D. Hoard, Eli Whitney, Barbara McClintock, Anna Botsford Comstock, Cynthia Westcoft, John Deere, Cyrus McCormick, P. G. Holden, Luther Burbank, Henry A. Wallace, and others.) As the result of your study and the descriptions on the preceding pages, what does it mean to be successful in an agricultural occupation? What factors account for the success of the person studied?

7. With your teacher, arrange to have several speakers appear to discuss the occupations in which they are engaged. Try to get people who are active in different agricultural occupations.

8. Compare the services rendered by a county agent or a veterinarian with those of a lawyer or a doctor. How do the services compare? What reasons, if any, do you have for rating one occupation higher than the others?

9. How did one or both of your parents (or some other persons whom you know well) get a start in their occupation? In what ways do their experiences provide suggestions for others who wish to get started?

10. Concepts for discussion:
    a. There are hundreds of occupations in agriculture.
    b. Some agricultural occupations require a college education, but most of them may be entered with high school or technical school education and training.
    c. The study of agriculture is an asset to persons in occupations dealing with rural people.
    d. Youth need to study occupations to learn as much about them as possible before choosing one.
    e. Money is only one of many rewards and satisfactions to be gained from the right job.
    f. If youth do not wish to leave their home areas, they will have to select from the kinds of jobs within commuting distances.
    g. Youth should think in terms of occupational flexibility and prepare for a group or cluster of occupations.
    h. Non-agricultural occupations require skills and abilities that are similar to those needed for agricultural occupations.
    i. There are agricultural occupations for nearly every kind of interest and nearly every level of ability.
    j. Many people have become famous as a result of their work in the field of agriculture.

REFERENCES

Hoover, N. K. *Handbook of Agricultural Occupations.* Interstate Publishers, Inc., Danville, IL 61832.

Irvins, L. S., and A. E. Winship. *Fifty Famous Farmers.* Macmillan, Inc., New York, NY 10022.

Roy, E. P. *Exploring Agribusiness.* Interstate Publishers, Inc., Danville, IL 61832.

Smith, M., J. M. Underwood, and M. Bultmann. *Careers in Agribusiness and Industry.* Interstate Publishers, Inc., Danville, IL 61832.

U.S. Department of Agriculture. *Career Service Opportunities in U.S.D.A.* U.S. Government Printing Office, Washington, DC 20402.

U.S. Department of Education. Occupational Information and Guidance Service. Washington, DC 20202. Pamphlets are available for various occupations.

U.S. Department of Labor. *Occupational Outlook Handbook.* U.S. Government Printing Office, Washington, DC 20402.

Chapter 5

Landscaping the Home Grounds

Well-kept homes with attractive surroundings for persons on farms and in cities add greatly to the satisfactions in life. Beautification of home grounds is possible even for families with modest incomes. Many changes can be made at small expense by members of a family if they become sufficiently interested to study the situation and learn to use good methods. The young people in a family can become skillful along these lines and thereby contribute to the enjoyment of the entire family.

OBJECTIVES

1. Identify the function and design purposes of plants located near the home.
2. Discuss how the areas surrounding the home can be improved.
3. Apply basic landscaping principles.
4. Demonstrate basic needs and care required of landscape plants.

What Is Included in Beautifying the Home Grounds?

When we refer to beautification of the home grounds, we include the area around the home that is available for lawn, trees, shrubs, and other plantings that contribute to the attractiveness of the home. However, the entire *homestead* may include other buildings and small areas of land for gardens and for other purposes. On the farm, we refer to the *farmstead*, which usually includes the home and its immediate surroundings, together with farm buildings, yards, driveways, gardens, orchards, and windbreaks.

In determining the overall plans for improving the home grounds, the family should consider the entire homestead or farmstead. In planning an

Fig. 5-1. Shopping plazas provide opportunities for persons in the landscaping business.

efficient and attractive arrangement, the family should take into account various buildings, driveways, windbreaks, and other features.

Plants located near the home should serve a useful purpose. For functional purposes, plants:

1. Provide shade or serve as a windbreak.
2. Eliminate or reduce erosion, sound (noise), and glare.
3. Direct people to the desired areas.
4. Define outdoor spaces for beauty and for specific uses.
5. Provide habitat and food for wildlife.

For design purposes, plants:

1. Provide a pleasing transition between buildings and the site on which the buildings are placed.
2. Focus attention on desirable features.
3. Screen unattractive features from view.
4. Highlight seasonal plantings and color.

If necessary, because of limited time and finances, home owners can distribute the labor and expense over a period of years by making a few changes each season. At the outset they should determine what changes are desirable and then develop detailed plans for bringing them about.

How Can the Surroundings of Homes Be Improved?

The starting place for an improvement program on the home grounds

Fig. 5-2. Landscaping and conservation are combined in this terraced lawn with extensive plantings of juniper. The wood chip mulch protects the soil and retains moisture around the shrubs. (Courtesy, Soil Conservation Service, U.S. Department of Agriculture)

should usually be with the exterior portions of the house itself. The house should be in a good state of repair and otherwise improved to the extent that finances will permit. Following this, all unsightly objects and structures near the house should be removed or plans should be made to screen them from view by special arrangements of trees and shrubs.

Before proceeding with improving the home surroundings, the family should make a plan. A sketch should be drawn to scale of the home grounds and the locations of buildings. Present trees and shrubs should be located on the sketch. Following this, the arrangement of the lawn areas and plantings should be shown as the family would like to have it. In arriving at plans, home owners should make a careful study of suggestions in this chapter and in various books and bulletins. The college of agriculture in each state provides helpful bulletins. By observing home grounds that are attractively arranged and landscaped in the community, home owners may get some ideas.

Improving the lawn

In the beautification of the home surroundings, a good lawn is essential. The lawn should be of moderate size in order to avoid excessive work in caring for it, with the lawn space being in proportion to the size of the house.

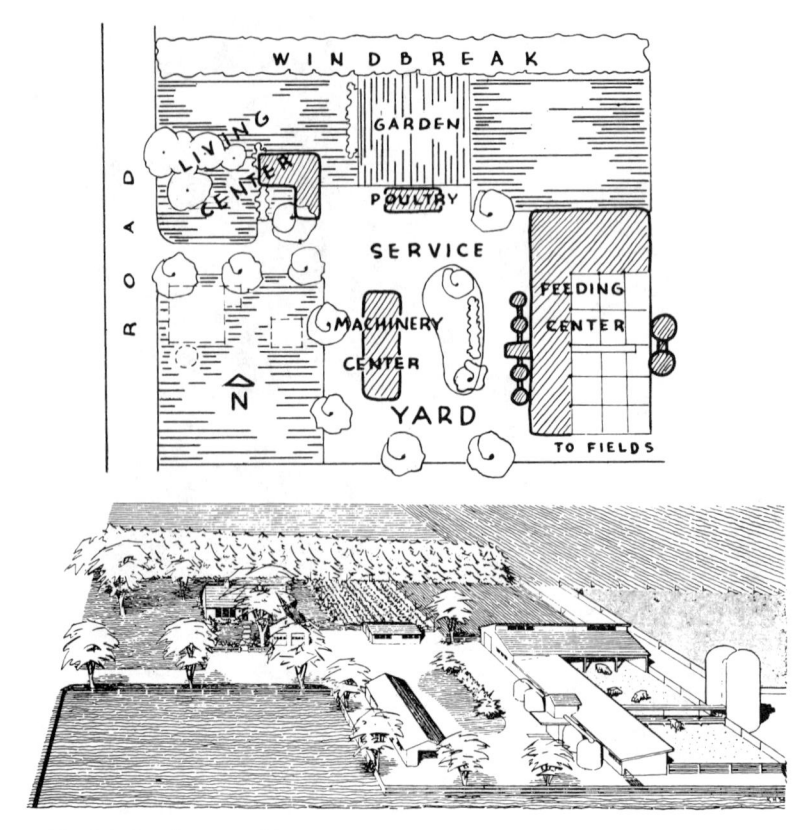

Fig. 5-3. A well-arranged farmstead showing the initial plan *(top)* and the place after carrying out the plan *(bottom)*. (From Circular 732, University of Illinois)

The surface of the lawn should be reasonably smooth, and it should be covered with a good stand of grass. In some cases, established lawns can be improved by reseeding where necessary after depressions have been filled. If the lawn is extremely poor, it will be best to start a new stand of grass after the soil has been plowed, thoroughly worked, leveled, and rolled. For reseeding or for general planting, Kentucky bluegrass, a cool-season grass, is frequently used in many portions of the United States. Other cool-season grasses are the fescues, the bent grasses, and the rye grasses. Mixtures of cool-season grasses are often used. Some warm-season grasses are Bermuda grass, carpet grass, centipede grass, zoysia grass, and St. Augustine grass.

A grass that is adapted to the area and that has the desired lawn qualities should be selected. Fine fescues do well in low-moisture conditions and require little maintenance. Bermuda grass and zoysia grass are both disease and drought tolerant. Bermuda grass is wear resistant and recovers well from general use. Kentucky bluegrass is a good general-purpose grass and is about average on most quality factors, as are the fescues. Kentucky

Landscaping the Home Grounds

bluegrass and the fescues are often sown together. Two or three varieties of grass, to form a blend, may provide a better lawn than just one variety. The college of agriculture in each state can recommend suitable grasses for lawns.

Most cool-season grasses are started from seeds. Seeds should be purchased from a reliable dealer and should carry labels that indicate freedom from weed seeds. Seeds should be scattered evenly, a small hand seeder being best for the job. These seeds may be covered by a light raking with a hand rake. After the raking, a straw mulch should be placed over the soil to prevent erosion and to conserve moisture. One 70-pound bale of straw will

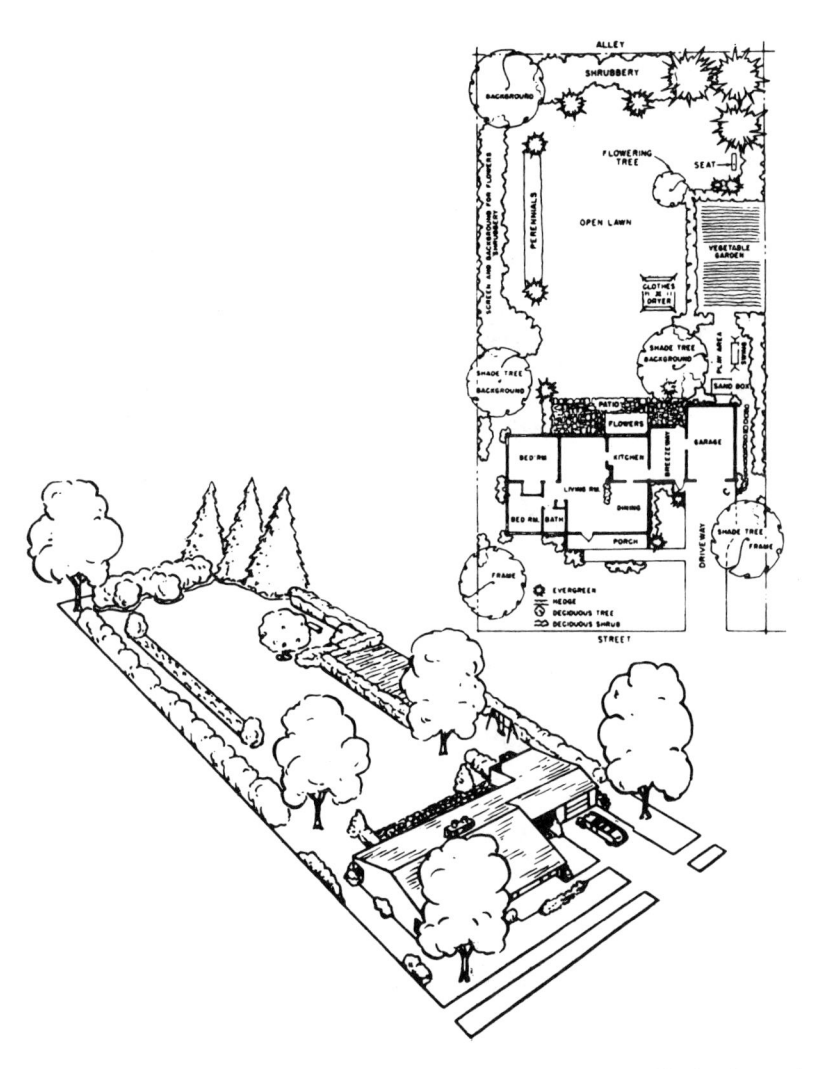

Fig. 5-4. The grounds of a city home are shown as they appeared in the plan and after the plan was carried out. (Courtesy, Illinois Department of Agriculture)

cover 1,000 square feet of area. Late summer or fall is usually the best time for seeding lawns. In reseeding portions of old lawns, the home owner should scatter grass seed in late winter.

Zoysia grass is a very durable grass with a very fine texture. Some home owners do not like zoysia grass because it does not become green until quite late in the spring and because its spreading tendencies make it difficult to control and to keep out of flower beds and gardens.

Fig. 5-5. A typical planting of English boxwood adds a finishing touch to a neat and attractively landscaped lawn in Maryland. (Courtesy, U.S. Department of Agriculture)

Some grasses, such as zoysia, are established by sprigging or plugging. Plugs may be purchased from nurseries. Each plug is about 2 inches in diameter and contains plants with soil. These plugs are set about a foot apart in the space where the lawn is being developed. As the plants develop, the entire surface of the ground becomes covered with grass. Bermuda grass is started in a manner similar to the procedure for zoysia grass. Portions of Bermuda grass plants are purchased and planted, and these portions develop and spread to form the lawn. Complete directions for starting these kinds of lawns can be secured from the nurseries where the plants are purchased.

A good fertilizer and liming program is necessary for a healthy, attractive lawn. The amount and kind of fertilizer and lime to apply depends on soil test results, grass species, climatic conditions, and mowing practices.

Nitrogen helps keep grass a deep green. Excess nitrogen will increase

Landscaping the Home Grounds

growth, which means more mowing, more watering, more dethatching, and possibly more insect and disease problems. Phosphorus is needed for root growth, and it is therefore important for establishing the lawn. Potassium is second to nitrogen in the amount necessary for growth. It promotes disease resistance, winter hardiness, drought tolerance, and use–traffic tolerance. While other minor nutrients are also needed, they are usually provided in fertilizers or are present in the soil.

Complete analysis fertilizers, those that contain nitrogen, phosphorus, and potassium, are the ones most frequently used. Some common analyses are 16-4-8, 5-10-5, and 10-10-10. The cool-season grasses are normally fertilized in the fall at a rate of 10 to 20 pounds of fertilizer per 1,000 square feet of lawn area. Fall is the season when much of the root development of the bluegrasses and fescues takes place. Two to 4 pounds of nitrogen and 1 to 2 pounds of potassium are generally recommended. Another application of fertilizer in the spring is often desirable. Application of nitrogen during hot weather is not recommended.

Warm-season grasses typically require slightly more fertilizer than cool-season grasses. Using a 16-4-8 analysis fertilizer at a rate of 10 to 20 pounds per 1,000 square feet in the spring and fall will provide 3 to 7 pounds of nitrogen and should keep the lawn growing well. It is always wise to have the soil tested and to check with the local vocational agriculture teacher or with a county extension officer regarding the best fertilizer program for a specific situation.

Newly planted lawns should be watered regularly and should be fertilized more frequently than is necessary for established lawns.

When applying fertilizer, the home owner should:

1. Apply the fertilizer when the grass is dry to avoid leaf burn.
2. Use a mechanical spreader to obtain an even distribution of the fertilizer. A uniform coverage is also more likely if two applications are made, with the second application being at a right angle to the first.
3. Water in all fertilizer applied thoroughly.

A helpful practice in lawn fertilization is to use fertilizer that becomes available gradually, rather than all at once. Nitrogen is coated with a special material to permit controlled release over a long period, thus providing the stimulation needed for a green lawn and for continuous growth for a longer period of time than is necessary for the regular farm fertilizers. For a general understanding of fertilizers and fertilization, see Chapter 16.

Many home owners have been attracted to the weed-and-feed lawn materials because of the ease of applying several materials with just one trip across the lawn. Recently, however, it has been discovered that ornamental shrubs and trees have been killed by the herbicides in these weed-and-feed mixtures. In lawns crowded with ornamentals, it is best to apply the weed control material and the fertilizer separately. Herbicides should be selected that, when properly applied, will not injure ornamental shrubs and trees.

Established lawns benefit from being rolled; the best time is in the spring after the ground becomes sufficiently firm that it will not pack when it is rolled. Rolling is done to firm the grass roots that may have been heaved by freezing.

In mowing lawns, some people make the mistake of setting the lawn mower to cut the grass extremely short. Food for the plant is manufactured in the leaves, and a weakened stand of grass may result if the leaf surface is seriously reduced by close cutting. Therefore, it is best to set the mower to cut the grass about 2 inches above the ground.

If the lawn is watered, the soil should be thoroughly soaked with suitable sprinklers, rather than given a light sprinkling. If only the surface is moistened, the plants may develop shallow root systems and be damaged severely after this layer of soil has dried out. Sprinkling attachments that distribute the water uniformly should be used.

In some situations, shady lawns may be a problem. If this is the case, shade-tolerant grasses, such as red fescue, should be planted. The shaded part of the lawn should also be fertilized more heavily than the rest of the lawn to reduce the competition for nutrients between the lawn and the trees. Watering the lawn frequently to a depth of about 6 inches will also help. Other practices of benefit to shaded lawns are removing all trees that add little or nothing to the landscape design, pruning trees quite heavily to allow more light to reach the lawn, pruning roots of trees when the roots are shallow and compete with grass for moisture and nutrients, and keeping the lawn free of leaves and other debris by raking or sweeping.

Providing ground covers

For some conditions, ground-cover plants may be better than lawns. On a dry or steep bank, in shady or rocky places, or between shrubs in foundation plantings, covers such as English ivy, pachysandra, myrtle, Ajuga, lilies-of-the-valley, plantain lilies, creeping juniper, and day lilies are all possibilities. Since more than 200 kinds of plants are used as ground covers, the home owner should be careful to select the best one for the site. Ground covers that root along the ground as they grow are the most effective for controlling erosion.

The best planting times for ground covers are fall and early spring to allow the root systems to develop. One plant for every 1 to 4 square feet will be needed. The closer the plants are set, the more quickly will the ground be completely covered by the plants. Care should be taken to free the area of weeds before the ground covers are planted. The addition of organic matter, fertilization, mulching, and watering will help assure good growth.

Proper depth of planting is essential, especially for areas that cannot be spaded.

Controlling insects and diseases

Many lawn problems are caused by insects and diseases. If a lawn is to be attractive, control measures must be taken.

Insects can cause the grass to turn brown and then die. Some insects attack plant roots; some eat both stems and leaves of plants; and some simply suck juice from the plants. Most of these kinds of damages can be prevented by proper treatment.

The larvae of several kinds of beetles, including the June beetle and the Japanese beetle, may cause extensive lawn damage. They feed on the roots of grass about an inch below the surface. Ant colonies damage lawns by destroying seed and by building large ant hills. Wireworms feed on roots and bore into the underground parts of stems. Billbugs feed on stems and leaves. A common problem is the sod web worm. Web worm damage shows up first as brown trails through the grass and then as larger brown areas. Some species damage crowns and roots as well as the leaves. Damaged sod will contain the web worm, dirty white to light brown in color with dark spots and about ¾ inch in length. The nearly white adult moth can be seen in the porch light at night or making zigzag flights when disturbed during the day. A nearly certain sign of web worms is the sight of many robins or blackbirds pecking at the turf. Lawns can be treated with diazinon, sevin, or spectracide. Other insects that feed on leaves and stems are armyworms, cutworms, and grasshoppers.

A few insects, such as aphids, leafhoppers, and chinch bugs, suck the juice from plants. Infestations must be fairly heavy before damage becomes visible.

There are many insecticides under a variety of trade names that can be used to control insects. Chlordane, diazinon, sevin, carboryl, dursban, malathion, and methaxychlor are all possibilities. Each should be applied according to the directions on the container. Spring and fall are the best times to treat lawns for grubs. Above-ground pests can be controlled by treatments any time during the growing season.

Diseases of lawns are usually caused by fungi. Fungus filaments can sometimes be seen on blades of grass. If the infection is severe, the spores of rust or smut may appear like a dust cloud in the air when the lawn is being mowed. Dollar spot, fairy rings, snow molds, powdery mildew, leaf smuts, slime molds, and helminthosporium are the most common. Yellowing, browning, coloring of a white to dusty gray and other abnormal colorations of the lawns are signs of disease problems. By following recommended cultural practices, selecting disease-resistant varieties, and applying fungicides according to the directions on the product labels, home owners can maintain healthy lawns, which is the best control for lawn diseases.

Controlling weeds

Problems with weeds can be avoided to a great extent by proper fertilizing, watering, and mowing. Some weeds can also be controlled by hand pulling. When these methods are unsuccessful, chemical control may be desirable.

For weeds such as crabgrass and goosegrass, a pre-emergence treatment (before the seeds germinate) is most effective. For crabgrass, this is when the forsythia is just past full bloom. Herbicides such as siduron (Tupersan), DCPA (Dacthal), and bensulide (Betasan) are recommended.

For post-emergence control of broadleaf weeds, some recommended herbicides are DCPA (Dacthal), mecoprop (MCPP), and 2,4-D.

Many other herbicides are available. All chemical treatment should be done with great care according to the manufacturer's directions.

Arranging drives and walks

The arrangement of the drive and walks is an important consideration in making the home grounds attractive. Since a broad expanse of grass has a pleasing effect, it is not desirable to have drives and walks through the central portion of a lawn. The drive leading to the house from the street or highway is best located at the side of the lawn, and it should lead to the side or rear of the house near the doors most used by the family. If walks are constructed from the house to this drive, no other approach to the street or highway is usually necessary.

The driveway should be surfaced with crushed rock, gravel, cement, asphalt, or any other material that provides a firm surface under all weather conditions. Walks that are used a great deal should be made of concrete, while those that are used but little may be made of smooth slabs of stone placed in the ground, which gives a stepping-stone effect.

Arranging plants

In planning the home grounds, home owners should carefully consider the locations for trees, shrubs, and smaller plants. Considerable help can be obtained from the general recommendations of experts along this line.

Some homes can be made more attractive by a few shrubs planted next to the houses to form what is known as the *foundation planting*. These shrubs should be grouped at the corners and angles of the house. Such plantings serve to overcome the bareness of the house foundation, and they tend to soften the effect of sharp corners and vertical lines. They also give

Landscaping the Home Grounds

the house the appearance of belonging with the rest of the surroundings. The plantings beneath the windows of the house should be of low-growing varieties of shrubs, and those at the corners should be of somewhat higher-growing varieties. It is not necessary to hide the entire foundation; in fact, most experts prefer to have parts of the foundation exposed to view. For portions of these plantings, certain types of evergreens may be appropriate.

For southern exposures, where shade for the home is desirable in the summer and sunshine in the winter, deciduous types of plantings should be selected.

Many modern homes have almost invisible foundations. For these houses, there is little or no foundation to screen from view; therefore, there is no need for special foundation plantings. Reducing or eliminating foundation plantings for newer homes can reduce landscaping costs and the labor needed for maintaining the lawn and the home.

The central portions of the front lawn should be kept free from plantings of shrubs and trees so that an open view is possible. Flowerbeds, curious stones, or other objects have no place in this open space because they detract from the desired effect. Tall-growing trees should be planted at the sides and rear of the house for purposes of framing and shading the house. Such trees should be located at least 25 feet from the house with spaces of 40 to 50 feet between them.

Shrubs may also be planted along the sides or boundaries of the lawn. For the most part, trees and shrubs should not be planted in straight rows.

Fig. 5-6. The foundations of most modern homes do not need to be screened from view.

Fig. 5-7. The shrubs create a pleasing effect by focusing attention on the entrance to the home.

Instead, they should be placed in irregular groups and clumps. This method of planting is most nearly like the arrangement that is found when trees and shrubs are permitted to grow naturally. However, if hedges are desired, the shrubs for this purpose can be set in straight rows.

The plantings of trees and shrubs at the rear of the house should also be carefully arranged. Portions of the grounds are often used for a variety of purposes, such as a playground, a place for drying clothes, a location for the garage, and, in some cases, a vegetable garden. The portions used for purposes that may not contribute to the general attractiveness of the surroundings should be screened from view by plantings of shrubs or trees. Photographs and drawings included in this chapter are examples of desirable practices in locating trees and shrubs.

Proper spacing should be provided between shrubs. A common mistake is to place young plants too close together so that, as they grow, they become crowded. Tall-growing shrubs should be planted 6 to 8 feet apart and low-growing shrubs, 3 to 4 feet apart. The first two or three years, the shrubs may be interplanted with annual flowers, which will add to the attractiveness of the plantings until the shrubs develop sufficiently to fill the spaces. In early and later years, annual and perennial flowers of various heights and kinds may be planted in front of the shrubs. Foundation plantings should be placed 4 or more feet from the house, depending on the growth habits of the shrubs planted.

Selecting trees and shrubs

In selecting the varieties of trees, shrubs, and flowers for their home grounds, home owners may want to consider the recommendations of their state college of agriculture. They may also want to observe neighborhood varieties that appear attractive and grow vigorously. While plants can be purchased from nurseries, some suitable plants can frequently be obtained from other home owners in the neighborhood, thereby reducing the cost considerably. Table 5-1 lists some suggested varieties.

Table 5-1. Suggested varieties of trees and other plants for use in homestead improvement[1]

TREES		
Trees for shading and framing the house		
American linden (basswood)	Honey locust	Silver maple
Ash	Norway maple	Sugar maple
Beech	Pin oak	Sweet gum
Flowering dogwood	Red maple	Sycamore
	Red oak	Tulip tree
	Scarlet oak	
Trees for accent purposes		
Amelanchier	Flowering crab	Purple-leaf plum
Birch (several varieties)	Flowering dogwood	Redbud
Blue spruce	Hawthorn	Russian olive
Bradford pear	Magnolia	Smoketree
English hawthorn	Mountain ash	White fir
Trees for windbreaks		
White pine	White spruce	Austrian pine
Norway spruce	Douglas fir	Red or Norway pine
	Scotch pine	

VINES		
Flowering vines		
Jackman clematis	Trumpet creeper	Chinese wisteria
	Japanese honeysuckle	
Vines for covering brick, stone, masonry		
Boston ivy	Bigleaf winter creeper	English ivy
Vigorous climbing vines with heavy foliage		
American bittersweet	Virginia creeper	Honeysuckle (several varieties)
Trumpet creeper	Sweet autumn clematis	Chinese wisteria

(Continued)

Table 5-1 (Continued)

SHRUBS

Shrubs for hedges

Five-leaved aralia	Amur privet	Rugosa rose
Japanese barberry	Morrow honeysuckle	Van Houtte spirea
Lemoine deutzia	Rugonis rose	American arborvitae

Shrubs for border plantings

a. Low-growing

Froebel spirea	Thunberg's spirea	Slender deutzia
Japanese barberry	Garden snowberry	Horizontal cotoneaster
	Coralberry	

b. Medium-growing

Spreading cotoneaster	Flowering quince	Bridalwreath or Van Houtte spirea
Lemoine deutzia	Rugosa rose	Snowhill hydrangea

c. Tall-growing

Weigela	Morrow honeysuckle	Mockorange (several varieties)
Euonymus (several varieties)	Tartarian honeysuckle	Beautybush (several varieties)
Forsythia	Lilac (several varieties)	
	Viburnum	

Shrubs for foundation plantings

Andora juniper	Spreading Japanese yew	Deutzia
Pfitzer juniper	Japanese barberry	Euonymus
Dwarf Japanese yew	Cotoneaster	Hydrangea
Lilac	Anthony waterer spirea	Rose

FLOWERS

Annuals

Aster	Calendula	Morning glory
Petunia	Sweet pea	Snapdragon
Ageratum	Zinnia	Marigold

Biennials

Foxglove	Canterbury bell	Pansy
Hollyhock	Sweet William	

Perennials

Delphinium	Iris	Hardy chrysanthemum
Veronica	Lily-of-the-valley	Goldenglow
Shasta daisy	Peony	Hardy aster
Phlox	Pyrethrum	

[1]These consist of varieties suited especially to the central and eastern portions of the United States, although many of those given are suited to other areas.

Fig. 5-8. An attractive backyard with border plantings that will soon screen out the homes in the background and provide privacy. Note curving lines and varying heights of plantings.

Raising flowers for home-ground beautification

In landscaping, flowers have a definite place, especially in border plantings along the margins of lawns. They also can be used appropriately with foundation plantings around the house. Flowers are most attractive with shrubs as backgrounds. Rarely, if ever, should flowers be placed in special beds out in open spaces in the lawn.

For color through the spring, summer, and fall, a mixture of annual and perennial flowers is best. Perennials require only moderate care and will renew themselves from the roots for many years. Most of them bloom in the spring or early summer. The following perennials, listed in approximate order of blooming, will provide flowers through most of the growing season: daffodils, bluebells, tulips, irises, peonies, delphiniums, phloxes, and chrysanthemums. In deciding where to plant, the home owner should consider the height of each plant when it matures. Some annuals that will support the colors provided by the perennials are alyssums, ageratums, asters, marigolds, morning glories, petunias, snapdragons, sweet peas, and zinnias.

Anyone with an especially strong interest in flowers may wish to make special arrangements when planting them. By careful study, an individual can determine the opening and closing times of some favorite flowers. The flowers could then be planted where they would be seen the most when they were open. An attractive arrangement would be to plant the flowers in the shape of a clock so that certain flowers would be specific hour markers because they open or close at that time.

The Swedish botanist Carolus Linnaeus identified the opening and closing times for some flowers as follows:

    6 a.m. Spotted cat's ear—opens.
    7 a.m. African marigold—opens.
    8 a.m. Mouse-ear hawkweed—opens.
    9 a.m. Prickley sow thistle—closes.
  10 a.m. Common nipplewort—closes.
  11 a.m. Star of Bethlehem—opens.
  12 m.   Passion flower—opens.
  1 p.m. Childing pink—closes.
  2 p.m. Scarlet pimpernel—closes.
  3 p.m. Hawkbit—closes.
  4 p.m. Small bindweed—closes.
  5 p.m. White water lily—closes.
  6 p.m. Evening primrose—opens.

How Can the Buildings and Plantings on Large Homesteads and Farmsteads Be Improved?

For farmsteads, small-scale farm layouts, and some other homesteads, the arrangement of buildings and additional plantings should be considered.

Arranging buildings

The arrangement of the buildings on a farmstead or a large homestead has a great deal to do with appearances and efficiency. The arrangement of farm buildings affects the amount of labor required in caring for the farm animals and for performing the other necessary farm chores. An especially desirable grouping, from the standpoint of saving labor, is to have the buildings placed around the sides of an open space, commonly called the service yard.

On farms, the matter of distance between buildings and their location in relation to each other should be carefully considered. For each 14½ feet that a person walks each day, he/she walks approximately a mile in a year, so it is best to plan for saving steps wherever possible. On the other hand, buildings should not be so close together that a fire in one might spread to the others. Ordinarily, the barns should be at least 100 feet from the house to avoid objectionable odors and to reduce fire hazards.

Landscaping the Home Grounds

Planting a windbreak

For a country home or a farmstead in the northern states, there should be a windbreak for protection during the winter season. This windbreak should be located on the sides that will give protection to the farmstead against the prevailing winds of winter.

Evergreen trees are preferred for windbreak plantings. Norway spruce and white pine are two varieties that will give good results under most conditions. Norway pine, Scotch pine, Austrian pine, and Douglas fir are also good trees for many portions of the country. Trees should be located at intervals of about 12 to 16 feet in rows about 16 feet apart. Two to four rows are usually recommended if the space permits.

A twin-row, high-density windbreak is now recommended for many homesteads. The windbreak planting is made up of three or four sets of double-row (twin-row) plantings, with the two rows of trees in each set

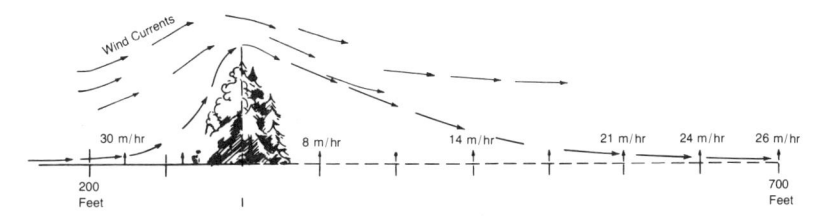

Fig. 5-9. The effect of tree plantings, for windbreaks, on wind velocity can be calculated from the above chart. (Courtesy, U.S. Department of Agriculture)

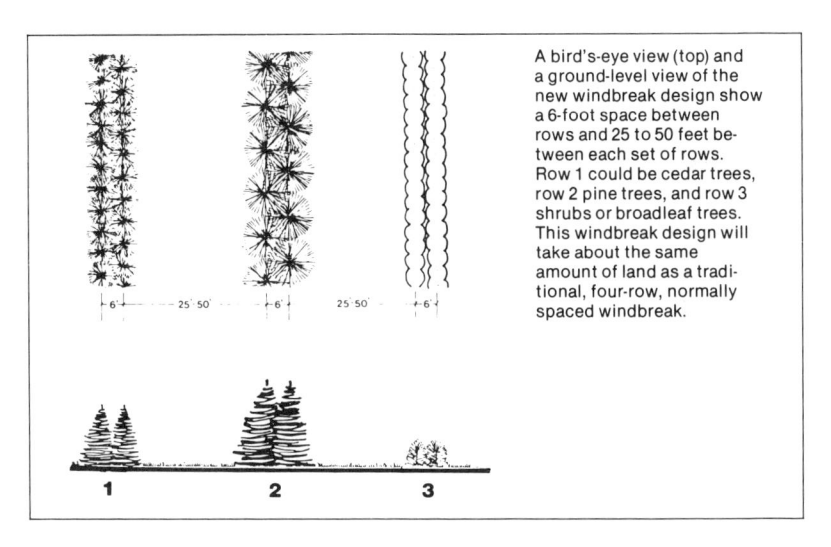

A bird's-eye view (top) and a ground-level view of the new windbreak design show a 6-foot space between rows and 25 to 50 feet between each set of rows. Row 1 could be cedar trees, row 2 pine trees, and row 3 shrubs or broadleaf trees. This windbreak design will take about the same amount of land as a traditional, four-row, normally spaced windbreak.

Fig. 5-10. New windbreak design. (Courtesy, *Soil and Water Conservation News*, U.S. Department of Agriculture)

being planted 6 feet apart, and the trees in the rows being placed 5 to 8 feet apart. From 25 to 50 feet of space is left between the sets of trees, wide enough to be farmed. Snowfall drops between the sets of trees, thus making it possible to plant the windbreak closer to the buildings than is possible with the traditional windbreak.

Using desert landscaping

For some areas, desert landscaping, including rock coverings, may be the most desirable way to beautify the home. The general principles for designing the home landscape are the same for desert landscaping as for other forms of landscaping. However, desert landscaping makes it possible to use the predominant plants of the area, such as cacti, to beautify the home.

Following guidelines for landscape planning

Thoughtful consideration of the following general guidelines for landscape planning will contribute greatly to the creation of a home environment that will be pleasing for many years. The home owner should:

1. By painting, cleaning up, and removing unsightly objects and materials, prepare the house and grounds for landscaping.
2. Carefully analyze the activities that will probably take place in the home and on the grounds—parties, picnics, games, gardening, bird watching, and similar activities.
3. Prepare drawings to scale that show all the present features and objects on the grounds, plus the kinds and locations of all new shrubs, trees, and other landscaping features.
4. Plan for all three dimensions of the area to be landscaped: the skyline, the ground space, and the boundary lines of the grounds.
5. Plan for the three major areas to be landscaped: the public area, the service area, and the living area.
6. Consider all grading, leveling, and additions of soil.
7. Take into account the cooling effects of the wind and the warming effects of the sun.
8. Use the natural features (trees, rocks, water) of the area to the greatest extent possible.
9. Hide or camouflage the undesirable or unattractive features of the home or grounds with plantings.
10. Maintain the home as the dominant feature of the grounds.
11. Carry out the dominant lines of the house by using tall plantings with tall houses, low plantings with low houses, etc.
12. By framing or using other arrangements of shrubs and trees, accent the desirable features of the house and grounds.

13. Select and locate each planting or landscaping feature in terms of its relationship to other plants and and landscaping features and its contribution to the overall effect.
14. Select plants that can be adapted to the particular soil and climatic conditions that exist.
15. Consider the growth requirements and growth habits of each plant when it matures.
16. By appropriately locating plants, fireplaces, and other landscaping features, keep open the largest possible area for use.
17. Direct drainage of water away from the buildings.
18. Eliminate or soften straight lines.
19. Provide for easy access to and from each of the landscaped areas.
20. By appropriately locating sidewalks, driveways, and other harsh features, make maximum use of available space.
21. Allow for privacy from the neighbors.
22. Avoid overlandscaping.
23. Provide for protection and beauty during all seasons of the year.
24. Through framing or other plant and shrub arrangements, accent desirable distant views.
25. Allow for general blending in of the house and grounds with the best features of the surrounding neighborhood.
26. Create as nearly a natural effect as possible.
27. Strive for a general effect and view of the house and grounds that is pleasing, whether viewed from inside the house, inside the grounds, or outside the grounds.
28. Establish a definite, logical order for accomplishing each part of the landscape plan that is in keeping with the time and money available.
29. Gear all planning to the amount of time and money available for proper long-time care and maintenance.
30. Try to involve every member of the family in the development of the landscape plan.

Transplanting trees and shrubs

Most trees and shrubs can be transplanted satisfactorily in either spring or fall. For a few varieties, spring is much preferable, while for certain others, fall is decidedly better. Where there is a great deal of freezing and thawing in the winter, spring is considered the best time for transplanting. If transplanting is done in the fall, it is best to wait until the leaves have fallen and the trees or shrubs are in a dormant state. When a tree or shrub is transplanted, it should be removed with considerable dirt and with a fair-sized clump of roots. Care should be taken to prevent the roots from drying out.

158   **AGRISCIENCE IN OUR LIVES**

In transplanting, individuals should keep several steps in mind. These include:

1. The hole should be dug sufficiently large so that the roots will not be twisted or crowded. It should permit the tree or shrub to be set with the roots as deep or slightly deeper than in its preceding location.

Fig. 5-11. Some wide-open spaces are pleasing to the eye and essential for picnics and other family recreational activities.

Fig. 5-12. Tall plantings should be used with tall houses.

2. While the hole is being dug, the layer of good top soil should be kept separate and placed at the bottom of the hole around the roots.
3. After the tree or shrub is put in place, some of the soil should be tamped around the roots with hands or feet.
4. Water should be poured around the plant and allowed to soak away.
5. The remainder of the soil should be placed around the plant, with a saucer-shaped pocket to catch water. If there is little rain, it is desirable to provide water at intervals until the tree establishes itself.
6. Pruning should be done to remove some of the branches entirely and to shorten others. This removal of top portions is necessary to balance the loss of some roots in transplanting.
7. Newly transplanted shrubs should be watered frequently. Sufficient water should be applied to soak the ground below the depth of the roots. Wetting only the surface of the ground will cause the roots of the shrub to grow upward to seek the water.

Pruning and caring for trees and shrubs

Trees and shrubs should be carefully pruned at the time of transplanting and at intervals thereafter to secure the desired shape and to remove the surplus growth. In all pruning work, each part removed should be cut back to a main branch or to the main trunk, with no stub remaining. If this precaution is observed, the wound will "heal" properly by the growth of new tissues over the surface.

The bottom limbs of young, growing trees must be removed over a period of years until the lowest limbs are the desired height above the ground. Tree limbs always remain at the same height, since upward growth takes place only at the tips of the branches.

In pruning shrubs with flowers, it is important to know the season in which the flowers appear and the age of the wood on which they grow. In general, shrubs should be pruned soon after their flowering period. Nearly all shrubs require some pruning each year. Usually some branches should be cut out entirely and some should be headed back. Maintenance pruning will result in more and better blooms, develop and keep the desired shape, revitalize plants, and remove diseased, dead, or injured wood. For flowering shrubs, about one-third of the older wood can be removed at the ground level. Too heavy pruning will result in a great deal of sucker growth. Forsythia, spirea, deutzia, and weigela should be pruned after they have finished blooming, since the flower buds are formed during the previous year and early pruning would destroy most of the bloom. Summer flowering shrubs and trees should be pruned in the early spring while the plants are still dormant. With the exception of shady-lawn problems, shade trees need only to have dead, diseased, and/or crowded branches removed. If trees need some additional pruning, it can be done sometime from February to

May. Some pine trees, such as the white pine, exude less sap when pruned at the candle stage of the new growth. Maples, elms, birches, and similar trees that exude a great deal of sap when cut in the spring can be pruned in the summer.

Fig. 5-13. Trees with rootballs ready for transplanting *(top)*. A tree placed in the hole prepared for it *(bottom)*. Note the depth and size of the hole.

HOW TO PRUNE

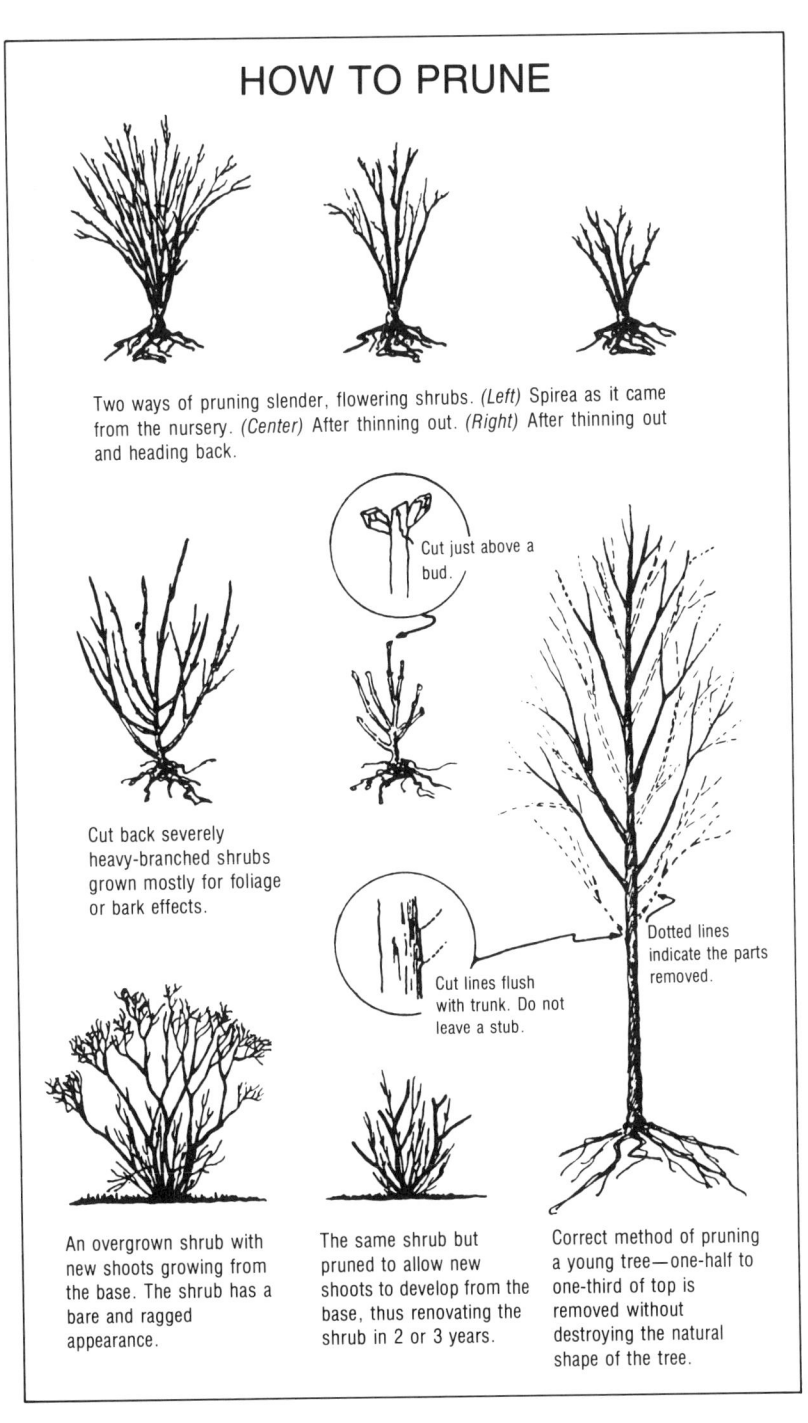

Two ways of pruning slender, flowering shrubs. *(Left)* Spirea as it came from the nursery. *(Center)* After thinning out. *(Right)* After thinning out and heading back.

Cut just above a bud.

Cut back severely heavy-branched shrubs grown mostly for foliage or bark effects.

Cut lines flush with trunk. Do not leave a stub.

Dotted lines indicate the parts removed.

An overgrown shrub with new shoots growing from the base. The shrub has a bare and ragged appearance.

The same shrub but pruned to allow new shoots to develop from the base, thus renovating the shrub in 2 or 3 years.

Correct method of pruning a young tree—one-half to one-third of top is removed without destroying the natural shape of the tree.

Fig. 5-14. Proper pruning of trees and shrubs is necessary for best effects. Note the detailed directions for pruning. (Courtesy, The University of Wisconsin)

Irrigating

Many plants indicate when they need water—by wilting, changing color, or growing more slowly than usual. Moisture needs vary according to the weather, the soil, the season, and the kind of plant.

Nighttime and early morning are the best times to irrigate. Newly transplanted plants will need more water than will established plants. In all cases, over-irrigation should be avoided. So that the amount of water to apply can be determined, the depth should be checked and the plant response observed. Moisture can be conserved by mulching.

Controlling pests

Most medium to large landscape plants need little protection from pests. It is best to concentrate on helping the plants grow well so that they can resist pest attack.

However, even healthy and well-cared for plants are subject to attack by pests and may need some help, such as washing aphids off trees, plugging nest areas of cutter bees, picking off caterpillars by hand, and cutting off and destroying tent caterpillar webs.

Insecticides should be used only when pest damage becomes excessive and other control measures are not effective. Care should be used in the selection and application of insecticides.

If a tree appears to be undernourished as indicated by poor leaf color, weak growth, and/or fungus or insect attack, fertilizer should be applied. Broadcasting is best for young trees, with the amount depending on the tree size.

What Are Some Possible Hobbies for People Who Raise Flowers and Other Ornamental Plants?

Two Nebraska farmers, the Sass brothers, became famous in producing new varieties of irises. They started this as a hobby many years ago. They became skillful in transferring pollen from the flowers on one plant to the flowers on other iris plants, a process known as *artificial pollination*. This process requires a great deal of skill and special knowledge. From many crosses of this type, they produced some new and outstanding varieties. Before long, what was once a hobby became an important industry to them.

Harry O'Brien started many years ago to develop a hobby in flower gardening and the growing of other plants for beautification of the grounds around his home in Ohio. He began writing articles about his hobby and occasionally appeared on the programs of garden clubs. He became

Landscaping the Home Grounds

nationally famous for writing articles about gardening and for giving lectures in different parts of the United States.

While few people will carry their hobbies as far as these persons, there are many opportunities for developing fascinating hobbies or specialties in connection with flowers and other ornamental plants. The following is a suggested list:

1. Growing flowers indoors.
2. Starting a flower garden with several varieties of annuals.
3. Raising biennial flowering plants.
4. Becoming an expert in raising one type of flower, or plant, such as dahlias, roses, irises, delphiniums, lilies, or cacti.
5. Learning to identify varieties of trees, shrubs, and flowers suitable for home-ground improvement.
6. Raising ornamental gourds.
7. Becoming an expert in arranging bouquets.
8. Raising ornamental evergreens.
9. Learning to maintain a good lawn.
10. Growing flowers for cut-flower arrangements.
11. Establishing a lawn with some special variety of grass, such as bent grass.
12. Raising flowers from spring-blooming bulbs, such as tulips, daffodils, and narcissuses.
13. Developing different types of terrariums under glass bowls.
14. Building and maintaining a small greenhouse for flowers and other plants.
15. Developing a picnic nook and constructing an outdoor fireplace or a barbecue pit.
16. Becoming skillful in propagating (multiplying) various flowering plants and shrubs.
17. Identifying and controlling various diseases and injurious insects of flowers, trees, and shrubs.
18. Becoming skillful in pruning the various shrubs so as to secure desired shapes and abundant bloom.
19. Starting an all-season flower garden.
20. Constructing a pool and putting water lilies and other appropriate plants in it.
21. Making a rock garden.

SUGGESTED ACTIVITIES

1. With others in your class, visit some attractively landscaped home. What are some of the features that make it attractive?
2. On Arbor Day or some other day in the spring, plan with your class to plant one or more trees on the school grounds. As a class,

decide on the variety, arrange for getting it, choose the location for it, transplant it, and prune it.

3. Make a drawing showing the location of the house and surrounding trees and shrubs on your home grounds at present. Use some appropriate scale, such as 1 inch to 20 feet. Using your ideas and advice from others, make a plan for your home grounds as you think they should be arranged. Give reasons for the changes you suggest. After getting ideas from your parents and others, start to make some of the plantings this year.

4. After studying information from various sources, including your observation of nearby growers of flowers, seed and nursery catalogs, and special bulletins and books, select flowers that you especially like. Classify them as annuals, biennials, and perennials, and complete the following.

	Season of blooming	*Height at blooming time*	*Color of bloom*
Annuals			
Biennials			
Perennials			

5. From the above list, make plans for growing flowers at home that will provide continuous bloom throughout the growing season.

6. With the other students in your class, plan an improvement program for your school grounds. Include in this plan the location for playground areas, trees, shrubs, and other features that you think should be included. Get suggestions from a landscape specialist. Have the plans approved by school authorities. If possible, make arrangements for securing trees and shrubs and plant them in the appropriate locations.

7. During the fall months, plan a flower show among members of your class. Have each bring a bouquet and be responsible for arranging it. Ask each student to learn the names of the flowers shown. Have an identification contest after some study of the flowers. Ask each student to rate all the bouquets for pleasing arrangement and harmonious selection of flowers, giving each a rating of excellent, good, or fair.

8. Organize a garden club among members of your class. For those interested in growing small flower gardens at home, make plans for judging the gardens the following fall.

9. Grow some flowering plants in the classroom. Make a schedule for taking turns in caring for them.

10. Concepts for discussion:
    a. Home grounds can be beautified at very little expense.
    b. Better results are usually obtained when plans of the home grounds as the family would like to have them are made.
    c. A good lawn is basic to beautifying the home grounds.
    d. Lawns should be fertilized in both early spring and early fall.
    e. In using herbicides, individuals must be careful to avoid harming ornamental plants.
    f. Ground-cover plants may be better than grass for some conditions.

g. Everything in the landscape, from driveways and sidewalks to shrubs and trees, should be located to achieve the best possible pleasing effect.
h. Flowers that will provide color from spring through fall should be selected.
i. The guidelines developed by landscape designers should be followed in planning for home beautification.
j. Most trees and shrubs are best transplanted in the spring or fall.
k. Trees and shrubs should be pruned when planted and as needed to maintain the desired shape.
l. Early flowering shrubs should be pruned after they finish blooming; summer flowering shrubs should be pruned in the spring while they are still dormant.
m. Cutting grass too short is harmful to the lawn.
n. When lawns and plants are watered, the ground should be soaked to a considerable depth so plant roots will grow down into the soil.

REFERENCES

Davis, W. B. *Landscape Trees for the Great Central Valley of California*. Cooperative Extension Leaflet 2580, University of California, Berkeley, CA 94720.

Hoover, N. K. *Approved Practices in Beautifying the Home Grounds*. Interstate Publishers, Inc., Danville, IL 61832.

Lieberman, A. S., and R. G. Mower. *Ground Covers for New York State Landscape Plantings*. Extension Bulletin 1178, New York State College of Agriculture and Life Sciences, Cornell University Ithaca, NY 14850.

Nelson, W. R., Jr. *Landscaping Your Home*. Circular 858, University of Illinois, Cooperative Extension Service, College of Agriculture, Urbana, IL 61801.

Sacamano, C. M., and W. D. Jones. *Ground Covers for Arizona Landscapes*. Cooperative Extension Service, College of Agriculture, The University of Arizona, Tucson, AZ 85721.

Scannell, R. J., and A. S. Lieberman. *Landscape Design for Residential Property*. Extension Bulletin 1099, New York State College of Agriculture and Life Sciences, Cornell University, Ithaca, NY 14850.

Schroeder, C., and H. B. Sprague. *Turf Management Handbook*. Interstate Publishers, Inc., Danville, IL 61832.

Vocational Agriculture Service. *Identifying and Controlling Lawn Insects*. VAS 5016, Vocational Agriculture Service, College of Agriculture, University of Illinois, Urbana, IL 61801.

Vocational Agriculture Service. *Turfgrass Diseases and Their Control*. VAS 5015, Vocational Agriculture Service, College of Agriculture, University of Illinois, Urbana, IL 61801.

Williams, T. G. (revised and edited by E. N. Weatherly). *Planning Your Home Landscape*. Cooperative Extension Service, College of Agriculture, University of Georgia, Athens, GA 30602.

Chapter 6

Growing Food for the Family

One of life's greatest pleasures is watching things grow. Many people, whether they live on the farm or in town, find this pleasure by having a garden. In addition to the pleasure of watching plants grow, gardeners can get great satisfaction from providing food for the family table. By using the proper methods, they can raise garden products of high quality. Youth often take part in gardening as a family project. Persons skillful in gardening are said to have a green thumb, but almost anyone can learn to garden well.

OBJECTIVES

1. Discuss the need for fruit and vegetables as a part of good nutrition.
2. Describe what is needed to raise vegetables.
3. Discuss the basic essentials of producing small fruits.
4. Discuss the basic essentials of producing tree fruits.
5. Discuss the basic science of vegetable and fruit preservation.

Why Should Vegetables and Fruits Be Raised?

Vegetables and fruits should be eaten abundantly because they are an important part of a healthful diet. Nutrition specialists indicate that the following should be eaten daily: (1) one or more servings of leafy green and yellow vegetables; (2) one or more servings of citrus fruits, tomatoes, strawberries, or some other good source of Vitamin C; (3) one or more servings of white potatoes or sweet potatoes; and (4) one or more servings of other fruits and vegetables. To understand how these fit into an adequate diet, see Fig. 2-10, which shows the Basic Seven groups of foods.

Some people think that vegetables and fruits can be purchased more cheaply than they can be raised. Usually this is not the case. Furthermore, if there is a homegrown supply available for each family, family members are more likely to eat sufficient amounts of these important foods.

From a fair-sized piece of land, vegetables and fruits may be produced that if purchased would cost many hundreds of dollars. Products from large

Fig. 6-1. Vegetables and fruits are one of the four basic food groups. Which of these vegetables and fruits can you grow at home? (Courtesy, U.S. Department of Agriculture)

gardens may also be sold. The actual expenses, such as seeds, plants, and fertilizer, may be kept at a low figure. Much of the labor can be performed by various members of the family in their spare time. By proper storage, canning, freezing, and drying, many of these products can be used throughout the year by the families who raise them.

Part of the savings from homegrown foods comes from the costs for storage, transportation, processing, packaging, and profits for foods purchased in a store. For every dollar spent on food, 25 cents or more goes for processing and packaging, with processing accounting for over 50 cents of the cost in some cases. About half of the price of highly processed foods, such as applesauce, canned tomatoes, and catsup, goes for processing. Of course, food processors can produce some items such as cottage cheese and frozen orange juice more efficiently and cheaply than an individual could.

In addition to providing a savings in the food bill and healthful foods, a home garden may yield vegetables and fruits higher in quality than those that could be purchased. Many of these, such as asparagus, peas, sweet corn, and strawberries, deteriorate if they are not used soon after they have been picked. Anyone who has not eaten peas and sweet corn cooked directly from the garden does not know the delicious flavor of these vegetables. The sugar changes to starch in these two vegetables soon after they have been picked if they are allowed to stand at high temperatures. Thus, much of the natural flavor is lost.

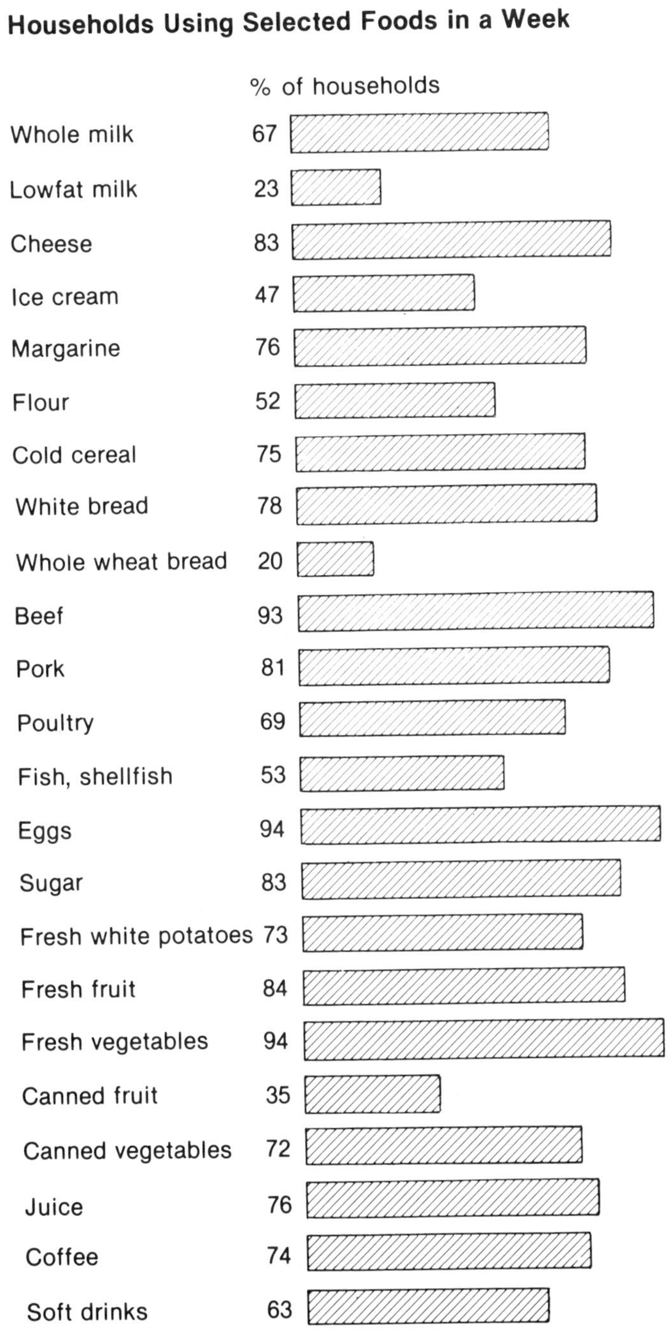

Households Using Selected Foods in a Week

% of households

Food	%
Whole milk	67
Lowfat milk	23
Cheese	83
Ice cream	47
Margarine	76
Flour	52
Cold cereal	75
White bread	78
Whole wheat bread	20
Beef	93
Pork	81
Poultry	69
Fish, shellfish	53
Eggs	94
Sugar	83
Fresh white potatoes	73
Fresh fruit	84
Fresh vegetables	94
Canned fruit	35
Canned vegetables	72
Juice	76
Coffee	74
Soft drinks	63

Fig. 6-2. U.S. households consume a wide variety of foods. (Courtesy, Nationwide Food Consumption Survey [48 states], U.S. Department of Agriculture)

Households Using Home-Produced Food in a Week

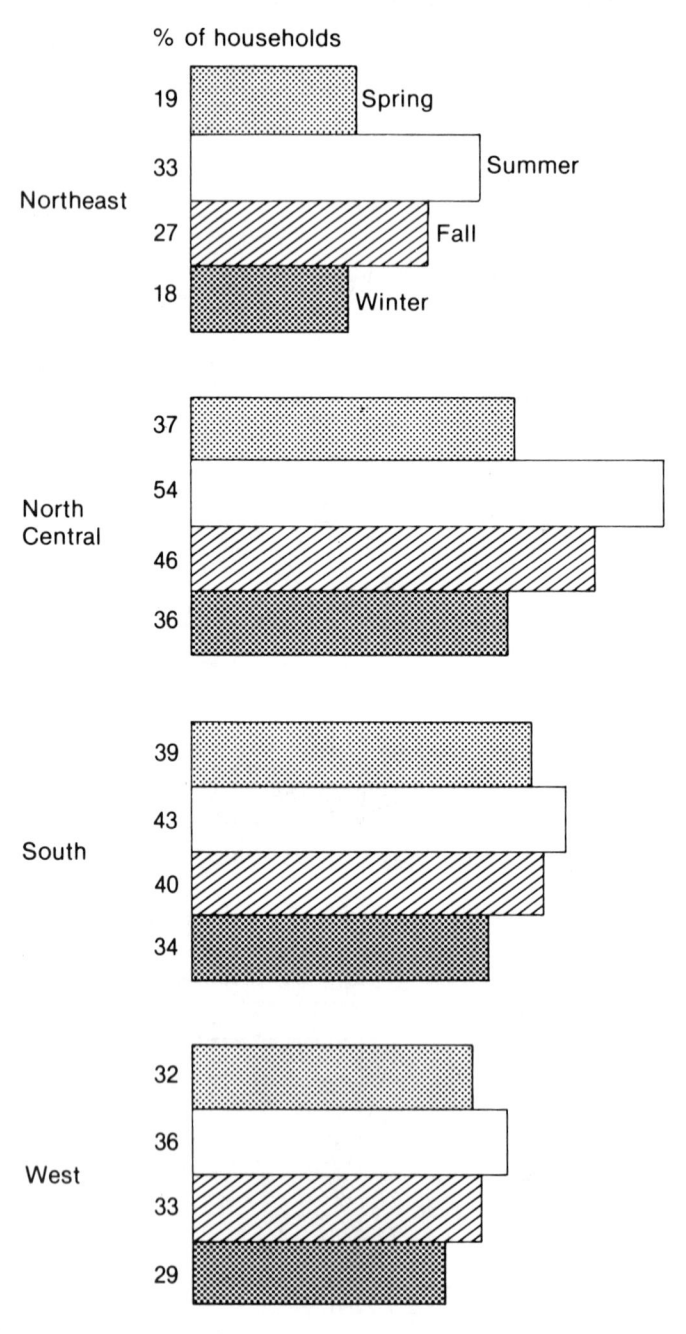

% of households

Northeast
- 19 Spring
- 33 Summer
- 27 Fall
- 18 Winter

North Central
- 37
- 54
- 46
- 36

South
- 39
- 43
- 40
- 34

West
- 32
- 36
- 33
- 29

Fig. 6-3. Many families in the United States produce a large part of their daily food needs. (Courtesy, Nationwide Food Consumption Survey [48 states], U.S. Department of Agriculture)

By taking an active part in gardening, young people can learn many interesting things about plant life. Gardening is a type of work through which youth can assume some responsibilities for the family living. Most of them will find considerable enjoyment in taking part in a family enterprise. In some communities, they have opportunities for entering garden clubs of various kinds.

In some cases, young people may raise vegetables and small fruits for sale and thus earn some money for themselves.

Where Should the Garden and Orchard Be Located?

On most places where land is available, it is possible to raise vegetables, small fruits, and even tree fruits. The location, or site, for growing these various products depends on the conditions of the individual homestead. The usual arrangement is to have tree fruits separate from vegetables and small fruits.

It is desirable to have the garden located on soil that drains well. If necessary, the soil should be tiled. A gradual slope is helpful for drainage; but the slope should not be so great that soil washing, or erosion, will take place. A southern slope is preferred in the North because the soil warms more quickly in the spring if this is the case.

The garden should be located on soil that is fairly loose and fertile. While much can be done to improve the soil by the use of animal manure and other fertilizers, it is advantageous to choose a place with average or better soil at the outset.

The garden should not be located near large trees, shrubs, or buildings because of the shading that results. Trees and shrubs also take a great deal of moisture and plant food needed by the garden crops.

The preferred location for the garden is near the home and a water supply. A garden near the home is convenient for working in it at odd moments and for making maximum use of the vegetables grown. A little water can improve the quality and yield of the garden produce as well as save the garden during dry spells. A garden next to the home is also easier to protect. Even though a fence is put up around the garden, being able to watch over it closely can help to limit the damage from rodents, birds, and other predators.

In some cases, if there is not sufficient space for a garden on the home place, nearby plots of ground may be rented for this purpose. If space for a garden is to be rented, it is wise to take into consideration the availability of water before the arrangements are concluded.

Which Vegetables Should Be Grown?

Beginners may wish to start with vegetables that are easy to grow. A few of these are the following:

English peas—good cool-weather plants with few insect problems.

Onion sets—cool-weather plants that can be used when small or full size.

Radishes—cool-weather plants.

Greens—kale, collard, leaf lettuce, cress, mustard, and spinich are all possibilities. These can be planted at the same time as cool-weather plants and will last most of the summer. Replanting can be done in late summer for a fall crop.

Tomatoes—transplant a month later than peas are planted.

Beans—snap or lima beans can be planted nearly any time; several times during the summer to keep a fresh supply.

Squash—summer yellow squash is a good choice.

Cucumbers—vines can be left to grow on the ground or can climb. Biggest problem is water.

Muskmelons—grow easily but take a lot of space.

Peppers—both green (sweet) and red (hot) peppers grow on small bushy-type plants.

Sweet corn—a favorite with most gardeners.

Fig. 6-4. Some of the items of equipment useful for gardening and working around the home grounds.

What Equipment Is Needed?

Garden equipment can be quite simple and inexpensive, or it can be elaborate and costly.

A hoe, spade, rake, and trowel are essential and, together with a measuring stick and planting string, provide the basic tools for gardening. For large gardens, a seeder and hand cultivator are desirable. Garden tractors, with the various attachments, can make gardening more enjoyable to the mechanically inclined and to those persons who have very large gardens. A garden hose long enough to reach all parts of the garden, plus a sprinkler attachment, will be needed if a source of water is available.

Some kind of duster or sprayer is also a necessity. Compressed air sprayers with a capacity of 1 to 5 gallons are the most satisfactory. Inexpensive plunger-type dusters with up to 3 pounds' capacity are also available and quite satisfactory. Crank-type dusters for large and small gardens are very easy to use.

What Are the Essentials for Raising Vegetables?

The vegetable garden, to be most successful, should provide an assortment of high-quality vegetables for use by the family throughout the growing season and during the winter months.

If their efforts in gardening are to be successful, home gardeners should become familiar with the best methods. The most important of these practices are discussed on the following pages.

Planning the garden

If someone were considering building a house, he/she would first spend considerable time in making plans for it. Careful planning is also necessary for success in gardening. Unless plans are carefully made, there will not be an assortment of vegetables throughout the year, and the most efficient use of the gardening space will not be obtained. The plans for a vegetable garden should be made during the winter months so that the seeds may be ordered and everything will be ready when the planting season arrives.

One of the first steps in planning is to select the *kinds* of vegetables to raise. The size of the garden will to some degree limit the kinds of vegetables to be raised. Preference should be given to the vegetables liked by members of the family, who often have prejudices against certain vegetables. These prejudices usually can be overcome if vegetables of good quality are raised.

Among the kinds of vegetables that should ordinarily be included are:

Fig. 6-5. Brussels sprouts really grow this way. (Courtesy, U.S. Department of Agriculture)

Fig. 6-6. A mature cauliflower plant.

Growing Food for the Family

(1) those in which the leaves and stems are used, such as asparagus, cabbage, celery, lettuce, rhubarb, and spinach; (2) those in which the roots or root-like portions are used, such as beets, carrots, onions, parsnips, potatoes, and radishes; (3) those in which the seeds are used, such as beans, peas, and sweet corn; and (4) those in which the fruit or fruit-like structures are used, such as cucumbers, peppers, pumpkins, squash, melons, and tomatoes.

Fig. 6-7. Acorn squash can provide many delicious meals.

After the kinds of vegetables have been decided upon, the *variety* or *varieties* of each should be chosen. The seed catalogs published by reliable garden seed dealers are helpful in a study of varieties. A beginner in gardening will do well to select varieties recommended in bulletins of the state college of agriculture. Ordinarily, only one or two varieties of each kind of vegetable should be selected. In growing some vegetables, the beginner should choose an early and a late variety. A number of those that are suitable are listed in Table 6-1.

After selecting the kinds and varieties of vegetables, the home gardener must decide how much space in the garden should be used for each kind. This is in part dependent upon the size of the garden and in part on the size of the family for which the garden is planned. A plan for the garden should be carefully drawn to show the location and space for each vegetable. The distances between rows and the vegetables that will follow those that occupy the ground only part of the season should also be shown on the plan. A suggested plan is shown in this chapter, although it is not expected that everyone will wish to have a plan just like the one shown.

Table 6-1. Types of varieties of common vegetables

Kind of vegetable	Types of varieties
Asparagus	Mary Washington, Waltham Washington
Beans, snap	
bush	Bush Lake 274, Contender, Kinghorn Wax, Top Crop, Tenderette, Tendergreen
pole	Challenger, Dade, Kentucky Wonder
Beans, lima	Dixie Butterpea, Henderson Bush, Jackson Wonder, King of the Garden, Thorogreen
Beets	Detroit Dark Red, Early Wonder, Ruby Queen
Broccoli	Waltham 29, Green Comet
Brussels sprouts	Jade Cross, Long Island Improved
Cabbage	
early	Copenhagen Market, Early Jersey Wakefield, Golden Acre
midseason	Marion Market, Red Acre
late	Danish Ballhead
Carrots	Danvers 126, Imperator, Nantes, Royal Chantenay
Cauliflower	Snowball, Snow King, Purple Head
Celery	Summer Pascal, Tendercrisp, Utah 52-70
Collards	Georgia, Vates
Corn, sweet	Butter and Sugar, Golden Cross Bantam, Illini Xtra Sweet, Jubilee, Seneca Chief, Silver Queen
Cucumbers	
pickling	Bravo, Chicago Pickling, National Pickling, Pioneer, SMR 18
slicing	Marketmore 70, Poinsett, Straight Eight, Sweet Slice, White Spine
Eggplants	Black Beauty, Black Magic, Early Beauty
Endive	Green Curled Ruffec, Salad King
Garlic	Creole, Elephant Large White Improved
Horseradish	New Bohemian
Kale	Dwarf Blue Scotch, Dwarf Green Curled, Siberian
Kohlrabi	Purple Vienna, White Vienna
Lettuce	
head	Buttercrunch, Great Lakes 659, Ithaca, White Boston
leaf	Black Seeded Simpson, Parris Island Cos, Salad Bowl
Muskmelons	Golden Crenshaw, Golden Beauty Casaba, Honey Dew, Hale's Best
Onions	Beltsville Bunching, Chieftan, Early Harvest, Granex, Yellow Sweet Spanish
Parsley	Moss Curled, Perfection

(Continued)

Table 6-1 (Continued)

Kind of vegetable	Types of varieties
Parsnip	All American, Hollow Crown
Peas	
edible pod	Dwarf Gray Sugar, Dwarf White Sugar, Sugar Snap
English	Alaska, Freezonian, Green Arrow, Little Marvel, Progress No. 9
Peppers	
sweet	California Wonder, Pimento, Yolo Wonder
hot	Hungarian Yellow Wax, Long Red Cayenne, Red Chili
Potatoes	Irish Cobbler, Katahdin, Kennebec, Red Pontiac, Sebago
Pumpkins	Cinderella, Connecticut Field, Jack-O-Lantern, Small Sugar
Radishes	Champion, Cherry Belle, Scarlet Globe, White Icicle
Rhubarb	Crimson, Ruby, Valentine, Victoria
Spinach	America, Bloomsdale Long Standing, Early Hybrid 7, New Zealand
Squash	Butterbar, Butternut, Early Prolific, Straightneck, Hubbard, Table Queen, Zucchini
Sweet potatoes	Centennial, Gold Rush, Jersey Orange, Nugget
Tomatoes	Better Boy, California 145, Columbia, Manapal, Small Fry, Supersonic, Tropi-Gro
Turnips	Golden Globe, Purple Top White Globe, Seven Top (greens)
Watermelons	Charleston Gray, Crimson Sweet, New Hampshire Midget, Rhode Island Red, Sugar Baby, Yellow Baby

The garden plan should be drawn to a suitable scale, such as 1 inch to 20 feet. With this scale, the outline for a garden that is 100 feet by 140 feet would be 5 inches by 7 inches. After the outline has been made, lines can be drawn to represent the rows, and the names of the vegetables can be shown for the various rows. In arranging the garden, the gardener should place the permanent crops, such as rhubarb and asparagus, along one side so they will not be damaged when the plot is being plowed. If some small fruits are to be part of the garden, space must be provided for them. The rows for the vegetables should ordinarily be placed the long way of the garden. For cultivation with a garden tractor, the rows should be about 3 feet apart.

In planning the garden, beginning gardeners should find the following suggestions helpful.

1. Group the early-season, quick-maturing crops together, starting at one side of the garden. These include lettuce, radishes, peas, beets, carrots, and early cabbages. Other crops can follow some of these as the early-maturing vegetables are removed.

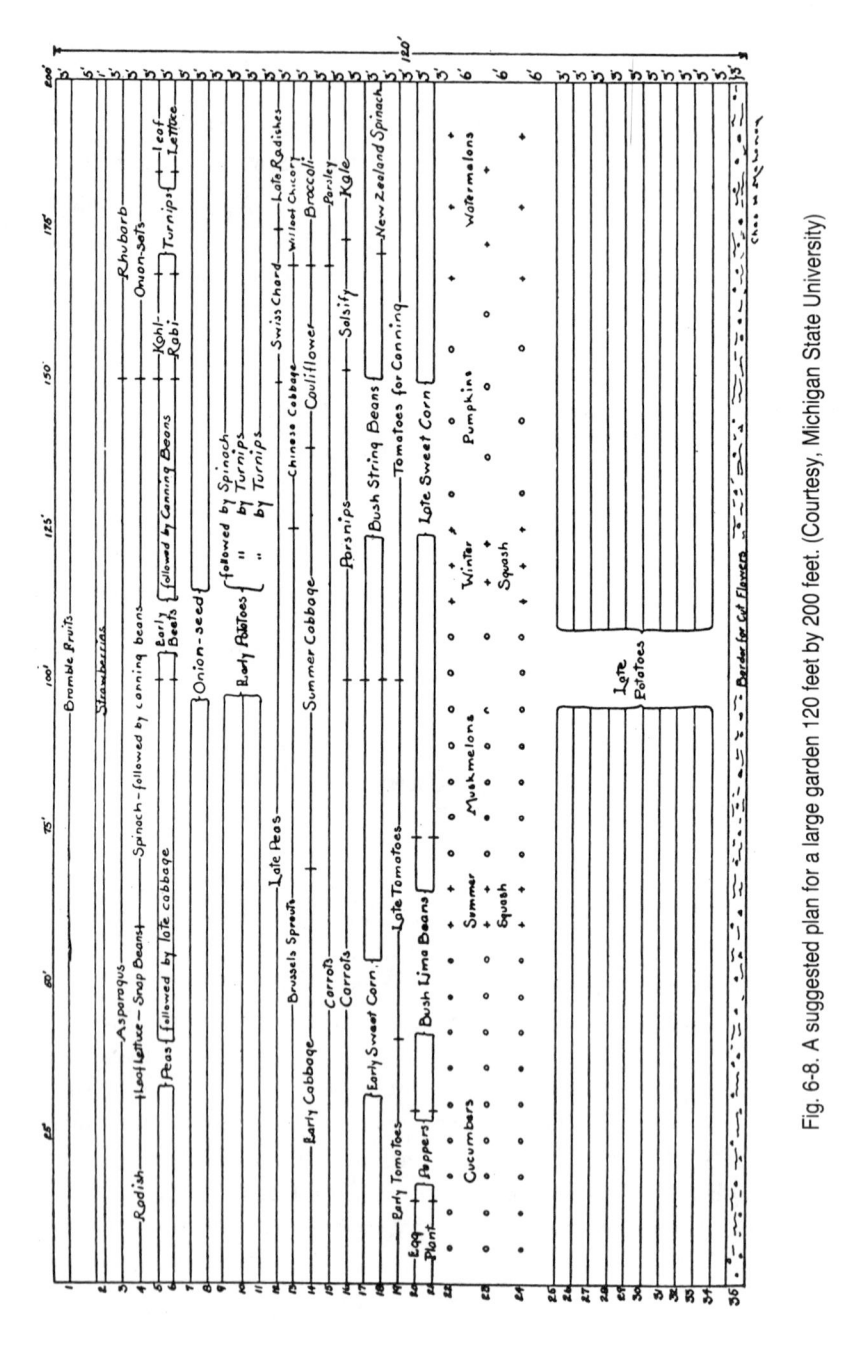

Fig. 6-8. A suggested plan for a large garden 120 feet by 200 feet. (Courtesy, Michigan State University)

2. Group the crops that take the entire season to mature. These include onions from seeds or small plants, broccoli, parsnips, chard, and salsify.
3. Group the tall-growing plants together in a place where they will not shade other plants. These tall plants include sweet corn, pole beans, and staked tomatoes.
4. Provide ample space for vine crops, such as squash and muskmelon, so that they will not interfere with other crops.
5. Select varieties of vegetables recommended for the particular state or region.

A *planting table* is a useful part of the garden plan. The table should show the kinds of vegetables, varieties, time of planting, and the distances between plants in the row. (See item 3 in the list of suggestions at the end of this chapter.)

If the garden is carefully planned, suitable amounts of each vegetable will be planted; the family will be provided with a fairly continuous supply of vegetables throughout the year; and the entire garden space will be kept in use throughout the season.

Seeds and plants should be purchased from a reliable firm. Perennial vegetables, such as rhubarb and asparagus, are frequently started from rootstocks. Sets or plants are frequently used for producing onions. Almost all other common vegetables, except potatoes, are started from seeds. Only a few early potatoes are ordinarily grown in a garden, but the potatoes for planting should be free from disease and of a recommended variety.

The amount of seed needed should be determined as accurately as possible, with allowances for some replanting. Table 6-2 will be helpful in determining the amount of seed or number of plants for the various vegetables and in computing the length of row for each of the vegetables for a family of four.

Preparing the seedbed and fertilizing the soil

Usually it is best to perform some of the operations for the preparation of the seedbed in the fall of the year. The first step is to remove all of the plant materials that are left on the garden.

If animal manure is available, applications should be made prior to plowing. Compost is also good for providing organic matter. The garden should be plowed rather deeply.

Compost adds organic matter to the soil and helps aerate the soil. It is also a good way to get rid of leaves, garden trimmings, and grass clippings. The compost can be made in a pile or in a contained area, such as a circle made with a snow fence. The compost is started from a 6- to 8-inch layer of plant material. For each 10 square feet of area, about a cup of a complete

Table 6-2. Seed and plant requirements of common vegetables for a family of four

Kind of vegetable	Amount/number per 100 feet of row	Amount/number for family of four
Asparagus	60 plants	60 rootstocks
Beans, bush	½ pound	1 pound
Beans, pole	½ pound	½ pound
Beets	2 ounces	2 ounces
Cabbage, early	60 plants	(50 plants) 1 packet
Cabbage, late	60 plants	(100 plants) 1 packet
Carrots	1 packet	1 packet
Celery	200 plants	(200 plants) 1 packet
Cucumbers	1 packet	1 packet
Lettuce, head	1 packet	(100 plants) 1 packet
Lettuce, leaf	1 packet	1 packet
Onions, seed	1 packet	2 packets
Onions, sets	1½ quarts	2–3 quarts
Parsnips	1 packet	1 packet
Peas, early	½ pound	2 pounds
Peas, late	½ pound	2 pounds
Peppers	50 plants	10 plants
Potatoes, early	6 pounds of tubers	12 pounds
Pumpkins	1 ounce	1 ounce
Radishes	1 ounce	1 ounce
Rhubarb	40 plants	10 roots
Spinach	1 ounce	¼ pound
Squash, summer	½ ounce	½ ounce
Squash, winter	1 ounce	1 ounce
Sweet corn, early	2 ounces	2 ounces
Sweet corn, late	2 ounces	3 ounces
Tomatoes, early	30 plants	10 plants
Tomatoes, late	50 plants	20 plants

garden fertilizer is spread on top of the plant material. Next, the pile is covered with 1 to 2 inches of soil. This layering of plant material and soil is repeated whenever material is available and until the pile is the desired height. Keeping the top of the pile level or slightly concave will help to catch rain water to provide the needed moisture. Some additional moisture will be needed if rainfall is light. The pile will be ready for use more quickly if it is turned with a spading fork every two months to admit air. It will take from four to six months for the compost to be suitable.

Plowing or spading should ordinarily be done in the fall, rather than in the spring, because the soil will have a chance to settle; the soil texture will be improved; the manure will have more time to decay; and the soil will usually dry out more quickly, so that earlier spring planting is possible.

The condition of the soil should be checked before the soil is plowed or spaded. If the soil crumbles when a handful of it is squeezed, then it is dry enough to work.

If the garden soil is very acid, lime should be added. Lime improves the texture of heavy soils and neutralizes the acidity, but its excessive use may be injurious to most garden crops. Consequently, it is not wise to add lime unless the soil is definitely acid, as shown by a reliable test of the soil. (See Chapter 16.) Some crops, including asparagus, beets, cabbage, cauliflower, lettuce, lima beans, muskmelons, peas, and spinach, benefit from the addition of lime if the soil is strongly acid. Most garden vegetables do best on soils that are slightly acid, so the amount of lime should be limited accordingly.

In most cases, it pays to add commercial fertilizers to gardens. These may be scattered uniformly over the surface and worked into the top layer of soil or broadcasted prior to plowing. A good general purpose fertilizer with a formula of 4-16-4, 5-10-5, 5-10-10, or 10-10-10 provides nitrogen, phosphorus, and potassium. This fertilizer may be added at the rate of 1 to 4 pounds per 100 square feet of surface.

If the garden is sufficiently large, it is desirable each year to plant a strip to clover, rye, or some other suitable crop that will be plowed under as

Fig. 6-9. A well-tended garden can provide many kinds of vegetables. Note that the asparagus tops have been permitted to grow so that they will be able to provide nutrients for the storage roots from which the edible spears will come the next year.

green manure. If a portion is seeded each season to this kind of crop, it may be possible to improve the soil over the entire garden in a period of a few years. Such crops may also be grown on space used the first part of the season for early-maturing vegetables.

Rotating the main crops, such as potatoes, tomatoes, etc., in the garden plot is desirable so that they will not be planted two years in succession on the same parts of the garden. Such a practice helps to control diseases.

Using a disk or a harrow, the home gardener should work the soil into a good state of tilth just previous to planting the vegetables. In small plots, the hand rake is suitable. Since many of the seeds are very small, the soil should be worked until it is quite fine and smooth.

The use of soil insecticides will help reduce soil insect problems. The granular form of the insecticides is easy to handle and can be applied along with the fertilizer. All soil insecticides should be applied according to the manufacturers' directions.

Planting the garden

Some vegetables are sufficiently hardy to be planted in the early spring as soon as the soil becomes dry enough. These cool-season or hardy vegetables include potatoes (early crop), onion sets and plants, onion seeds, radishes, lettuce, peas, spinach, and cabbage. Crops that should be planted a little later include beets, carrots, and late potatoes. Others that should not be planted until all danger from frost is past are sweet corn, tomatoes, peppers, beans, squash, cucumbers, watermelons, eggplants, and muskmelons.

The seeds of some vegetables, such as celery, early cabbage, cauliflower, tomatoes, and peppers, are usually started early in *hotbeds* or *coldframes* so that the plants will be fairly large when the weather is sufficiently warm to plant them in the open garden. For a small garden, these plants are usually purchased.

The depth at which seeds should be planted in the garden soil will depend somewhat on the type of soil and the size of seed. A general rule is to plant the seeds at a depth four times the largest diameter of one seed. The seeds should be planted shallower than this in heavy soils and deeper in light soils. Directions for depth and thickness of planting are usually printed on the packages of seeds.

Planting and later care will be easier if straight rows are marked out in advance of making the furrow openings for the seeds. Wooden stakes and strong twine will aid in this task. Shallow furrows for small seeds can be made by drawing a hoe handle along the twine or string. For deeper furrows, the corner of a hoe will serve well.

The plants that have been started earlier or purchased should be *transplanted* into the garden at the proper time. Plants such as tomatoes and peppers should not be transferred to the garden until all danger of frost is past.

Cabbage plants may be transplanted early. The main or late crop of cabbage may be started by rather thickly planted seeds in a small space in the garden. The resulting plants may then be transplanted in the garden in sufficient time to mature before fall.

In the transplanting process, as much soil as possible should be left on the roots of the plants so that the tiny root hairs will not be stripped from the roots. The plants should be set in holes at a slightly greater depth than they were growing, and a small amount of soil should be packed around the roots. Some water should then be added to settle the soil around the roots. After the water has soaked away, the remainder of the hole should be filled with soil.

Cabbage, lettuce, onion, and some other plants can be hardened to withstand frost; others, such as tomatoes and peppers, cannot. Hardening also helps the plants withstand chilling, drying from winds, moisture shortage, and high temperatures. Withholding water and lowering the temperature to slow the rate of growth for a period of two weeks before planting is the best way to harden the plants. About a week before they are to be transplanted, plants in beds or flats should be blocked out with a large knife. Blocking, or cutting the roots, causes new roots to form quickly near the plants to make recovery from transplanting easier.

Cultivating the garden

The soil in the garden should be cultivated carefully after the vegetables have started to grow. The main object in cultivating is to kill the weeds. Cultivation also helps to conserve the moisture and to permit air to enter the soil more readily. This helps to make plant food materials available to growing plants. Thus, it seems desirable to cultivate the soil often enough to keep out the weeds and to keep the soil loose. Cultivation should be shallow to avoid injury to the roots of the vegetables.

In order to save labor in a large garden, the gardener should use a tractor-drawn cultivator as much as possible. Small tractors suitable for garden work save considerable labor. It is usually necessary, however, to use a hand hoe or some other tool next to the rows. Some hand weeding and thinning in the rows is also desirable. Where much of the work is done by hand, a wheel cultivator may be used.

Some gardeners may prefer to mulch their gardens in May or June after the weeds have been cleaned up. Straw or peat moss placed between the rows and around the plants to a depth of 2 to 4 inches will help control weeds and conserve moisture. Black plastic may be used for tomatoes, eggplants, cucumbers, melons, and squash. For transplanting, it is easier to put down the plastic first and set the plants through the plastic. It is best not to use mulches during rainy seasons and in wet areas.

Controlling diseases and insects of garden crops

It is important to know how to control the serious diseases and insect pests of vegetable crops.

Insect pest control begins with the use of soil insecticides as already discussed. Eliminating any accumulations of weed and crop wastes will also help, since such accumulations are good breeding places for insects. Rotenone, pyrethrum, sevin, and malathion are good dusts for controlling most common garden insects. Rotenone will control cabbage worms, cucumber beetles, Mexican bean beetles, and squash vine borers. Malathion is used for aphids and cabbage worms. Sevin is used to control corn earworms, blister beetles, corn borers, potato bugs, and several other pests.

Disease control may also be a problem, especially in places where gardens have been located for a period of years. Some of the more common diseases are yellows and black rot of cabbage; asparagus rust; bean bacterial blight, mosaic, and root rot; black spot of beets; blight, scab, and blackleg of potatoes; wilt, leaf spot, mosaic, and anthracnose of tomatoes; the mosaic and wilt diseases of vine vegetables.

For most gardeners, control efforts for diseases will be limited to buying treated seed, buying disease-resistant varieties of vegetables; removing and destroying weed and crop residues; rotating the garden plot and the locations of the various kinds of plants in the garden; and controlling insects to

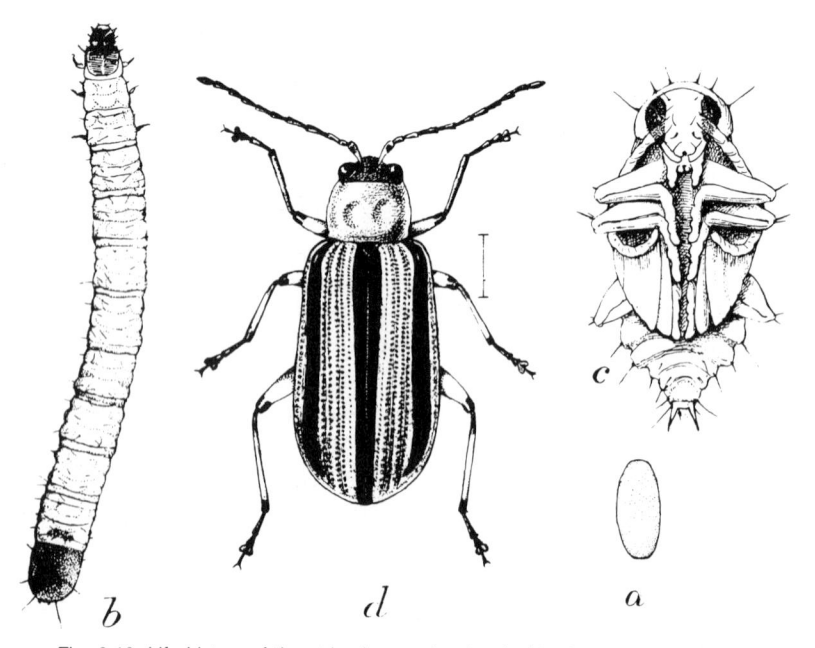

Fig. 6-10. Life history of the striped cucumber beetle (*Acalymma vittata* [F]). (Courtesy, U.S. Department of Agriculture)

Growing Food for the Family 185

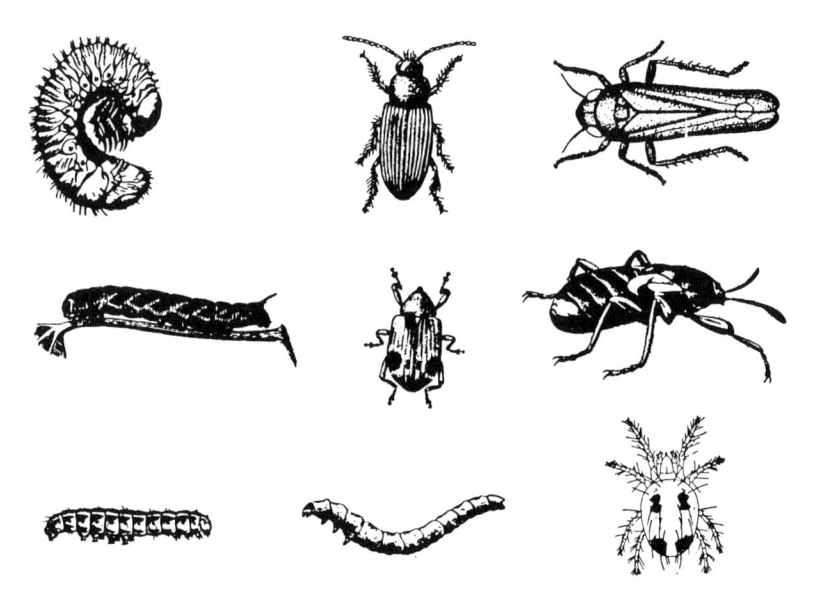

Fig. 6-11. Some garden pests. *(From top to bottom: left row)* White grub, hornworm, armyworm; *(center row)* seed corn beetle, strawberry weevil, corn rootworm; *(right row)* leafhopper, chinch bug, spider mite.

keep down the spread of disease. It is also best to stay out of the garden when the plants are wet with rain or dew. For the dedicated gardener, there is a variety of seed treatments and sprays available, such as captan, maneb, zineb, and mancozeb. In each case, the directions on the package should be followed with extreme care.

Harvesting the vegetables

Food fresh from the garden or ripened on the vine provides the flavor treat that keeps many people raising food in a garden year after year. The secret to the best flavor is harvesting the crop at the right time.

There are many ways to determine the best time to harvest, depending on the characteristics of each vegetable. While only by experience can a person become an expert, Table 6-3 lists general guidelines for some of the popular vegetables.

What Are the Essentials for Producing Small Fruits?

A garden is not complete unless it provides small fruits in addition to the vegetables. The kinds of small fruits that should be included depend on the space available, the locality, and the tastes of the family. Strawberries,

Table 6-3. General guidelines for harvesting common vegetables and melons

Kind of vegetable or melon	Time to harvest
Asparagus	When spears are 6 to 8 inches high.
Beans, lima	When beans are firm and bright and pods are full but still green.
Beans, snap	When beans have just started to grow inside the pods and pods snap easily.
Beets	When skin is smooth and bright and roots are about 2 inches long. Small- to medium-sized beets are best.
Broccoli	When heads are still green and compact and before flowers show color.
Brussels sprouts	When plants are bright green and have tight heads.
Cabbage	When heads appear to be full and heavy.
Cantaloupes	When they are beginning to be yellow-orange and they separate easily from the vine.
Carrots	When they are medium size, firm, brittle, and bright orange-yellow.
Cauliflower	About two weeks after tying the leaves and the heads are white and compact.
Celery	When stalks are medium size and easy to snap.
Collards	When plants are bright green and have small midribs.
Corn, sweet	About two weeks after silk starts to turn brown when kernels are shiny and still in tender milk stage.
Cucumbers	When they reach desired size and are a dark shiny green.
Eggplants	When fruit is dark purple and before they reach full size.
Lettuce, head	When heads are desired size and firm.
Lettuce, leaf	When leaves are large enough and before seed stems appear.
Onions	When they reach desired size for early use; when tops are falling over with some green color left for storage.
Peas, English	When pods are not filled and peas are tender, sweet, and bright green.
Peas, Southern	When pods are nearly filled and starting to turn purple or yellow.
Peppers	When they reach desired size and are firm.
Potatoes, white	When vines die; for early use, when large enough.
Potatoes, sweet	When most of the tubers are 2 to 3 inches in diameter.
Radishes	When they reach desired size and roots are still firm and tender.
Rhubarb	When stems are 8 to 15 inches long.
Spinach and Swiss chard	When leaves are crisp and green.
Squash, summer	When they reach desired size, are a bright color, and have tender shell.
Squash, winter	When rind is hard and the proper color.
Tomatoes	When fruit is right color, uniform, and firm.
Turnips	When they are of medium size and are firm.
Watermelons	When ground spot turns yellow.

Raising strawberries

Strawberries are grown frequently by home gardeners. A part of the regular garden makes a good location for the strawberry bed. Since strawberries can be expected to bear satisfactorily for only two years in one location, it is usually desirable to start new plantings every two years at least. After the plants in the new location become productive, the old patch may be plowed under and the space used for vegetables again. The rows of strawberries should be placed parallel with the rows of vegetables so that the same cultivation methods can be used. Areas where tomatoes, potatoes, peppers, and eggplants have been grown within three years should be avoided to reduce the possibility of verticillium wilt.

In selecting varieties of strawberries, home gardeners should consider recommendations of their state college of agriculture. Some possible choices are Catskill, Earliglow, Guardian, Mastadon, Ozark, Red Chief, Sparkle, Sunrise, and Surecrop.

Strawberry plants should be purchased from a reliable nursery so that vigorous and disease-free plants will be secured. As soon as the plants arrive, they should be "heeled in" if they cannot be set out immediately. This process consists of loosening the bundles, placing the plants close together with the roots in a trench a few inches deep, covering the roots with soil, and applying water freely to the soil around the roots.

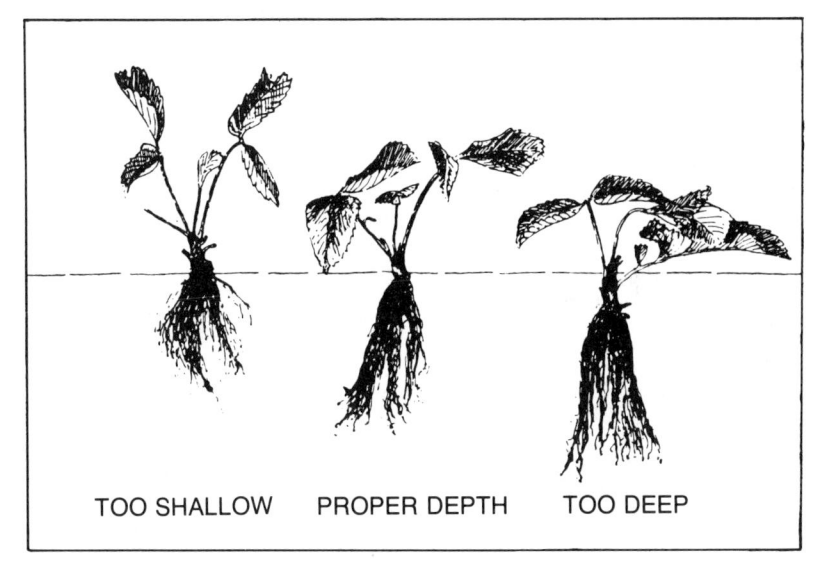

Fig. 6-12. Correct and incorrect depths to transplant strawberry plants. The crown of the plant should be at the ground level *(center)*. (From Circular 453, University of Illinois)

The *matted-row* system of growing strawberries is most commonly used for spring-bearing varieties. In this system, the plants are set 18 to 24 inches apart in rows 4 feet apart. During the season, the plants send out runners in all directions. These take root and form new plants so that each row develops a matted growth for a width of about 2 feet. During the first year, the flowers are pinched off to encourage root establishment and plant growth. After the first year, all new runners should be removed. The new plants bear fruit the following year.

Many people prefer the hill system, with plants spaced 6 to 7 inches apart, because of the greater ease of using tools for weed control. With the hill system, runners are removed each 7 or 10 days. Only four to six runners are allowed to grow from each plant.

In setting strawberry plants, the gardener should insert them in the soil at the proper depth. The *crown*, which is the region from which the roots and leaves arise, should be located at the surface of the ground. A spade or a trowel may be used to make the openings in the soil for the roots. The soil should be firmly packed around the roots. Unless the soil is very moist, it is desirable to add water to the soil around each plant. Spring is ordinarily considered to be the best time for starting a strawberry bed. The newly set plants should be given careful cultivation. Some hand weeding may be necessary between the plants.

Fig. 6-13. It's fun picking and eating strawberries like these. (Courtesy, U.S. Department of Agriculture)

In the northern states, the strawberry patch should be *mulched* for protection during the winter season. Straw, hay, or other similar materials may be used for mulch. These materials are best put on the rows to a depth of 2 or 3 inches in the early winter, before the ground freezes hard. Mulching prevents injury to the roots of the plants from extreme cold and from "heaving" of the soil, which would take place as a result of alternate freezing and thawing. The spring growth of the plants is also retarded by mulching, so that blooming will take place after danger from late frosts is past. Part of the mulch should be removed in the spring as soon as a few leaves from the strawberry plants begin to appear through it. The material may be placed between the rows, where it helps to keep down the weeds and to conserve moisture. Leaving some of the material on the rows helps to keep the developing berries from touching the ground.

After one year of bearing, a strawberry patch may be renewed for an additional crop. One method is to remove the plants from the centers of the rows where the oldest plants are found and thin out the plants in the portions that remain. This should be done soon after the bearing season ends. In some places, gardeners can maintain the beds for up to five or more years. To do this, the leaves are cut off immediately after harvesting; a pound of fertilizer, such as 10-10-10, is spread on each 25 feet of row, and the leaves and mulch are spaded or tilled into the soil. The beds are left at a width of 15 to 18 inches. Water is provided to stimulate new growth.

Strawberries are attacked by the tarnished plant bug and the strawberry bud weevil. Weekly sprays of malathion can be used for control during the blossom period before harvesting starts. A coarse netting can be purchased to protect the berries from birds.

Raising raspberries and blackberries

In many parts of the United States, raspberries and blackberries are grown successfully. Plantings of these small fruits usually remain productive for several years. Therefore, if they are included in the regular garden space, they should be planted along one side.

There are three common kinds of raspberries: black, red, and purple.

Some recommended varieties of raspberries and blackberries are as follows: (1) black raspberries—Allen, Jewel, and Bristol; (2) red raspberries—Newburgh, Cherokee, Hilton, Taylor, and Heritage; (3) purple raspberries—Cumberland; and (4) blackberries—Darrow, Ranger, and Raven.

In selecting varieties of these small fruits, gardeners should follow the recommendations from their state college of agriculture.

The best time for starting a new planting of raspberries or blackberries is spring. The plants can best be obtained from a reliable nursery, although they may be secured from old plantings, provided they are free from serious diseases. Red raspberries and blackberries produce *suckers* at some distance from the plants. These may be used for new plantings if each is carefully dug with a small portion of the root from which the sucker is developed. Purple and black raspberries form new plants by a process known as *tip layering,* in which the tips of the canes develop roots where they come in contact with the soil. Gardeners can speed up this process by covering the ends of some of the canes with soil several weeks previous to the time that new plants are desired. The newly developed roots, along with portions of the canes, can then be removed and planted.

Fig. 6-14. *(Top left)* Heritage fall-bearing red raspberries. *(Top right)* Jewel black raspberries. *(Bottom left)* Trailing variety of blackberries. *(Bottom right)* Erect variety of blackberries. (Courtesy, U.S. Department of Agriculture)

The usual method of spacing raspberries and blackberries is to set the plants at 2- to 4-foot intervals in rows about 8 feet apart. The new plants should be set somewhat deeper than they were growing originally. The soil should be carefully packed around the roots, and water should be added if the soil is dry.

After a patch has been started, it is desirable to cultivate occasionally between the rows. This will check the growth of weeds and will prevent the spread of plants outside the rows. After the patch has been established, the spaces around the plants can be mulched with layers of straw or other similar materials.

Pruning is necessary for raspberries and blackberries to grow well. In pruning these plants, the gardener should keep in mind that the fruit is borne on canes that grew the previous year. These canes die after producing one crop. It is best to remove these fruiting canes after the crop has been gathered, since they may be diseased.

The new canes of black and purple raspberries and blackberries should be pruned during the year of growth; the tips of these new canes should be cut or pinched off when the canes reach a height of 2 or 2½ feet. This practice will cause the canes to branch out and become stocky. The following spring, before growth starts, these side branches should be cut back to a length of 10 to 12 inches. The weak canes should be removed at that time leaving four to six strong canes per plant.

Red raspberries do not require as much pruning as the other raspberries. The fruiting canes should be removed after the crop is gathered. The following spring, the young canes should be cut back to a height of about 4 feet before growth starts. The weaker canes should be removed entirely, leaving about four to six vigorous canes per plant, if in hills. If in continuous rows, canes should be thinned to about 6 inches apart.

Raising currants and gooseberries

A few currant and gooseberry bushes may be grown. These will remain productive for several years and may be placed in the border plantings of the lawn or elsewhere. Among the best varieties of the currants for most conditions are Perfection, Wilder, and Red Lake. Downing and Poorman are varieties of gooseberries commonly recommended.

Currant and gooseberry plants should be secured from a nursery and then planted at 4- to 6-foot intervals.

Pruning consists of removing some of the weakest of the new canes and some of the old canes that have borne fruit for several years. About 6 to 10 strong canes should be left per bush. The pruning should take place before the spring growth starts.

The chief *insect pests* of the currant and gooseberry are currant worms and aphids. Currant worms are greenish with numerous black spots. The worms feed rapidly and may eat all of the leaves from the bushes unless they are destroyed; spraying or dusting with rotenone is an effective control measure. Application of a solution of nicotine sulfate effectively controls the aphids.

Anthracnose and leaf spot can be serious diseases. Zineb is often used for control.

Raising grapes

Home gardeners usually prefer to grow grapes for the table rather than for drying or wine making. Grapes are best located at one side of the garden in rows that will permit easy cultivation. If space is limited, a few grape vines may be trained on a fence, against the sunny side of a building, or on an arbor or a pergola.

The local climate is the major factor in determining the variety of grape to be planted. For the cooler climates, Concord, Iona, Niagara, and Golden Muscat are often chosen. These are known as eastern varieties, sometimes called slip-skin varieties because the skin separates so easily from the pulp. Exotic, Pierce, Ribier, Golden Muscat, Thompson Seedless, and Red Malaga do well in the intermediate and warmer areas.

Grapes are fairly easy to grow. In the cooler climates, they rarely need water. Two or three waterings may be needed in warm climates. The most troublesome disease is mildew. To prevent mildew, a fine sulfur dust can be used throughout the growing season. The most troublesome insect, the grape leafhopper, can be controlled with malathion.

Fig. 6-15. The vines in this vineyard are pruned "umbrella" style, with two canes projecting from each side at the top of the plant.

Grapes planted in fertile soils ordinarily do not need to be fertilized. If plants show signs of poor growth, it is best to consult with the local vocational agriculture teacher or county extension agent.

Grapes may be started from plants that have been purchased. They may also be started from *cuttings* or by *layering.* Cuttings consist of pieces of the canes or branches containing about four buds each. These are set in the soil with all buds covered except one. During the season, some of these take root and can be transplanted to a permanent location. In the layering process, one of the canes that has been left attached to a vine is covered with soil. The end is left uncovered. During the season, roots are formed on the covered portion, and the following spring the new plant is ready for transplanting.

The plants of grapes should be set at intervals of about 8 feet apart.

Grapes are *pruned* more severely than any other of the common small fruits. The usual method of pruning is the four-cane Kniffen system. For this, a two-wire trellis is needed, the first wire of which is about 2½ feet from the ground and the second about 2 feet above the first. The pruning should be done previous to the time that growth starts in the spring. Four canes of the previous year's growth are left. These are the fruiting canes, and they should be cut off or "headed back" so that 10 to 15 buds are left on each. In addition, four renewal spurs should be left. These are best located near the main trunk, with two buds on each, where they produce fruiting canes for the following year. (See Fig. 6-16.)

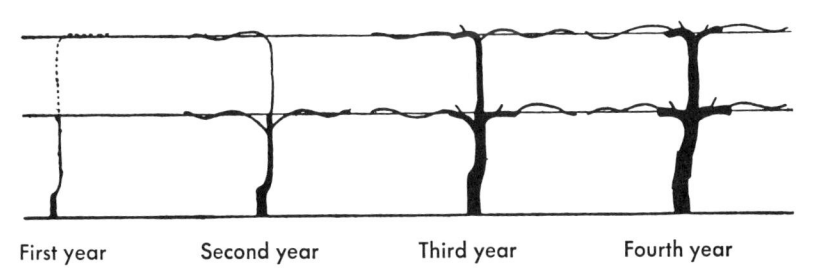

First year   Second year   Third year   Fourth year

Fig. 6-16. Grapes must be pruned properly to secure good fruit. The above diagram shows the method for pruning in successive years under the four-cane Kniffen system.

What Are the Essentials for Raising Tree Fruits?

In addition to a garden that produces an abundance of vegetables and small fruits, people with sufficient space may have a few fruit trees or a small orchard for producing fruits. These may include apple trees of summer, fall, and winter varieties and a few cherry and plum trees. In many areas, pears and peaches may also be grown successfully. Fruit trees must be sprayed to control insects and diseases. Because of the labor involved, many people may decide not to raise them.

Selecting kinds and varieties

In planning to raise tree fruits, individuals should select varieties of apples and other tree fruits that are sufficiently hardy to withstand the climatic conditions of their particular section of the United States. Furthermore, they should get varieties of fruit of the choicest quality possible for given conditions. For these reasons, they should learn the varieties that are recommended by their state college of agriculture. In this way, they can take advantage of any new developments. For example, the Geneva Agricultural Experiment Station in New York recently released *Liberty*, a new disease-resistant apple variety. Home gardeners will need to spray only for insects. Local nurseries will carry the varieties that grow best in the area.

Dwarf (size-controlled) trees are recommended for the home garden as well as for commercial production. An acre of land will handle 1,000 dwarf apple trees as compared to 35 standard-sized trees. Dwarf trees use sunshine better; produce excellent fruit; are easy to spray, prune, and harvest. They also begin to produce fruit at an earlier age than do standard-sized trees. Dwarf trees should be about 6 to 8 feet in height.

Some of the kinds of dwarf fruit trees now available are peach, plum, apple, pear, cherry, apricot, and nectarine. Pruning can be used to further limit tree size. Other factors affecting size are the kind of rootstock, the richness of the soil, the size of the variety, and the earliness of production. Early production tends to reduce the tree size.

For best results, varieties and kinds of fruit should be selected to fit the growing conditions. It is especially important to obtain fruit from a reliable nursery.

Planting the trees

As soon as the young trees are received from the nursery, they should be heeled in if they cannot be set out at once.

It is important to set the trees about 30 feet apart so that there will be plenty of room between them for spraying and other work when the trees grow larger. Dwarf trees may be set at intervals of about 15 feet.

As soon as the locations of the trees have been determined, the holes should be dug. The hole for a tree should be large enough for all the roots without crowding and deep enough for the tree to be set slightly lower than it stood in the nursery row. After the broken and damaged ends of the roots have been cut off, the tree should be set in the hole with the roots spread out. Some of the top soil should be thrown around the roots and thoroughly tramped. Water should be added to settle the soil around the roots. The hole can then be filled with soil. Care should be taken in the planting process so that the roots are not exposed to the air long enough to become dry.

Pruning fruit trees

The first pruning of a tree is done immediately after it has been planted in its permanent location. The object of this pruning is to give the tree the desired shape. Usually, trees that have several lateral or side branches are purchased. From three to five lateral branches should be left and the other branches should be pruned with smooth cuts near the trunk so that the wounds will "heal" readily by growth from the margins of the scar. The lowest branch that is left should be about 18 inches above the ground, and the others should be uniformly distributed, so that no two branches of the same size will develop at the same place on the trunk. If this is done, splitting that takes place at weak crotches of larger trees can be avoided.

After a young tree has been trimmed to the desired shape, some additional pruning should be done from year to year to remove branches that interfere with each other. Each kind of fruit tree requires special pruning. Information on pruning can be secured from bulletins from the state colleges of agriculture.

Caring for the soil

The soil around fruit trees may be cultivated frequently enough to keep grass and weeds from growing. A layer of straw, hay, or sawdust may be placed around the trunks of the trees to prevent growth of grass and weeds. To discourage rodents, a 6-inch layer of fine stone or cinders should be placed around the tree trunk. A hardware cloth, ¼-inch mesh, placed around the trunks will help protect the trees. The wire cloth should reach the lowest branches and be 6 inches or more in diameter. Using herbicides to control weeds is preferable to using mechanical methods. Pre-emergence herbicides are used before the weeds appear; post-emergence herbicides are used after the weeds can be seen. For weeds such as crabgrass, goosegrass, pigweed, and lamb's quarters, the herbicides diuron, simazine, dichlobenil, and diphenamid are used. Paraquat, a contact herbicide, will kill small established annual weed grasses, broadleaf weeds, and some perennial weeds. In using chemicals to control weeds, individuals should follow the latest recommendations.

Controlling insect pests and diseases of tree fruits

In order to obtain good fruit, individuals should spray fruit trees regularly and carefully to control insect pests and diseases.

Directions for spraying fruit trees are available from the state colleges of agriculture. Some of the concerns that sell spray materials provide a complete one-package or all-purpose mixture for spraying or dusting home orchards, which is effective for controlling most insects and diseases. To control insects and diseases of fruits, growers should apply the proper materials at recommended times and with equipment that distributes the materials to all parts of the trees. In some communities, persons with spray rigs may be hired to spray fruit trees.

Some common diseases that must be controlled are scab, black spot, powdery mildew, and rust. Serious insect pests include caterpillars, aphids, mites, scales, leaf rollers, leafhoppers, and cankerworms.

Raising citrus trees

Citrus trees can be grown also, but only in the warmer, low frost-risk areas. Growers should select varieties that will do well in their particular area, that will meet their interests, and that will fit in the available space.

Planting and caring for citrus trees is not difficult. The grower should start with small trees, about two years old, and plant them in the spring after all danger of frost has passed. The trees should be planted according to procedures followed for any other balled and burlaped (B&B) tree. The rootball on each tree should be about 10 inches in diameter. The trunks should be whitewashed or wrapped so that they will not sunburn. The trees should be carefully watered, and they should be pruned to remove sucker growth and dead limbs. Applying fertilizer is not usually necessary. Fruit should be picked and used as it matures. Keeping the trees healthy is the best way to control insects and diseases.

How Should Vegetables and Fruits Be Stored for Winter Use?

In order to obtain the most benefit from raising vegetables and fruits, home gardeners may want to store considerable quantities for winter use.

Canning is a common method of storing some kinds of vegetables and fruits. Over one-third of the households in the United States can some vegetables and fruits each year. To avoid spoiled food, the home canner should use jars made for canning, use new lids or disks each time, and cook according to instructions, such as those found in USDA home canning publications. Bulging lids, odors, leaks, molds, and spurting liquids are signs of food spoilage; however, botulinum toxin, which can be fatal, is invisible.

Some kinds of vegetables can be satisfactorily stored with only a small amount of special preparation if the proper conditions are provided. A special vegetable room in the basement, with a window for ventilation, is a very desirable arrangement. The temperature in a good vegetable room should

Fig. 6-17. Not all garden produce is put into the freezer. Here, some heated fruit is being packed loosely into jars. (Courtesy, U.S. Department of Agriculture)

be kept at 40°F or slightly less. The air must be able to circulate, and it should be kept moist so that the vegetables will not become dry and shrivelled. In a storage place of this sort, most root crops, such as carrots, can be kept in moist sand, bunches of celery can be placed with the roots in moist soil, and potatoes can be stored in open crates or bins. Apples and pears will store well in baskets or other containers under these conditions.

Vegetables, such as squash, pumpkins, and sweet potatoes, which require a dry storage place, can be kept satisfactorily in an ordinary room or in the basement. Onions should be stored in a cool, dry place.

In some communities, freezer-locker plants are available. Some families are utilizing this type of storage for crops such as peas, lima beans, greens, string beans, and peaches. These products are processed, packed, and frozen while fresh. Many families are securing home freezers for storing certain vegetables, fruits, meats, and other food products.

Some vegetables can be stored in the garden where they grew. When the ground begins to freeze in the fall, the vegetables should be covered with a heavy mulch of straw, hay, or leaves. The mulch prevents freezing of the vegetables and makes digging easier in the winter.

One of the oldest methods for preserving food is by drying. Depending on the kind of food, enough water is removed to reduce water content to between 5 per cent and 25 per cent. In hot, dry climates, foods will dry in a few days. In other climates, foods can be dried in the kitchen oven with drying trays, an oven thermometer, and a small fan. Special food dehydrators can be purchased from several sources. Only the best vegetables and fruits should be dried. It is better to overdry than to underdry. Dried food should be placed in tightly sealed freezer-style plastic bags immediately after drying and then placed in a cool, dry, dark room for storage.

SUGGESTED ACTIVITIES

1. With the help of one or both parents, estimate the amount and value of the vegetables and fruits produced by your family during the past year. Estimate the cost of the vegetables and fruits that were purchased. How many kinds of vegetables and fruits were raised last year at your home? Which kinds might be produced in increased amounts?

2. Secure bulletins on gardening from the college of agriculture in your state. Also, secure seed catalogs from one or two firms that sell garden seeds. After studying these and information from the preceding chapter, make a list of kinds and varieties of vegetables you believe you could raise in your home garden or a garden of your own.

3. Make a plan for your home garden or a garden of your own in which the drawing is made according to suitable scale. On this plan, show the locations of the rows for the various vegetables, the succeeding crops (if any), and the distance between rows. Also make a planting table for this garden, which may be arranged as follows:

Kind of vegetable	Variety	Time of planting	Distance between plants in row	Amount of seed

For this table, use bulletins on gardening that you have obtained from the college of agriculture in your state. Also, have at least one seed catalog available for studying the characteristics of the various vegetables.

4. What diseases and insect pests damaged vegetables in your home garden during the past year? How might each be controlled?

5. Assume responsibility for raising vegetables and small fruits for use by your family.

6. Raise some vegetables and small fruits and open a roadside stand to earn some money during the summer months.

7. Bring specimens to school of some of the varieties of apples that are grown successfully in your area. Be prepared to identify them for the class. Using bulletins from your state college of agriculture and from other sources, make a list of the varieties of the various tree fruits that can be successfully grown in your area.

8. Report to your class on how vegetables and fruits are stored in your home. Give suggestions on how the storage conditions might be improved.
9. Arrange for a fruit grower or some other person who understands pruning to show your class how to prune fruit trees.
10. Concepts for discussion:
    a. Vegetables and fruits are an important part of a healthful diet.
    b. High-quality foods can be produced in a home garden more cheaply than they can be purchased in a grocery store.
    c. Gardening can be fun.
    d. The garden should be located near the home, near a water supply, and on soil suitable for growing vegetables and fruits.
    e. Equipment for gardening can be quite simple and inexpensive.
    f. The garden should be carefully planned to include the number and kinds of vegetables and fruits desired.
    g. Garden crops should be grouped in the garden according to time of maturity and height of the mature plants.
    h. A well-prepared seedbed will help to produce high yields of healthy vegetables and fruits.
    i. Most garden crops do best on soil that is slightly acid.
    j. In most cases, it will pay to use some fertilizer.
    k. Main garden crops should be rotated so they are not planted two years in succession in the same parts of the garden.
    l. Some vegetables can be planted earlier in the year than others.
    m. The depth at which seeds should be planted depends on the type of soil and the size of the seed.
    n. In transplanting, as much soil as possible should be left on the roots of the plants.
    o. The main purpose in cultivating is to kill weeds.
    p. For most gardeners, the best disease-control effort will be to use sound cultural practices.
    q. Gardeners will need to use good cultural practices, insecticides, and hand picking to control insect pests.
    r. There are several kinds of small fruits that home gardeners can grow.
    s. Considerable space and labor are required for growing tree fruits.
    t. It costs less to eat at home than it does to eat out.
    u. The greatest benefit from a garden is obtained when some vegetables and fruits are stored for use in the winter.

REFERENCES

Evans, B. R., A. L. Stacey, and Danny Gay. *Insects and Diseases of Home Vegetable Gardens.* Bulletin No. 835, Cooperative Extension Service, College of Agriculture, University of Georgia, Athens, GA 30602.

Lee, J. S., and D. L. Turner. *Introduction to World AgriScience and Technology.* Interstate Publishers, Inc., Danville, IL 61832.

Oebker, N. F., and P. M. Bessey. *Vegetable Varieties for Arizona.* Cooperative Extension Publication Q337, College of Agriculture, The University of Arizona, Tucson, AZ 85721.

Swaider, J. M., G. W. Ware, and J. P. McCollum. *Producing Vegetable Crops.* Interstate Publishers, Inc., Danville, IL 61832.

Topoleski, L. D. *The Home Vegetable Garden.* Cornell Information Bulletin 101, Cooperative Extension Service, New York State College of Agriculture and Life Sciences, Cornell University, Ithaca, NY 14850.

True, L. F., and S. Fazio. *Citrus Trees in the Home Garden.* Cooperative Extension Service Publication Q39, College of Agriculture, The University of Arizona, Tucson, AZ 85721.

U.S. Department of Agriculture. *Dwarf Fruit Trees.* Leaflet No. 407, Agricultural Research Service, U.S. Government Printing Office, Washington, DC 20402.

U.S. Department of Agriculture. *Food for Us All.* Yearbook of Agriculture: 1969, U.S. Government Printing Office, Washington, DC 20402.

U.S. Department of Agriculture. *Gardening for Food and Fun.* Yearbook of Agriculture: 1977, U.S. Government Printing Office, Washington, DC 20402.

U.S. Department of Agriculture. *Growing Vegetables in the Home Garden.* Home and Garden Bulletin No. 202, U.S. Government Printing Office, Washington, DC 20402.

U.S. Department of Agriculture. *Keeping Food Safe to Eat.* Home and Garden Bulletin No. 162, U.S. Government Printing Office, Washington, DC 20402.

Wott, J. A., and J. Chamberlain. *Recommended Vegetable Varieties.* Cooperative Extension Service Publication HO-101, Purdue University, West Lafayette, IN 47907.

Chapter 7

Conservation—The Wise Use of Natural Resources

"Come and get it!" "Take all you can!" "Cut out and get out!" From pioneer days until recent times, many people in the United States have looked upon forests, soil, wildlife, and other natural resources as theirs for the taking. Little thought has been given to the future, except possibly to assume that "there will always be plenty left."

OBJECTIVES

1. Define conservation.
2. Discuss soil loss through erosion.
3. Discuss conserving water resources.
4. Outline the hydrologic cycle.
5. Discuss forest and coastal wetlands preservation.
6. Describe basic wildlife management techniques.

Here and there, individuals questioned the recklessness with which nature's gifts were being used and wasted. However, for the most part, their voices were drowned in the rush to "conquer the wilderness" and to "get while the getting is good." The United States has been blessed with an abundance of resources that furnish the necessities of life—food, shelter, and clothing—as well as many luxuries. Perhaps no other nation has been so ruthless and shortsighted in the use of these resources.

Millions of acres of once fertile land have become totally worthless for farming, and a still larger area is rapidly reaching the point of exhaustion. The vast forests have been depleted until there is less than one-fifth of the original area of productive timber, and the good timber is disappearing more rapidly than it is being replaced by new growth.

Many species of game and other forms of the wildlife of streams, prairies, and forests have been greatly reduced; and certain species of animals once plentiful have disappeared entirely or are nearing extinction. Rivers have become muddy. Serious floods have devastated some regions and dust storms have occurred in others. Many streams have been reduced to dry runs; increasing numbers of wells are failing to provide water; and serious droughts are prevalent in some regions. In no small way, humans have been responsible for these conditions. Instead of working with nature, they have in many cases destroyed gifts that required centuries to produce.

Fig. 7-1. "Cut out and get out!" Note the poor cutting practices and the lack of small growth. (Courtesy, U.S. Forest Service)

The widespread effects of the environment of human activity is nowhere more evident than in the acid rains that are becoming more serious with each passing year. The rain that falls in many areas of the world is made acidic by the sulfur dioxide and nitrogen oxides put into the atmosphere from industrial plants and other sources. As a result, there are already thousands of lakes in the northern part of the United States so acidic that fish and other forms of aquatic life cannot survive in them. These lakes have been classified as dead lakes. Canada, parts of the United States, and parts of other countries have lost much of their sport fishing industries because of the acid rain problem. West Germany fears that most of its forests are being killed by acid rain, and many other countries have similar problems.

In retrospect, we realize *now* that our natural resources should have been used with more regard for the future. Under the same circumstances, would the present generation treat our natural resources better than previous generations treated them?

What Is Conservation?

Some think that conservation consists of setting aside natural resources and hoarding them for future use. In a broad sense, *conservation is the wise use of natural resources in order to benefit the largest number of people now and in the future.* Conservation also means *use without waste,* or with the smallest waste possible.

In this chapter we are concerned primarily with the conservation of the natural resources that are most closely associated with agriculture and land use. These resources include *soils, water, forests, and wildlife.* If properly conserved, water in streams and under the ground will be available for an indefinite period. Soils, forests, and wildlife, under proper protection and planning, can be maintained or even increased and improved if methods known to be practical are utilized.

Fig. 7-2. Nature has provided many gifts to make life more interesting and enjoyable. This scene is in the Columbia National Forest in Washington. (Courtesy, U.S. Forest Service)

Why Should Conservation Be Practiced?

A few years ago, one farm leader said, "Damage to the land is important only because it damages human lives. The whole purpose of conservation goes back to that fact." What did he mean?

Some reasons for conserving soils, water, forests, and wildlife are as follows:

1. Millions of people get part or all of their income from lumbering and related industries and from fishing, hunting, and trapping. These people provide many valuable products for the food, clothing, and shelter of the entire population. Trees help to determine the climate we have all around the world. One large tree can provide cooling equal to the output of 10 room-size air conditioners.
2. Some forms of wildlife directly benefit farmers and thus are indirectly important to everyone. It has been estimated that birds save millions of dollars annually by consuming harmful insects. Some snakes eat field mice and other harmful rodents. A single bull snake may save a farmer many dollars per year. A large portion of the food of skunks consists of insects.
3. Forests, bodies of water, and various forms of wildlife contribute to the recreation and enjoyment of many people. Millions of tourists visit the various states annually, largely to enjoy the natural surroundings. These tourists spend large amounts of money that benefit the residents of the states visited.
4. The wise use of natural resources is necessary for national security. Only as the United States remains strong, with ample reserves of resources for periods of emergency, can it be secure.
5. The present generation has an obligation to future generations to continue the supply of natural resources of soils, water, forests, and wildlife. Do you agree?
6. If any one of these four resources is used carelessly or unwisely, the others may be seriously affected. For example, the depletion of for-

Fig. 7-3. Floodwaters erode irrigated land in Idaho. (Courtesy, Soil Conservation Service, U.S. Department of Agriculture)

ests might lead to increased flooding, and this in turn to increased soil erosion, to decreased crop yields, to decreased income to farmers, and to decreased prosperity to towns and cities.

The importance of conservation to both urban and rural people is illustrated in a large area in Illinois. A dam was constructed across a river to supply water for the homes, business firms, and industries in a fairly large city. A large lake formed behind the dam provided a reserve supply of water. However, after a few years, it was discovered that soil was gradually accumulating in the lake. This soil was washed into the lake from hundreds of acres of land that drained into the area above the dam. In due time, the capacity of the lake was reduced by one-fourth, and the situation has become increasingly serious.

In addition, the loss of fertile top soil is a serious problem to farmers. In order to solve this problem, rural people and urban people must work together. Thus, *"conservation is everybody's business."*

In 1976, the collapse of the Teton Dam provided another vivid example of the importance of sound water and flood control practices to everyone. When the dam collapsed, it released rushing torrents of water up to 60 feet deep, flooding over 128,000 acres of farmland and destroying homes, offices, equipment, animals, and crops. Many people lost their lives. More than 300,000 acres of irrigated land were left without a source of water. In addition to the millions of dollars needed to clean up and to repair the damage, there was an uncounted loss in income to the entire area.

Those fortunate enough to live near the open country have excellent opportunities to study the soil, water, trees, and wildlife and to learn how these resources are being misused and how they may be conserved. The great out-of-doors is our laboratory for many interesting experiences.

Why Is Soil Erosion a Serious Problem?

For the United States as a whole, the greatest loss in *soil fertility* comes about through erosion. It is estimated that at least 5.5 billion tons of soil are washed out of fields and pastures every year. Enough top soil erodes from the land into the Mississippi River alone in one year to build an island 1 mile long, ¼ mile wide, and 200 feet high. Each June the waters of the Palouse River in eastern Washington may carry away about 3 million tons of fertile top soil. In less than 100 years, this region has lost enough top soil to cover an area eight blocks square and eight stories high. According to USDA estimates, at least 25 million acres of once good land have been lost; about 280 million additional acres have been damaged seriously; and about 775 million acres have been eroded to some extent.

Since the early pioneer days, people in the United States have been

Fig. 7-4. Flooded farmlands indicate the need for controlling water when and where it falls.

wasteful in the use of their greatest natural resource, the soil. The following statement was written in 1747 by a man who lived in Connecticut:

> When our forefathers settled here, they entered a land which probably never had been ploughed since the Creation; the land being new, they depended upon the natural Fertility of the Ground, which served their purpose very well, and when they had worn out one piece they cleared another without any concern to amend their land, except a little helped by the Fold and Cartdung.

While George Washington was President of the United States, he included the following statements in a letter to a friend:

> Our lands, as I mentioned in my first letter . . . were originally very good; but use and abuse have made them quite otherwise. . . . We ruin the lands that are already cleared, and either cut more wood, if we have it, or emigrate into the Western Country.

Most of our ancestors came from European lands where soil erosion was not a serious problem because the rains were gentle and the water readily absorbed into the soil. Consequently, the early settlers handled soil much as they had in the "old country."

Even today, there are some farmers who are extremely careless or indifferent about the use of the soil on their farms. Some say, "What's the difference to me? This farm will last as long as I want it." On the other hand, increased numbers of farmers are realizing that they have a responsibility for maintaining or improving the soil on their farms during their lifetimes.

Although the loss of soil and soil fertility affects farmers, it also affects everyone else. J. N. "Ding" Darling, a great cartoonist interested in conservation, once said, "Soil is our bread and butter, our beefsteak, the clothes we have on our back." If we understand the meaning of this statement, we can understand why the soil has been called our nation's basic heritage.

History reveals that some ancient civilizations became extinct or degraded because they were wasteful and careless in the treatment of their soil. Thus, in portions of the world, soil that was once fertile is now wasteland or supports only a fraction of the people once there. This is true of regions in Mesopotamia (now Iraq), Israel, Egypt, China, and several other countries. Much of the "promised land" described in the Bible as "flowing with milk and honey" is now nonproductive and barren. If people in the United States will become alert to the need for conserving soil and will do something about it, this fate need not befall this country. In portions of Europe, evidence of what good farming can do to preserve the soil is found in fields that have been cultivated for a thousand years or more. These fields are still highly productive. A permanent type of agriculture is possible if proper methods are used for improving and conserving the soil.

Almost everyone is probably aware that water may cause serious erosion on sloping land. However, water may cause erosion on bare land that is nearly level. Raindrops splash away the soil. The process is called splash erosion. Raindrops strike the bare soil like tiny bombs and loosen soil particles that bounce upwards, as shown in Fig. 7-5. The pounding raindrops seal up the pores and prevent the water from soaking into the soil. This causes the formation of pools and ponds, which finally overflow, carrying soil away.

There are many signs indicating that soil is being washed away. Some of these signs are:

1. The soil in the streams and rivers after a rainstorm.
2. The soil that fills roadside ditches.
3. The cuts or ruts in the fields after a heavy rainstorm.
4. The huge gullies that have been formed by running water in some fields.
5. The eroded roadside banks. This is especially noticeable where new roads are being constructed.
6. The fields abandoned because the top soil has been washed away and crops can no longer be produced on them profitably.
7. The lakes, streams, and rivers that are being gradually "choked" with silt and mud.

Although water erosion is the most serious type of erosion, in many sections of the United States, soil may be carried away by the wind. In 1934, millions of acres in parts of Oklahoma and nearby states were blown all the way to the Atlantic coast and beyond by vast dust storms. This area once supported great herds of buffalo. At that time a covering of native grasses

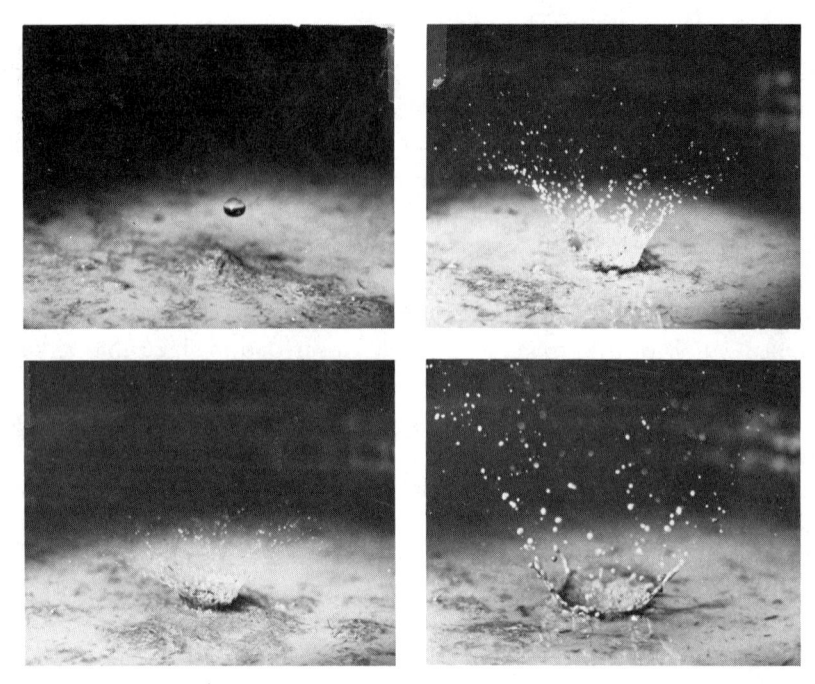

Fig. 7-5. A falling raindrop hits the ground with the force of a small atomic bomb. When it strikes the soil, the force splashes soil particles as much as 2 feet high and 5 feet from the spot where the raindrop hit. (Photo for Soil Conservation Service by U.S. Naval Research)

Fig. 7-6. This severe erosion along the highway could have been prevented. (Courtesy, U.S. Department of Agriculture)

Conservation—The Wise Use of Natural Resources

anchored the soil so that the winds could not disturb it. Later, much of this land was plowed to raise wheat. In a period of serious drought, there was no plant covering to hold it in place; and serious wind erosion resulted. This area was brought under partial control by improved cropping practices, but in 1954 and 1955, more than 13 million acres in seven states were damaged by wind erosion and about 19 million additional acres were in condition to blow. In the summer of 1983, extreme drought conditions covered sections of the Corn Belt and some southern states. This kind of weather is a reminder that only continuous sound cropping practices can keep the land in condition to prevent serious wind erosion problems.

Many people may actually see the signs of wind eroding the soil and not realize it. The huge dust storms that are described in the newspapers are not the only indicators of wind erosion. Some of the common signs of wind erosion are:

Fig. 7-7. Winds up to 80 miles per hour in Sully County, South Dakota, cause soil loss from wind erosion. (Courtesy, Soil Conservation Service, U.S. Department of Agriculture)

1. The dust in the air over cultivated fields on windy days.
2. The dust that collects around windows that are not airtight.
3. The dirt on top of snow along the roadside next to fields where crops grow.
4. The dirt that collects on plants, cars, houses, and other objects.

Because soil losses are of national concern, some steps have been taken by the federal government to improve the situation. The Soil Conservation Service was established in 1935 by an act of Congress to "provide permanently for the control and prevention of soil erosion." Through county Agricultural Stabilization and Conservation Service committees, the federal government has made payments to many farmers to help defray the cost of applying soil conservation practices to their land. Agricultural colleges, county agricultural agents, high school vocational agriculture teachers, and others are aiding in various ways to reduce soil losses.

Preventing soil losses

Conservation, when applied to soil, means use without serious deterioration. Nature's method was to grow protective grasses on the prairies and dense forests on rough areas. Under these conditions, soil losses from water erosion were kept at a minimum. Furthermore, such a system preserved a balance between plants and animals and aided greatly in the control of water.

Fig. 7-8. This grass waterway is helping prevent the formation of a gully where the water collects and runs off the field. (Courtesy, Soil Conservation Service, U.S. Department of Agriculture)

Conservation—The Wise Use of Natural Resources

Various methods or practices are helpful in preventing losses of soil that occur because of water erosion. The method to use depends upon conditions; in some cases, two or more methods should be used in combination. The most important of these practices are:

1. Treat land according to its needs and use it in accordance with its capabilities. Some land is suited to tilled crops and to a variety of other uses. Some land should remain in hay and pasture crops most or all of the time. Some land should be kept permanently in trees.
2. Use a system of cropping suited to each farm. In most cases, grasses and legumes should be included in a rotation, since these are more effective than row crops in holding the soil.
3. Construct grass waterways in natural "draws" or drainageways in cultivated fields. These spread out the water that collects and allow it to flow slowly off the land without damaging the soil.
4. Cultivate sloping fields and plant row crops on the contour. This means farming around the slopes, rather than up and down. The furrows and the marks formed by various tillage implements serve as small terraces or dams that catch and hold the water, thus preventing the soil from being washed away.
5. Use strip cropping on some sloping fields. In this system, strips of various crops are planted on the contour, with alternate strips of legumes and grasses, small grains, and row crops. Erosion on row crops is checked by the noncultivated crops.

Fig. 7-9. Strip cropping can be used alone or in conjunction with contouring. Strip cropping pays big dividends in saving the soil and in increasing yields. (Courtesy, Soil Conservation Service, U.S. Department of Agriculture)

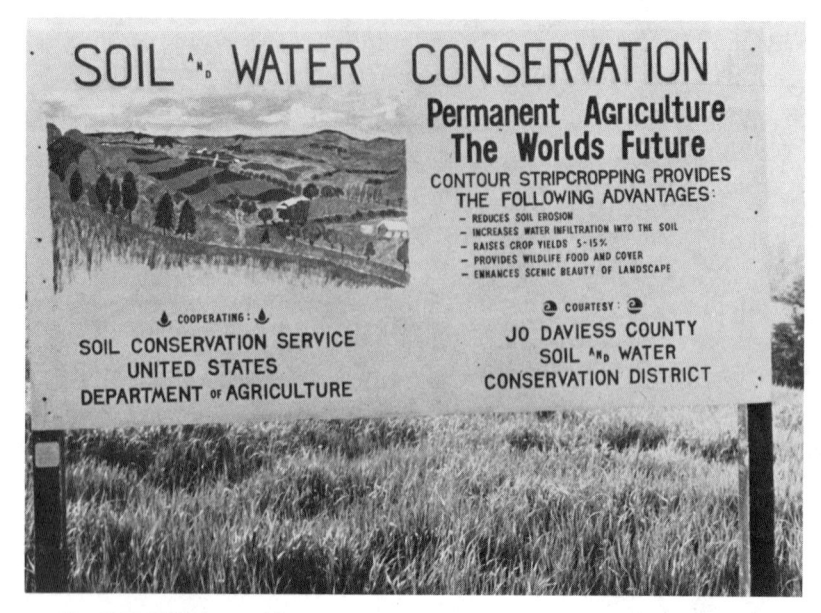

Fig. 7-10. This sign explains the advantages of contour strip cropping. (Courtesy, *Soil and Water Conservation News*, U.S. Department of Agriculture)

6. If the slopes are quite steep but not too steep for cultivation, construct terraces at intervals on the hillside. Terraces are also needed for long, gradual slopes. These terraces retard the flow of water, thereby allowing most of it to soak into the soil. For water not immediately absorbed, terraces act as eave troughs and conduct the surplus water slowly to one side of the field or to sodded areas that prevent washing.
7. Plant cover crops on fields from which cultivated crops can be harvested. Cover crops such as rye, vetch, winter wheat, and some others hold the soil in place during the winter season.
8. On portions too steep for cultivation, plant trees or grass as a permanent type of vegetation that will prevent excessive soil washing and water run-off.
9. Under appropriate conditions, construct farm ponds. These aid in controlling erosion, preventing floods, and providing water for livestock and other uses. They are also places for recreation.
10. Use conservation tillage farming. With conservation tillage, farmers employ various techniques to leave the crop residues on the surface of the ground to protect the soil and to preserve moisture. In tilling, they use subsoilers and fluted coulters to loosen the soil for plant root development.

Because of the concern about sediment pollution, many state offices of the Soil Conservation Service, as well as state agencies, have published guides for controlling erosion and sediment on construction sites. The guides contain descriptions of various erosion control measures and how-to information for establishing vegetation and for designing and installing structures. The guides are used by property owners, land developers, city planners, engineers, county officials, and state officials.

Wind erosion may be controlled by one or more of the following methods:

1. Planting crops in strips, with row crops alternated with small grains, grasses, or legumes.
2. Planting row crops at right angles to the prevailing winds.
3. Using shelter belts of trees to break the force of severe winds.
4. Keeping land subject to serious wind erosion in grass or trees.

How Can the Land Be Reclaimed?

Much of the damage to the land as a result of surface mining can be corrected by proper management. In many states, surface mining sites have been turned into fertile fields, and the research continues on ways to do this better. For example, researchers discovered a native grass *(Panicum clandestinum)* growing on surface-mined land in the Northeast in acid soil. This plant, called *deertongue,* was studied, and 20 strains were blended to produce Tioga deertongue. This grass grows on acid mine sites under nearly any moisture condition with little or no help. It can also be used to reduce erosion in woodland plantings. Tioga deertongue is a warm-season perennial that grows 1 to 3 feet in height and has leaves 1 inch wide and 4 to 8 inches long that are shaped like a deertongue.

How Can Water Resources Be Conserved?

"You never miss the water until the well runs dry." This old proverb indicates how we tend to take our water supply for granted. Underground water, from which numerous cities and farms get their main supply, is being seriously depleted in many places. This is because we are removing large amounts of water without maintaining the conditions that would replenish them.

Some cities that must depend directly on water from snow and rain are also in a similar situation. At one time, the great reservoirs in the Catskills that provide much of the water supply for New York City were nearly empty. Some U.S. cities that utilize water from small streams and artificial lakes are at times threatened by shortages due to periodic droughts.

Every living thing requires water. Water is the life blood of the soil, for plants cannot grow without it. So-called deserts support some forms of plant and animal life only when some moisture is present. To produce 1 pound of dry matter in a corn plant, 365 pounds, or over 45 gallons, of water is required. To produce just 1 bushel of corn, about 5,000 gallons of water is necessary; thus, a yield of 100 bushels on 1 acre requires 500,000 gallons. The human body is two-thirds water. The apple is 87 per cent water, and a beef cow is about 75 per cent water. Individuals use considerable water for their various personal needs. Industries require vast amounts of water.

Controlling loss of the water supply

When the first colonists came here, most of the continent was blanketed by forests or grasses. Under these conditions, much of the water that fell in the form of rain or snow was absorbed by the soil. As forests were removed and grasslands were plowed, nature's cover was disturbed and increasing amounts of water flowed over the surface toward streams and rivers. Much soil was lost in the process. The top soil went first, and since it absorbs water more readily than the subsoil, the process became increasingly serious as time went on.

It is doubtful that all floods could be prevented, even if all native vegetation and the original soil could be replaced. However, floods are probably much more serious in regions where the native vegetation has been removed, with the result that increased quantities of water rush quickly

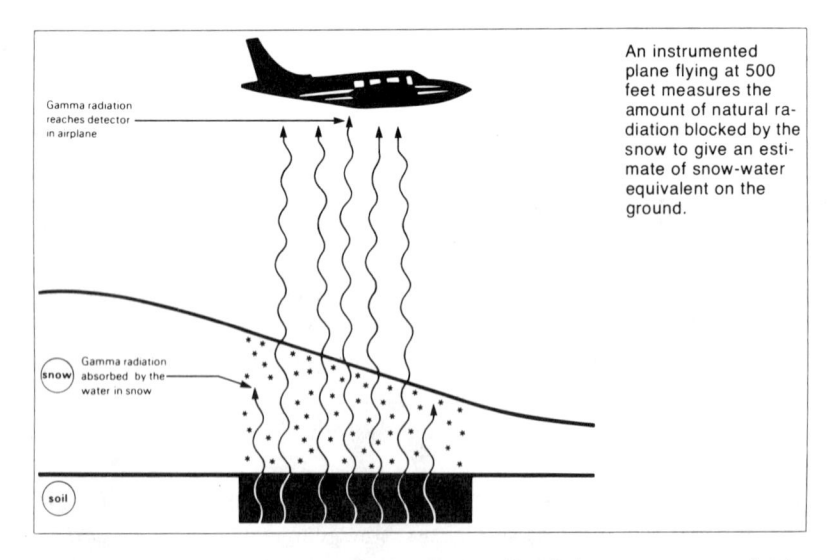

Fig. 7-11. Meteorologists are using a supersensitive Geiger counter to predict the extent of flooding. Snow and soil moisture weaken the natural radiation from the earth; thus, by measuring the radiation, meteorologists can determine the amount of snow and water. (Courtesy, National Oceanic and Atmospheric Administration)

over the surface toward streams and rivers instead of being absorbed and released gradually. Not only may floods become more serious, but droughts may occur more frequently because the surface vegetation has been disturbed, thus increasing the losses of surface soil and thereby reducing the amount of water absorbed by the soil.

Continued pumping of water from the ground for irrigation and drinking can also lead to other problems. At Las Vegas, it has caused the land surface to drop nearly 4 feet. In addition, huge cracks up to 3 feet wide have appeared in the ground and have damaged some of the buildings.

The problems for farmers with dwindling water supplies are especially difficult. Farmers use about a third of the water withdrawn from surface and ground supplies. About 98 per cent of this water is used for irrigation on over 51 million acres of land. Of the 178 million acre-feet of water used for irrigation in 1975, only two-fifths were consumed by crops. (An acre-foot is enough water to cover 1 acre of land 1 foot deep, or about 326,000 gallons.) As water is pumped from the ground, the water levels continue to decline, making it ever more costly to pump the water and, in many cases, so reducing the supply of water that there is not enough to irrigate the land, and it can no longer be farmed.

Some scientists indicate that the water supply in this country has changed very little. However, there are more people, and each person is using increased amounts of water. Irrigation and manufacturing firms are also using increased amounts of water.

Multiple reuse of water is now a fact of life in many places, since water consumption for all uses equals or exceeds the recoverable stream flow and ground-water supply.

In a study of water conservation, it is important to understand the water cycle, sometimes referred to as the *hydrologic cycle*. The following features should be noted in Fig. 7-12, which shows this cycle.

1. Water gets into the air chiefly by evaporation from land and bodies of water, by the evaporation of rain and snow before it reaches the ground, and by moisture given off from the leaves of plants by a process known as *transpiration*. One large tree gives off as much as 150 gallons of water each day during the summer months.
2. Moisture-laden air becomes cooler, causing the moisture to condense. This falls in the form of rain or snow. (Some moisture may collect in the form of dew on objects that have cooled during the night, much as a pitcher "sweats" on a hot day when it is filled with cold water.)
3. Moisture falls on the earth's surface. Some runs off, but normally a large amount is absorbed. Of the portion absorbed, some sinks deep into the soil and rock layers where it is stored as *ground water*, the top of which is called the *water table*.

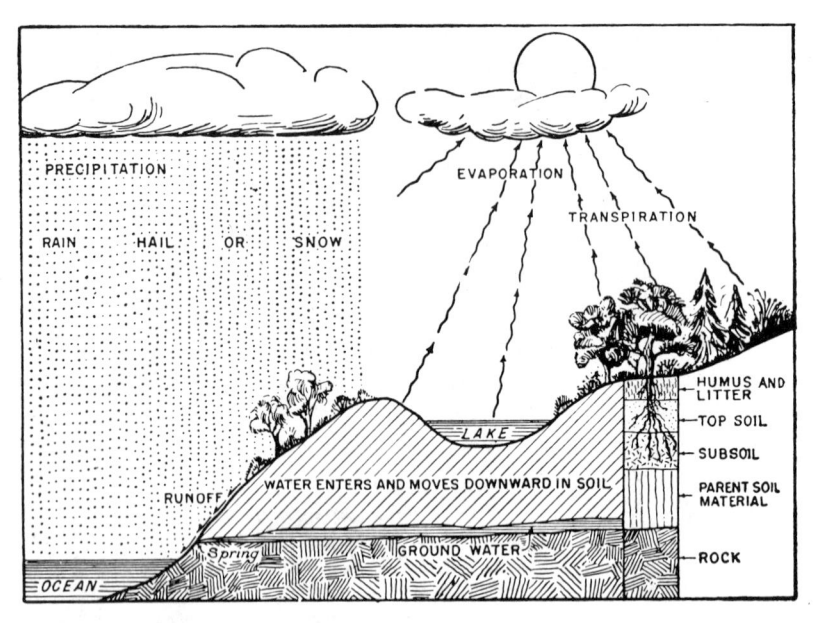

Fig. 7-12. The movement of moisture in all of its forms from earth to sky and back again is called the water cycle, or the hydrologic cycle. (Courtesy, U.S. Department of Agriculture)

4. If the rate of rainfall is faster than the water can be absorbed, part of the water runs off; this is called *surface run-off*. It finds its way into creeks and rivers and finally into lakes and oceans. If run-off is excessive, floods may occur. This is the portion of the water that causes soil erosion.
5. Water evaporates from the surface of the land, from plants, and from bodies of water, thus completing the cycle.

Humans have disturbed the supply of water from underground sources. When the surface vegetation is changed, less water soaks into the soil to replenish the underground supply. Wells drawing on underground water may be increased in number and used to the extent that water from this source is pumped out faster than it is replenished. In some places, because of these conditions, the capacities of wells have been seriously reduced, or the wells have failed entirely.

Increasing soil water

People can do a great deal to increase and conserve soil water. Some of the methods recommended for preventing soil erosion aid in the absorption of rainwater by the soil. Grasses and other cover crops, contour cultivation, terraces, and reforestation are important in this process.

Belts of trees planted at right angles to the direction of prevailing winds may aid in preventing excessive evaporation from the surface. They are also useful in areas where there is a small amount of rainfall. The use of straw and other plant materials for surface mulches helps to prevent evaporation and to check run-off of water from sloping fields. The term *trashy cultivation* is used to describe this practice in wheat fields of the West and elsewhere.

Action taken to control brush and weeds can reduce water losses and bring land back into productivity. In West Texas, ranchers brought a dried-up creek back to life by controlling the brush growing in the watershed. Mesquite and other undesirable brush grew to such an extent that these plants used all the water and the creek simply dried up and remained dry for 40 years. After the mesquite and brush were removed and the rangeland was planted to grass, the creek again flowed with water for livestock and a local town water supply.

Controlling streams and rivers

Streams and rivers should be considered in water conservation. Various practices on the land that cause water to soak into the soil aid in preventing overflows or floods at certain seasons and low water levels at others. Dams across streams at various points hold back excess water during periods of heavy rainfall, and this water can be released during periods of limited rainfall.

In some places in the United States, huge dams have been built to aid in flood control and to provide water for cities, for electrical power, and for irrigation. These dams also tend to prevent floods by holding back large quantities of water during heavy rains and rapid thaws of snow.

If water is properly controlled as it falls, much damage to valuable land and property can be avoided.

Controlling water pollution

The actual loss of water through run-off and evaporation is not the only major problem in maintaining an adequate water supply. Much water is also lost because it is so polluted it cannot be used.

Certain fish, especially trout, require clear, cool water and shaded banks. Many streams, which have become loaded with silt or contaminated with sewage and wastes from factories, have become depleted of many of the most desirable species of fish. Furthermore, such streams are less scenic and less desirable for recreation. Some polluted streams may become a menace to human health.

Some forms of pollution, such as coal washings and oil, are so bad that the water cannot be cleaned and purified for human use. Such water is too often returned to streams and lakes to pollute more water and kill fish, other wildlife, and plants. The use of detergents has also created water pollution problems. There is no known way to remove all of the detergents from the water. In some places, enough detergent remains in "cleaned and purified" water to cause foaming in streams. River tugboats churn up huge quantities of "suds" because of the detergents left in the water when the water is returned to the streams. In some places, foaming occurs when water is turned on at faucets in homes.

Some of the numerous ways in which water becomes polluted are from:

1. Trash and garbage being thrown into the water.
2. Factory wastes being drained into streams, rivers, and lakes.
3. City sewage being drained into the streams.
4. Soil being washed into the waterways from the fields, cutover river banks, and construction projects.
5. Vacationers putting unburned fuel from motor boats, camp wastes, garbage, etc., into the water.
6. Fish and plants dying in the streams because of other forms of pollution.
7. Streams being used by farm animals.
8. Stream beds being filled with undesirable plant growth because the water flow is too little to keep the stream beds clean.

Some of the actions that can be taken to keep water clean are:

1. Keep waste and trash out of the streams, rivers, and lakes.
2. Keep domestic animals away from water to be used by people.
3. Support legislation and its enforcement to prevent dumping of industrial wastes into streams, rivers, and lakes.
4. Support legislation and its enforcement requiring cities and towns to treat all sewage before putting it into rivers and streams.
5. Support legislation and other efforts to control soil erosion.
6. Support and encourage research designed to improve ways for treating and purifying water before it returns to the streams.

How Can Forests and Woodlots Be Preserved?

>Woodman, spare that tree!
>Touch not a single bough!
>In youth it sheltered me,
>And I'll protect it now.

Thus, the poet George Pope Morris tells us that trees are our friends. It has already been emphasized that trees play an important part in the conservation of water and in the prevention of erosion. Under favorable conditions, there is a deep layer of humus on the forest floor that acts as a sponge in soaking up water. The trees and undergrowth serve to break the fall of heavy rains, thus aiding in the process of water absorption by the soil. Root masses permeate the soil to considerable depth. These conditions practically prevent erosion, even on steep slopes, so that the rate at which soil is lost is usually less than the rate at which soil is formed on the forest floor.

Conserving forests

Forests are being cut much faster than they can reproduce. At the present, in the United States, about one-fourth of the total land area is forestland. Of this, about 500 million acres are suitable for timber crops. About one-fifth of the commercial forestland is government owned. Much of this is in national forests under the supervision and protection of the U.S. Forest Service. In addition, many of the states have state forests. Most of the remainder is privately owned. About 40 per cent of the land in private commercial forests consists of wooded areas on farms.

If the present acreage were properly managed, no additional space would be needed to provide lumber and other forest products. If proper methods of cutting trees were used and if small trees were allowed to remain, forests would perpetuate themselves by natural processes under most conditions. In some cases, however, in order for forests to be restored on

Fig. 7-13. Collecting sap, from which maple syrup and maple sugar are made, from the sugar maple tree.

eroded or burned-out land, small trees of desired varieties must be transplanted. Joint efforts of private companies and the U.S. Forest Service have reduced the number of fires and have improved methods of cutting and reforestation.

Forest fires cause millions of dollars of loss each year. Over 90 per cent of forest fires start because people are careless. Greater care by everyone who visits forests would prevent most of these losses.

Young people who are interested in careers in forestry should study the opportunities and qualifications for these kinds of jobs.

Establishing woodlots and forest areas

On many farms, there are plots of land that should be used entirely for growing trees. In some southern states, one-half or more of the farmland is wooded. These trees can be considered a farm crop. From them, farmers obtain cordwood, pulpwood, sawlogs, fence posts, maple syrup, turpentine gum, Christmas trees, nuts, and other products. In addition, wooded areas help conserve soil and water, and they can be used for recreational activities.

In value, forest products rank twenty-second among all farm crops, but they would rank higher if proper cutting methods and other good practices were used on the wooded areas. A carefully managed woodlot is practically self-perpetuating. This calls for selective cutting of trees as they reach maturity and for the protection of small trees of various sizes to provide for replacements. Ordinarily, permitting livestock to graze in farm woodlots is not advisable, since the damage usually offsets any gains. A woodlot with trees

Fig. 7-14. A U.S. Forest Service plane drops a slurry on a fire in the Ozark National Forest, Arkansas. (Courtesy, U.S. Department of Agriculture)

and undergrowth of varying heights provides a continuous supply of food and cover (protection) for many forms of wildlife.

About half the timber going into houses in the United States comes from small private forests. State laws and conservation district programs in Washington, Idaho, Oregon, and northern California are being used to improve woodland management and to maintain the productivity of farm forests. The laws and programs are designed to reduce soil erosion to the lowest possible amount. As a result, the rivers and streams are clearer, wildlife has benefited, timber production has increased, and grazing for cattle has improved.

In starting and maintaining woodlots of forest areas on farms, farmers may need to transplant trees or plant seeds. Oak, hickory, and walnut trees may best be started from the nuts. Sugar maple, Norway (red) pine, white pine, Scotch pine, spruce, ash, sycamore, beech, yellow poplar, and basswood are other trees frequently found in woodlots in the North Central states. For tree belts planted to control "blow-out" areas, black locust, jack pine, Norway (red) pine, and Scotch pine are frequently used. For Christmas trees, the Scotch pine is the best-selling tree species. However, many people like white pine, red pine, jack pine, Austrian pine, Douglas fir, Norway spruce, and red cedar.

In some states, trees for woodlots and reforestation in general can be secured at a small cost from the state agricultural colleges or the state conservation departments.

In many states, some high schools sponsor school forests. Days are set aside when students transplant trees and apply various practices that are recommended for forest development and wildlife management. These forests provide practical experience for the students and demonstrate approved forest practices to the community. In what other ways are school forests of value to students?

Conserving coastal wetlands

The preservation of Virginia's barrier islands, a 42-mile chain of nine islands along the Delmarva Peninsula and the only undeveloped chain of barrier islands along the nation's East Coast, is a major project of The Nature Conservancy. Other examples of the work of this nonprofit organization are a cypress swamp in South Carolina and a wildlife range in New Mexico. It has been involved in nearly 1,500 preservation projects, from tropical rain forests in Hawaii to trout lakes in the Adirondacks. The barrier islands are more than an example of beauty in nature. This area, where the land meets the sea and the salt water meets fresh water, is economically important also. Almost all the finfish and shellfish harvested by the eastern shore's seafood industry spend part of their lives in coastal wetlands like these. Oysters, clams, and crabs spend their entire lives in the marsh estuary system, feeding off particles of deteriorating marsh grass. The aquatic life of these wetlands and sand dunes provides feeding, resting, and nesting areas for over 200 species of waterfowl and shorebirds. The marshes are the state's only breeding sites for the forester's tern, willet, and laughing gull. Royal terns summer on Virginia's barrier islands and winter in Peru. Canada geese and a variety of ducks winter here and summer farther north. Both research and public recreation will be provided for in the long-term plans for the area. This is a good example of how preserving an area in its natural state is of benefit to both people and wildlife. Are there areas of this kind near your home?

How Can Wildlife Be Conserved?

When the colonists first came to this country and for many years following, there was an abundance of wild animals. These were hunted for furs, food, and even for sport, often without regard for the future. In time, the natural habitats of these animals were disturbed as forests were cut, swamps were drained, and large areas of land were taken over for farms and ranches and suburban development. The cutting of firewood, the impoundment of water, the pollution of water, and the overuse of hiking trails have also contributed to the destruction of wildlife.

As a result, the American bison, or buffalo, which once numbered an estimated 75 million, became almost extinct by 1890. The American elk, which once numbered millions, were reduced to a few thousand. The bighorn sheep and pronghorned antelope met the same fate. Many buffaloes were killed only for their hides, bighorn sheep for their horns, and elk for two teeth, which men wore on their watch chains. To some degree, these species have been built back up by proper conservation methods, but such measures came too late to save some species. For example, the passenger pigeon, Labrador duck, great auk, Carolina parakeet, dodo, New Zealand quail, Eskimo curlew, Stellar's sea cow, Barbary lion, and heath hen are extinct. Since 1900, more than 40 animals have become extinct. This same fate threatens the whooping crane, California condor, bald eagle, prairie chicken, American alligator, cheetah, wolf, Gila monster, rhinoceros, walrus, giant panda, and several species of whales and tigers.

Probably, we should not be too critical of the early settlers, as they did many creditable things to develop a growing nation. Then too, it should be recognized that eliminating densely forested areas helped some forms of wildlife. Deer and quail, for example, have increased in numbers because of this and protective game laws.

In spite of adverse conditions, the annual return from wildlife in the form of meat and fur is considerable. In other words, certain forms of wildlife may be considered to be valuable "crops" to be harvested the same as corn, wheat, and potatoes.

An interrelationship and interdependency exists between many living things in nature. Fish, birds, and game cannot be considered separately from the conservation of trees, other vegetation, and water. If they are to be maintained or increased, streams must be kept clean. The planting of shrubs and trees makes conditions more favorable for certain forms of wildlife. In addition, trees add to the beauty of the countryside, provide an income in their own right, and aid in the conservation of soil and moisture. Some kinds of wildlife, in turn, protect the forest and crops from various enemies, and so the endless cycle continues. Thus, we again have evidence of the interrelationship between soils, trees and other vegetation, water, and wildlife. The serious disturbance of one is likely to affect the others unfavorably. Improvement of one will frequently lead to improvement of others.

Several methods helpful in the conservation of wildlife are as follows:

1. *Reducing take or harvest.* There are game laws that restrict the season for hunting and the number of certain species of game and fur animals that can be taken per day or per person. The purpose of these laws is to provide for the continuance of sufficient breeding stock. The harvest must be adjusted to the crop produced, as determined by game specialists who are familiar with the conditions. Some species, such as certain song birds and other animals needing special protection, have no open season in some states, which means that it is illegal to hunt them at any time. In some situations, however, increased fishing and hunting are encouraged in order to prevent overpopulation of wildlife.
2. *Controlling natural enemies.* This is a complicated process, since reducing certain species may upset nature's balance along unexpected lines. Furthermore, it is important to understand which species are really harmful under various conditions. Red foxes, for example, may destroy some birds, but they also consume field mice and other harmful animals. Nearly every species of wildlife has value to some group of people. If we help nature, a desirable balance usually will be maintained.
3. *Establishing refuges and sanctuaries.* By this means, in certain areas, wildlife are protected from hunters and poachers. Millions of acres are now included in these areas for wildlife, with additional acres for waterfowl. Birds that migrate during certain seasons have stop-off stations throughout the country where they may find food and shelter in safety. These areas are primarily under the supervision of the Bureau of Biological Survey of the federal government. Small acreages on many individual farms also furnish valuable refuges.
4. *Artificially restocking areas.* Fish hatcheries are maintained by

Fig. 7-15. Refuges, such as this one, provide safe environments for wildlife.

federal and state governments, and at intervals various streams and bodies of water are restocked. Pheasants and some other forms of wildlife are sometimes reared and released in certain regions, although this practice has questionable value after a species has become established. Restocking of already overpopulated ponds and land areas is not a good practice.

5. *Providing good habitats for wildlife.* If sound land use is practiced and good soil practices are used, conditions are most favorable for many kinds of wildlife. Shelter areas furnish cover for animal life. These may be provided by shrubs, small trees, and vines that grow in gullies, ravines, fence rows, and odd corners. In some cases, desirable species of plants, such as multiflora rose for hedges, may be started in these areas.

6. *Preventing water pollution.* Some cities and industrial firms have been extremely careless in the disposal of sewage and other wastes. In many cases, these materials have been dumped directly into streams and lakes. The resulting pollution has killed countless fish and other forms of wildlife that inhabit streams and lakes. Most states have passed laws prohibiting these practices, but in many cases these laws are not enforced. For example, the chemical wastes from one industrial firm in Illinois were allowed to flow into a nearby river. As a result, all of the fish, some weighing up to 50 pounds each, and many other forms of water life were killed for a distance of 100 miles or more downstream. Although the pollution was ultimately

Conservation—The Wise Use of Natural Resources

stopped, it will take several years to restore the fish and the wildlife in the stream.

7. *Protecting endangered species.* In addition to laws against hunting endangered species, there are laws that make it illegal to sell articles made from these species. A New York State law barring the sale of articles made from endangered species such as the alligator and the vicuna was held to be constitutional. The law challenged was the Mason and Harris law that bans importation into the state and sale in the state of articles made with the skins of mammals, birds, amphibians, and fish threatened with extinction.

8. *Carefully using chemicals to control pests.* The need for care in the use of pesticides is dramatized in books such as Rachel Carson's *Silent Spring* and Lewis Herber's *Our Synthetic Environment*. Many examples are given of how mistakes in the use of pesticides have helped destroy birds, fish, and other wildlife. Examples are also given of how reproduction in wildlife may be affected by the long-time buildup of chemicals in the soil and water.

9. *Establishing wildlife care centers and wildlife rescue squads.* In some states, wildlife care centers have been established where injured and sick wildlife can be taken for treatment and care until the wildlife are able to return to their natural habitats. In Florida, there is a wildlife rescue squad that responds to calls about injured wildlife.

The concern people have for the preservation of wildlife is shown by the success of the "Return a Gift to Wildlife" programs. In many states, cit-

Fig. 7-16. An elk calf about three weeks old. (Courtesy, U.S. Forest Service)

izens can direct that specified amounts of money from their income tax refunds be withheld as donations to be used to improve the management of fish and other wildlife.

However, as is pointed out elsewhere in this book, today's agriculture is dependent upon pesticides for controlling diseases, insects, and weeds that would otherwise destroy the crops and livestock needed for human consumption. Pesticides are, in fact, one of agriculture's major weapons in the fight against hunger and starvation. In addition, pesticides are used to destroy the insect carriers of human diseases. Without pesticides, there would likewise be little defense against the weeds, insects, and disease pests that attack gardens, lawns, shrubs, flowers, and shade trees.

Pesticides and other chemicals are also used to help maintain wildlife. The lamprey, which has nearly destroyed the Great Lakes fish and fishing industry, is being brought under control through the use of chemicals. The Dutch elm disease is being fought with chemicals. Many other uses of chemicals help to preserve wildlife. Temporary damage to some forms of wildlife from the use of pesticides for agricultural purposes will be more than offset by the benefits to wildlife.

Biological controls for pests are helpful, and research in this area is continuing. However, these usually do not provide a sufficiently high degree of control to eliminate the need for chemical controls—and they cannot be developed quickly enough.

Through stronger regulations restricting the use of pesticides, and through new developments in biological controls, the danger to wildlife from the use of chemicals can be minimized. If other favorable conditions, such as those that have been discussed, are maintained, much game and other wildlife will flourish even where most of the land is used for farming. Some of the practices previously discussed for conserving the soil provide these conditions. About 70 per cent of the wild fur crop in the United States is trapped on farmland. This includes muskrat, skunk, opossum, raccoon, mink, fox, and weasel.

In some cases, land owners have taken steps to control hunting by forming game-management cooperatives. Hunting of game animals on these farms is controlled. Only responsible persons are allowed to hunt on the land in these areas, and the number of hunters is controlled. In some cases, fees are charged for hunting privileges.

Determining harmful species

As previously mentioned, determining with certainty which species of animals are harmful and which are beneficial is not easy. Skunks in general are beneficial because they consume harmful insects. In addition, pelts from these animals have considerable value. An occasional skunk may kill chickens, but this is not sufficient reason for killing all skunks. Of all species of

Fig. 7-17. Part of a stream in eastern Illinois has been set aside as an experimental area studied jointly by the Illinois Natural History Survey and the Illinois State Department of Conservation. Anglers report their catches and the number of hours they fish. These fish biologists weigh, measure, and record the catches. The data collected are studied and are used later in the formulation of management recommendations designed to improve fishing in Illinois streams. (Courtesy, Illinois Natural History Survey)

animals, perhaps snakes, owls, and hawks are most frequently destroyed without regard for their value. Most snakes are harmless and a few, such as bull snakes and pine snakes, destroy field mice and other harmful rodents. Of the owls, only one is "bad," the great-horned owl. Only 4 of 14 species of hawks feed on poultry and song and game birds. Hawks deserving little protection are the goshawk, the sharp-shinned hawk, Cooper's hawk, and the duck hawk.

A person should learn to identify the harmful snakes, which include rattlesnakes, coral snakes, water moccasins, and a few other species. These and other animal outlaws, such as starlings, English sparrows, and rats, should be fair game for the hunter.

Enjoying wildlife without killing

Many people enjoy birds, squirrels, and other forms of wildlife in their natural settings without hunting them. Some people enjoy photography and thus do their hunting with a camera. Others encourage favorable conditions for the multiplication of some desirable species of birds, such as wrens, bluebirds, martins, and wood birds. They put out food in times of serious scarcity, provide various types of protective cover, and construct nesting boxes.

How Can Plans Be Developed for the Better Use of Land?

As mentioned earlier in this chapter, good land use is important in the conservation of good soils, water, forests, and wildlife. It is important for individuals to recognize that for the best interests of everyone, each acre should be put to its best use and protected in appropriate ways. Someone has said, "Most of the land is good if you know what to do with it."

The problem of land use is an important one in the planning of cropping programs for individual farms. Similarly, wise land use should be considered for large areas and regions. Just as land may be used improperly on an individual farm, it may be used unwisely in entire regions. Some large areas should be used primarily for reforestation, other areas for grazing purposes. Perhaps the old Indian who strode onto a field of freshly plowed prairie sod in one of the Dakotas had a viewpoint of land use that many white persons have failed to understand. Looking sadly at the furrow slices, he slowly shook his head and muttered: "Him wrong side up." What do you think he meant?

Utilizing scientific findings

In many ways, scientific investigations have contributed to the better use of land. Scientists have had a hand in the development of crop rotations, cultivation methods, terraces, dams, and fertilizing practices. In addition, specialists in plant breeding are producing plants better adapted for soils subject to erosion or drought. Being developed are strains of trees that grow rapidly and produce high yields. Methods have been developed for using legumes and grasses to make silage. These methods thus contribute to soil conservation and provide excellent feed for livestock. Livestock feeding experts are studying methods for using larger amounts of pastures and roughages in the rations of farm animals.

Developing a balanced program on each farm

Each farm should be planned carefully with respect to livestock and crop production. In the development of plans that also take into account the best uses of land, the following should be considered:

1. The various portions of a farm should be used for the purposes for which the land is best suited.
2. Cropping systems should be adapted to the community and to the land.
3. Livestock and feed production should be balanced so that as many animals as possible can be raised on homegrown feeds.
4. Foods should be raised for family use.
5. The use of the land should be planned in relation to possible effects on nearby farms.

If these factors are properly considered, farmers should be able to provide for the conservation of soils, water, forests, and wildlife. If farmers plan programs for their farms carefully, they and their families, as well as future generations, will benefit by increased returns.

Specialists from various agencies can help farmers develop plans for their farms. These include extension specialists from the state agricultural colleges, county extension agents, high school vocational agriculture teachers, and soil conservation specialists and other representatives of various federal agencies.

Thousands of farmers and other land owners throughout the country are becoming interested in developing plans to better conserve the resources on their farms.

Fig. 7-18. Windbreaks of trees and shrubs in the North Dakota Great Plains area protect the soil from wind erosion and beautify the countryside. (Courtesy, U.S. Department of Agriculture)

Planning the use of land by districts and larger areas

For the greatest progress in planning the use of land, the cooperative effort of many farmers is necessary, since drainage and run-off water from one farm may affect adjacent farms. Assistance by outside experts can be better utilized if many farmers in a given area are interested in planning. Because various species of wildlife may use several farms in the course of a year, cooperative effort is desirable.

In every state, laws have been passed making it possible for the farmers in a given community to organize soil conservation districts so that they may cooperate most effectively among themselves and with various state and federal agencies in conserving soil. The responsibility for such a district rests on the local people, and a majority must be in favor of it before a district can be formed. After such an organization has been formed, each individual farmer may secure the help of specialists in planning a suitable farming program. At the end of a recent year, farmers and ranchers had organized 2,922 soil conservation districts, in which most of the nation's farmland was included.

In certain states, notably Wisconsin, progress has been made in rural zoning. Under state law, a county may vote to restrict further settlement in areas unsuited for farming. The restricted zones are used for forests and recreational purposes.

Perhaps the most extensive project in planning to date is that located in the area drained by the Tennessee River and its tributaries. It covers an area that includes portions of several states in which several million people live. Under a federal act, the Tennessee Valley Authority (TVA) was created. The construction of several large dams, generation and distribution of electricity at reasonable rates, soil improvement by fertilizers and legumes, control of erosion, improvement of forests, and development of local industries with low-cost electricity have been carried out.

One of the big problems in the United States is that each year about a million additional acres of land are required for the expansion of cities, highways, and airports and for other public and private uses. Much of this land was highly productive and such land is a limited resource.

When he was Secretary of Agriculture, Henry A. Wallace said, "The whole matter of planning to bring about better land use is new in the United States. How to use the land to the best advantage for present and future generations is one of the most important problems that confronts us."

SUGGESTED ACTIVITIES

1. What are some of the ways in which the natural resources of soil, trees, and wildlife are still used unwisely in your community? How do you account for the failure to use them properly? What meth-

Conservation—The Wise Use of Natural Resources

ods are used for conserving them? What other conservation practices are needed?

2. Each of the following phrases indicates chains of relationships that represent the effects of upsetting the balance of nature. (For example, in the first, depletion of forests leads to increased run-off of water, etc.) Explain each series and carry each farther, if possible. Work out some additional chains.
   a. *Depletion of forests*—decreased water entering soil—lowered water level in soil—drying up of wells and springs.
   b. *Improper tillage of soil*—increased amounts of soil carried by water—increased soil in stream—deposits of soil behind dams—decreased water-holding capacity of dams—decreased water power.
   c. *Removal of shrubs and trees*—decreased birds—increased insects—damaged crops—lowered income for farmers.
   d. *Destruction of coyotes on the western plains*—increased number of jack rabbits—increased destruction of crops—greater loss to farmers. (Note: It is estimated that 62 large jack rabbits consume as much grass as one 1,000-pound cow.)

3. Take a field trip to study the evidences of soil losses in some area near your school. What types of losses are most serious? Are there gullies in the fields? Is the top soil gone from hillsides? Are streams muddy? Why do you think these losses have occurred? What are the effects on the people in that community? What might be done to remedy the situation?

4. Take a trip to some farm where various practices are in use for conserving the soil. What methods are used? How does each method aid in preventing erosion?

5. After a heavy rain, get a sample of water from a muddy stream. At the same time, get a sample of water from a faucet. Put each sample in a separate glass jar. After the samples have stood for several days, explain the differences you see.

6. Make a survey of wooded or forested areas in your community. What products are obtained? What varieties of trees are most common? What methods are used to maintain these woodlands? What suggestions for improvement would you offer?

7. With other members of your class, consider the possibility of starting a school forest. Usually, the first step is to write to the conservation department in your state to determine the procedures for securing land and trees. That department can also give helpful suggestions on methods of securing and transplanting small trees.

8. Keep a record of the species of wildlife seen on some area of land near your home or your home farm, if you live on a farm. Indicate the kinds, numbers of each seen, date seen, and other information about actions and habitats.

9. If arrangements can be made, plan a visit with your class to a soil conservation district or demonstration project. Usually, this can be done with the help of the vocational agriculture teacher, county extension agent, or someone in the Soil Conservation Service.

10. With your teacher and the rest of the class, plan excursions of various types related to conservation. These might include field trips to study wild flowers, native trees, shrubs, and animal life. Per-

haps trips can be taken to a national or state forest, a forest ranger station, a lumber mill or a paper mill, and a farm where maple syrup is made.

11. Through firsthand observation and from a study of printed materials, become an expert on the home, food, and various habits of one or more forms of animal life in your region.

12. With the rest of your class, build a library of books and bulletins on the subject of conservation. Various agencies may be contacted for literature. (See list at end of chapter.)

13. Construct a bird feeding station outside a window of the classroom. Provide food in the form of suet, seeds, etc. Note the kinds of birds using it and the food eaten by each.

14. Study the life of some great naturalist, such as Thoreau, Burroughs, Muir, or Audubon, and make a report to your class.

15. Make a list of rules that you think every good citizen should follow relative to removing trees and shrubs, building camp fires, observing game laws, cooperating with land owners when hunting or fishing on their farms, and using a gun safely.

ORGANIZATIONS INTERESTED IN CONSERVATION

Agencies not connected with the federal government

American Forestry Association
1319 18th Street, N.W.
Washington, D.C. 20036

American Museum of Natural History
(publisher of *Natural History*)
Central Park W. at 79th Street
New York, New York 10024

Boy Scouts of America
P.O. Box 61030
Dallas–Forth Worth Airport
Dallas, Texas 75261

Camp Fire, Inc.
230 Camp Fire Road
Chappaqua, New York 10514

Conservation Foundation, The
1717 Massachusetts Avenue, N.W.
Washington, D.C. 20036

Council for Planning and Conservation
Box 228
Beverly Hills, California 90213

Environmental Action, Inc.
1346 Connecticut Avenue, N.W.
Washington, D.C. 20036

Fund for Animals, Inc.
Suite 731
140 West 57th Street
New York, New York 10019

Girl Scouts of the U.S.A.
830 Third Avenue and 51st Street
New York, New York 10022

Izaak Walton League of America
1800 North Kent Street
Arlington, Virginia 22209

National Audubon Society
950 Third Avenue
New York, New York 10022

National Geographic Society
(publisher of *The National Geographic Magazine*)
17 and M. Streets, N.W.
Washington, D.C. 20036

National Parks Association
1701 18th Street, N.W.
Washington, D.C. 20009

National Recreation and Parks Association
1601 North Kent Street
Arlington, Virginia 22209

National Wildlife Federation
1412 16th Street, N.W.
Washington, D.C. 20036

Student Conservation Association
Box 550
Charleston, New Hampshire 03603

Water Pollution Control Federation
2626 Pennsylvania Avenue, N.W.
Washington, D.C. 20037

Wilderness Society, The
1901 Pennsylvania Avenue, N.W.
Washington, D.C. 20006

Wildlife Management Institute
709 Wire Building
Washington, D.C. 20005

Wildlife Society, The
5410 Grosvenor Lane
Bethesda, Maryland 20814

World Wildlife Fund (U.S.)
1601 Connecticut Avenue, N.W.
Washington, D.C. 20009

Agencies connected with the federal government

Federal Power Commission
Washington, D.C. 20585

Tennessee Valley Authority
Director of Information
Knoxville, Tennessee 37900

U.S. Department of Agriculture
Washington, D.C. 20250
- Agricultural Stabilization and Conservation Service
- Extension Service
- Forest Service
- Soil Conservation Service

U.S. Department of Education
Washington, D.C. 20202

U.S. Department of the Interior
Washington, D.C. 20240
- Bureau of Land Management
- Bureau of Mines
- Bureau of Reclamation
- Fish and Wildlife Service
- Geological Survey
- National Park Service
- Oil and Gas Division

Agencies in various state governments

Department of public instruction or state department of education

State college of agriculture

State department of conservation

State department of agriculture

REFERENCES

Bosworth, D. A., and A. B. Foster. *Approved Practices in Soil Conservation.* Interstate Publishers, Inc., Danville, IL 61832.

Conservation Chart (depicts the conservation of resources in two imaginary communities). Sport Fishing Institute, 608 - 13th Street, N.W., Suite 801, Washington, DC 20005.

Donahue, R. L., R. H. Follett, and R. W. Tulloch. *Our Soils and Their Management.* Interstate Publishers, Inc., Danville, IL 61832.

Hurt, D. R. *The Dust Bowl: An Agricultural and Social History.* Nelson-Hall, Inc., Chicago, IL 60606.

Kircher, H. B., D. L. Wallace, and D. J. Gore. *Our Natural Resources and Their Conservation.* Interstate Publishers. Inc.. Danville IL 61832.

Lee, J. S., and D. L. Turner. *Introduction to World AgriScience and Technology.* Interstate Publishers, Inc., Danville, IL 61832.

"Price of Abundance, The." (Color film on wise land-use decisions, 29 min., 16 mm.) Shawnee Resource Conservation and Development Area, 1305 North Yale Avenue, Rural Route 6, Box 255, Marion, IL 62959.

Schroeder, M. J., and C. C. Buck. *Fire Weather.* USDA Agricultural Handbook No. 360, U.S. Government Printing Office, Washington, DC 20402.

U.S. Department of Agriculture. Slide sets available:

"4-H Wildlife Project and Demonstration, The"

"Hunt Safely"

"Making Plaster Casts of Animal Tracks"

"Sharing Our Land with Wildlife"

"Some Game Birds of North America"

"Wildlife for All"

U.S. Department of Agriculture. *Soil Conservation.* U.S. Government Printing Office, Washington, DC 20402.

U.S. Department of Agriculture. *Soil and Water Conservation News.* U.S. Government Printing Office, Washington, DC 20402.

U.S. Department of the Interior. *River of Life.* Conservation Yearbook 6. U.S. Government Printing Office. Washington. DC 20402.

Chapter 8

The Mechanization of Farming

The substitution of machinery for hand labor is one of the main reasons for the rapid strides made in increasing the farmer's productivity per unit of work in recent years. Farm jobs that required the work of many people and many days can now be done by fewer people in a fraction of the time. The story of the mechanization of farming is one of the most interesting and dramatic parts of the history of agriculture.

OBJECTIVES

1. Discuss the development of agricultural mechanization.
2. Demonstrate basic safety precautions when using machinery or power equipment.
3. Discuss what is ahead in mechanized farming.
4. Identify modern farm machinery.

For centuries, farming was an occupation in which brute strength and backbreaking toil were necessary to eke out an existence. "The Man with the Hoe," a famous painting by Millet, a noted French artist, dramatizes this side of farming. Inspired by this painting, Edwin Markham, a well-known American poet, wrote a poem by the same title in which he included the following lines:

> Bowed by the weight of centuries he leans
> Upon his hoe and gazes on the ground,
> The emptiness of ages in his face,
> And on his back the burden of the world.

For 5,000 years or more, the hoe was the primary tool for preparing the soil. It was the forerunner of the plow. At first, a crooked stick, a ram's horn, or a sharp rock served as a kind of hoe. Some centuries before the birth of Christ, people learned to use metal and from it fashioned the blades for crude hoes to prepare the soil. The hoe as it is known today is a cultivating tool, and it is still used widely in cultivating crops in many countries.

The nail, invented 5,000 years ago, is still the primary fastener for materials ranging from wood to metal to masonry. The square or flat-shaped nail with its rough edges has a greater holding power than a modern wire nail. Up to 200 years ago—when machines were invented to cut nails from cold iron, nails were hand forged by blacksmiths.

How Has Mechanization Affected Farming in the United States?

The Pilgrims and other early settlers tilled their fields with hand tools. Until about 150 years ago, little progress had been made over the hoe type of farming.

During the past century, startling changes have taken place in the development and use of farm machinery and power equipment on farms in the United States. In fact, more progress has been made during this period than in all the thousands of years since the first attempts at tilling the soil were made. In the broad sweep of these developments in the United States, three main periods stand out, one prior to 1850 and the other two in the century and more that has followed. These are described briefly in the following paragraphs, but it should be recognized that there was considerable overlapping of the periods.

The human- or hand-power age. To a large degree, hand tools and hand power characterized farming in the United States up to about 1850. The hoe and other crude tools were used in tilling the soil. Corn was planted by hand in hills, and small grain was broadcast by hand. The sickle, the scythe, and the cradle scythe were used to harvest wheat. About 1800, wooden moldboard plows that were pulled by oxen or horses were developed. About this time, some other crude machines, also pulled by horses or oxen, were invented. The steel plow and the horse-drawn reaper were invented during the 1830's, but they were not adopted widely until about the middle of the last century.

The animal-power era. During the period of 1850 to about 1910, oxen, horses, and mules were used to pull plows and other tillage implements invented prior to the middle of the century and after. Although oxen were used as draft animals on some farms during the early part of this period, they were gradually replaced by horses and mules, which were faster moving and easier to handle. Horse-drawn reapers and improved types of grain harvesters came into use. Animal power operated threshers and other types of stationary equipment through the use of tread mills and sweeps. During the latter part of this period, gang plows, combines, and other large types of

The Mechanization of Farming

Fig. 8-1. The cradle scythe was a labor-saving device of the hand-power age in U.S. farming.

Fig. 8-2. Horses provided a chief source of power on farms in the United States during the last half of the nineteenth century and the first part of the twentieth century.

machines were pulled by horses or mules in multiple hitches. About the time of the Civil War, stationary steam engines were used for operating threshers and some other types of equipment. In a limited way, steam engines were used for pulling plows and some other implements. However, this type of power was never widely adopted for this use because the engines were heavy and hard to maneuver.

The mechanical-power era. About 1900, the internal-combustion engine, commonly known as the gasoline engine, first operated crude types of "traction engines" on a few large farms. These were gradually improved and the word *tractor* came into use. Over the years, tractors have been made lighter in weight with more payload, ease of operation, and general usefulness. Small, medium-sized, and large tractors have been developed for varying conditions on farms. Special implements have been designed for use with tractors. In recent years, rubber tires have become standard on tractors and on many other types of farm machinery.

With the coming of trucks, as well as tractors, to farms, horses decreased rapidly in number. In a recent year, there were nearly 4.8 million tractors on farms in the United States; many farms are entirely without horses, at least for draft purposes.

During recent years, electricity has been widely used for lights and some of the power on farms in the United States. Many remarkable developments have occurred and others are in the offing, as described later in this chapter.

One of the most spectacular developments of modern times has been the application of the airplane to farming. Although its use is still limited, its possibilities have been demonstrated on widely diverse jobs such as spraying and dusting crops to control insects and diseases, broadcasting grass and legume seed, seeding rice and some other grains, broadcasting fertilizers, and spreading chemicals to kill weeds and brush on rangeland. It has also been used for patrolling forests and carrying persons to fight forest fires, hunting coyotes and other animals that destroy livestock, taking aerial photographs of farmlands for use in soil conservation planning, checking fences and cattle on the range, and carrying doctors and medical supplies to isolated rural families. In a recent winter, carrying hay in the "operation haylift" made it possible to save the lives of some cattle and sheep stranded on the range without feed because of heavy snowfall. Because of its dramatic aspects, this use of airplanes was widely publicized.

Effects of mechanization

Through mechanization and application of power to farming, plus increases in yields per acre and per animal from better methods of farming, the United States has become a land of plenty. These developments in producing food and fiber helped the Allies win World War II.

Farm Productivity

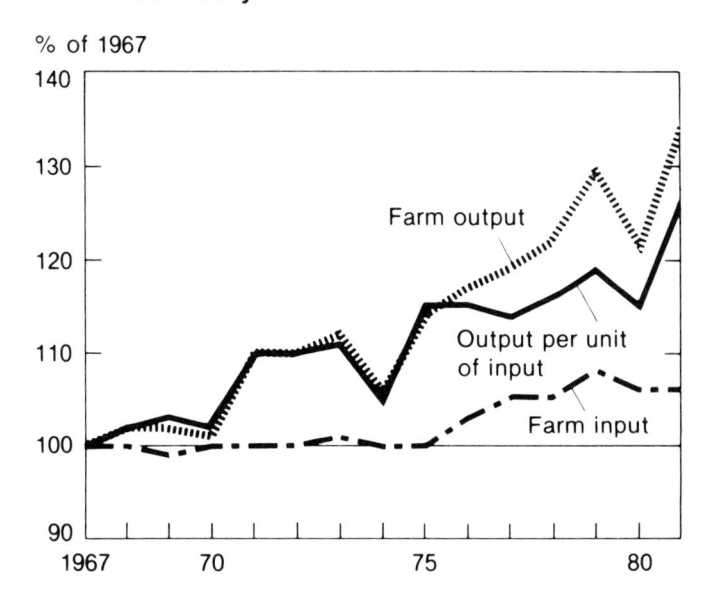

Fig. 8-3. Farming efficiency, as shown by the amount of farm production per unit of farm input, continues to increase. (Courtesy, U.S. Department of Agriculture)

Reducing hand labor needed for crop and livestock production. In the past century, mechanization and the application of other scientific developments to farming have increased the output per farm worker. From 1950 to 1980, the hand labor required for farming went down about 70 per cent, while the production per person went up nearly 570 per cent. The increase in production was greater for crops than for livestock because crop production has been mechanized to a greater extent than livestock production and because per unit production increases for crops have exceeded those of livestock. Thus, with mechanization increasing the amount of work each person can do, fewer people are needed in agricultural production. Fifty-five years ago, 27 per cent of the total population was still on the farm; now only about 2.5 per cent remains.

In 1955, labor made up 32 per cent of the cost of farming; by 1990 it was only 8.6 per cent. The mechanical power and machinery cost of farming, on the other hand, have continued to increase.

Mechanization has greatly reduced the number of hours of hand labor needed for crop and livestock production. For example, the number of hours of hand labor needed in 1950 was 15.1 billion; in 1980, it had been reduced to 4.27 billion. Tables 8-1 and 8-2 show the changes in the hours of hand labor per crop acre and per crop production unit for 1935–39 and for 1976–80, and Tables 8-3 and 8-4 show the changes in the hours of hand labor per animal unit and per animal product for the same two time periods.

Use of Selected Farm Inputs

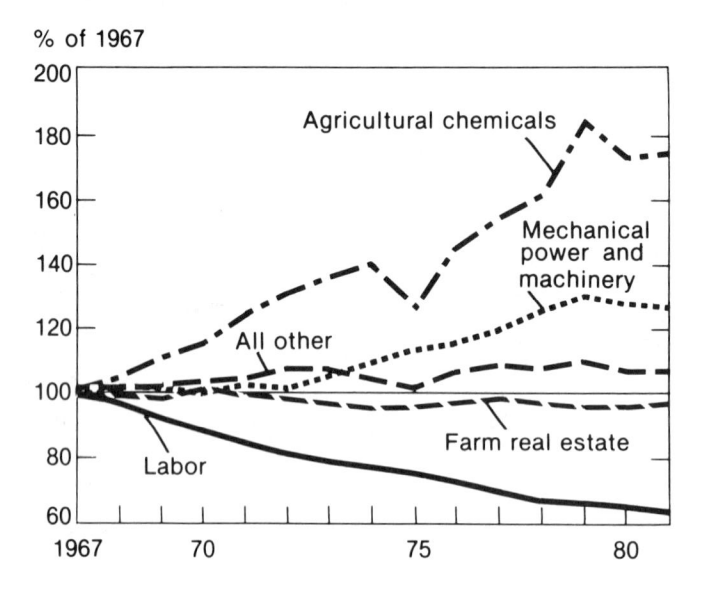

Fig. 8-4. Labor use has declined since 1967, while the use of mechanical power and machinery has increased nearly one-third and agricultural chemicals over two-thirds. (Courtesy, U.S. Department of Agriculture)

Almost everyone is in favor of these trends. However, there are some disadvantages as well as advantages to the mechanization of farming. Some of the advantages are as follows:

1. The drudgery and backbreaking toil in farming have been greatly reduced.
2. Production and income per person engaged in farming have been markedly increased.
3. Workers formerly engaged in agriculture have entered other occupations for producing devices and conveniences that make possible high living standards for all people.
4. Acreages formerly needed to produce feed for horses and mules are available to produce food for humans. Many people can be fed from the 60 million or more acres thus released since 1910.
5. Our nation has greater security because a smaller percentage of its people are needed to produce the necessary agricultural products and a larger percentage are available to provide industrial products and services needed in peace or war. It should be recognized, however, that some of this people power is used to produce machinery and other materials for farms; in reality some of these workers have replaced persons who formerly worked on farms.

The Mechanization of Farming

Table 8-1. Hours of labor per crop acre, 1935–39 and 1976–80

Crop	Hours of labor 1935–39	Hours of labor 1976–80
Corn	28.1	3.5
Cotton	99.0	7.0
Hay	11.3	3.4
Potatoes	69.7	36.8
Sorghum	13.1	3.8
Soybeans	11.8	3.6
Sugarbeets	98.0	25.0
Tobacco	415.0	248.0
Wheat	8.8	2.8

Source: U.S. Department of Agriculture.

Table 8-2. Hours of labor per crop unit, 1935–39 and 1976–80

Crop unit	Hours of labor 1935–39	Hours of labor 1976–80
Corn, 100 bu.	108.0	4.0
Cotton, 1 bale	209.0	7.0
Hay, 1 ton	9.1	1.5
Potatoes, 1 ton	20.0	3.0
Sorghum, 100 bu.	102.0	7.0
Soybeans, 100 bu.	64.0	12.0
Sugarbeets, 1 ton	8.4	1.2
Tobacco, 100 lbs.	47.0	12.0
Wheat, 100 bu.	67.0	9.0

Source: U.S. Department of Agriculture.

Table 8-3. Hours of labor per animal unit, 1935–39 and 1976–80

Animal unit	Hours of labor 1935–39	Hours of labor 1976–80
100 chickens	30	11
100 broilers	25	0.5
100 layers	221	53
1 dairy cow	148	42

Source: U.S. Department of Agriculture.

Table 8-4. Hours of labor per hundredweight of animal product, except for chickens, 1935–39 and 1976–80

Product	Hours of labor 1935–39	Hours of labor 1976–80
Beef cattle	4.2	1.3
Chickens	9.0	2.8
Broilers	8.5	0.1
Chickens, 100 eggs	1.7	0.2
Hogs	3.2	0.5
Dairy cattle	3.4	0.4
Turkeys	23.7	0.4

Source: U.S. Department of Agriculture.

6. Farm work can be done quickly when weather and soil conditions are most favorable. Crops can be planted, tilled, and harvested in a fraction of the time formerly required.

Some of the disadvantages are as follows:

1. Farmers must have more capital than formerly to engage in farming, primarily because of the large investments in machinery, tractors, and other equipment. The average investment in production assets used in agriculture has now reached about $405,167 per farm and $264,513 per farm worker. Much of this investment is in machinery and equipment. In effect, through the adoption of modern equipment, farmers have substituted capital for labor.
2. Farmers must have a larger and more stable farm income than formerly because they must purchase tractor fuel, electricity, and replacements for mechanical equipment.
3. Some persons displaced from farming may find difficulty in establishing themselves elsewhere and adjusting to other types of work.
4. Farms have increased in size in order to make the greatest possible use of equipment. The resulting decrease in number of farms has decreased opportunities for individuals to become established in farming.
5. Farmers need to have considerable mechanical ability and skill to maintain and operate modern farm equipment effectively.
6. Farmers must have a continuous supply of tractor fuel and electricity if they are to maintain production of essential farm products. Difficulties may arise in periods of war or at other times if supplies of fuel and electricity are reduced or diverted to other uses.

Level of Assets and Debt

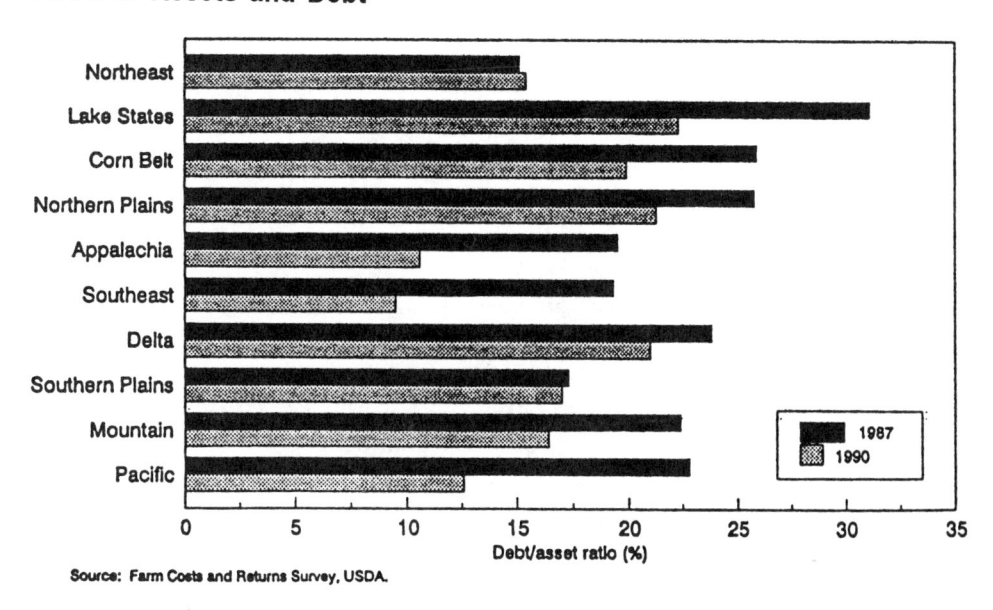

Fig. 8-5. There has been a substantial decline in the average debt/asset ratio of commercial farms between 1987 and 1990 in all regions except the Northeast, Delta, and Southern Plains. (Courtesy, U.S. Department of Agriculture)

7. Operators of large farms have benefited more than small operators because the per acre cost is higher for providing the necessary equipment for small layouts. To be most economical, farm machines must be operated more hours per year than many small farmers can operate them.

Great changes in farming have occurred as a result of mechanization. Farmers, as never before, have become businesspersons who need to sell large quantities of products at favorable prices in order to buy and operate the necessary equipment. Increasingly, mechanization makes it possible for rural people to enjoy the benefits of a mechanical and power age. They cannot go back to an "ox-cart" or a "horse-and-buggy" stage of farming. Most of them would not want to go back, even if they could.

What Changes Have Taken Place in Equipment Used for Producing Crops?

As indicated earlier, tremendous strides have been made in the mechanization of farming during the past century.

The cotton picker is considered one of the great labor-saving machines of all time. First commercially produced in 1941, it does the work of nearly 2 million hand pickers. A two-row picker will pick as much cotton as 50 hand pickers.

An experimental lettuce harvester "feels" the head and if it is mature, triggers a cutting blade. It could cut labor by one-half.

A hay cuber chops and presses hay into animal bite-size pieces, resulting in more economical transport, less waste, and reduced storage requirements (one-third to one-half as much space as bales). It makes possible the total mechanization of hay handling.

A second generation harvester for processing tomatoes has an electronic sorter that eliminates 15 to 20 hand sorters now needed on commercial harvesters.

The tobacco harvester is probably one of the most talked about pieces of machinery around. Adoption of a mechanical tobacco harvester, plus removal of barriers to larger production units, reduced labor required from 415 hours per acre in 1935–39 to an estimated 248 hours per acre in 1976–80, a reduction of about 40 per cent.

A cling peach harvester, with a crew of three or four workers, can harvest from 6,000 to 10,000 pounds in an hour and from 4 to 5 acres in a day. This compares with a crew of about 20 workers picking the same amount by hand.

In a recent year in Delaware, 12 harvesters took care of 1,200 acres of asparagus at about 6 acres per machine-hour, replacing 400 workers. The interest in mechanized harvesting there grew after 1969, when 5,000 acres in Delaware and New Jersey went unharvested because of a shortage of workers.

Dates on palms 30 to 40 feet high are now mostly harvested by cutting the bunches and lowering them to a shaker trailer. The fruit is shaken into bulk bins. Labor requirements have been reduced by 80 per cent, and harvest costs have been cut in half—down to about 20 per cent of total production costs.

While great advances have been made in the mechanization of farming, the most modern equipment for saving labor has been adopted on the larger farms and on farms where conditions are otherwise favorable for its use. One of the trends that has changed this situation is the development of mechanized equipment for small farms.

Equipment for tilling the soil

The development of tillage tools down through the thousands of years since people started to till the soil is a fascinating story. Some of the first plows were crudely made from forked sticks and were pulled by people. As long as 5,000 years ago, in some parts of the world, primitive plows were

The Mechanization of Farming

pulled by oxen. Actually, there was little improvement over the crude hoes and forked stick types of plows until about 160 years ago. Even today, on about half of the farms of the world, crude wooden plows are still in use.

Up to 1830 in the United States, a few plows with wooden moldboards had been developed that were designed to turn over a layer of soil. These gradually replaced the older types that merely stirred the soil. Some moldboard plows with metal points were adopted in the New England states during the Revolutionary War, where they were known as "minuteman" plows. Metal parts were added to these plows, but the first plow with an all-metal moldboard of cast iron was patented in 1797. Farmers were slow to adopt it, partly because they feared the iron would poison the soil. In the Middle West, wooden moldboard plows with cast iron shares were used to turn the prairie sod, but these plows would not scour[1] in the black, loamy soil after the original prairie sod had been broken. In 1837, John Deere, a blacksmith, built one of the first steel plows with a smooth, polished moldboard that scoured well in prairie soils.

All of these early plows were of the "walking" type; that is, the operator walked behind the plow and guided it by grasping the handles while oxen or horses pulled it. About the time of the Civil War, the first "riding" plow was developed. This plow, called a "sulky" plow, which was mounted on wheels and pulled by two to four horses, had a seat on which the operator could sit.

Fig. 8-6. This picture of slaves drawing a crude plow is one form of evidence that the tilling of the soil is an ancient occupation. (Courtesy, J. I. Case Company)

---

[1] Scouring is the cleaning action of the soil as it passes over the wearing surfaces of the plow. This action keeps the wearing surfaces of the plow clean and shiny so that friction is reduced, and the soil can slide along the plow surface and be turned over. When a plow will not scour, rust and wet soil stick to the wearing surfaces.

The operator could plow as much as 2 acres per day, as compared to about 1 acre with the walking plow. A few years later, two-furrow plows, called "gang" plows, appeared; these were pulled by six to eight horses and could plow 3 or 4 acres per day. Today, three- and four-furrow tractor-drawn plows that can plow an acre or more an hour are in common use. Larger plows and tractors are used in some parts of the United States.

Fig. 8-7. A no-till plow invented by the Hardens. A spike-toothed slot-filler wheel follows the spring-loaded coulter and subsoil shank to fill the slot cut by the subsoil to prevent seeds from dropping down too far. (Courtesy, Soil Conservation Service, U.S. Department of Agriculture)

Implements for working the soil after it has been plowed have changed greatly. "Brush" harrows consisting of portions of trees with the branches attached were used by the pioneers. Later, various types of harrows, or "drags," were made with wooden spikes, or "teeth," set in wood. These were replaced by iron parts. The disk harrow, developed later, had circular blades that made it possible to cut through roots and clods. Spring-tooth harrows and other kinds of equipment also came into use for working up the soil.

Until the early 1800's, the hand hoe prevailed as the method of cultivating row crops, such as corn and tobacco. A single-shovel, plow-type cultivator pulled by one horse was developed about 1820. Many improvements were made, including the development of a walking cultivator pulled by a team of horses. Later it was displaced by the riding type. The earlier models of these were of the one-row type; later, two-row cultivators were used in some parts of the country. In recent years, tractor-drawn cultivators of two-row to eight-row types have come into common use.

Many other types of implements for tilling the soil and cultivating various crops are now in use, as discussed in Chapter 17.

Equipment for planting crops

Broadcasting by hand, as portrayed by the sower of Biblical times, was used in the United States as the chief method of planting wheat and other small grains until less than a century ago. This method continued in extensive use until the Civil War when farm labor became scarce. Earlier, various types of grain drills and seeders had been developed, and these were improved and adopted extensively during and after the Civil War. Some of these seeded and covered the grain in one operation. The end-gate type of broadcasting seeder attached to the rear end of a wagon was another type that was widely used. Grain drills with disk-type furrow openers came into use before 1900. Today, with a 12-foot tractor-drawn drill, one person can seed 50 to 60 acres per day at an accurate rate and at uniform depth. In the wheat-growing areas, two or more of these drills are often hitched together to increase the daily acreage. Some modern drills are equipped with attachments for seeding legume and grass seed and for distributing fertilizer, so that grain and grass can be seeded and fertilizer spread in one operation.

The hoe was the common method of planting corn until about 1850. Hand-type, hill-drop corn planters were developed first, and some crude types of horse-drawn planters appeared during the first half of the last century. One type of two-row planter, developed about 1860, made it possible to plant corn in hills. One rider drove the horses and another pulled a lever to drop the corn on cross marks previously made with a sled-type marker. Later, a mechanism for dropping corn in hills was developed for corn plan-

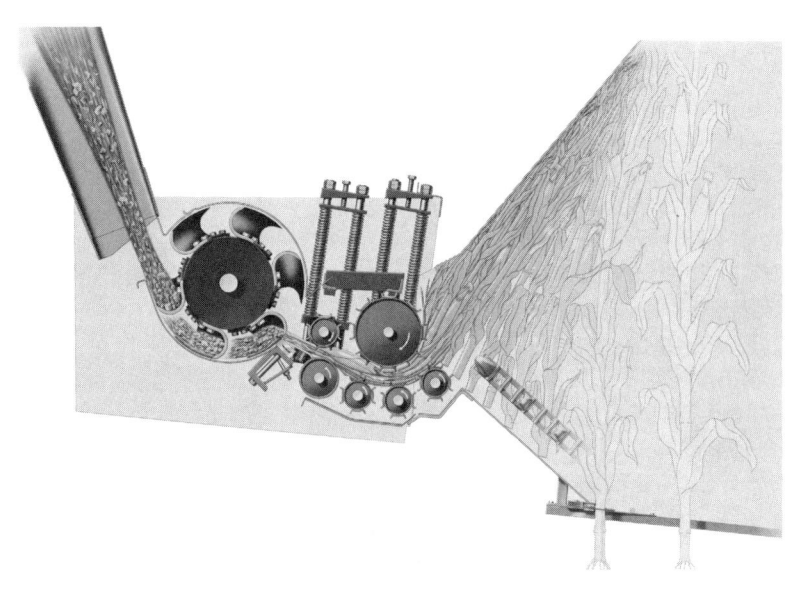

Fig. 8-8. A diagram of the flow of corn through an ensilage cutter. (Courtesy, Allis-Chalmers Manufacturing Company)

ters. It was tripped by a wire with knobs at the desired intervals. Today, tractor-drawn planters of two-row to eight-row types are generally used. Some of these have fertilizer attachments that drop the fertilizer into the soil at the desired places along the rows. Recently, attachments have been added to planters for applying insecticides and weed-killing chemicals to the soil.

Equipment for harvesting crops and threshing grain

An interesting drama could be written on the changes that have taken place in harvesting and threshing wheat from Biblical times to the present. Sickles made from flint were first used to cut grain about 3,000 B.C. Then came the sickle with a curved metal blade. This was followed by the scythe and the cradle scythe. All of these were hand methods of cutting, and the grain was then bound by hand into sheaves, or bundles. One of the first successful attempts to construct a horse-drawn machine for cutting grain was the McCormick hand-rake reaper of 1831, as shown in Fig. 8-9. This reaper was followed by various improved reapers, which culminated in the self-binder that cut the grain and bound it into bundles with twine, tied by a special type of mechanical knotter. As early as 1880, horse-drawn combined harvester-threshers were perfected to cut and thresh the grain in one operation. These became known as combines. At first, these were used only on large wheat farms; but today tractor-powered combines of various sizes and designs are widely used throughout the United States.

The methods of threshing wheat likewise passed through a series of dramatic changes. Some of the early methods were threshing the cut grain by the treading feet of animals, by rollers drawn over the cut grain, or by

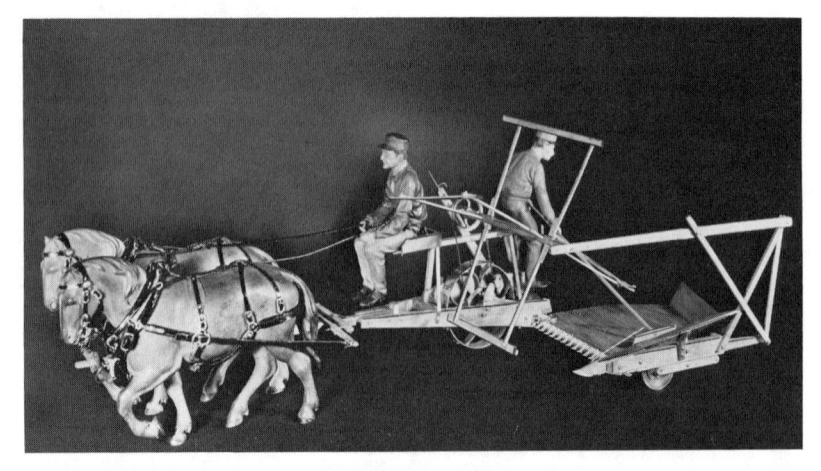

Fig. 8-9. An early model of the McCormick grain reaper. The person in the rear raked off the grain with a hand rake. (Courtesy, International Harvester Company)

The Mechanization of Farming

Fig. 8-10. The steam engine *(top)* and threshing machine *(bottom)* used to make quite a team for separating the grain from the straw.

the use of a hand flail. All these methods required the additional steps of shaking the wheat from the shattered straw and then winnowing to separate the wheat from chaff. The latter step consisted of placing the wheat and chaff in a container, which was raised so that the contents were poured out

Fig. 8-11. This vertical bale elevator *(top)* and bale carrier *(bottom)* make putting hay in the barn easy. (Courtesy, Sperry New Holland)

slowly. The chaff and pieces of stems were blown away by the wind, and the heavy kernels of wheat fell into a container on the ground. These primitive methods are still used in many parts of the world.

Although some simple hand-powered threshing machines were designed prior to 1850, horse-powered threshers did not come into common use until after the mid-century. Power for these machines was provided by horses on tread mills or hitched to sweeps that were connected with the thresher by belts or a tumbling-rod mechanism. Later, steam engines came into use, and by 1900, large steam engines and huge threshing machines were widely used. The gasoline-powered tractor appeared during the first part of the century, and by the 1920's, this type of power had practically displaced the steam engine for operating threshers. Combines that harvest and thresh small grains and some other crops have gradually displaced most threshing machines throughout the United States.

As recently as 1930, most of the corn was picked by hand, one ear at a time, from bundles of corn or from the standing stalks. Some mechanical pickers had been developed prior to that time, but they were only moderately successful. Today, self-propelled corn pickers are standard equipment on Corn Belt farms and in many other parts of the country. One person operating a two-row picker can husk more corn than a dozen good hand pickers. Picker-shellers and corn combines are recent developments.

As already discussed, the cotton picker is a good example of mechanizing the harvesting of crops, with a one-row picker being able to harvest as much cotton per hour as 50 hand pickers. Although machine pickers are increasing rapidly, much cotton is still picked by hand because many fields are too small or rough for mechanical pickers and because hand labor is cheap in some areas.

Methods of making hay have also changed greatly. The scythe for cutting, the hand rake for gathering into windrows, and the hand fork for pitching onto a load or stack were typical tools used in the haymaking process a century ago. Horse-drawn mowers and rakes came into use, and later the hay loader had a part in reducing the hand labor required to make hay. The buck rake mounted on a tractor, the field baler, and the field forage harvester (or field chopper) are some of the modern machines from which farmers may choose, depending on the conditions and the preferences of the individual farmer.

How Has Electrification Affected Farming?

Most farms now have electricity supplied from central power stations. From 1970 to 1980, the average amount of current used per farm grew from 917 kilowatt-hours to 1,313 kilowatt-hours. There are about 400 different farm uses of electricity, although no single farm has this many. The follow-

Energy Used in Agricultural Production

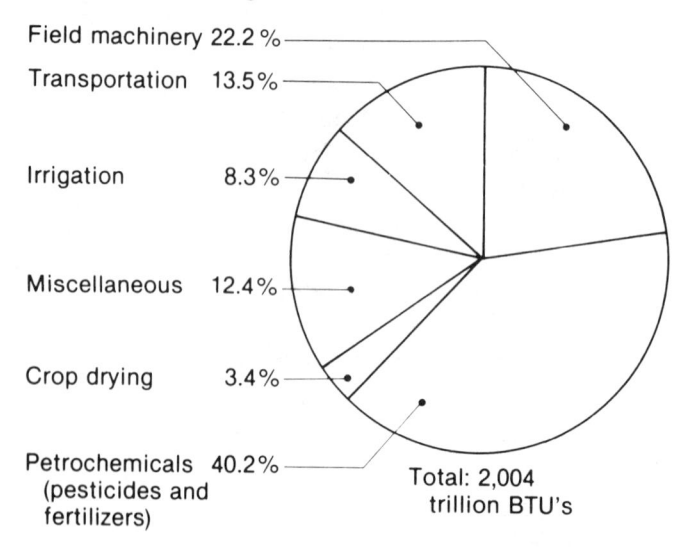

Field machinery 22.2%
Transportation 13.5%
Irrigation 8.3%
Miscellaneous 12.4%
Crop drying 3.4%
Petrochemicals 40.2% (pesticides and fertilizers)

Total: 2,004 trillion BTU's

Fig. 8-12. Farmers consumed about 2.5 per cent of the 80 quadrillion BTU's of energy. (Courtesy, U.S. Department of Agriculture)

ing are only a few of these uses. What additional ones have you observed on farms and in farm homes in your section of the country?

Barn cleaner	Household uses of many kinds
Bench grinder	Irrigation
Chick brooder	Lamb brooder
Clippers	Lawn mower
Corn dryer	Lighting
Corn grader	Mechanized feeding equipment
Corn sheller	Milk cooler
Cream separator	Milking machine
Drainage	Motors for many uses
Egg candler	Paint sprayer
Egg grader	Pig brooder
Electric fence	Portable drill
Electric welder	Poultry debeaker
Feed grinder	Power saw
Feed mixer	Refrigeration
Grain elevator	Silo unloader
Hay drying equipment	Soldering iron
Hay hoist	Stable cleaner
Heater for livestock waterers, etc.	Time switch for poultry lights
	Vacuum groomer
Heaters	Ventilators
Hoists	Water heater for milkhouse
Home pasteurizer for milk	Water system
Hotbed heating system	Water trough deicer

The Mechanization of Farming

With several hours of hand labor performed by electric power for a few cents, electricity has replaced much hand labor on some farms. Thus, much hand work has been eliminated, and farm life has been made more pleasant.

What Changes Have Been Made in Farm Buildings and Farmstead Equipment?

Early farmers had crude stables and barns that provided mere shelters for livestock. Often, machinery was left outdoors; even today this is fairly common.

Modern farmers are finding that suitable farm buildings, as well as equipment of various kinds, are profitable investments. New designs in farm buildings are making them more durable and more useful for various purposes. Buildings that can be adapted at a small expense for use for various kinds of livestock have been designed. Labor-saving features are making it possible to reduce the hand labor in caring for livestock.

One example of the effects of buildings and equipment in saving labor is shown by the reduction in human work-hours required per dairy cow per year. In 1909, as a U.S. average, over 135 work-hours were required per

Fig. 8-13. Badger Pow-R-Trac Feeder for dairy cattle. (Courtesy, Badger Northland, Inc.)

cow per year for hand milking and other chores done largely by hand. With new types of barns and milking equipment and mechanical devices for many other tasks, less than half this time is now required.

New materials and methods for constructing farm buildings have come into use. Laminated wood rafters, steel frames, galvanized and aluminum sheets for sides and roofs, and new kinds of plywood are available. Concrete is used widely for floors and for pavements for barnyards and feeding floors. Pole-type supports are used to reduce the costs of some shelters suitable for livestock, machinery, and hay storage. One-story barns are replacing barns with haylofts. Also available are some partially or completely prefabricated farm buildings that may be assembled quickly.

Large buildings are being constructed for the storage of many kinds of farm machinery used on modern farms. Many farms have well-equipped shops for repairing farm machines and for constructing needed equipment.

New types of grain storage, equipped with power-driven elevators, are being constructed. In some structures for storing grain and hay, equipment has been installed for artificially drying the stored products so that spoilage can be prevented.

Electricity is widely used in farm buildings for lighting and for operating various appliances, such as barn cleaners, feed conveyors, automatic ventilators, and automatic waterers.

Many types of livestock equipment now have features that reduce hand labor. Silos are equipped with mechanical devices for removing the silage at the top. Some silos are designed and equipped so that the silage is removed mechanically at the bottom and carried to the feed troughs without hand labor. Trench silos and bunker-type silos contain self-feeding devices that allow the animals to poke their heads through to get at the feed. As they push against the gate with their shoulders, the gate moves toward the feed and keeps it within reach of the animals. Thus, they eat their way into the silos with little or no handling of the stored feed by the farmer. Some barns are designed so that hay can be eaten by livestock in a similar manner.

Some kinds of livestock equipment are almost completely automatic, thus eliminating most of the hand labor and some of the head work. Automatic watering systems provide ample water to livestock at all times. Automatic machines grind and mix feeds in desired proportions and elevate the feed to overhead bins from which it flows by gravity to feed bunks where livestock eat it. In some poultry buildings, time clock devices are connected with poultry feeding installations so that the chickens are fed automatically at the desired times of the day.

One ingenious person perfected an automatic cattle feeder with a time clock that turns on a horn at regular intervals. The cattle learn to "come and get it." After an interval of time that permits the cattle to arrive, an electrical device meters out the correct amount of feed into the trough from which the cattle eat.

The combined buildings and equipment on a farmstead should be suited to the kinds of crops grown, the types of livestock raised, the labor supply available, and the farm income. Careful planning is needed to keep costs at an economical minimum. In Chapter 12, additional information is given on buildings and other equipment for livestock.

Plans and other information for farm buildings can be secured from the state colleges of agriculture and from commercial concerns.

How May Farm Machinery and Power Equipment Be Operated Efficiently and Safely?

To get the most from the large investment in machinery and power equipment on farms, farmers must have considerable mechanical ability and skill. Machine life can be increased and more efficient service secured if machinery is properly used and adjusted and given good care. Today, many persons skilled in demonstrating, servicing, and repairing tractors and other farm machinery are employed by farm equipment establishments.

Selecting farm machinery and power equipment

In modern farming, good equipment is a profitable investment if it is selected carefully in terms of the needs of each farm. Machines of desirable kinds and sizes must be carefully chosen so that the total investment is kept within reasonable bounds.

Even with careful selection, thousands of dollars are required to provide the equipment for a typical farm in the United States. Some jobs on the average farm do not justify the investment required if the individual farmer were to own the necessary equipment. In such cases, it is frequently possible to hire the jobs done by individuals who own the necessary equipment; this is often called custom work. Other arrangements may be made, such as two or more farmers going together to purchase some of the less-used types of equipment. Thus, farmers must figure carefully in deciding whether or not to purchase a piece of machinery and in determining the size most suitable for their individual conditions if they decide to buy one.

When farmers compare the cost of purchasing a machine with the cost of custom work, they should consider the following factors of cost in machine ownership: depreciation, insurance, interest, repairs, taxes, and the risk of ownership. In addition to these factors, labor costs and other operating costs, such as fuel and oil, must be considered. If the costs for custom work do not exceed the costs of ownership more than a reasonable amount for the custom operator's profit, it would probably be wiser to hire the work done.

One of the more common mistakes made in selecting farm equipment is to choose equipment that is larger than needed for the amount of work to be done. The saving of time in performing a particular operation with an oversized machine may not be sufficient to pay for the added costs of ownership, particularly if there is no way to make profitable use of the time saved.

Operating and adjusting

Tractors may use more fuel than necessary if they are improperly operated and poorly adjusted. Combines and corn pickers leave a great deal of grain in the field, and plows require more power and do a poor job of plowing if they are not kept in good operating condition. Minor adjustments may be all that is necessary in many cases.

The following practices aid in reducing the fuel consumption of farm tractors: (1) keep the carburetor in proper adjustment, (2) use the fuel for which the engine is designed, (3) use machines which utilize as nearly as possible the full capacity of the tractor, (4) for light loads, shift to the highest gear possible and thus reduce the engine speed, (5) have machines in good running order to keep drafts at a minimum, (6) use the tractor and machines when soil conditions are most suitable, and (7) shut off the engine rather than let it run idle when power is not needed.

Operator's manuals or instruction books for tractors and other major pieces of farm equipment are provided by the manufacturers. Operators should study these and follow the instructions carefully.

Servicing and repairing

Keeping tractors and other farm machines in good running order involves lubricating properly. This prolongs their lives and reduces the cost of operation. The operator's manual should be followed with respect to the kinds of oils and greases and the frequency of lubrication. Minor repairs should be made as needed. Much of this work may be done by the operators. Owners of automobiles frequently have their cars lubricated and serviced in garages because it is convenient to take their cars there at intervals. Owners of tractors and farm machines must do most of this servicing on their farms. However, for making major repairs and adjustments, they must secure the services of skilled mechanics.

Developing a farm shop

In order to repair farm machinery and construct farm equipment, many farmers have well-equipped shops on their farms in which they can do this work. A special building may be constructed for this purpose, but more frequently the shop is located in one end of a machine shed or in a garage. A source of heat is desirable so that the shop will be comfortable to work in during cold weather. Usually, winter is a slack season on the farm, and farmers have time available for repairing and servicing farm equipment.

Using safety precautions

Many accidents occur on farms. The increased use of machinery and power equipment leads all other causes of accidents to farm people. Some of these result in loss of hands or limbs, or even in death. Most accidents are due to carelessness and could be avoided.

Using the following safety rules from the Farm Equipment Institute will help to eliminate the major causes of accidents with farm machinery.

1. Keep all shields and guards in place on power shafts, belts, and chains.
2. Stop machine before adjusting and oiling it.
3. Disconnect power before cleaning mechanism when it becomes clogged.
4. Keep hands, feet, and clothing away from power-driven parts of machine.
5. Keep off implement unless seat or platform is provided. Keep others off.

What's Ahead in Mechanized Farming?

As discussed in earlier portions of this chapter, farm equipment and types of power on farms have changed greatly in the past century or so. These changes have been greatest in recent years and are continuing. Periods of war and national defense help speed up changes. This was true in the Civil War, in World War I, and in World War II, largely due to the shortage of labor on farms and the need for increased production of farm products. In recent years, high costs of labor and other costs of production have had a similar effect.

Improvements in farm machines are continually being made to increase their efficiency and particularly to reduce hand labor. These changes are coming so rapidly that new developments may become common practice or out of date in a short time. One of the developments, now seen on many farms, is the combining of various operations in one machine. This has been done in the case of the combine for harvesting and threshing wheat and other small grains. Another is the use of a field sheller for corn that husks and shells corn in one operation, as previously mentioned. The shelled corn is put in special types of bins equipped with dryers that reduce the moisture content quickly, so that the corn does not spoil. Another development is the

Fig. 8-14. Corn is being cut, chopped, and blown into a trailing forage box. (Courtesy, Badger Northland, Inc.)

use of a grain drill that in one trip over the land does a complete job of preparing the seedbed, planting grain, and applying fertilizer.

A machine called a mulch planter can be used to plant corn. In one operation, this machine prepares the seedbed in narrow strips in a field of legumes, distributes fertilizer, and drops the seeds. With this machine, it is possible to grow corn on soil subject to erosion and to plant corn in successive years on the same field, while building up the humus and fertility level of the soil.

Attachments of various kinds may be quickly mounted on tractors. These reduce the total investment in machinery and increase the efficiency of the equipment.

Equipment has been developed to irrigate crops in areas that have insufficient rainfall. It has been used for special crops, such as potatoes and small fruits; for some field crops, such as corn; and on pastures, with promising results. Special pumps, lightweight pipes that can be quickly un-

The Mechanization of Farming

coupled and easily carried to a new location, and spray nozzles that can spread the water over wide areas have been designed for this purpose.

Some automatic irrigation systems have been tried experimentally. When the moisture content of the soil falls to a certain level, controls buried in the soil automatically turn on the water; when the soil has received sufficient water, the controls automatically turn the water off.

Electricity is being used in many ways, thus bringing *automation* to many farm tasks. Machines grind, mix, and convey feeds. Installed in dairy barns, pipelines connected to milking machines carry the milk to bulk tanks where it is cooled automatically to the desired temperature. At the touch of a button, silo unloaders remove silage from the tops of silos and drop it into conveyors that carry the silage to feed bunks without manual handling. The feeding of grain and hay to dairy cattle also has been almost completely mechanized on some farms. The same is true for feeding poultry, as previously explained. Barns are equipped with ventilation systems that automatically maintain a uniform temperature.

In some establishments, eggs are cleaned, graded, and packaged primarily by automation. Irradiation is being used to control insects in grains and to prevent sprouting of stored potatoes.

The development of a mechanical and power age on the farm has been comparable to that in the city. Only time will show what use may be made of atomic energy and solar heat in farming. New developments and discoveries

Fig. 8-15. Artificial drying of crops makes the harvest season a little less under the control of the weather.

will undoubtedly find a place in farming and displace much of the mechanical and power equipment now used.

SUGGESTED ACTIVITIES

1. Arrange a panel discussion with some of your classmates in which you discuss the pros and cons of the mechanical age in farming.
2. Talk to a farmer or a retired farmer who has had many years of farm experience. Ask for a description of the type of equipment used on the farm when the farmer was a child. What were the advantages and disadvantages of those times, as the farmer sees them now?
3. Visit a farm machinery dealer in your community and observe the new machinery on display. Arrange to have the person in charge answer some of your questions.
4. List the uses of electricity on some farm with which you are familiar. Describe the different devices for which electricity is used.
5. Visit a well-equipped livestock farm and note the various buildings and equipment for livestock.
6. Make a list of accidents from farm machinery that you have heard or read about. What caused each accident? How might the accidents have been avoided?
7. Describe some farm machine you have seen that represents a new design. In what ways is it different from its predecessors?
8. Concepts for discussion:
    a. Farming has changed from a hand-tool to a power-equipment age.
    b. The reduction in labor needs through mechanization is still going on.
    c. Production increases have been greater for crops than for livestock because of greater mechanization for crops and because of greater per unit production increases.
    d. Increases in farming efficiency have been as great as or greater than those in other industries.
    e. Farming is still the greatest single employer of workers among industries.
    f. Land that was used to produce feed for horses is now being used to produce food for people.
    g. As mechanization of farming increases, the capital required for farming increases.
    h. Mechanization of farming has led to an increase in the size of farms.
    i. Mechanization of farming and an increase in the size of farms have led to increased demand for skilled labor and management and decreased need for unskilled labor.
    j. As mechanization leads to increased centralization of farming, food production becomes more subject to action by labor unions, which provide transportation and processing of both farm produce and the materials needed to produce food.
    k. In general, the larger the farm machine, the greater the amount of work one person can do with it in a given amount of time.

The Mechanization of Farming

MODERN FARM MACHINERY PHOTO QUIZ

The following pictures are of farm machinery you might see on farms and in fields as you drive through the country. The pieces of equipment shown here are only a few of the many types available from the various companies.

The pictures are grouped according to the company manufacturing the machinery and supplying the photographs—Allis-Chalmers Manufacturing Company, John Deere, International Harvester Company, J. I. Case Company, and Sperry New Holland. How many of these pieces of equipment can you identify? The names are listed by number in the order of their appearance at the end of the series of photographs.

AGRISCIENCE IN OUR LIVES

1

2

3

4

5

6

The Mechanization of Farming 263

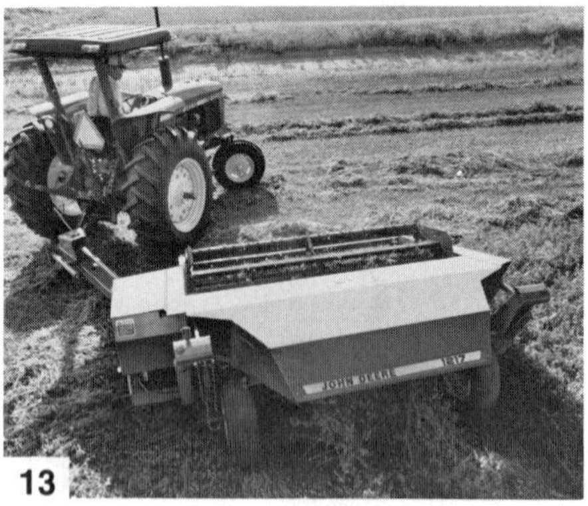

13

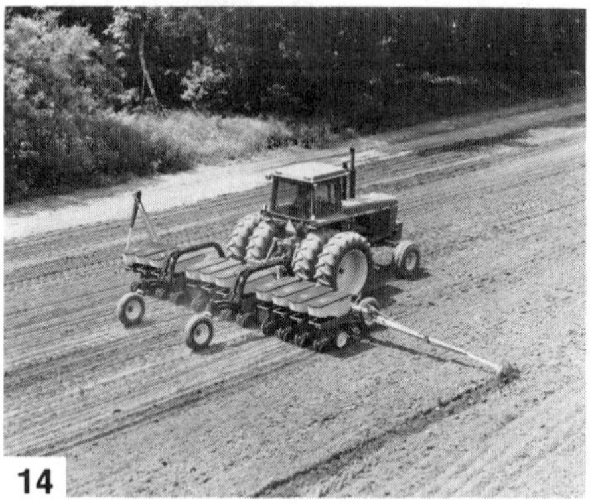

14

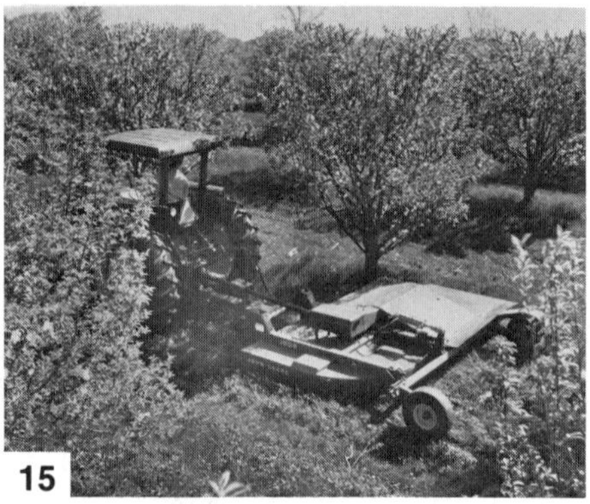

15

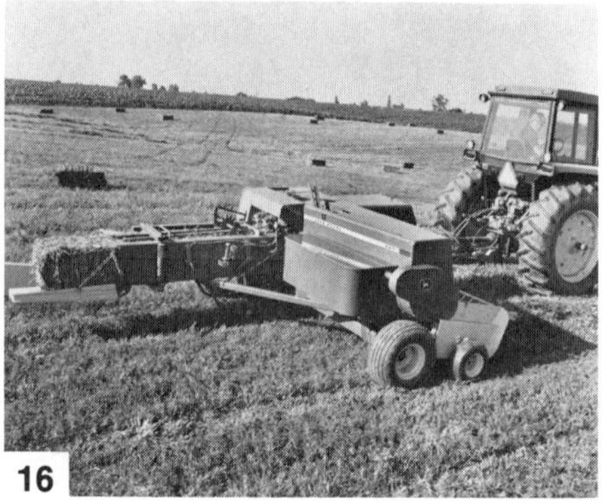

16

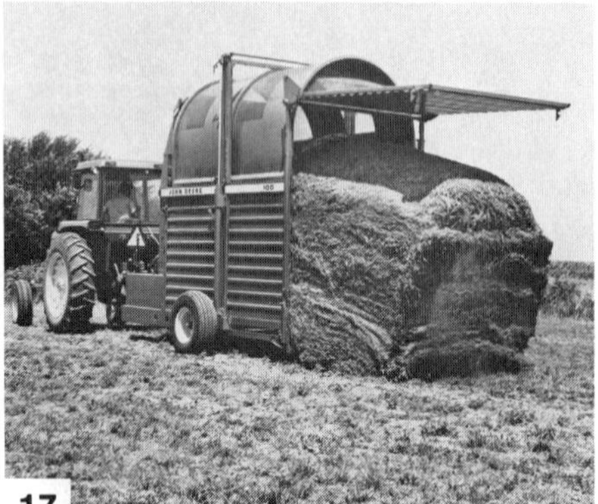

17

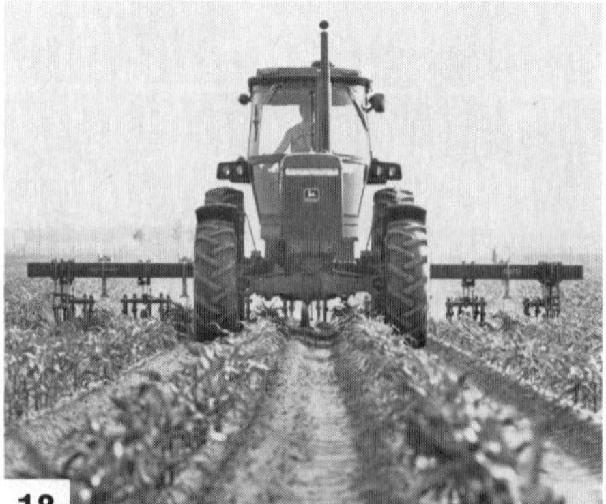

18

The Mechanization of Farming

266   AGRISCIENCE IN OUR LIVES

25

26

27

28

29

30

The Mechanization of Farming

NAMES OF PHOTO-QUIZ MACHINERY

Allis-Chalmers Manufacturing Company

1. M3 Gleaner combine in soybeans.
2. Tracto-mounted Lasco Lightning Weeder—an electrical discharge charge system with 50 kilowatts of electrical power.
3. Model 8070 tractor with seven-bottom moldboard plow.
4. Model 880 four-row cotton harvester.
5. N7 Gleaner combine equipped for corn.
6. Operator's cab on modern tractor.
7. Grain combine with header pivots for hillside harvesting.
8. Model 1400 chiselvator-style tine tiller, 40 to 60 feet in width.

John Deere

9. Mulch tiller with spherical disk blades and heavy-duty chisel standards.
10. Model 520 integral grain-soybean drill for planting 7- and 10-inch rows.
11. Twelve-row 50 series header.
12. The largest of John Deere's new "Titans" line of combines, the turbocharged 8820, with up to 45 per cent more productive capacity than the 7700, the previous top of the John Deere combine line, harvesting eight rows of soybeans.
13. Model 1217 mower-conditioner.
14. Twelve-row planter.
15. Model 1008 rotary cutter.
16. Model 467 baler.
17. Model 100 stack wagon that can build haystacks 10 by 7 by 8 feet or stover stacks weighing up to 1½ tons.
18. Model 825 cultivator mounted on tractor.

International Harvester Company

19. Model 1480 Axial-Flow combine harvesting rice.
20. Model 5288 tractor, 160 horsepower, with disk.
21. Cotton harvester.

J. I. Case Company (A Tenneco Company)

22. Model 4890 tractor with forage harvester.
23. Model 1190 tractor with three-bottom plow.
24. Model 2090 tractor with eight-row cultivator.
25. Model 4690 tractor with tiller leaving mulch on surface.

Sperry New Holland

26. Pivot-tongue windrower in Sudan grass.
27. Baler with bale thrower.

28. Manure spreader.
29. Automatic bale wagon unloading.
30. Round baler.
31. Self-propelled forage harvester in corn.
32. Combines in Oregon grass-seed crop.
33. Combines in Texas red wheat.
34. Loading hay into a tub grinder.
35. Feed grinder-mixer.
36. Unrolling a round bale of hay.

REFERENCES

Burke, S. R., and T. J. Wakeman. *Modern Agricultural Mechanics.* Interstate Publishers, Inc., Danville, IL 61832.

Phipps, L. J., and C. L. Reynolds. *Mechanics in Agriculture.* Interstate Publishers, Inc., Danville, IL 61832.

U.S. Department of Agriculture. *Agricultural Statistics, 1992.* U.S. Government Printing Office, Washington, DC 20402.

U.S. Department of Agriculture. *Contours of Change.* Yearbook of Agriculture: 1970, U.S. Government Printing Office, Washington, DC 20402.

Chapter 9

Characteristics of Farm Animals

Domestic animals have long been considered a prized possession. In the Bible, the word *cattle* included all livestock, i.e., cows, goats, sheep, camels, horses, donkeys, etc. The word *cattle* comes from a Latin word meaning "capital" or "property." In early times, a person's wealth was measured in terms of the number of livestock owned.

Animals have always contributed in many ways to the human existence. Many of the heroes of the past became so known because of their hunting ability. As animals were tamed or domesticated, dependence on them became even greater.

OBJECTIVES

1. Identify how farm animals are beneficial to mankind.
2. Discuss the scientific classification of farm animals.
3. Discuss basic animal behavior and traits.
4. Describe the characteristics of the principal breeds of livestock.

How Are Farm Animals Beneficial?

Farm animals are the prime example of how people have been able to use animals for many purposes in daily life.

The happiness and comfort of individuals are greatly dependent on domesticated animals. In everyday life, people are continually coming in contact with their contributions. both directly and indirectly.

Animals as sources of materials for clothing and protection

Animals furnish many materials from which clothing is made. Ancient peoples used hides and furs for much of their clothing and in many cases for making tent-like structures in which to live. Most of these materials at that time were secured from wild animals. Many changes have been made in clothing since that early period. Skins and furs, a considerable part of which now comes from domesticated animals, are still used extensively in clothing. Leather made from the hides of cattle and some other animals is used for shoes and other articles of wearing apparel.

A great deal of clothing is made from cloth that contains wool from sheep. Wool was very important in the very earliest civilizations. Fabrics of wool have been unearthed in the ruins of villages inhabited by the Swiss Lake Dwellers about 10,000 years ago. The Babylonians made wool cloth as early as 4,000 B.C. Wool and sheep are mentioned many times in the Bible. Woolen cloth is a good insulation for the body, since it keeps the cold out in winter; under conditions of intense heat, it keeps the heat away from the body. Each fiber of wool is covered with fine scales that lock together when spun into yarn and woven into cloth. The best grade worsteds are made from fine, long fibers. Woolens and felt are made from short fibers. Various animals in addition to sheep provide fiber for cloth. These include Angora goats raised in parts of the United States, Kashmir goats and camels in some countries in Asia, and llamas and alpacas in some countries in South America.

Some wild animals are now being raised in confinement for the production of fur. These include mink, silver foxes, beavers, and chinchillas. Chinchillas were nearing extinction in their native habitat in South America when a few animals were brought to the United States and allowed to multiply in captivity. The number has increased sufficiently to produce furs for commercial use.

Animals as beasts of burden

For generations, horses and oxen relieved much drudgery. Oxen were probably used for pulling heavy loads before horses were used. Oxen are still used extensively for draft purposes in India, China, Italy, and some other parts of the world. Tractors have replaced most of the work horses on farms in the United States. Even on cattle ranches, horses are used less frequently than formerly. Many ranches have been fenced to confine the cattle. Ranch hands spend about as much time in jeeps and station wagons as on horseback.

Horses were extensively used in the past for transporting people and their belongings from place to place. Trains, automobiles, and airplanes

Fig. 9-1. Oxen are a curiosity in the United States, but not in some parts of the world. Note the yoke and attachments for pulling a load.

now do this. However, horses still render some valuable services and, in addition, are sources of considerable pleasure. Some people like to ride them as a means of recreation. Horse races are watched with enthusiasm by many people. In some sections of this country and other countries where there are no improved roads, horses are still used for transporting people from place to place. When storms prevent regular mail delivery, horses are still called upon to deliver the mail, just as they were used by the famed Pony Express.

Animals as sources of food

Many foods are derived from animals. Milk has often been called nature's most perfect food. People have used milk, butter, and cheese longer than there has been recorded history. Excavations in Switzerland revealed cheese-making equipment used several thousand years ago by the ancient Lake Dwellers. The Old Testament of the Bible includes many references to the use of milk, butter, and cheese. Records from India, about 2,000 B.C., show that butter was used as a food and as an offering to the gods. Cows, worshipped as sacred animals in that country, roamed in cities and elsewhere, as they do today. Ice cream is a major dairy product today, but it is not new. About 1296 A.D., Marco Polo took an ice cream recipe back to Italy from China. Ice cream was first made in the United States about 1777.

About 400 varieties of cheese are made in various parts of the world. Many of these are available to us. Milk is the favorite beverage and ice cream the favorite dessert of most young people. Every young person should have at least a quart of milk per day in some form or other. Every adult should consume at least a pint. Cows furnish most of the milk supply in this country, although some is provided by milk goats. The dairy cow has been called the "foster mother" of the human race. However, in some parts of the world, milk is also secured from the yak, water buffalo, sheep, goat, camel, and reindeer.

Chickens furnish eggs, which are a valuable part of our diet. Meats from several kinds of farm animals are important as food. The animals most commonly used for this purpose are beef cattle, hogs, poultry, and sheep. The meat from the mature sheep is called *mutton,* and that from the young sheep is called *lamb.* What is meat from hogs called? From older beef cattle? From younger beef cattle?

In most nations, foods from animal sources comprise an important part of the people's diet. In 1990, in the United States, consumption of meat, poultry, and fish averaged about 185.0 pounds per person. In addition, the average person ate large amounts of dairy products. Some types of vegetarians consume some animal products, such as milk, butter, and eggs. Some persons who consume little or no food from animal sources may be improperly nourished.

By-products from meat animals

Parts of animal bodies that were formerly wasted are now made into many valuable products at large meatpacking houses. The modern packinghouses are making considerable profit from these materials, which are classed as *by-products* of the packing industry.

The number of by-products is so extensive that only a few can be listed here. Hoofs, horns, and bones are made into combs, knife handles, buttons, glue, gelatin, and many other articles. Some important medical products are made from portions of the carcasses. For example, the pancreatic glands of millions of cattle and swine are needed to supply insulin for persons with diabetes. Liver extract is widely used by people with anemia. Tankage and meat scraps, valuable as protein feeds for livestock, are also by-products of the packing industry. Materials that cannot be used for more valuable purposes are converted into fertilizers. Thus, nearly every part of each animal is utilized in some way in large meatpacking establishments.

One strange by-product is pig heart valves. Every year, thousands of people in this country undergo heart valve transplants and many more thousands undergo the operation worldwide. The valves are taken mostly from pigs weighing less than 80 pounds, which means that most of the valves must be imported from countries where pigs are slaughtered at lighter weights. The valves from pig hearts are excellent replacements for human heart valves because they are durable, resistant to infection, and not readily rejected by the human body. There are, of course, many other ways in which animals contribute to human health.

Classification of Farm Animals

The domestic animals of major economic importance belong to Phylum X, Chordata, of the Animal Kingdom. The phyla of the Animal Kingdom are:

Phylum I: Protozoa
Phylum II: Porifera
Phylum III: Coelenterata
Phylum IV: Platyhelminths
Phylum V: Nemathelminths
Phylum VI: Echinodermata
Phylum VII: Annelida
Phylum VIII: Mollusca
Phylum IX: Arthropoda[1]
Phylum X: Chordata

Phylum X, Chordata, is further divided into the following classes:

Class I: Pisces (fish)
Class II: Amphibia (frogs, toads, etc.)
Class III: Reptilia (turtles, snakes, alligators, etc.)
Class IV: Aves (all birds)
Class V: Mammalia (bears, dogs, horses, cattle, humans, etc.)

Cattle, horses, sheep, and pigs fall into Class V. The animals in this class are characterized by being warm-blooded, having hair or fur on their bodies, bearing young, and nourishing them from mammary glands.

The particular order within Class V to which cattle, horses, sheep, and pigs belong is the Order Ungulata. The animals in this order have strong teeth for grinding their food and have hoofs with one or more toes such as the hoof of horses and the split hoof of cattle. The following complete classification for the horse is illustrative of how animals are grouped.

Kingdom: Animal
Phylum: Chordata
Class: Mammalia
Order: Ungulata
Family: Equidae
Genus: Equus
Species: caballus

---

[1]Phylum IX, Arthropoda, will be discussed in more detail in Chapter 19, "Insects—Friends or Foes?"

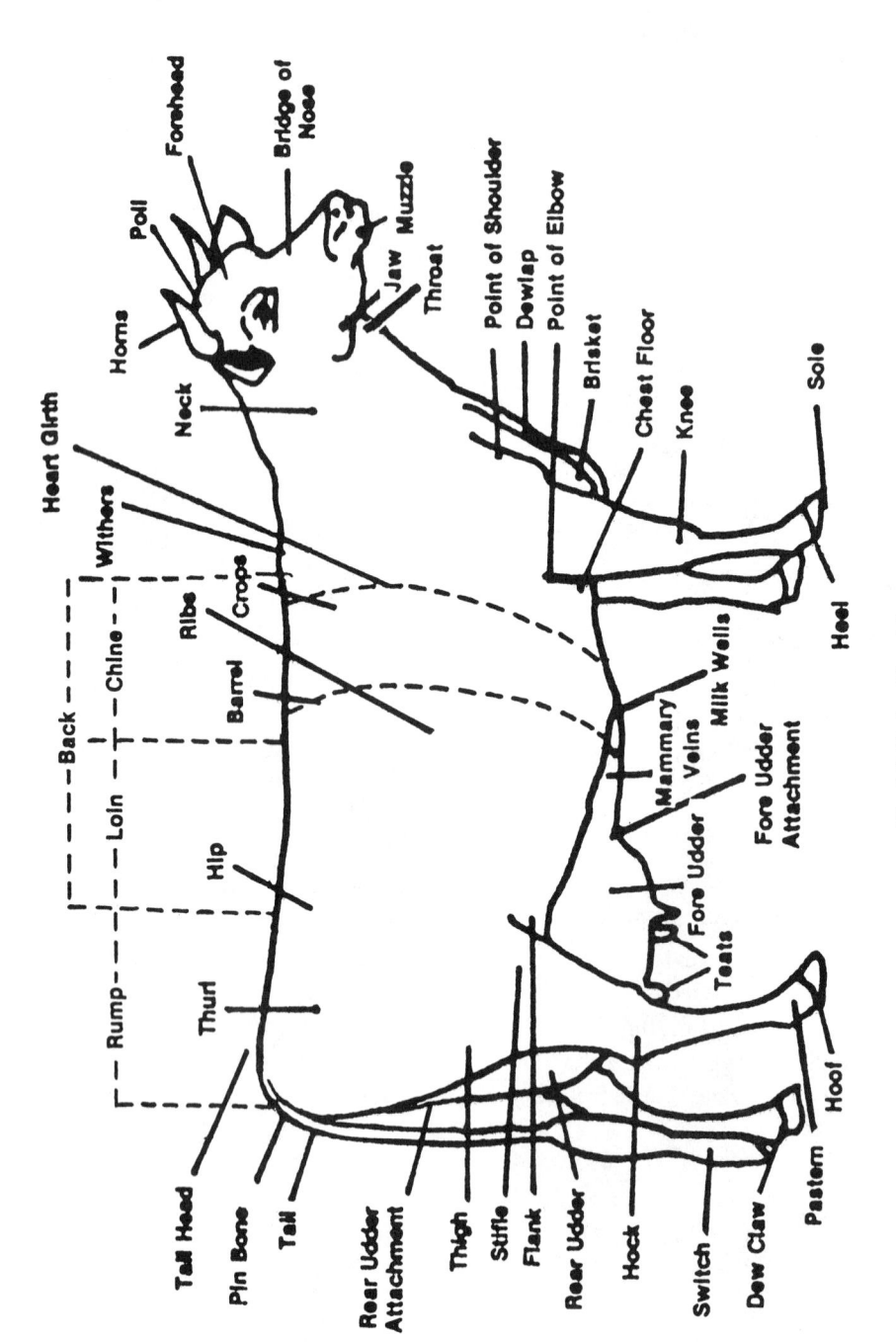

Fig. 9-2. Parts of the dairy cow.

Characteristics of Farm Animals

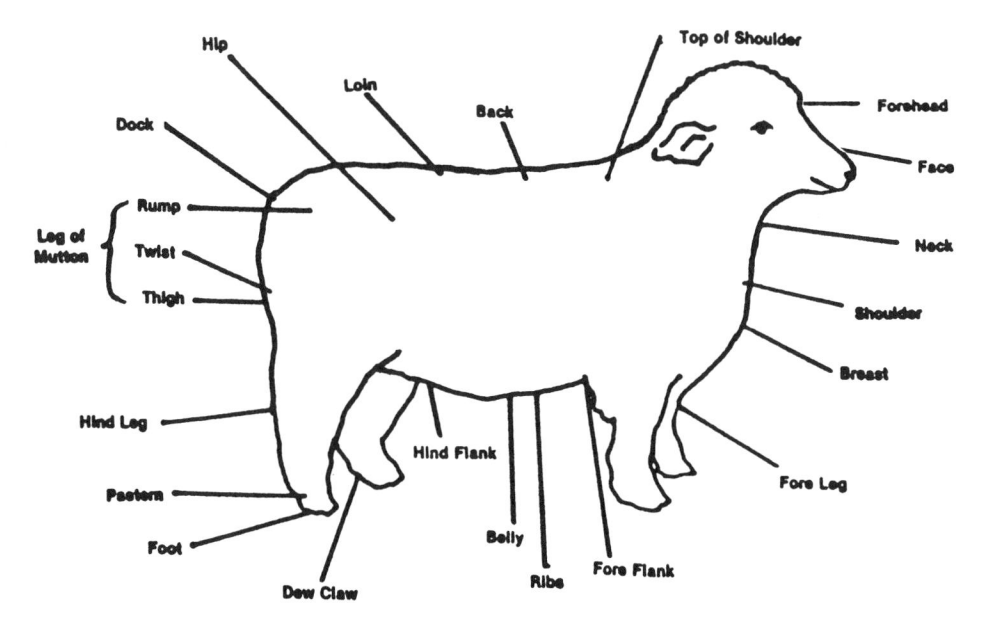

Fig. 9-3. Parts of the sheep.

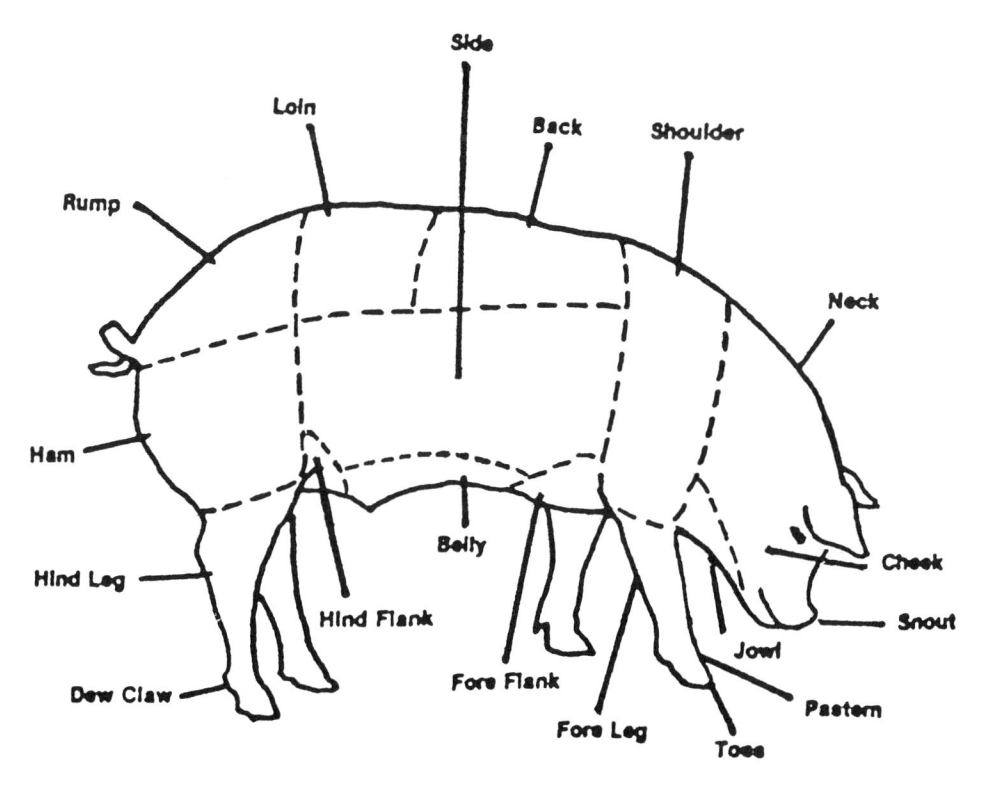

Fig. 9-4. Parts of the hog.

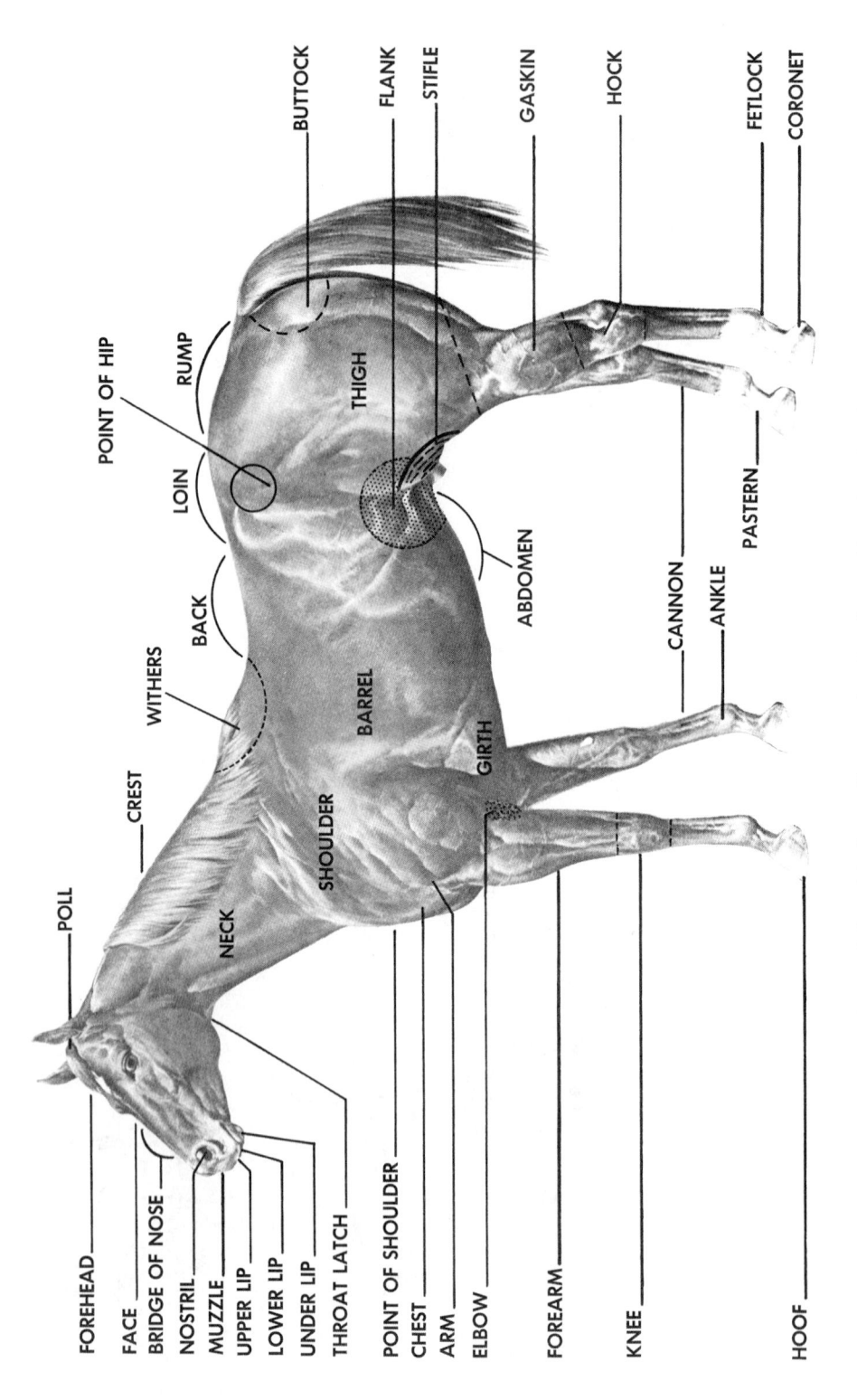

Fig. 9-5. Points of the horse. (Courtesy, American Quarter Horse Association)

Why Are Domesticated Animals Important to Successful Farming and Ranching?

Many farmers and ranchers in the United States keep enough farm animals to consume most or all of the crop products raised on their farms and ranches. These farmers obtain their incomes largely from the sale of animals and animal products. This system of farming is called *livestock farming*. The sale of livestock and livestock products comprises about half of each dollar of income to farmers in the United States.

If good farm animals are properly raised and marketed, they usually bring a greater return for the crops that they consume than could be obtained if the crops were sold directly on the market. It has often been said that hogs are "mortgage lifters" and that "prosperity follows the dairy cow." Sheep are sometimes called the "golden hoof" of farming. These and other terms used by farmers indicate their high regard for livestock and suggest the importance of livestock in farming.

It is more difficult to maintain the fertility of the soil by grain farming than by livestock farming. By farmers' feeding the crops to livestock, large percentages of the food materials that the plants contain are returned to the soil in the form of manure. Livestock farmers also raise legumes, such as clover and alfalfa, for feed. These crops benefit the soil, as described in Chapter 16.

Livestock farmers may use rough and hilly land for pasture. Over 660 million acres of land in the United States are in grass. Much of this land cannot be used for other purposes. People do not eat grass, but livestock do; and they convert it into meat, milk, wool, and other valuable products for human use. In other words, farm animals can be looked upon as factories that convert feed of little direct value to people into products useful to them.

What Are the Important Kinds of Livestock in the United States?

The United States is the leading country of the world in livestock production. It ranks first in beef production, although both India and the Soviet Union lead the United States in total cattle numbers. The United States also leads in the production of milk and cheese. China and the Soviet Union surpass the United States in the number of hogs. The United States ranks fourteenth in number of sheep. If both number and quality of all kinds of livestock are considered, no other country equals the United States.

If we were to study the distribution of the kinds of farm animals in the United States, we would find considerable variation among the states. In 1982, the top 10 states ranked in the order of *total number of beef cattle* on

U.S. Cattle Inventory

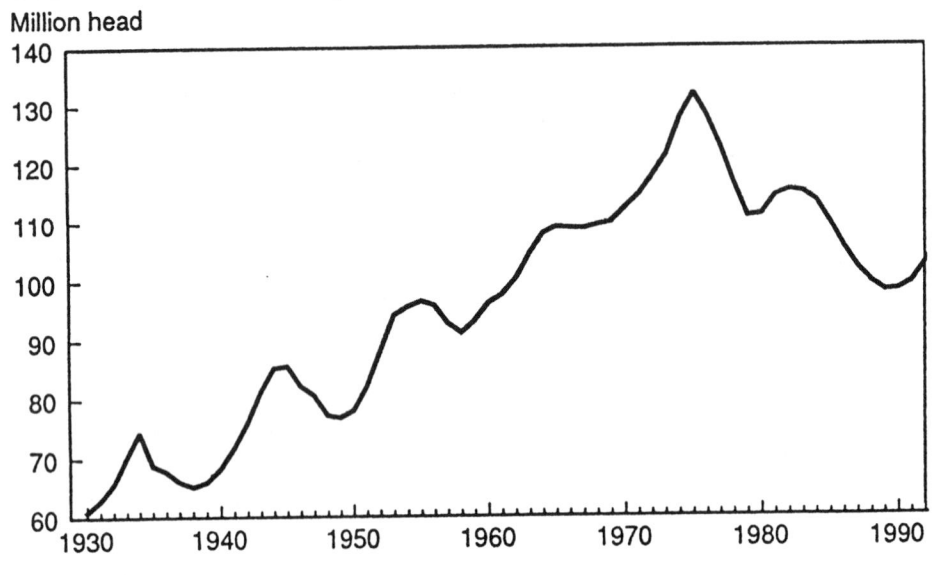

Fig. 9-6. The total number of beef cattle on farms continues to trend upward, . (Courtesy, U.S. Department of Agriculture)

farms were Texas, Oklahoma, Missouri, Nebraska, Kansas, Iowa, Montana, South Dakota, Florida, and California. In 1981, the 10 states with the most cows kept for *milk production* were Wisconsin, California, New York, Minnesota, Pennsylvania, Michigan, Iowa, Ohio, Texas, and Missouri, in that order.

The top 10 states in *swine production* in 1981 were, in order, Iowa, Illinois, Minnesota, Indiana, Nebraska, Missouri, Ohio, North Carolina, Kansas, and South Dakota. Parts of all these states, except North Carolina, are located in the Corn Belt. About two-thirds of all hogs in the United States in 1981 were in the first six states.

Texas led all states in *sheep production* in 1982. The other top states, ranked in order, were California, Wyoming, South Dakota, Colorado, Utah, Montana, New Mexico, Oregon, and Idaho.

In *broiler production* in 1981, the 10 leading states were Arkansas, Georgia, Alabama, North Carolina, Mississippi, Maryland, Texas, Delaware, Colorado, and Virginia. In *turkey production* in 1981, North Carolina, Minnesota, California, Arkansas, Missouri, Virginia, Texas, Iowa, Indiana, and Wisconsin led the country, in that order.

The decrease in numbers of horses and mules over the last 35 years has been one of the greatest changes in U.S. agriculture. In 1959, horses and mules on farms and ranches numbered only about 3 million as compared to about 26 million in 1920; and the downward trend continues. Tractors and trucks have largely replaced horses and mules as sources of power on farms in the United States. The decrease in numbers of horses and mules is less pronounced in some of the southern states where farms are small. However, it looks as if "Old Dobbin" is on the way out in most parts of the United States, except on farms not suited to tractors and on ranches. Horses in considerable and increasing numbers are used for pleasure purposes.

On some farms and ranches, other kinds of animals also are raised. Some farmers raise large numbers of ducks, especially in the eastern states. A few raise fur animals, such as silver foxes, mink, and chinchillas. Some specialize in saddle horses and light horses for racing. What unusual kinds of farms are found in your area?

What Are Some Interesting Habits and Traits of Farm Animals?

The people who get the most enjoyment from their associations with farm animals have usually developed an understanding of their habits and traits. Through study and observation, everyone can learn many interesting features of farm animals.

Farm animals differ in their habits of eating and in their "equipment" for eating. A cow has thick and relatively inflexible lips, but she has a long and rough-surfaced tongue with which she gathers the grass and pulls it into her mouth. A horse's tongue is not adapted for such a purpose, but its lips are especially flexible so that it can "crop" (bite off) the grass close to the ground with its teeth. The upper lip of a sheep is divided in the middle, which makes it possible for the sheep to nibble grass close to the surface of the ground. Cattle and sheep have front teeth on the lower jaws only. They have tough, pad-like gums in the front parts of the upper jaws.

Cattle and sheep eat rapidly and swallow their feed with little chewing when they first take it into the body. The feed is later *regurgitated* (brought back into the mouth) and carefully rechewed. This process is often referred to as chewing the cud. Animals that chew their cuds are called *ruminants*. Several hours are spent each day by such animals in this process. Cows, for example, spend about 8 or 9 out of every 24 hours in actual grazing, with the remainder spent about equally divided between chewing their cuds (ruminating) and just resting. They stand up about 16 hours out of each day. How many hours each cow actually sleeps seems to be a mystery. Some scientists believe they sleep little if at all, and the same may be true of sheep.

Animals have far more intelligence than many people think they have, and this is particularly true of domesticated animals. Dogs are usually ranked at the top of animals in intelligence and cats are second. It may surprise us to know that the lowly pigs are really quite intelligent, and some sci-

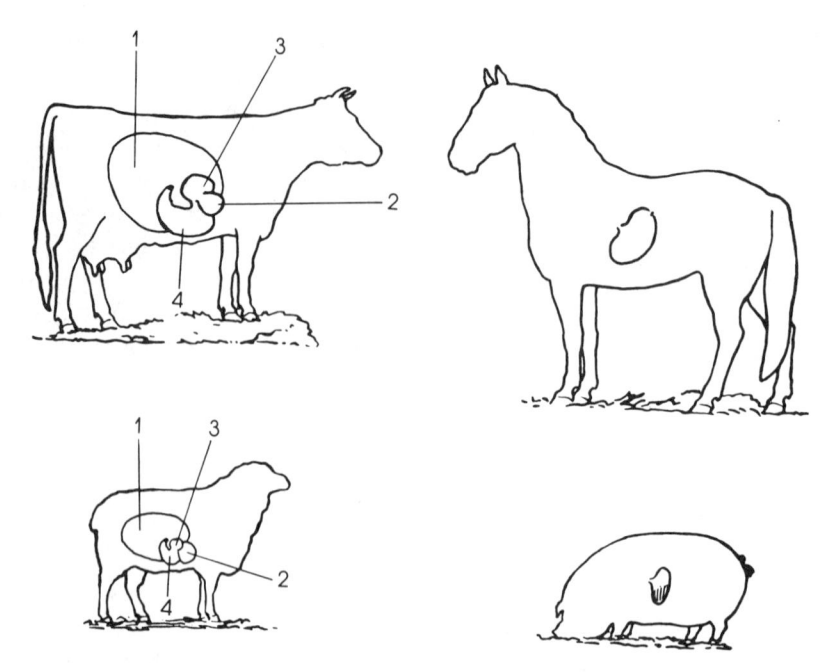

Fig. 9-7. Farm animals differ in their stomach capacities. Cattle and sheep are ruminants. They have stomachs with four compartments that can handle large quantities of bulk feeds. Horses and pigs are nonruminants. The stomach capacity of the average cow is 266 quarts; sheep, 31 quarts; horse, 19 quarts; and hog, 8 quarts. (Courtesy, Michigan State University)

entists believe they are the smartest of the hoofed animals. Most pigs are sent to market at a young age; thus, they do not have a chance to show their "talents." Other domesticated animals in order of intelligence are mules, horses, goats, cows, sheep, and chickens. Individual animals of any one kind vary widely in their levels of intelligence.

Cattle are able to eat large quantities of coarse feed, such as grass and hay. This is made possible by a large paunch, one of the four stomach compartments possessed by each cow. The digestion of this feed is a complicated process consisting of chewing and rechewing, soaking, churning, chemical action, and the work of tiny organisms. The inside of a cow's stomach is an interesting place, at least to some scientists. They have studied the processes in it by making a small opening directly into the paunch, or rumen, of a live cow. Without discomfort to the animal, they have studied the movements of the stomach as well as the contents. They have found that the paunch (or rumen) is a large fermentation vat teeming with billions of microorganisms including bacteria, yeasts, and protozoa that aid in breaking down the feeds and in forming proteins, certain vitamins, and other nutrients.

The stomach of a horse or a hog is similar to a human stomach. A horse has a large caecum (at the same place as the appendix of a human), which is an important part of the digestive system, as it provides extra capacity for considerable hay and other bulky feeds. A chicken has a crop in which the food is stored and softened. From there it passes to a modified stomach, then to the gizzard, where it is ground finely, and then to the intestines.

There is still considerable mystery about how milk is manufactured by a cow. The main "factory" for this process is located in the udder. Nutrients are carried in the blood to the tissues of the udder in which the milk is made. An estimated 400 pounds of blood pass through the udder for every pound of milk produced. The so-called "milk veins" in front of the udder are really vessels that carry some of the blood from the udder. Arteries that carry blood from the heart to the udder, as well as additional veins that carry blood from the udder, are located in the body of a cow. Between milkings, the udder of a producing cow fills with milk. At milking time, a cow does not "give milk"; it has to be taken from her! If the udder is massaged by hand or by machine, a message is carried to the cow's brain and relayed to the pituitary gland located at the base of the brain. This gland releases a hormone into the bloodstream, which causes muscles in the udder to squeeze the milk out of the milk glands into "milk cisterns" in the udder. Thus, milk is put under pressure in the udder. All that is necessary for getting the milk from the udder is to squeeze the teats with the hands or with a milking machine, and the pressure in the udder forces the milk out. This let-

Fig. 9-8. Bob's Pansy Girl, a registered Ayrshire cow owned by D. R. Shores, Skowhegan, Maine. As a six-year-old, she produced 20,240 pounds of milk, with a butterfat test of 6 per cent. (Courtesy, Ayrshire Breeders' Association)

down of the milk by a cow is largely an involuntary process. If a cow is frightened during the milking process, she may hold up her milk. This also is an involuntary process caused by a hormone from the adrenal glands that checks the action of the muscle in the udder, which would otherwise force the milk from the tissues where it is made.

Cattle and sheep are especially *gregarious,* meaning that each tends to stay with other animals of its kind. This trait probably traces back to the period previous to domestication when animals of each kind banded together for protection against enemies. In sheep, this trait is known as the *flocking instinct.* Of what value is this trait to the western sheep raiser?

In the protection and care of their young, farm animals show many interesting traits. A cow hides a newborn calf in the grass or brush, just as a deer hides her fawn. A mother hen may ruffle her feathers and attack with her beak if she senses that her chicks are in danger. A ewe may stamp her feet and at times charge threateningly at a stranger who approaches her lamb, though at other times she is quite timid. A cow may use her head and horns to protect her calf if the occasion seems to demand it. Some people could profit from the examples of courage, patience, and faithfulness that animals show in caring for their young.

The activities of farm animals at rest vary. Some animals, like the horse,

Fig. 9-9. Cattle tend to stay together in herds. This herd of Angus is grazing on a lush New England pasture. These cattle would be equally at home in other parts of the country. (Courtesy, American Angus Association)

rest and even sleep while standing up, although they lie down part of the time. Horses get up with their front quarters first, while cows do just the opposite. These actions have been explained as traits acquired by evolution and survival of the fittest during the early stages of the development of these animals. One explanation is that the small ancestor of the horse developed in grassy country. When alarmed, it raised up on its front feet to lift its head above the grass as quickly as possible to look around. On the other hand, cattle descended from animals that grew up in areas covered by low trees. Therefore, when surprised, a cow would get up with hind feet first so that she could keep her head close to the ground as long as possible to look underneath the tree branches and see the cause of alarm.

Many of the traits mentioned were especially useful to the ancestors of farm animals, before the period of domestication, in protecting themselves from their enemies. After hundreds and even thousands of years of domestication, some of the instincts of animals have persisted, though under most conditions they are of little or no value today.

Horses do not actually pull a load to which they are hitched. In reality, they push against their collars, from which the tugs, or traces, extend to the implement being pulled.

An interesting characteristic of some livestock is the tendency of herds or flocks to have a special organization, with animals in the group having different levels of authority. The position of each animal is won by force. For example, through fighting with others in the flock, hens arrange themselves into a "peck order," with a "boss hen" that has the right to peck all others at the top. Thus, hen A (the top boss) pecks B but is pecked by none; both peck C; and all three peck D; and so on to the bottom hen that is pecked by all others in the flock but pecks none. A hen's rank determines to a major degree her rights for food, space, nests, and other requirements. Actually, the hens near the top often produce more eggs because they have a better opportunity to eat, while the others tend to stay in the background. Cows have a "bunt order," with a boss cow at the top. To some degree, horses in a group have a rank, which is established through kicking and biting.

An alert livestock producer can tell animals apart, though they are alike in color, by peculiarities in the way each stands or holds its head, its friendliness or lack of it, and in other ways. Thus, animals have individual personalities just as people do.

Animals are protected from excessive heat and cold in various ways. In the winter, cows and horses exposed to cold grow thick, long coats of hair. All animals protect themselves from the hot sun by seeking the shade when it is available. On most parts of their skin, horses have sweat glands that provide a good cooling system. The sweat glands of cows are on the muzzle, which is moist when the animal is in good condition. Brahman cattle, however, have some sweat glands in the skin on various parts of the body; and the cooling surface is increased by loose folds of skin. Because of these and other characteristics, Brahman cattle can withstand heat up to 95°F without discomfort. Other breeds of cattle are affected unfavorably at temperatures above 75°F. Hogs have no sweat glands; thus, they seek water or mud in order to keep cool. A dog uses its tongue and mouth as a cooling system. In hot weather, a dog keeps its mouth open and moves its tongue to increase the evaporation of moisture.

What Are the Characteristics of the Principal Breeds of Livestock?

Most of the common breeds of livestock have been in existence for many years, but a few breeds are of recent origin. Most of the breeds now found in the United States originated in other countries and were imported to this country. The native homes of many breeds are the British Isles and some other European countries.

In raising livestock, a farmer should select a breed that is adapted to the purpose for which the animals are to be kept. For milk production, the farmer should choose a dairy breed, not a beef breed. Furthermore, the farmer should choose a dairy breed that produces a type of milk suited to the market demands in a given situation. Under some conditions, milk with a low butterfat content might be produced and sold with greater profit. Dairy breeds differ considerably with respect to the average butterfat content of the milk.

The farmer should also select a breed that is adapted to the conditions under which it is to be raised. If going into extensive cattle production in the western states, a farmer would probably choose a beef breed. Furthermore, Herefords probably should be selected, because animals of this breed are especially hardy under range conditions. If raising a large number of chickens primarily for egg production, the farmer should choose a breed accordingly.

Regardless of which breeds they choose, livestock breeders will be most successful if they keep only the best animals. There are good and poor animals in every breed. Careful breeders will use the best individuals to improve their herds and will discard the poorest individuals. So, it is possible for farmers to build up profitable herds of any breed.

Many successful livestock farmers are raising farm animals that have been improved to the extent that definite breed characteristics are clearly indicated. These animals are either *purebreds* or *grades*. Whether a given animal is a purebred or a grade cannot always be determined by its appearance, but for purposes of identifying that animal with a specific breed, such a distinction is usually unnecessary. If it is necessary, the owner can be asked. In most cases, the owner will have a *registration certificate* from the breed registry association to certify that a particular animal is a purebred.

If both parents of an animal are purebreds of the same breed, the animal itself is a *purebred*. The term *purebred* thus signifies that all ancestors of the animal in question have been members of the same breed back to the time the breed was founded. Sometimes a purebred animal is called a "thoroughbred" or a "full-blood," but neither is the correct term. If only one parent, usually the sire, is a purebred, the animal is called a *grade*.

A *crossbred* animal is one whose sire and dam were both purebreds, but of different breeds. In some cases, as many as three breeds are used in crossbreeding. However, in each generation, a purebred sire is used.

A *scrub* is an animal that shows no indication of good breeding. Its ancestors were nondescript and usually inferior.

Any animal, regardless of whether it is a purebred, a grade, or a scrub, can have a *pedigree,* because this term simply refers to a list of the ancestors or the family history of an individual. However, only the pedigree of a purebred animal is recognized by any of the breed associations, and only purebred animals are eligible for registry in a breed association.

In selecting a breed of any kind of livestock, the livestock producer should become familiar with several breeds in order to make a good choice.

Breeds of dairy cattle

Dairy cattle, in general, may be distinguished from beef cattle by their lean, angular appearance and by their well-developed milk-producing organs. Most of the breeds have horns, but dairy producers usually remove them for safety reasons.

The breeds found in the largest numbers in most communities are the Holstein, Jersey, and Guernsey. Occasionally, other breeds such as the Ayrshire and the Brown Swiss may be found.

Holstein-Friesian. The Holstein-Friesian breed (usually called Holstein) is the most widely distributed of all the dairy breeds in the United States. This breed was developed in the Netherlands more than a thousand years ago. Although some Holsteins were imported as early as 1621, the present breed traces back to about 1861. The cattle are usually black and white (sometimes red and white), and they are the largest of all the dairy breeds found in the United States, with the bulls weighing about 2,200 pounds and the cows about 1,500 pounds. In production, Holstein cows average larger quantities of milk with a lower percentage of butterfat than many cows of other breeds. The average butterfat test for their milk is about 3.4 per cent.

Jersey. The Jersey breed is second only to the Holstein in numbers in the United States. The Jerseys were brought to this country about 1800 from

Fig. 9-10. Holstein bull, 3H629 Limestone Standout Strephon 1597697, with a 1983 Predicted Difference of +$206, +1,890 pounds of milk, −0.15 per cent butterfat, and +45 pounds of butterfat. An excellent bull with 8,551 daughters in 2,485 herds. (Courtesy, Eastern AI Cooperative, Inc.)

the Island of Jersey in the English Channel. Jerseys are very versatile—they are good grazers, they tolerate hot temperatures and do well in cool climates. They mature early and have a long life. Their color varies from a light to a dark fawn, with or without white markings. Jersey cows are the smallest in size of all the important dairy breeds, with bulls weighing about 1,500 pounds and cows about 1,000 pounds. While they produce less milk than other breeds of dairy cows, their percentage of butterfat is higher than for any other breed. The average test for butterfat is about 4.9 per cent.

Guernsey. The Guernsey breed was imported to the United States about 1831 from the Island of Guernsey in the English Channel. Guernseys vary in color from a rich golden or light fawn to a reddish fawn, with varying degrees of white markings. The cows are small, but somewhat larger than Jerseys in size, with the bulls weighing about 1,800 pounds and the cows about 1,100. They average slightly larger quantities of milk than Jerseys and slightly lower butterfat, about 4.6 per cent.

Ayrshire. The Ayrshire breed, whose native home was Ayr County, Scotland, first arrived in the United States about 1822 by way of Canada. The accepted breed color is white and red, with the red ranging from light to deep cherry to mahogany or brown. Mature bulls weigh about 1,850 pounds and cows from 1,200 to 1,600 pounds.

Brown Swiss. The Brown Swiss breed originated in Switzerland and was brought to this country in 1869. The cows range from a light to a dark

Characteristics of Farm Animals

Fig. 9-11. Basil Lucy Minnie Pansy *(top)*, appraised Excellent—91—is the current leading living lifetime producer for milk and butterfat for the Jersey breed. To date, she has produced 264,857 pounds of milk and 12,857 pounds of butterfat. She is owned by William H. Diley & Sons, Canal Winchester, Ohio. Rocky Hill Favorite Deb *(bottom)*, appraised Excellent—91—is the current milk and butterfat champion of the Jersey breed for one lactation with 35,881 pounds of milk and 1,923 pounds of butterfat. She is owned by Walebe Farms, Inc., Collegeville, Pennsylvania. (Courtesy, The American Jersey Cattle Club)

Fig. 9-12. Cleverlands Nances Judy Jean *(top)*, appraised Excellent—92—is the only Guernsey cow ever to have produced over 30,000 pounds of milk for five lactations. She is owned by Mr. Ray Orisio, Weedburn, Oregon. Her Cow Performance Index is eighteenth in the breed (+367). Her daughter, Lone Palm TH Jeans Judy *(bottom)*, appraised Very Good—88—has the highest Cow Performance Index of the Guernsey breed (+464). (Courtesy, The American Guernsey Cattle Club)

Fig. 9-13. Brown Swiss bull, 7B638 Ken IR Elegant Reward 171140, with a 1983 Predicted Difference of +$208, +1,272 pounds of milk, +0.15 per cent butterfat, and +71 pounds of butterfat. This bull has 99 daughters in 57 herds. (Courtesy, Eastern AI Cooperative, Inc.)

brown in color. It is one of the heaviest of the dairy breeds, with the bulls weighing about 2,000 pounds and the cows about 1,400 pounds.

Other breeds. Other dairy breeds found in smaller numbers in the United States include the Red Danish, Dutch Belted, and Red Poll. The Milking Shorthorn breed is discussed under the section that follows.

Breeds of beef cattle

The breeds of beef cattle commonly found in most communities are the Shorthorn, Hereford, and Aberdeen Angus. Their descriptions, as well as those of other breeds, follow.

Shorthorn. The Shorthorn breed has been present in the United States since 1783. Its native home was Durham and York counties in England. Shorthorns are widely distributed in the United States. The individual animals are red, white, or any combination of these, such as spotted or roan. Mature bulls weigh about 1,800 to 2,000 pounds, and mature cows weigh from 1,100 to 1,500 pounds. Polled Shorthorns have the same basic characteristics, except horns. Milking Shorthorns are able to give considerable milk, in addition to producing fairly good beef. They are often referred to as *dual-purpose* cattle because of these qualities. Milking Shorthorn bulls usually weigh between 2,000 and 2,500 pounds, while the cows usually weigh between 1,400 and 1,800 pounds.

Fig. 9-14. Two excellent-type Milking Shorthorn cattle. Kingsdale Lady Elegance 42nd 373834 *(top)* produced 25,270 pounds of milk and 792 pounds of butterfat in 365 days at twice daily milking. Korncrest Pacesetter 368707 *(bottom)* is the breed's highest PTI sire. (Courtesy, American Milking Shorthorn Society)

Characteristics of Farm Animals

Hereford. Herefords, whose native home is Hereford County, England, were first brought to the United States about 1817. Herefords are the most numerous of all beef cattle in the United States. Like Shorthorns, they are red and white, but their markings are quite uniform. The head is always white; thus, they are often called "white faces." White is usually found on the underline, along the back, on the switch, and on the lower parts of the legs. The remainder of the body is a rich shade of red. Herefords normally

Fig. 9-15. Hereford bull *(top)* and cow *(bottom)*. Herefords are the most numerous of the beef cattle breeds. (Courtesy, The American Hereford Association)

have horns; Polled Herefords have no horns. The average weight of Hereford bulls is 1,800 to 2,100 pounds; for cows, it is 1,100 to 1,500 pounds.

Aberdeen Angus. Cattle of the Aberdeen Angus (usually called Angus) breed were first imported from Aberdeen County, Scotland, in 1883 by a Kansas breeder. This breed is second only to the Hereford breed in numbers. Angus cattle can be identified by their short, black hair and absence of horns. However, since individual red cattle occasionally occur, some breeders have established Red Angus herds. Both Aberdeen Angus and Red Angus bulls weigh about 1,600 to 2,000 pounds; cows weigh about 1,000 to 1,400 pounds.

Brahman. The Brahman is the oldest of all breeds of beef cattle. It has been bred in India for thousands of years. The American Brahman breed was developed in the southern part of the United States from 266 bulls and 22 females of humped cattle of several strains imported from India between 1854 and 1926. Brahmans are used many times to cross with other beef breeds, particularly in the southern states. They vary in color, from a very light gray to red to almost black. They can withstand heat better than most breeds and are resistant to flies, ticks, and certain diseases. The weight of the average bull usually ranges from 1,800 to 2,400 pounds. The average cow weighs about 1,200 to 1,800 pounds.

Santa Gertrudis. The Santa Gertrudis is a relatively new breed of cattle, having been developed on the King Ranch in Texas. It originated from crosses of Brahman bulls on Shorthorn cows, followed by careful selection and breeding within these crosses. The Santa Gertrudis is cherry red. Santa Gertrudis cattle are very large—bulls may weigh between 2,200 and 2,600 pounds, and cows between 1,400 and 1,800 pounds.

Charolais. The Charolais is one of the oldest of the French breeds. In 1930, a Mexican imported 2 bulls and 10 heifers. In 1931, he imported 8 more bulls and 29 females. The first Charolais to be imported to the United States arrived at the King Ranch in Texas in 1936. In 1981, there were over 32,000 registered Charolais in the United States. Charolais cattle are creamy white. Like the Santa Gertrudis, Charolais cattle are quite large, with bulls weighing between 2,000 and 2,500 pounds, and cows between 1,250 and 2,000 pounds.

Red Poll. The Red Poll, like the Milking Shorthorn, is a popular dual-purpose beef breed. Its native home was the eastern coastal area of England. It was brought to the United States in 1873. Some breeders use it in a typical cow-calf herd for meat production. It varies in color from light red to a very dark red and has a white tail switch. Mature bulls weigh somewhere between 2,000 and 2,200 pounds, and cows between 1,430 and 1,700 pounds.

Other breeds. The Devon, also from England, has a deep red to pale chestnut color. The Galloway, from Scotland, is black with soft, wavy hair.

Fig. 9-16. A 2,350-pound Angus bull *(top)* and a 1,350-pound Angus cow with her calf *(bottom)*. These cattle are outstanding examples of the Angus breed. (Courtesy, American Angus Association)

Fig. 9-17. Brahman bull *(top)* and cow *(bottom)*. (Courtesy, American Brahman Breeders Association)

Characteristics of Farm Animals

The Scotch Highland, from Scotland, may be any of several colors—black, red, brindle, yellow, or silver. The Beefmaster, Brangus, and Charbray were all developed in the United States. The Beefmaster is usually red, but the color may vary. The Brangus may be a light gray or red, but the predominant color is black. The Charbray is usually a creamy white. Other notable breeds are the Simmental from Switzerland; Limousin, Maine-Anjou, and Blonde d'Aquitaine from France; and Chianina from Italy.

Fig. 9-18. Excellent-type Charolais bull *(top)* and cow *(bottom)*. (Courtesy, American-International Charolais Association)

Breeds of hogs

In the past, the common breeds of hogs were divided into the lard type and the bacon type. This well-defined classification has almost disappeared due to the efforts of the various breed association to produce a meat-type animal, somewhere between these two types, that would yield meat more to the liking of the consumer.

Fig. 9-19. Ideal Duroc boar *(top)* and gilt *(bottom)*. (Courtesy, United Duroc Swine Registry)

Characteristics of Farm Animals

Of the breeds, the most common are the Duroc, Poland China, Chester White, and Hampshire. The Berkshire and the Spotted are also found in some communities. The principal difference between the first four breeds is color. The Duroc is solid red; the Poland China is all black, except for small amounts of white on the feet, head, and other parts of the body; the Chester White is all white; and the Hampshire is black with a white belt around the body at the shoulders. The Berkshire is marked very similarly to the Poland

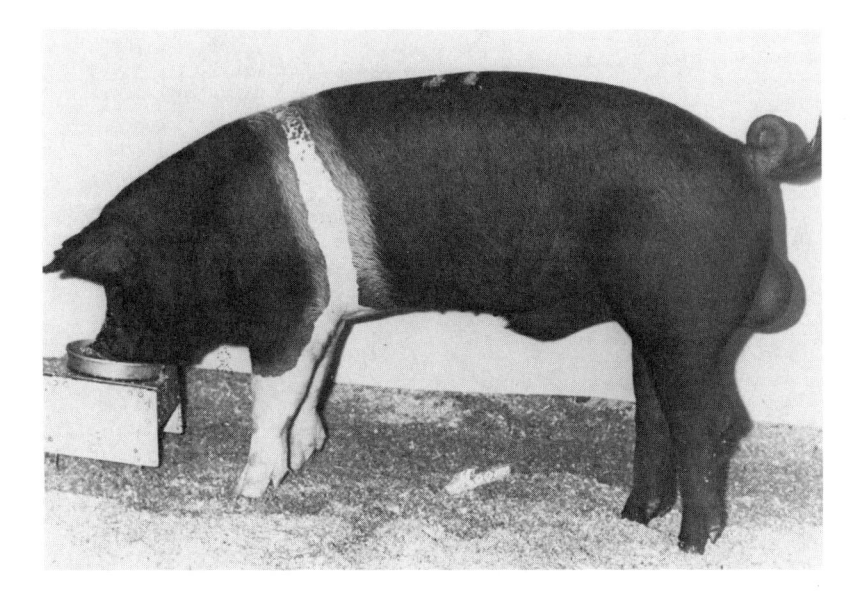

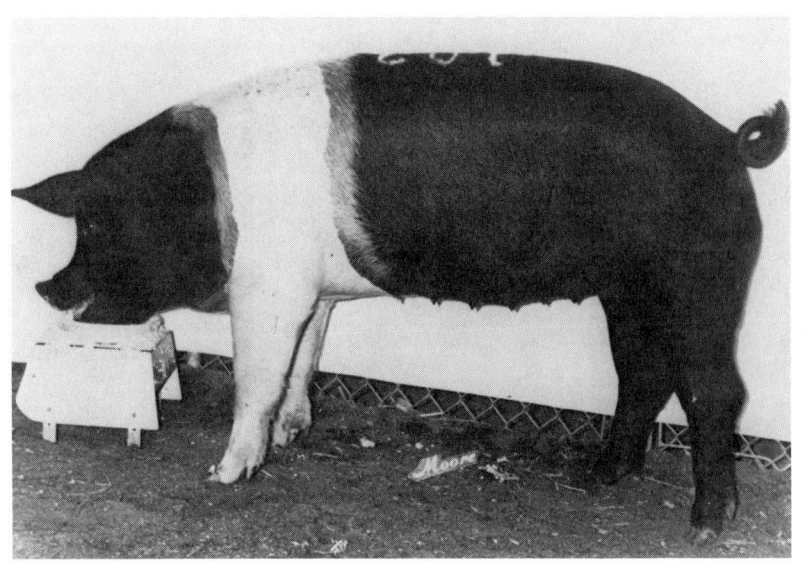

Fig. 9-20. Modern-type Hampshire boar *(top)* and gilt *(bottom)*. (Courtesy, Hampshire Swine Registry)

China, but it may be identified by the short, upturned nose and erect ears. The Spotted is spotted black and white.

Other breeds are the Yorkshire and the Tamworth. The Yorkshire is white, like the Chester White, and the Tamworth is red, like the Duroc. However, unlike the Chester White and the Duroc, they have long, narrow bodies and erect ears. The Tamworth has a long, narrow face, while the Yorkshire has a dished face.

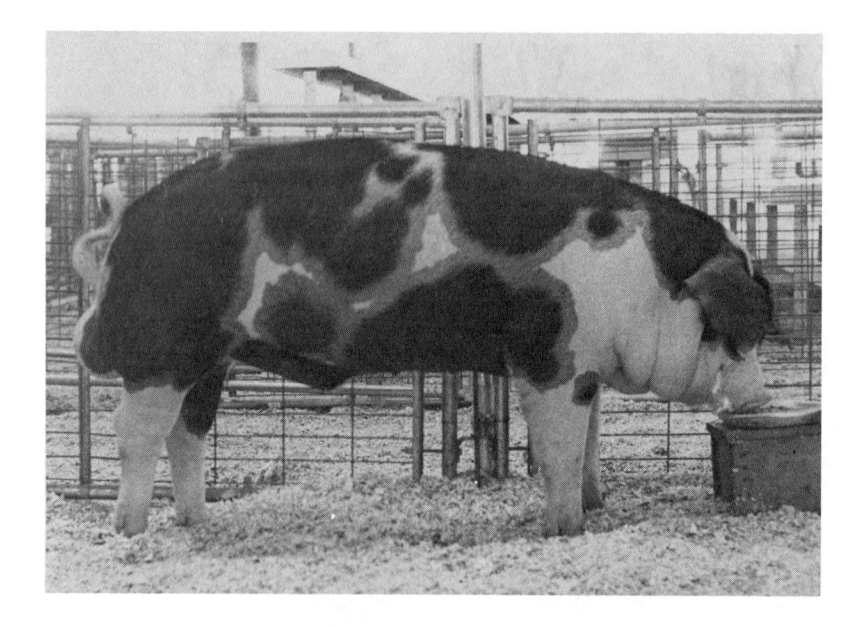

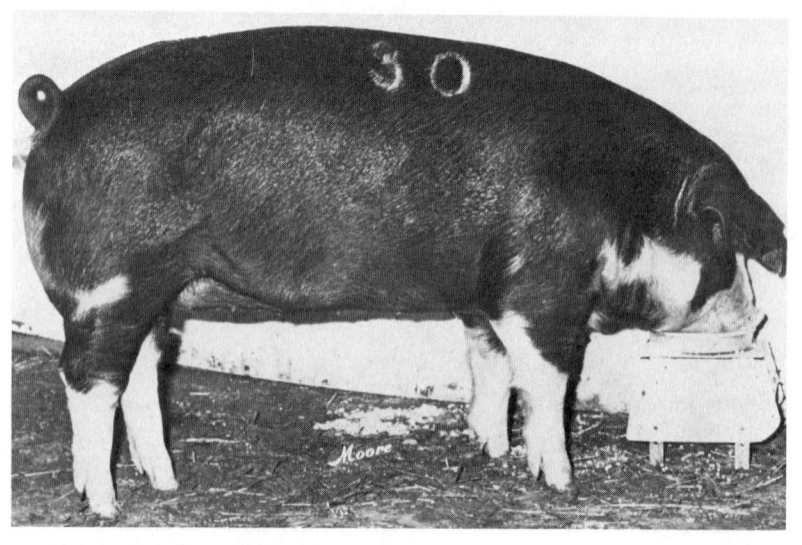

Fig. 9-21. Outstanding Spotted boar *(top)* and gilt *(bottom)*. (Courtesy, National Spotted Swine Record, Inc.)

Characteristics of Farm Animals

The native homes of these common breeds are as follows: (1) Duroc, New York and New Jersey; (2) Poland China, Butler and Warren counties, Ohio; (3) Chester White, Chester County, Pennsylvania; (4) Hampshire, probably in Hampshire County, England; (5) Berkshire, Berkshire County, England; (6) the Spotted, chiefly in Indiana; (7) Yorkshire, Yorkshire County, England; and (8) Tamworth, Stafford County, England.

In recent years, several new U.S. breeds have been developed through

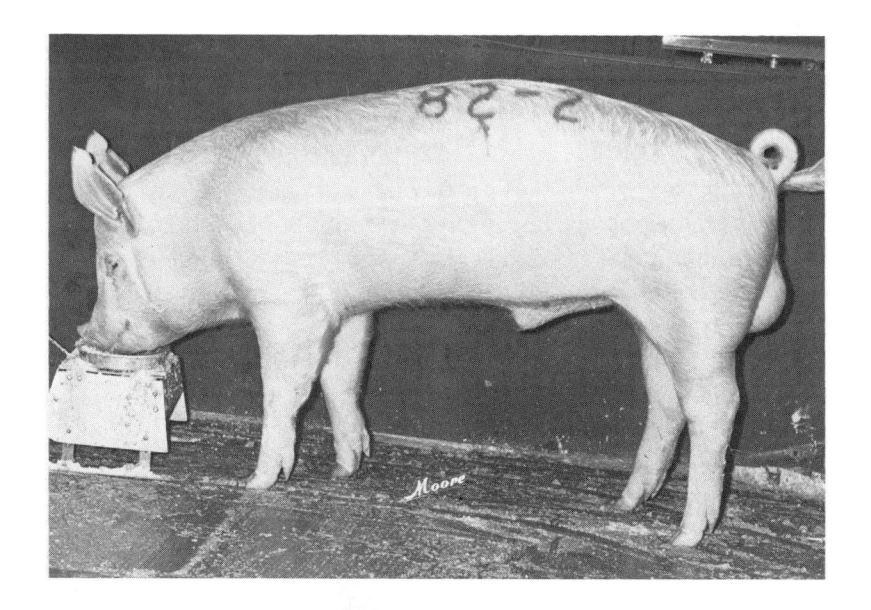

Fig. 9-22. Modern-type Yorkshire boar *(top)* and gilt *(bottom)*. (Courtesy, American Yorkshire Club, Inc.)

crosses of various breeds followed by careful selection. These new breeds include Minnesota No. 1, Minnesota No. 2, Minnesota No. 3, Beltsville No. 1, Beltsville No. 2, Maryland No. 1, San Pierre, and Palouse.

Nearly all hogs now raised commercially are crossbred. The main reason for raising any purebred hogs is to supply boars for crossbreeding.

Breeds of horses

Horses, in general, may be divided into two groups or types: (1) light horses and (2) heavy or draft horses. Light horses are used for pulling carriages, racing, riding, and other purposes where a light load is involved and either speed or style is desired. Draft horses are used for pulling heavy loads at a walk.

The number of draft horses on farms has decreased greatly during recent years. Draft breeds are harder to differentiate between than the breeds of most other kinds of farm animals. However, if we observe closely, we can soon learn to become reasonably skillful in identifying the breeds. Breeds of draft horses include the Percheron, the Belgian, the Shire, and the Clydesdale. Representatives of these breeds are not numerous today in the United States.

Fig. 9-23. An interesting contrast in breeds of horses—a Shetland Pony colt and a Percheron stallion. (Courtesy, Iowa State University)

The native homes of the common draft breeds are as follows: (1) Percheron, the La Perche district of France; (2) the Belgian, Flanders region in Belgium; (3) the Clydesdale, southern Scotland, through which flows the Clyde River; and (4) the Shire, Lincoln and Cambridge counties in England, the name probably coming from the word *shire,* which refers to the counties of England.

Light horses are kept for pleasure purposes both by urban and rural people. They are also used on farms and ranches for herding cattle and sheep and for inspecting these animals over large areas. The breeds of light horses include the Thoroughbred, the American Trotter or Standardbred, the American Saddlebred Horse, the American Quarter Horse, the Tennessee Walking Horse, the Welsh Pony, the Hackney Pony, and the Shetland Pony.

The Thoroughbred is used as a fast running horse with a jockey in the saddle. It originated in England. The American Trotting Horse, or Standardbred, developed in the United States, is used primarily as a harness racehorse, hitched to a light cart, or sulky. The American Saddlebred Horse, Quarter Horse, and Tennessee Walking Horse are breeds that also originated in the United States. The American Saddlebred Horse is primarily used as a pleasure horse, but it can also be used as a stock horse because it is capable of travelling long distances. For horse shows and pleasure riding, it is often trained for five gaits—the walk, trot, canter, rack, and

Fig. 9-24. The Appaloosa in action at the Sixteenth National Appaloosa Horse Show in Boise, Idaho. (Courtesy, Appaloosa Horse Club, Inc.)

either the running walk, fox trot, or show pace. The Quarter Horse, also used as a pleasure and a stock horse, got its name because of its great speed for a quarter of a mile or less. The Tennessee Walking Horse got its name from the running walk, which made it suited for use on large plantations. The Shetland Pony, whose native home is the Shetland Islands, is the smallest breed of horses. The Appaloosa is a riding horse of great endurance. It is a descendant of Spanish horses and was first bred in the United States by the Nez Percé Indians in Idaho and Washington. The Appaloosa has very distinctive characteristics: the eyes are encircled with white; mottled skin is prominent about the nostrils; the hoofs are vertically striped; and the coat is spotted in various patterns. Sometimes the entire coat is spotted, and sometimes there is just a frosting of white over the hips. The Appaloosa is used for stock work, pleasure and trail riding, field hunting, jumping, rodeo work, endurance and distance riding, cutting and reining, combined training and dressage, and racing.

Many of the horses used on ranches are descendants of the wild horses that originated from horses brought to this country by the Spaniards about four centuries ago. They are often called *mustangs* or *bronchos*. Frequently, they are crossed with some of the light breeds.

Breeds of sheep

Breeds of sheep may be more difficult to identify than the breeds of some other kinds of livestock. However, this should make the art of identification all the more fascinating.

Breeds of sheep may be divided into two groups: *mutton type* and *fine-wool type*. The mutton breeds produce wool of considerable value, but they have been especially improved from the standpoint of the quality of meat obtained from them. The breeds of the fine-wool type are used for meat purposes also, but they have been developed especially for the production of heavy fleeces with fine fibers. The mutton-type breeds may be further divided into the long-wool class and the medium-wool class, in accordance with the wool characteristics, indicated by the names of the classes.

The medium-wool mutton types are usually found outside the range states. These include the Shropshire, Oxford, Hampshire, Dorset, Southdown, and Cheviot breeds. Some of the mutton long-wool breeds, such as the Lincoln and Cotswold, may also be found. However, on the western ranges, breeds of the medium- and fine-wool types, such as the Merino, Rambouillet, Corriedale, and Columbia, predominate.

The Shropshire, Hampshire, Oxford, and Dorset, as well as some other breeds, are natives of counties of England with corresponding names. These breeds, along with some others, are frequently called the "down" breeds because they were developed in the regions of low hills, called downs, in central and southern England. The Southdown, for example, was developed in a region of these hills in southern England known as the South Downs.

Some of the medium-wool mutton types—the Cheviot, Columbia, Corriedale, Dorset, Hampshire, Montadale, Oxford, Shropshire, Southdown, and Suffolk—are described in the following paragraphs.

Cheviot. The Cheviot is an English breed. It is a very stylish and beautiful breed, characterized by a black nose. It is one of the smaller breeds, with rams averaging about 175 pounds and ewes 125 pounds. Its fleece is quite long and dense, 4 to 5 inches a year, but light, with rams shearing 8 to 10 pounds and ewes 6 to 7 pounds a year.

Columbia. The Columbia, a crossbreed of Lincoln rams and Rambouillet ewes, was developed around 1912 at the King Ranch in Wyoming. Bred to fit the rigorous mountain range conditions, it has longer legs than most mutton breeds and a more rugged constitution. It is a large breed, with rams weighing about 200 to 275 pounds and ewes about 150 to 200 pounds. Its wool is 4½ to 5 inches in length, and its fleece weighs 12 pounds or more.

Corriedale. The Corriedale is a dual-purpose sheep, bred for both wool and mutton. Crossbred from Lincoln and Leicester rams and Merino ewes, it was brought to the United States in 1914 from New Zealand. It is white with dark points. Rams weigh about 180 to 225 pounds and ewes 125 to 185 pounds. Rams will produce an average of 15 to 25 pounds of wool each year, and ewes will produce about 10 to 15 pounds. The fleece is dense and of good length with a well-defined crimp.

Fig. 9-25. Lambs being judged in the show ring. (Courtesy, New York State College of Agriculture and Life Sciences, Cornell University)

Fig. 9-26. Stylish Polled Dorset ram *(top)* and ewe *(bottom)*. (Courtesy, Continental Dorset Club, Inc.)

Dorset. Both the Horned and the Polled Dorsets are English breeds. The Horned Dorset ram has a large, corkscrew-shaped horn; the ewe has a much smaller horn that turns outward and downward. Both the Horned and the Polled Dorset breeds are fairly large, with rams weighing 175 to 250 pounds and ewes 125 to 175 pounds. The fleece is light—rams produce 8 to 12 pounds a year and ewes 6 to 8 pounds a year.

Hampshire. The Hampshire, originating in England, dates back to the 1880's for flocks in the United States. It can be identified by its black or dark brown head, face, and legs. It is one of the most popular of the mutton types, as well as being one of the largest. The rams exceed 275 pounds in weight, while the ewes weigh 180 to 200 pounds. A fleece yields on the average of 7 to 8 pounds.

Montadale. The Montadale, a crossbreed of Columbia ewes and Cheviot rams, was developed in the United States for mutton production. Its distinguishing characteristic is that its head is free of wool. Rams weigh 200 to 250 pounds and ewes 150 to 200 pounds. The average fleece weight for a ewe is about 10 to 20 pounds.

Oxford. The Oxford, an English breed, can be identified by its dark gray or brown face, ears, and legs. It is also noted for its huge size—rams weigh 250 to 350 pounds and ewes 175 to 250 pounds. Its heavy fleece weighs 10 to 15 pounds, with fibers 3 to 5 inches long.

Fig. 9-27. A small flock of excellent-type Hampshire sheep. (Courtesy, American Hampshire Sheep Association)

Shropshire. The Shropshire, an English breed, is distinguished by having a dark face and a complete covering of wool from its nose to feet. It is medium in size, with rams weighing 175 to 250 pounds and ewes 135 to 175 pounds. The average fleece yield is 8 to 10 pounds.

Southdown. The Southdown is one of the smallest and oldest English breeds. It has excellent mutton qualities. Rams weigh 165 to 225 pounds and ewes 125 to 160 pounds. The year's fleece yields 5 to 8 pounds.

Suffolk. The Suffolk has very distinguishing characteristics. Its polled head, ears, and legs are jet black, covered by short, fine hair. It is the largest mutton-type medium-wool English breed, with the rams weighing 250 to 325 pounds and the ewes 160 to 250 pounds.

Notable long-wool mutton breeds are the Cotswold, Leicester, Lincoln, and Romney. These are described as follows.

Cotswold. Some people believe that the Cotswold is the oldest of all breeds of sheep, dating back nearly 600 years in England. Its white wool may have gray specks or a bluish cast. Rams weigh about 250 to 300 pounds and ewes 175 to 225 pounds. The wool has long, wavy ringlets or curls with strands from 8 to 14 inches in length. A year's growth of wool will weigh 10 to 20 pounds.

Leicester. There are two types of Leicesters—the *English* and the *Border*. They both originated in England. The English Leicester is the smallest of the long-wool breeds. Rams usually weigh 250 pounds or more and ewes 180 pounds or more. The face and legs are covered with hair. The fleece shears an average of 10 pounds. The Border Leicester is larger and longer, has less wool on the face and legs, has whiter hair, and has a shorter fleece than the English type. Its fleece will usually weigh 8 to 10 pounds.

Lincoln. The Lincoln is an English breed of sheep with several notable characteristics. It is the largest breed of sheep—rams weigh 250 to 375 pounds and ewes 200 to 275 pounds. It is covered with a long, coarse fleece, ranging from 10 to 12 inches in length, which is the heaviest fleece of any mutton breed. A ram will shear 20 pounds or more and a ewe 12 to 16 pounds.

Romney. The Romney differs from the other English long-wool breeds in several ways. It is smaller—rams weigh about 225 to 250 pounds and ewes 175 to 200 pounds. It is more rugged; thus, it does well in cool climates. It is shorter-legged. Its fleece is denser and finer—it is usually 4 to 6 inches long and yields 12 to 18 pounds annually.

The fine-wool breeds include the American and Delaine Merino and the Rambouillet, which are described below.

Merino. There are many different types of Merino sheep, but the most popular breeds in the United States are the American and the Delaine

Characteristics of Farm Animals

Fig. 9-28. Suffolk ram *(top)* and ewe *(bottom)*. (Courtesy, National Suffolk Sheep Association)

Merino. Both types originated in Spain. Most rams have horns, although there are some polled, while the females are all polled. Both types are noted for having fleece of outstanding quality. American Merinos are separated into two classes—Class A and Class B, which are lacking in mutton qualities. Class A Merino rams weigh 130 to 170 pounds and ewes 90 to 130 pounds. Their wool usually measures 1½ to 2 inches in length and usually shears 25 to 30 pounds for rams and 15 to 25 pounds for ewes. Class B Merino rams usually weigh 150 to 175 pounds and ewes 90 to 140 pounds. Their fleece usually measures 2½ to 3 inches in length. Rams usually produce 25 pounds in a year, and ewes 15 pounds. Class C, or Delaine, Merinos produce good mutton, unlike the American Merinos. They have shorter legs and weigh more than Class A and Class B sheep. Rams should weigh 150 to 225 pounds and ewes 90 to 150 pounds. Their fleece is also denser—3 to 5 inches in length; it yields 10 pounds for rams and 11 pounds for ewes.

Rambouillet. The Rambouillet had its origins in France and Germany from Spanish Merino parent stock. Rams may be polled or horned. Ewes are hornless. It is the largest of the fine-wool breeds, with rams weighing 225 to 275 pounds and ewes 140 to 200 pounds. The fleece is dense, with fibers 3½ inches in length. A year's growth for a ram is 15 to 25 pounds of wool; for a ewe, it is 10 to 18 pounds.

Other breeds. These include the carpet–wool type Black-faced Highland from Scotland, which is a horned breed, and the fur-type Karakul from Asia, best known for its black or brown lamb pelts, which are used to make "Persian lamb" furs. Karakul rams have horns, but the ewes are hornless. Some of the more recently developed breeds are the Romeldale, the Southdale, the Polypay, the Tailless, and the Targhee.

Breeds of chickens

Many breeds of chickens have been developed. The breeds differ from each other primarily in body shape and size. Within some of these breeds are several varieties, which are differentiated chiefly by color of feathers and, in some cases, kind of comb. The standards for the different breeds are set by the American Poultry Association.

Each breed belongs to a class that indicates where the breed originated—the American Class, the Asiatic Class, the English Class, or the Mediterranean Class. The American Class includes the Plymouth Rocks, Wyandottes, Rhode Island Reds, Rhode Island Whites, New Hampshires, Jersey Giants, Dominiques, Javas, Lamonas, Buckeyes, Chanticlers, and Hollands. The Asiatic Class has the Brahmas, the Cochins, and the Langshans. In the Mediterranean Class are the Leghorns, Minorcas, Spanish, Blue Andalusians, Anconas, and Buttercups. The English Class includes the Dorkings, Redcaps, Cornish, Orpingtons, Sussex, and Australorps.

Fig. 9-29. A flock of Rhode Island Red chickens. (Courtesy, U.S. Department of Agriculture)

Most of the purebred chickens in the United States belong to one of four breeds. These breeds are New Hampshire, Plymouth Rock, Rhode Island Red, and Leghorn. In these breeds, the skin and shanks are yellow. The typical color of New Hampshires is chestnut red with some black. There are no separate varieties of New Hampshires. There are several varieties of Plymouth Rocks, but the White is the most common variety. The most common variety of Rhode Island Reds is the Single Comb. Members of this breed have plumage that is brownish red with some black. The most common variety of Leghorns is the Single Comb White. This breed is smaller than the other three named above, and it is kept primarily for egg production. The other three breeds are sometimes called *general purpose* because they are usually kept for both eggs and meat. Leghorns lay white-shelled eggs; the other breeds indicated above lay brown-shelled eggs.

Many of the chickens raised in recent years have been developed from special breeding stock in which crossbreeding and other special methods of mating have been used to produce strains for specific purposes. In producing them, some of the older breeds have been used. Various breeding methods have made it possible to secure especially efficient strains for both meat and eggs, depending on the purpose the poultry raiser has in mind.

The Plymouth Rocks, New Hampshires, and Rhode Island Reds are breeds of U.S. origin. Plymouth Rocks were developed primarily in the

Fig. 9-30. The famous Beltsville Small White turkeys. (Courtesy, U.S. Department of Agriculture)

state of Massachusetts and probably derived their name from the historic rock in that state. The other two breeds were developed in the states whose names they bear. The native home of the Leghorn is Italy; the name is derived from the port of Leghorn.

Chickens are raised in flocks ranging in size from a few birds to thousands of birds. The particular breeds that predominate differ in various states and sections within states.

Some farmers in the United States raise other kinds of poultry, such as ducks, turkeys, and geese. These are raised principally for meat. Each of these kinds of poultry has several breeds from which to choose.

Merely selecting a breed of farm animal does not insure success in livestock production. Success depends on selecting profitable animals within a breed, providing them with good care, feeding them properly, and keeping them healthy. These phases are discussed in the chapters that follow.

SUGGESTED ACTIVITIES

1. Make a list of articles of your clothing that are made entirely or partially of animal products. From what kind of animal was each derived?

2. Make a list of the foods you ate the past week. Which of these came from animal sources?

3. What kinds of farm animals are most important in your community? How do you account for these animals instead of other kinds being raised?

4. Study the kinds of domestic animals raised in some foreign country. What unusual animals do you find? What unusual uses are made of these and other animals?

5. Report to your class on the topic: "Animals I have owned and raised." Emphasize especially the interesting habits and traits that you have noticed.

Characteristics of Farm Animals

6. Make a study of the traits of some domesticated animals with which you are familiar. How do these animals protect their young and themselves? What type of social order seems to prevail for each kind of livestock?

7. While riding through the country, note the breeds of animals on the farms you pass. Make a note of the number of kinds and breeds of livestock that you see. To what extent do you note that some breeds are popular in certain communities? How many different breeds can you identify on one such trip?

8. Make a special study of the origin and important characteristics of your favorite breed of some kind of livestock. Who are the outstanding breeders of the present day? What are some of the noted sires and families of this breed? Where could you secure breeding stock of good quality?

9. Concepts for discussion:
   a. Animals provide clothing.
   b. Animals are a source of power and transportation.
   c. Animals are a source of food.
   d. Nearly every part of an animal can be used to produce some product.
   e. Every animal has characteristics that make it different from every other animal.
   f. Farm animals can bring a greater return for the crops that they consume than could be obtained if the crops were directly sold on the market.
   g. Much of the land in this country is best used for grazing by livestock.
   h. Livestock manure is useful for maintaining soil fertility.
   i. The use of horses and mules as a source of farm power is decreasing, while the use of horses for recreation is increasing.
   j. The regions of the United States differ with respect to the kinds of livestock raised.
   k. Animals have interesting habits, just as people do.
   l. Some animals appear to have a great deal of intelligence.
   m. The kind of food an animal eats is determined largely by the kind of digestive system it has.
   n. Some animals tend to stay together in flocks or herds, which makes them more manageable.
   o. Many animal characteristics have developed as a result of the conditions under which the animals grew.
   p. Animals have individual personalities by which they can be identified.
   q. Animals have various means of protection against the heat and cold.
   r. Each breed of livestock has identifiable physical characteristics that are transmitted to its offspring.
   s. Breeds of livestock should be selected for the purpose for which they are best adapted.
   t. Animals of the same breed transmit their breed characteristics to their offspring.
   u. The pedigree of an animal refers to its ancestors.

v. Animals that have digestive systems adapted to handling roughages and that regurgitate their feed for chewing a second time are called ruminants. Other animals are called nonruminants.

BREED REGISTRY ASSOCIATIONS

Dairy cattle associations

American Guernsey Cattle
  Club, The
2105J South Hamilton Road
P.O. Box 27410
Columbus, Ohio 43227

American Jersey Cattle Club,
  The
2105J South Hamilton Road
P.O. Box 27410
Columbus, Ohio 43227

American Milking Shorthorn
  Society
1722JJ South Glenstone Avenue
Springfield, Missouri 65804

Ayrshire Breeders' Association
2 Union Street
Brandon, Vermont 05733

Brown Swiss Cattle Breeders
  Association of U.S.A., The
P.O. Box 1038
Beloit, Wisconsin 53511

Dutch Belted Cattle Association
  of America, Inc.
Box 358
Venus, Florida 33960

Holstein-Friesian Association of
  America
P.O. Box 808
Brattleboro, Vermont 05301

Beef cattle associations

American Angus Association
3201 Frederick Boulevard
St. Joseph, Missouri 64501

American Brahman Breeders
  Association
1313 La Concha Lane
Houston, Texas 77054

American Breed Association,
  Inc.
306 South Avenue A
Portales, New Mexico 88130

American Chianina Association
Box 159
Blue Springs, Missouri 64015

American Hereford Association,
  The
715 Hereford Drive
Kansas City, Missouri 64101

American-International Charolais Association
1610 Old Spanish Trail
Houston, Texas 77054

American Maine-Anjou
  Association
564 Livestock Exchange
  Building
1600 Genesee Street
Kansas City, Missouri 64102

American Murray Grey
  Association
1222 North 27th Street, Suite 105
Billings, Montana 59107

American Polled Hereford
  Association
4700 East 63rd Street
Kansas City, Missouri 64130

American Red Poll Association
Box 35519
Louisville, Kentucky 40232

American Scotch Highland
  Breeders' Association
Box 81
Remer, Minnesota 56672

Characteristics of Farm Animals

American Shorthorn
  Association
8288 Hascall Street
Omaha, Nebraska 68124

American Simmental
  Association, Inc.
1 Simmental Way
Bozeman, Montana 59715

Beefmaster Breeders Universal
GPM South Tower, Suite 350
800 N.W. Loop 410
San Antonio, Texas 78216

Devon Cattle Association, Inc.
Box 628
Uvalde, Texas 78801

International Brangus Breeders
  Association, Inc.
9500 Tioga Drive
San Antonio, Texas 78230

Red Angus Association of
  America
4201 I-35 North
Denton, Texas 76201

Santa Gertrudis Breeders
  International
P.O. Box 1257
Kingsville, Texas 78363

Texas Longhorn Breeders
  Association of America
3701 Airport Freeway
Fort Worth, Texas 76111

Swine associations

American Berkshire Association
601 West Monroe Street
Springfield, Illinois 62704

American Landrace
  Association, Inc.
Box 647
Lebanon, Indiana 46052

American Yorkshire Club, Inc.
P.O. Box 2417
West Lafayette, Indiana 47906

Chester White Swine Record
  Association
Box 228
Rochester, Indiana 46975

Hampshire Swine Registry
1111 Main Street
Peoria, Illinois 61606

Inbred Livestock Registry
  Association
Route 4, Box 207A
Noblesville, Indiana 46060

National Hereford Hog Record
  Association
Route 1, Box 37
Flandreau, South Dakota 57028

National Spotted Swine Record,
  Inc.
110 West Main Street
Bainbridge, Indiana 46105

Poland China Record
  Association
368 West Douglas, Box B
Knoxville, Illinois 61448

Tamworth Swine Association
R.R. 2, Box 36
Winchester, Ohio 45697

United Duroc Swine Registry
1803 West Detweiller Drive
Peoria, Illinois 61615

Light horse associations

American Morgan Horse
  Association, Inc.
Oneida County Airport
  Industrial Park
Box 1
Westmoreland, New York 13490

American Paint Horse
  Association
P.O. Box 18519
Fort Worth, Texas 76118

American Quarter Horse
  Association
2736 West 10th Street
Amarillo, Texas 79168

American Saddlebred Horse
  Association
929 South Fourth Street
Louisville, Kentucky 40203

Appaloosa Horse Club, Inc.
P.O. Box 8403
Moscow, Idaho 83843

Arabian Horse Registry of
America, Inc.
3435 South Yosemite Street
Denver, Colorado 80231

Palomino Horse Association,
Inc., The
P.O. Box 324
Jefferson City, Missouri 65102

Pinto Horse Association of
America, Inc.
7525 Mission Gorge Road,
Suite C
San Diego, California 92120

Pony associations

American Shetland Pony Club
P.O. Box 435
Fowler, Indiana 47944

Pony of the Americas Club, Inc.
1452 North Federal
P.O. Box 1447
Mason City, Iowa 50401

Welsh Pony Society of America,
Inc.
Box 2977
Winchester, Virginia 22602

Draft horse associations

Belgian Draft Horse
Corporation of America
P.O. Box 335
Wabash, Indiana 46992

Clydesdale Breeders
Association of the United
States
Route 1, Box 131
Pecatonica, Illinois 61063

Percheron Horse Association of
America
P.O. Box 141
Fredericktown, Ohio 43019

Sheep associations

American Cheviot Sheep
Society
R.R. 1, Box 100
Clarks Hill, Indiana 47930

American Corriedale
Association, Inc.
P.O. Box 29C
Seneca, Illinois 61360

American Hampshire Sheep
Association
Route 10, Box 345
Ashland, Missouri 65010

American Shropshire Registry
Association, Inc.
P.O. Box 1970
Monticello, Illinois 61856

American Southdown Breeders'
Association
Route 4, Box 14B
Bellefonte, Pennsylvania 16823

American Suffolk Sheep Society
55 East 100 North
Logan, Utah 84321

Columbia Sheep Breeders
Association of America
P.O. Box 272
Upper Sandusky, Ohio 43351

Continental Dorset Club, Inc.
Box 506
Hudson, Iowa 50643

National Lincoln Sheep
Breeders' Association
R.R. 6, Box 24
Decatur, Illinois 62521

National Suffolk Sheep
Association
P.O. Box 324
Columbia, Missouri 65201

Goat associations

American Dairy Goat
  Association, Inc.
P.O. Box 865
Spindale, North Carolina 28160

American Goat Society, Inc.
Route 112, Box 112
DeLeon, Texas 76444

REFERENCES

Ensminger, M. E. *Animal Science*. Interstate Publishers, Inc., Danville, IL 61832.

Ensminger, M. E. *Animal Science Digest*. Interstate Publishers, Inc., Danville, IL 61832.

Ensminger, M. E. *Horses and Horsemanship*. Interstate Publishers, Inc., Danville, IL 61832.

Ensminger, M. E. *The Stockman's Handbook*. Interstate Publishers, Inc., Danville, IL 61832.

Hunsley, R. E., and W. M. Beeson. *Livestock Judging, Selection and Evaluation*. Interstate Publishers, Inc., Danville, IL 61832.

Lee, J. S., and M. E. Newman. *Aquaculture: An Introduction*. Interstate Publishers, Inc., Danville, IL 61832.

Vocational Agriculture Service. *Breeds of Beef Cattle*. VAS 1024a, Vocational Agriculture Service, College of Agriculture, University of Illinois, Urbana, IL 61801.

Vocational Agriculture Service. *Breeds of Sheep*. VAS 1049, Vocational Agriculture Service, College of Agriculture, University of Illinois, Urbana, IL 61801.

Vocational Agriculture Service. *Breeds of Swine*. VAS 1045, Vocational Agriculture Service, College of Agriculture, University of Illinois, Urbana, IL 61801.

Vocational Agriculture Service. *Dairy Cattle Breeds*. VAS 1046, Vocational Agriculture Service, College of Agriculture, University of Illinois, Urbana, IL 61801.

Chapter 10

Improving Herds and Flocks—Biotechnology

The animals raised on farms and ranches are descendants from animals that at one time existed in the wild state. Many changes have been made in these animals since they were first domesticated. Improving domesticated animals in ways that would make them more useful for the various purposes for which they are kept has been particularly important.

OBJECTIVES

1. Discuss the domestication of animals.
2. Determine the scientific basis for livestock improvement.
3. Explain how characteristics of chromosomes and genes determine makeup of body cells.
4. Apply biotechnology to improving herds and flocks.
5. Describe the four methods used for breeding stock selection.

How Were Animals First Domesticated and Improved?

The domestication of some kinds of animals was begun in the remote past, long before the time of any written records. While the exact time period is not definitely known, there is evidence from inscriptions and sketches on stone tablets, crude models of various sorts, and other remains of ancient civilizations indicating that some kinds of animals have been domesticated for several thousands of years. In the Old Testament of the Bible, many references are made to shepherds and their flocks and to milk and milk products.

Primitive people for a long period of time hunted animals in the wild state for food. As a more settled mode of life was adopted and the human population increased, it became more difficult to secure wild animals for food. This probably led to practices such as herding the animals together and driving them to favorable feeding and watering places. By such means, a more constant and reliable source of food was established. These and similar practices were the early beginnings of domestication.

Primitive people probably first learned that certain animals could be tamed readily when some of the young were captured and reared. It is thought that dogs were the first animals to be completely domesticated, probably as long ago as 10,000 B.C. It is believed that dogs were used as food during the early period of domestication, since bones of prehistoric dogs that show teeth and knife marks made by humans have been discovered. Probably later, primitive people found that dogs could track down other food animals and thus were worth more as providers of food than as a type of food. At some time during the early period of domestication, dogs were found to be useful as guards and companions. At a later date, dogs were used to pull loads, to assist in herding other animals, and to do other things.

All the important groups of farm animals (including sheep, cattle, horses, swine, and chickens) were undoubtedly domesticated several thousand years ago. The exact order of domestication is in doubt, but sheep were probably one of the first and horses one of the last. In Iraq, a prehistoric village, inhabited at least 7,000 years ago, has been uncovered. Evidences of domesticated animals are the bones of sheep, goats, swine, oxen, and horses. Another village, perhaps 10,000 years old, produced relics showing that the people were simple hunters of game animals with few or no domesticated animals. What do these findings seem to indicate?

Camels and elephants have been tamed and used for many centuries in some parts of the world. The supply of elephants for domestic use is still obtained largely through the capture of individuals in the jungles of India and Africa, the reason for this being that it is difficult to rear the young in captivity.

The turkey is a native of North America, and wild turkeys are still found in some places in this country. Turkeys are the national fowl for Thanksgiving dinners because the Pilgrims served wild turkey at the first Thanksgiving dinner in 1621. However, as a domesticated fowl, the turkey came to us in a curious manner. Soon after the discovery of North America, Spanish explorers took some of these birds from the continent back to Spain, and from Spain the birds spread into other parts of Europe. After being domesticated, some of them were brought back to the United States about 1800. One explanation of the name *turkey* is that by mistake people thought the bird came from the country of that name. Others believe the name was derived from the peculiar call of the turkey, which is something like "turk, turk." Commercial U.S. production dates back to the 1920's when domestic turkeys were few and expensive. These early turkeys were bred for colorful plumage rather than for meat. In the 1940's, the U.S. Department of Agriculture produced the Beltsville Small White, a broad-breasted small turkey designed for the family table. Turkeys are raised in every state, with North Carolina, Minnesota, and California producing the most in 1981. The wild turkey can still be found in 21 states: those south of the Ohio River, plus Pennsylvania, Maryland, Illinois, Missouri, Arkansas, Louisiana, Oklahoma, Texas, Colorado, New Mexico, and Arizona.

One of the interesting developments of recent years is the partial domestication of some kinds of fur-bearing animals such as the silver fox, mink, beaver, and more recently the chinchilla, which was imported from South America.

Before humans attempted to domesticate and improve animals, some types of improvement took place among these animals in the wild state. This improvement came about because of the hardships and dangers to which these animals were exposed. Within each species of animals, there were some individuals in each generation that were more rugged than others, some more fleet of foot, some better able to ward off the attacks of other animals, some better adapted to exist on certain types of feed, or some better fitted to withstand other types of hardships. As a result, those that were able to survive under these conditions became the parents of the next generation. By such a process, continuing for many generations, extremely hardy races of animals gradually came into existence, while some species became extinct. This process is called *natural selection*, because nature was responsible for it. It is sometimes referred to as "survival of the fittest." Why are these words appropriate?

The process of natural selection is illustrated well in the development of the horse during the period previous to domestication. Scientists believe that the horse descended from an animal about the size of a small dog that had several toes on each foot. During a period covering millions of years, many changes took place in the structure of these horse-like animals, the most important of which were the great increase in size and the development of one large, hoof-covered toe on each foot. It is interesting to know that some of the skeletal remains of these prehistoric horses, which have been helpful to scientists in tracing this development, have been found in some parts of the United States. It is believed that the first true horses developed on the North American continent millions of years ago. Some of these early horses found their way to other parts of the world when the North American continent was connected with Asia by a land bridge. For a long period of time the horse was extinct in North America. The early explorers re-established the horse in North America.

As people worked with animals in the early period of domestication, they probably noticed that some were better suited for specific purposes than others. They probably observed that some cows gave more milk than others; that some cows fattened more readily than others; that some horses were faster than others; that some sheep had better wool than others; and that all animals tended to vary in many other respects. After a period of time, people began to realize that many of the desirable characteristics, as well as some undesirable ones, tended to be transmitted from parents to offspring. As a result, a crude start was made in the selection of good individuals for breeding stock. Thus, some breeders selected the cows that gave the most milk; others selected the cows that fattened more easily. Some selected the large draft horses that could carry or pull heavy loads; others selected the light horses that could travel rapidly. Some selected sheep for their heavy fleeces; others selected them for their meat-producing qualities. In similar ways, definite attempts were made to develop other types of animals that could be adapted for other special purposes.

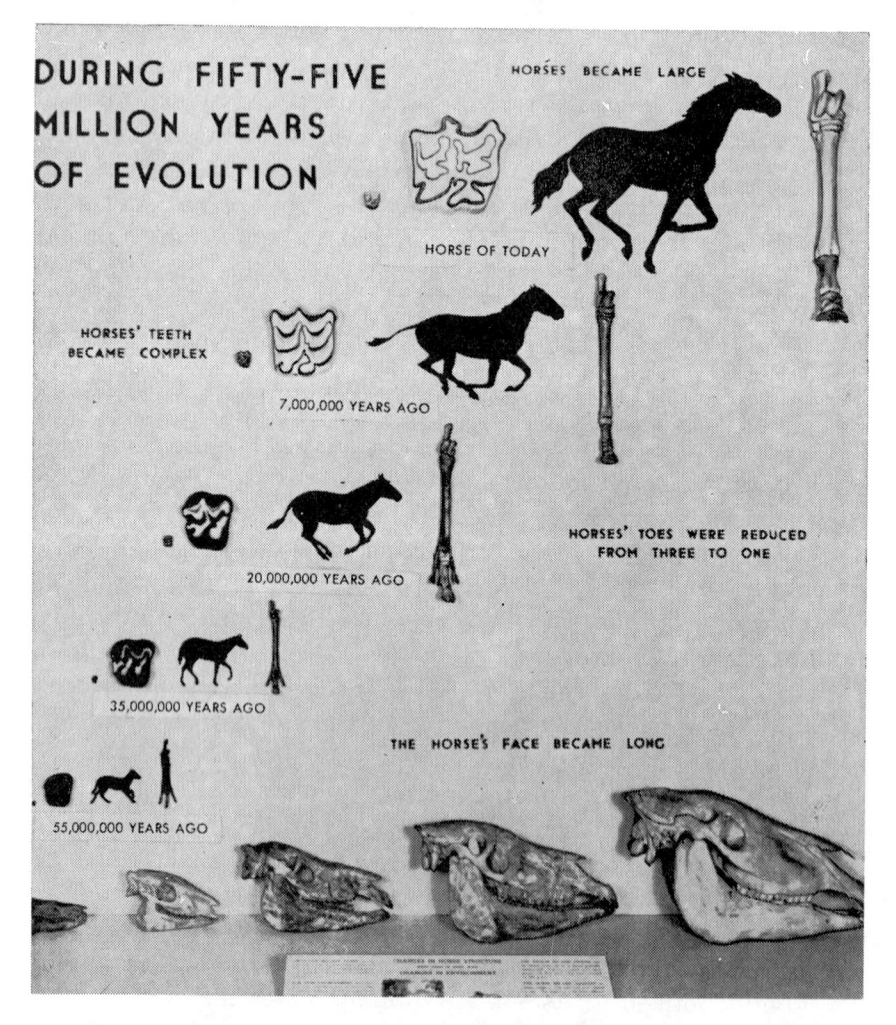

Fig. 10-1. Our domestic animals came from wild species that went through many changes. Some stages in the development of horses are shown, beginning with small three-toed animals 55 million years ago. (Courtesy, Chicago Museum of Natural History)

Sheep, cattle, pigs, horses, and goats were first brought to this country by the early Spanish explorers. These animals were raised in various settlements, often in connection with the early Spanish missions. Some of the stray horses, hogs, and cattle from these settlements reverted to the wild state. The longhorn cattle, which came up from Mexico into the territory later called Texas, originated from this source. The mustangs, or wild horses, of the West got started in the same way. The wild hogs found in some parts of the country are descendants of hogs first imported by early colonists.

In addition to the early importations from Spain, livestock were brought to the eastern seaboard by the early settlers from other countries. These shipments occurred at intervals over a long period of time. Most of the present breeds of livestock reached the United States in this manner.

The Arabian horse is believed to be the first breed of livestock to be domesticated. Many centuries ago, the tribes of Arabia tamed it, and in the process, improved its speed, stamina, intelligence, and beauty of form. Many breeds of farm animals have been developed, some by livestock breeders in some parts of Europe and, more recently, in the United States.

Why Is It Desirable to Improve Farm Animals?

In many parts of the United States, the most profitable farms and ranches are those that raise considerable livestock. The most prosperous of these livestock raisers give special attention to the improvement of their herds and flocks along the particular lines for which each kind of livestock is kept.

Usually a farmer sells milk from dairy cows on the basis of the amount of milk and percentage of butterfat in the product. Therefore, the production of a cow is usually measured in terms of the amount of milk and percentage of butterfat produced in one year. Some groups of farmers have organized Dairy Herd Improvement Associations, which employ individuals to keep production records for each cow in the herds of the members of the associations.

The improvement in dairy cows can be seen in what has happened to the average production per cow. In 1957, there were 19.8 million producing cows on farms. These cows averaged 6,303 pounds of milk per cow per year. In 1980, there were only 10.8 million producing cows on farms, but the average production per cow had increased to about 11,875 pounds of milk per year. As a result of this higher production per cow, total milk production increased from 124.6 billion pounds of milk in 1957 to 128.0 billion pounds in 1980.

The same kind of improvement can be seen in egg production figures. In 1940, for each chicken on farms as of January 1, about 18 eggs were produced. In 1980, there were 242 eggs produced for each chicken on farms as of January 1. These figures indicate a great increase in production that exceeds 250 eggs per chicken per year in good flocks.

The superior producers, or performers, are usually the most profitable to raise. This is true of beef cattle, dairy cattle, chickens, hogs, sheep, and horses, though the purposes for which each of these kinds of animals is kept are different. Brood sows that raise large litters of fast-gaining pigs are usually the most profitable. Ewes that raise fast-growing lambs and those that grow heavy fleeces of good-quality wool return the greatest profits. There are some strains of meat animals that gain faster and produce 100 pounds of gain on considerably less feed than other strains kept under similar conditions; here also, the profits are in favor of the best producers.

There are numerous possibilities for livestock improvement, as can be noted when records made by exceptional individual animals are analyzed. A few cows have produced more than 150 pounds of milk a day, their own weight in milk in less than two weeks, and over 20 tons of milk during a year. One cow produced 42,805 pounds of milk in a year; another produced 1,614 pounds of butterfat in a year. As a five-year-old, Beecher Arlinda Ellen, an Indiana cow, gave 55,661 pounds of milk, an all-time record. During peak days of production, she gave 95 quarts of milk per day. When hitched to a light racing cart on which a driver was seated, a horse ran a mile in less than two minutes. A team of draft horses pulled a load equal to nearly 30 tons on a wagon on pavement. A sow produced a litter of 18 pigs

Fig. 10-2. A ewe bred to produce a litter of lambs rather than the usual one or two. (Courtesy, *Agricultural Research*, U.S. Department of Agriculture)

A U.S. No. 1 Hog Carcass Yields More Lean Meat and Less Lard Than a No. 4

Wholesale Yields of Average 165 lb. Slaughter Hog Carcass*

Cuts	U.S. No. 1	U.S. No. 4
	pounds	
Four major lean cuts:		
Ham	37.6	28.3
Loin	32.0	23.4
Picnic shoulder	15.7	12.2
Boston butt	12.9	9.5
Total	98.2	73.4
Other cuts:		
Spareribs	5.9	4.8
Belly	21.3	26.1
Jowls	4.0	5.5
Neck bones	2.6	1.9
Front feet	2.7	2.2
Lean trim	9.6	8.7
Total	46.1	49.2
Lard	15.9	30.6
Hind feet, tail, skin, etc.	4.8	11.8
Total carcass weight	165.0	165.0

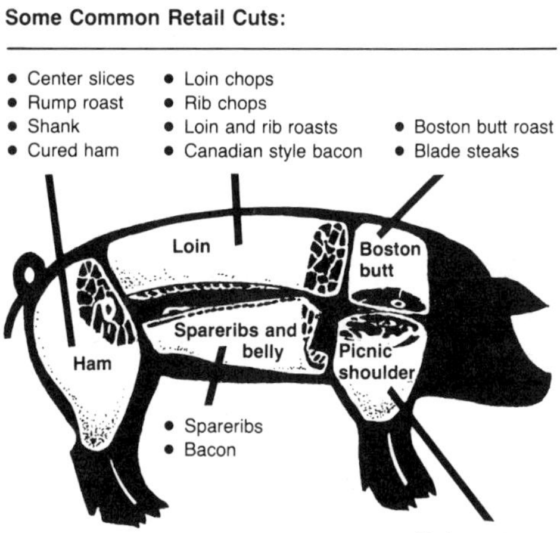
Some Common Retail Cuts:
• Center slices
• Rump roast
• Shank
• Cured ham
• Loin chops
• Rib chops
• Loin and rib roasts
• Canadian style bacon
• Boston butt roast
• Blade steaks
Loin
Boston butt
Ham
Spareribs and belly
Picnic shoulder
• Spareribs
• Bacon
• Picnic roasts
• Arm steaks
• Cured picnic

* Based on USDA cutting and trimming methods.

Source: AMS Market News, USDA.

Fig. 10-3. The better the type of hog, the greater the proportion of meat cuts the consumer wants. (Courtesy, *Farmline*)

that weighed 882 pounds at 8 weeks of age and a litter of the same number reached 5,036 pounds at 180 days of age. A hen produced 359 eggs in a year. Some chickens raised for meat purposes have reached an average of 3 pounds at less than eight weeks of age.

These records represent the upper limits of livestock improvement, at least as of a recent date; but new records will probably be established. Of course, these records are far above those that most livestock breeders can hope to attain with their herds and flocks, because individuals of such extremely high abilities are rarely found, and the conditions under which such records are made are usually far different from those that can be provided on most farms. However, when we compare these records with the averages on our farms, we realize that many of our farm animals are farther down the scale than is desirable under any conditions.

What Is the Scientific Basis for Livestock Improvements?

In the production of farm animals, everyone should realize that both *heredity* and *environment* influence the performance of each animal, just as they influence the performance of plants. Heredity refers to characteristics that are transmitted to an individual from its parents. Thus, a calf from Hereford parents inherits the white face. Likewise, a hen inherits its egg-laying capacity. (This capacity may or may not be for high production.) A colt has tendencies toward size, strength, or speed, depending on its inheritance. Homing pigeons inherit the tendency to find their way home after being transported long distances. Bird dogs inherit the tendency to "point" when they are near game.

A seed of a plant develops as the result of fertilization that takes place when a tiny pollen granule is joined with an ovule in the flower. The resulting seed contains the materials that determine the characteristics of the plant that grows when the seed is planted. Similarly, each animal has its origin in the union of two tiny cells. One of these cells is produced by a female and one by a male, and the cells unite following the mating process. The developing young of farm animals spend a portion of their early life in the bodies of their mothers, except for poultry, whose young develop in eggs outside the bodies of their mothers.

Chromosomes and genes determine characteristics

The characteristics of an animal are determined by the chromosome makeup of its body cells.

Each body cell contains many pairs of chromosomes. When a cell divides or duplicates itself to produce growth (mitosis or mitotic division), *each* chromosome in the cell also divides or duplicates itself so that the new cell formed will have exactly the same kind and number of chromosomes (and chromosome pairs) that the parent cell contained. This kind of cell division takes place millions of times in order to produce an adult animal.

Fig. 10-4. Each cell contains many pairs of chromosomes.

Some of the cells produced are for reproduction. Cell division to produce sperms and eggs for reproduction (meiosis) is different from cell division for growth. When sperm (in the male) or eggs (in the female) are formed in the reproductive organs, the chromosome pairs in the cells from which the sperm or eggs are formed split, with each chromosome of a pair going to help form a different sperm or egg. Each sperm or egg formed will contain a chromosome from every chromosome pair in the reproductive cell. Thus, the sperms or eggs contain only half the number of chromosomes contained by the body cells. When a sperm from the male fertilizes the egg of the female, one new cell is formed and the pairs of chromosomes are re-

Improving Herds and Flocks

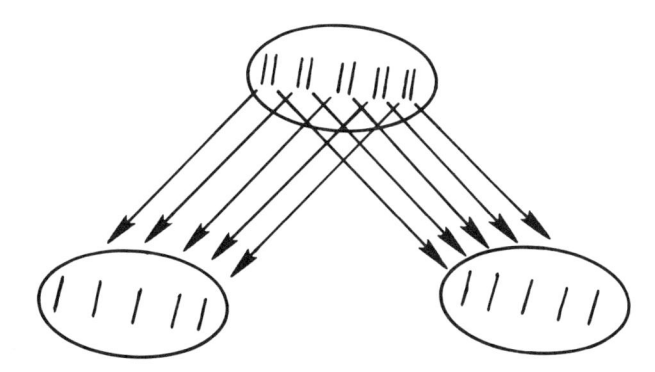

Fig. 10-5. The chromosome pairs split, forming cells with half the original number of chromosomes. These new cells can reproduce themselves just as other body cells reproduce themselves.

established by the pairing of the chromosomes from the male with those of the female. Each animal begins as this single cell formed when the sperm of the male unites with the egg of the female, one half of each pair of chromosomes coming from the male and the other half coming from the female.

All of the pairs of chromosomes are alike with the one exception of the pair of chromosomes associated with sex determination. The chromosome that determines the sex of the offspring appears in either the male or the female, but not in both. One of the sex-determining chromosomes will bring about the development of a male, the other will bring about the development of a female. In mammals, the male carries the sex-determining chromosome; in birds, the female carries the sex-determining chromosome. There are some characteristics linked to the sex chromosome in such a way that these characteristics appear in one sex only.

The chromosomes contain many genes that are the actual carriers of the characteristics of each kind of animal. There are genes or combinations of genes for each characteristic, and each type of gene appears on the chromosomes of both parents. Thus, the genes are in pairs just as the chromosomes are in pairs. As the new cell reproduces itself to grow into the new animal, each chromosome reproduces itself so that each additional cell formed has the same hereditary or genetic makeup as the cell from which it was formed. This means that each cell produced from the cell formed when the sperm and egg united has the same kinds and number of chromosomes carrying the same types of genes. Since, in each case, each parent contributes half of the hereditary material that goes to the offspring, the distinctive characteristics of the parents are duplicated in the offspring.

The development of the various parts of the body is also provided for in the cell. The chromosomes contain material that directs the development of cells so that, as the cells multiply, differentiation occurs and the various parts of the body develop as they should.

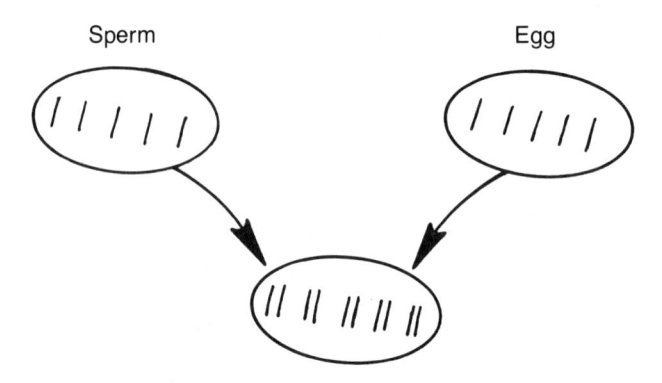

Fig. 10-6. In fertilization, a single cell is formed when sperm and egg unite and the chromosome pairs are re-established.

An interesting aspect of heredity is that the offspring from given parents may vary considerably from their parents and among themselves. No one is exactly like either of his/her parents, although there probably are many similarities. The parents may have brown eyes and their children may have blue eyes. They may both be tall, while their children may be shorter than they, although in that case the children are likely to be above average in height. Brothers and sisters may differ among themselves in certain characteristics.

In the flock of a certain poultry breeder, two hens were selected; one laid 237 eggs and one laid 240 eggs in a year. These hens were mated to the same rooster. The former had daughters that averaged 247 eggs and only one that produced fewer than its mother. The second hen had daughters all of which laid fewer than 200 eggs, though these offspring received the same feeding and care as those from the first hen. Obviously, the first hen had a better inheritance to pass on to her offspring than the other hen. In other words, she was more *prepotent* for high egg production, as scientists would say.

In some dairy herds, an occasional cow has daughters nearly all of which are high producers, while other cows have produced few or no daughters of merit.

These differences have been accounted for by geneticists. First, even though the chromosomes of a pair look alike, each of them may carry slightly different hereditary material, and since the pairs split to form sperm or eggs, there are many possibilities for differing combinations of chromosomes in the sperm or eggs. For example, if all the chromosome pairs could be put into a definite order before they split to form sperm or eggs, then as the first pair split, one chromosome would go to help form one sperm or egg and the other chromosome would go to help form another sperm or egg. Beginning with the second pair of chromosomes, however, more than one combination would be possible since pure chance would determine which half of the second pair would join each half of the first pair. Thus, if there were only four pairs of chromosomes in a cell, the following would be possible:

Pairs of chromosomes: A1, A2/B1, B2/C1, C2

Possible combinations of chromosomes when pairs split to form sperm or eggs are:

A1, B1, C1
A1, B1, C2
A1, B2, C1
A1, B2, C2
A2, B1, C1
A2, B1, C2
A2, B2, C1
A2, B2, C2

Of course, only two of the eight possibilities could occur from the division of the one cell in the illustration. If A1, B1, C1 happened to be one combination, the other combination would be A2, B2, C2; and a combination of A1, B1, C2 would result in a second combination of A2, B2, C1. The great number of sperm produced would, however, result in the opportunity for every possible combination of chromosomes to occur many times.

Each pair of chromosomes added to the number of pairs in a cell doubles the number of possible combinations of chromosomes for forming a sperm or an egg and adds that many more opportunities for differences among offspring of the same parents. Cattle have about 30 chromosome pairs in a cell. How many combinations of chromosomes could cattle have? Each of the many combinations of possibilities in the female egg could unite with any one of the same combinations of possibilities in the male sperm, resulting in a great number of variations in characteristics of offspring.

Another way in which differences are caused in offspring of the same parents is by the exchange of hereditary material between two chromosomes of a chromosome pair in a reproductive cell. If this occurs, then the chromosome sets formed when the pairs split to form a sperm or an egg would be slightly different with regard to hereditary material from the chromosome sets formed from chromosome pairs in which no exchange of hereditary material occurred.

Mutations may also occur. The genetic makeup of a plant or an animal is extremely stable. However, once in a while an accident that actually changes a gene and creates a new plant or animal may occur in nature. Scientists have been able to cause mutations by the use of X-rays and chemicals.

Other differences can be caused by the effect of environment. Sunlight, the seasons of the year, food, water, and other environmental factors can cause differences to appear in offspring of the same parents. Thus, a cow may inherit a tendency for high production, but if fed poorly, it will not produce milk or butterfat in large quantities. A son of a fast horse may inherit tendencies toward speed, but if he is poorly trained or improperly fed, he may never win a race. On the other hand, unless a cow has inherited a capacity for high production, no amount of feeding and care will make her into a high-producing cow. Thus, good heredity and good environment are both necessary.

As an example of the manner in which certain characteristics are inherited, let us consider the nature of the offspring that will result if cattle of the Aberdeen-Angus breed are mated to those of the Hereford breed. The calves from these matings will have white faces and black bodies but no horns. From which breed was each of these characteristics inherited? If these offspring are mated with other offspring of the same sort, there will be individuals that again show horns, the red color, or other characteristics in various combinations that were covered up for one generation. The characteristics that may be transmitted in a hidden manner are called *recessive,* and those that always show if present at all are called *dominant*. Thus, the white-faced characteristic is dominant, and the red color and horns are recessive in the example given.

To illustrate still further the color characteristic of cattle, let us suppose that animals are generally black, but occasionally a red animal is born. Black is the dominant color and red is recessive. If the symbol *B* is used for the dominant black and *b* for the recessive red, the manner in which the color would be transmitted is as follows:

1. If each parent carried only the gene for black, all offspring would be black and would carry only the gene for black.

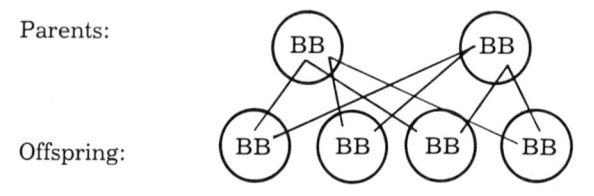

    Parents: BB × BB
    Offspring: BB BB BB BB

2. If each parent carried only the gene for red, all offspring would be red and would carry only the gene for red.

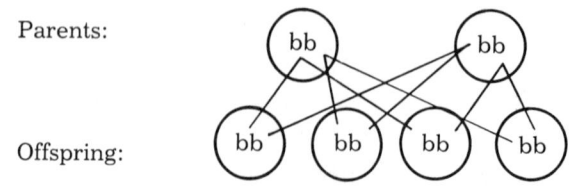

    Parents: bb × bb
    Offspring: bb bb bb bb

3. If one parent carried the gene for black and one parent carried the gene for red, all offspring would be black but would carry the genes for red and black.

Improving Herds and Flocks 331

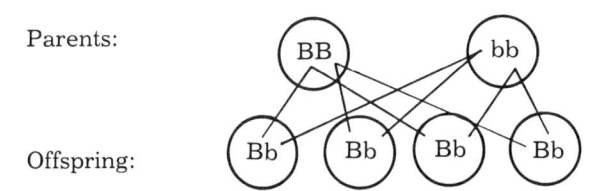

Parents: BB — bb
Offspring: Bb Bb Bb Bb

4. If each parent were black and carried the gene for red, one-fourth of the offspring would be black and would carry only the gene for black; one-half of the offspring would be black and would carry the genes for red and black; and one-fourth of the offspring would be red and would carry only the gene for red.

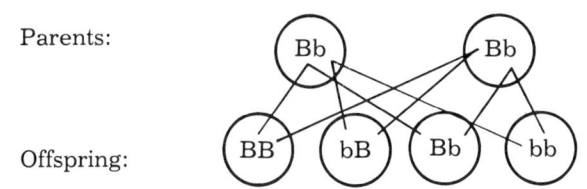

Parents: Bb — Bb
Offspring: BB bB Bb bb

5. If one parent carried only the gene for black and the other parent the genes for red and black, all offspring would be black, with half of them carrying only the gene for black and half the genes for both red and black.

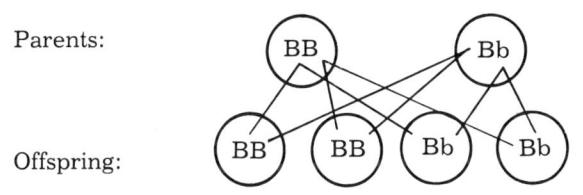

Parents: BB — Bb
Offspring: BB BB Bb Bb

6. If one parent carried only the gene for red and the other parent the genes for red and black; half of the offspring would be red and would carry only the gene for red; half of the offspring would be black and would carry the genes for both red and black.

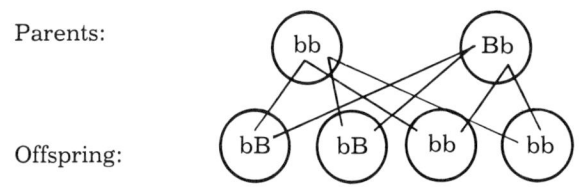

Parents: bb — Bb
Offspring: bB bB bb bb

With many of the characteristics important in animal production, more than one kind of gene is involved. For example, the characteristics of gaining weight rapidly, of producing many eggs, of producing many pounds of milk are the result of the combined effects of many genes. Since it would take many millions of animals to obtain every possible combination of genes, the improvement of livestock through breeding is based mainly on production records for individual animals and herds.

Occasionally, in breeding livestock, extremely undesirable features may occur in the offspring. For example, in beef cattle, dwarfed and malformed calves may occur. This trait of dwarfism is recessive and occurs in offspring of normal parents only if genes for it are transmitted by both parents to one of their offspring.

The whole process of animal improvement is very complicated, and livestock breeders must be extremely skillful when trying to improve their herds and flocks.

Gregor Mendel, an Austrian monk, was responsible for discovering many of the peculiarities of inheritance in plants. Since then, it has been demonstrated that inheritance in animals is similar in many respects to that in plants. Although breeders of farm animals had made considerable progress in livestock improvement prior to the discovery of the writings of Mendel in 1900, there was but little understanding of the reasons for certain breeding practices until Mendel's work became known. Later developments in the science of animal breeding made possible rapid improvements in some herds and flocks.

How Can Livestock Be Improved?

One of the greatest challenges to livestock breeders is to improve their herds and flocks along the lines for which these animals are kept. Thus, good dairy farmers desire a high average production of milk and butterfat per cow in their herds. Chicken raisers strive for a high average egg production per hen. Swine raisers seek to secure a large output per sow in terms of number of pigs raised per litter, weight of litter at weaning, and rapid, economical gains from birth to market. Sheep producers strive for a high lamb production per ewe and heavy fleeces of good quality. Beef cattle raisers strive to secure a high calving rate and to produce calves that gain rapidly and economically.

Many livestock raisers set goals in terms of average production per animal in their herds and flocks. To reach these goals requires the use of good feeding and care methods. Records of production must be well kept and effectively used. Every effort should be made to locate good parents capable of producing offspring that are also good producers.

The following are examples of levels of production that can be reached under favorable conditions. Because some livestock producers can secure higher levels than these, they should set their goals higher.

Average number of lambs per 100 ewes .... 125 lambs
Average weight of fleece per ewe (medium wool) .... 9 pounds
Average weight of fleece per ewe (fine wool) .... 14 pounds
Average number of pigs raised per litter .... 9 pigs

Milk Production and Prices

Dollars per cwt

Farm Price, All Milk

Billion Pounds

Milk Production

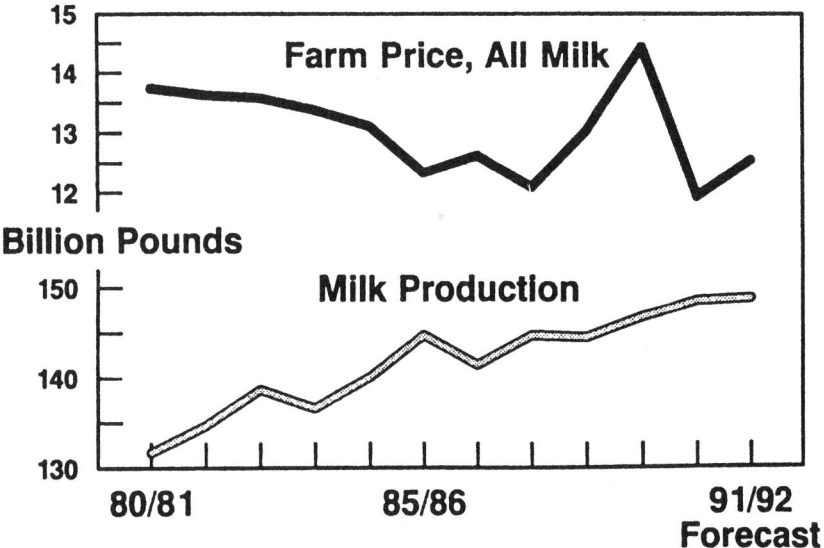

80/81   85/86   91/92 Forecast

Fig. 10-7. Increased production of milk per cow, thus maintaining total production even with decreasing numbers of cows, is due in part to improvement through breeding. (Courtesy, U.S. Department of Agriculture)

Average weight of litter of pigs at 56 days .... 300 pounds
Age at which hog weighs 200 pounds .... 5½ months
Average number of eggs per hen per year .... 240 eggs
Age at which chickens for meat weigh 3 pounds .... 8½ weeks
Number of pounds of butterfat per cow
 per year (2 × milking) .... 500 pounds
Number of pounds of milk per cow per year .... 15,000 pounds
Calving percentage in a beef herd .... 95 per cent
Weight of beef calves at six months of age .... 500 pounds

One way in which farm animals have been improved has been through the development of breeds. Most breeds were developed within the last few hundred years, some within quite recent years. Each of these breeds was developed by the selection of similar type individual animals for breeding stock. Selection was continued for many generations, until the animals had fairly uniform characteristics. After a period of time, the breeders of this group of animals formed a breed association and prohibited the use of outside breeding stock for the production of animals that would be eligible for registry in their association.

Although there are many people who argue about the relative merits of various breeds, the success of livestock raisers depends less on the particular breeds they choose than on the animals they select and on the other methods they use for improving their herds or flocks. Livestock raisers should select animals capable of high production, regardless of the breed. For their breeding herds, they should select males and females capable of transmitting the desired qualities of their offspring.

One method used by livestock breeders to improve their livestock is to purchase *purebred* sires of outstanding merit. Such a sire is mated to the females in a herd or a flock, thus contributing to the improvement of the next generation. Continued improvement is dependent on the selection of good females and males in successive generations.

Some breeders of purebred livestock follow the practice of *inbreeding* to a greater or lesser degree. In this practice, closely related animals are mated. By the use of inbreeding, it has been possible to obtain greater uniformity in the offspring and to make the transmission of certain desirable characteristics more certain. In the development of many breeds, for example, the practice of inbreeding was extensively used to bring about greater uniformity in color, form, and other characteristics. Inbreeding should not be used with farm animals of average or inferior quality because certain undesirable features may crop out and become firmly fixed along with some desirable characteristics.

Crossbreeding is being practiced with livestock more than formerly. This consists of using parent stock from two or more different breeds. For example, a Duroc boar might be mated with a Poland China sow. Frequently, the offspring from such a cross are more vigorous and gain more rapidly than purebred litters. In some cases, three-way crosses of hogs are used with success. In this case, sows from the crossbred litters are mated to boars of a third breed.

Crossbreeding is used with considerable success with most kinds of livestock, including hogs, beef cattle, dairy cattle, and meat types of chickens. Experiments in crossbreeding dairy cattle have shown that the offspring produced more milk and butterfat than their purebred dams. When these crossbred heifers were bred to a bull of a third breed, the heifers from these three-breed crosses produced still more milk and butterfat. To secure these results, however, livestock breeders used proved sires. In order to secure good results from crossbreeding, livestock breeders must select good males and females capable of transmitting high production to their offspring.

The mule is an outstanding example of increased vigor from crossbreeding. To produce a mule, a male member of the ass family is crossed with a female member of the horse family. The mule is especially hardy and can withstand adverse conditions much better than horses. However, mules are sterile, which means that they do not reproduce themselves.

Inbreeding combined with crossbreeding is now being used extensively to improve chickens. The first step in this process is to produce inbred lines

Fig. 10-8. A mule colt and its mother. A mule is a hybrid. (Courtesy, U.S. Department of Agriculture)

by mating close relatives for several generations. Only the best of these inbred lines are used. As a second step, some of these inbreds are crossed. In combining four inbred lines, for example, Inbred A is crossed with Inbred B, and Inbred C is crossed with Inbred D. As a third step, offspring of these two crosses are mated (see Fig. 10-9). The eggs resulting from this cross are hatched to produce hybrid chicks, which are purchased by farmers and commercial poultry raisers who maintain laying flocks.

Some inbreeding followed by crossing of inbred lines has been tried with swine with some beneficial results. This may be an important method for improving these and other kinds of livestock, just as it has been with chickens.

Artificial breeding, or artificial insemination as it is sometimes called, is being used with some kinds of livestock. This is a technique of breeding in which the semen from a good sire is transferred to females by artificial methods. By this means, the services of a good sire may be used on many more females than is possible by natural mating. So far, this method has been used most extensively with dairy cows, but its use is being extended to other kinds of livestock.

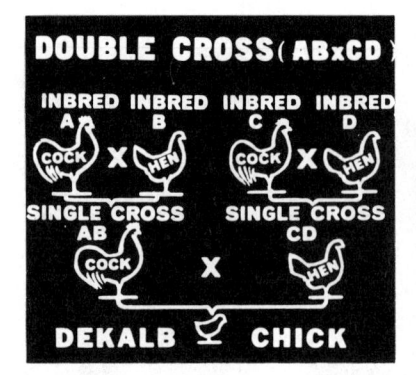

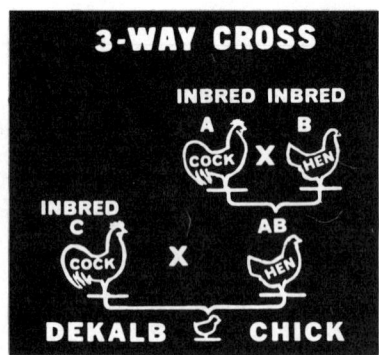

Fig. 10-9. The use of inbreeding, followed by crossing of inbred lines, has led to improvements in chickens. Superior inbreds are crossed to produce hybrids. DeKalb Chix result when two or more inbred lines are crossed. Double cross DeKalb Chix are produced when four inbred lines are combined, as shown in the diagram on the left, DeKalb Chix can also be produced by combining three inbred lines *(right)*. Crossbred chickens (sometimes called hybrids) are simply crosses of two standard breeds, not inbred crosses. (Courtesy, DeKalb Agricultural Association, DeKalb, Illinois)

How Can Profitable Livestock Be Selected?

Success in breeding livestock is dependent on selecting parent stock capable of transmitting desirable characteristics to their offspring. The selection of this parent stock calls for a great deal of skill. Four general methods may be used in various combinations for breeding stock selection. These include selection by (1) appearance or type, (2) pedigree, (3) performance or production, and (4) kinds of progeny produced.

Appearance or type. For a long time, livestock breeders have selected animals that are pleasing to the eye. Some people believe that the animal is an accurate indication of its producing ability. Actually, as with people, "Beauty is only skin deep." Cows of good dairy type may be disappointing as producers, and cows that are equally good in type may vary widely in production. The type of a meat animal is not an accurate indication of its gaining ability or of the quality of the meat in its carcass. Thus, looks are deceiving in determining the qualities of greatest importance in breeding livestock. Even so, most breeders are interested in placing some emphasis on type; but alert breeders give attention to additional factors.

Pedigree. A pedigree is a record of an animal's ancestry. It is possible to determine the names of the ancestors of a purebred animal and perhaps to find out something about their production. This information is available from the registry association of a particular breed.

Animals with good ancestors are more likely than otherwise to develop into good individuals and to have offspring that are likewise good. However, this is not always the case. A dairy cow, for example, may have superior blood lines as indicated by her pedigree, but she may be mediocre as an individual or she may have inferior offspring.

The production records of the close relatives of an animal under consideration merit the most attention in a pedigree, if the pedigree shows this information completely and accurately. The production of the mother and grandmother, their offspring, and the offspring of the sire and grandsire is of some significance. The production of the full sisters and half-sisters of the male or female under consideration is especially valuable in predicting the ability to transmit high production.

Performance or production. The real test of the value of an animal is how well it produces or performs when given proper feed and care. How good a dairy cow is as a producer of milk and butterfat can be determined from records of her production. Similarly, egg records can provide an accurate basis for ascertaining each hen's production. Race horses are tested by timing them on the race track. A machine known as a *dynamometer* has been used to determine the pulling ability of a horse or team. Wool from a given sheep can be weighed and checked for quality. The performance of hogs and beef cattle can be calculated by their rate of gains and the feed requirements per pound of gain.

Kinds of progeny. An animal's true worth as a breeder is determined by the kind of progeny (offspring) that it produces. Hogs similar in appearance may differ in their ability to transmit these qualities to their offspring. The daughters of some high-producing hens may be good layers, while the daughters of other high-producing hens may be only average producers. This may be true even though the hens are mated to the same rooster. Similar situations exist for other kinds of livestock. It is evident that careful records and observations must be made of the parent stock and their offspring to determine the true worth of the parent stock.

After a sire has had sufficient offspring to indicate his transmitting ability, he is called a *proved sire*. A sire may be proved good or poor, depending on whether his offspring are superior or inferior. The selection of males and females for breeding purposes according to the performance of their offspring is the most important step in a breeding program.

The use of records to prove a dairy bull is shown by the following example. In this case, the sire is a good transmitter of production, as shown by the high average production of his daughters and by their producing more than their mothers.

	Yearly production		Per cent butterfat in milk
	Milk	Butterfat	
Eighteen daughters (21 records)	14,102 lbs.	506 lbs.	3.6
Mothers (67 records)	11,999 lbs.	394 lbs.	3.3
Increase	2,103 lbs.	112 lbs.	0.3

A possibility in livestock breeding is to develop animals that are naturally resistant to certain diseases. For example, some experiments have shown that certain strains of hogs are resistant to brucellosis. Some strains of cows are resistant to mastitis. Some strains of chickens are resistant to leucosis. Because the improvement of animals along these lines is complicated and costly, only a beginning has been made.

Improving dairy cattle

Good dairy producers seek to develop herds consisting of cows that produce large amounts of milk and butterfat. Breeders of purebred cattle may give considerable attention to type as well as to production.

Selecting for type. Many studies have shown that there is little or no relation between the type of dairy cows and their ability to produce milk and butterfat. Consequently, it is not possible to develop a high-producing herd merely by selecting animals on the basis of their appearances. However, most breeders of purebreds consider it important to give some attention to type. Likewise, a dairy producer who is selecting a foundation heifer or cow will usually want to secure an animal of approved type.

A dairy cow of desirable type is lean and angular in appearance. She has a large, symmetrical udder. Her body is deep in the region of the chest. She has a large middle or barrel, as indicated by the spring of ribs and depth of body. Her back is straight and her rump is broad and level. Her face is moderately long and her muzzle is broad. She is large for her breed.

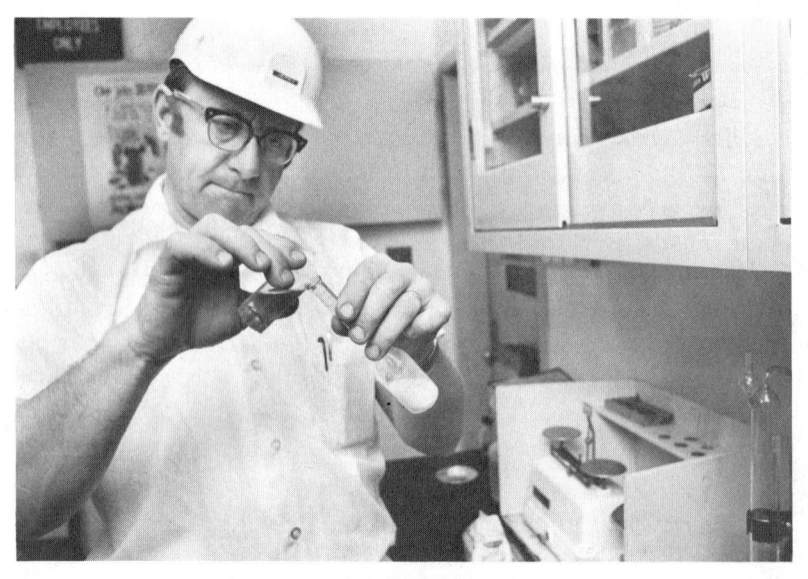

Fig. 10-10. Testing milk for butterfat.

Selecting high-producing cows. The only method for determining accurately the producing ability of a dairy cow is to keep records of her production for a period of a year or more. Records for several years, or lifetime records, for each cow are especially valuable. The records should include the amount of butterfat as well as the amount of milk she produced. Until Babcock discovered an accurate method of testing milk, there was no practical way to determine accurately the amount of butterfat produced by each cow.

In selecting a cow for a breeding herd, a dairy producer should give her production consideration. A good cow, when mature, should produce 500 pounds or more of butterfat per year on twice-a-day milking. If she has daughters of producing age, these should be high producers. Usually, however, such cows are sold only at high prices, so the only alternative is to purchase a promising heifer.

A desirable way to secure a dairy heifer for breeding purposes is to select a daughter of a cow that is herself a high producer and has daughters that are high producers. Other close relatives should also be high producers. The sire of the heifer should be a bull that has proved himself by having several daughters that are high producers, as explained earlier. If not a proved sire himself, he should be a son of a proved sire. Complete pedigrees include much of this information.

Selecting a dairy sire. A dairy bull who is himself a good proved sire, as shown by high-producing daughters, as this is the most accurate indication of his true worth as a sire, should be selected. If a proved sire is not available, it is desirable to secure a son of a good proved bull from a high-producing cow who is a daughter of a good proved bull. It is well to note the production of full sisters of the bull and of other close relatives. Many farmers who select bulls for breeding by artificial insemination base their choice partly on the Predicted Difference (PD) for milk yield of the progeny of the bulls. A PD of 500 pounds means that cows sired by these bulls will give an average of 500 pounds more milk in a 305-day lactation than their herdmates of the same breed.

Improving a herd. As indicated previously, many breeders of dairy cattle are able to secure the services of good sires by using insemination. In a recent year, 5 million dairy cows in the United States were bred by artificial insemination. By this means, each sire was used to breed an average of nearly 2,000 cows annually, and some outstanding sires were used to breed 10,000 or more cows per year. Trained persons are hired to perform the insemination process, with semen secured from establishments where outstanding bulls are kept.

An owner of a dairy herd should save heifers for replacements from high-producing cows that transmit their good qualities to their offspring.

Dairy producers have recently adopted a new system for evaluating

Fig. 10-11. Princess Breezewood R A Patsy 3816059 (VG), owned by Gelbke Bros., Vienna, Ohio. She is a "National Butterfat Champion—Regardless of Age, Breed, or Milking Frequency." Her record, at five years and two months for a year's milking, was 36,821 pounds of milk and 1,866 pounds of butterfat. (Courtesy, Holstein-Friesian)

Fig. 10-12. This Brown Swiss cow shows the results of improvement through breeding. Arbor Rose Mac Ruby 174924, classified as Excellent, was Senior and Grand Champion, National Show, and Senior and Grand Champion, Pacific International. Her 365-day production records with two times per day milking include 23,211 pounds of milk at 4.4 per cent butterfat and 30,360 pounds of milk at 4.4 per cent butterfat. (Courtesy, The Brown Swiss Cattle Association of America)

dairy animals called *linear classification*. The system consists of giving a *numerical score* for *each* of the *traits* included in the evaluation. This system will make it possible to have a score, based on an animal's appearance, that may indicate the animal's productivity and longevity. This system of evaluation may help with the selection of animals for herd improvement.

Improving chickens

Poultry raisers who keep flocks to produce eggs are interested in securing pullets capable of high production.

Securing good producing stock. Most poultry raisers today secure baby chicks from commercial hatcheries. Reliable hatcheries provide chicks that come from breeding stock of proven merit for high production. Today, some poultry breeders have developed high-producing strains in breeds such as White Leghorns and New Hampshires. Some hybrid chickens of improved laying ability are being developed through inbreeding and crossbreeding methods, as explained earlier in this chapter. Likewise, hatcheries provide chicks that gain rapidly and economically for meat production.

Broiler Production and Prices

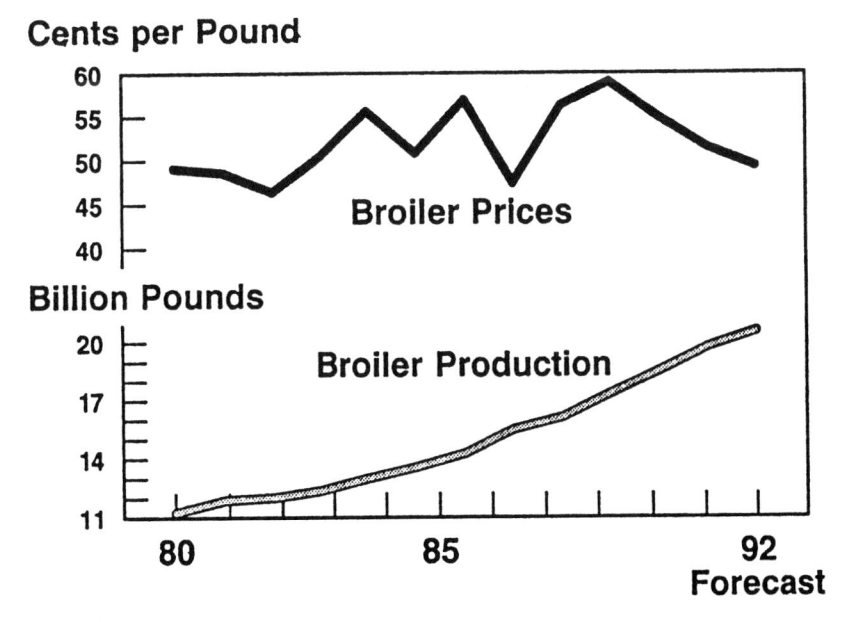

Fig. 10-13. As a result of good breeding and other practices, broiler production continues to rise. (Courtesy, U.S. Department of Agriculture)

Fig. 10-14. In selecting hens for a breeding flock, some top-flight poultry raisers use trap nests and record the production for each hen, as shown above. (Courtesy, U.S. Department of Agriculture)

Individuals who buy baby chicks should be sure to get them from hatcheries that secure eggs from flocks bred for high egg production and free from *pullorum*, a disease that may otherwise be transmitted through the eggs.

In some hatcheries, specially trained persons sort chicks by sex. Poultry raisers who prefer pullets may purchase chicks of that sex; or if males are preferred, they may purchase them.

Selecting high-producing hens. Good poultry raisers check their flocks at frequent intervals and eliminate the poor producers. This process is called *culling* the flock. The average person can learn to cull with considerable accuracy after a little instruction and practice. Culling should be done throughout the year. Many good poultry raisers cull their flocks whenever the eggs produced per day total less than half the number of hens.

Improving hogs

Hog breeders strive to select breeding stock that will produce large litters of fast-gaining pigs of approved type. Much emphasis is now being placed on meat-type hogs.

Selecting gilts and sows. Gilts from litters of eight or more pigs raised

Improving Herds and Flocks

that reach a weight of 300 pounds or more at 56 days of age and also gain rapidly after that age are preferred. The gilts selected should be large for their age. A good gilt weighs 200 pounds or more at six months of age. In selecting a sow, a hog breeder should consider the kinds of litters she has raised. These litters should measure up to the standards just described.

The pedigree of each gilt or sow should show close ancestors that have good-producing ability in number of pigs raised and fast-growing qualities.

In type, a desirable gilt or sow has a moderately long, deep body. She has a strong, uniformly arched back. She is moderately wide and uniform in width from front to rear. Her rear quarters are well developed. Her feet and legs are straight with strong, upright pasterns. Each sow or gilt should have 12 or more properly developed teats. The littermates of a sow or gilt and the offspring of a sow should be uniformly good in type and size. A gilt or sow should have a quiet disposition.

Selecting a boar. The hog producer should choose a boar that comes from a good litter. In type, the boar should be larger for his age and show more ruggedness than a gilt. He should be a purebred. If possible, he should have already sired good litters. His mother should be a high producer and his father should have sired litters of good type with fast-gaining qualities. Other individuals in his pedigree should also have shown the ability to produce large litters of fast-growing pigs. He should be secured from a breeder who has a healthy herd.

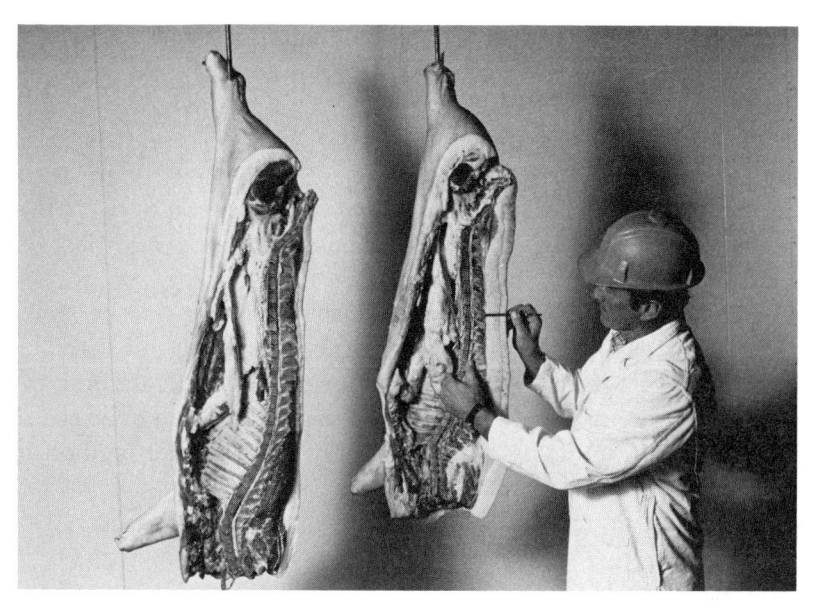

Fig. 10-15. The carcass from a lean line of pigs *(left)* has little backfat compared to the carcass being measured *(right)*. (Courtesy, *Agricultural Research*, U.S. Department of Agriculture)

Fig. 10-16. At 56 days, this little fellow will be weighed, along with the rest of the litter, to determine his rate of growth. (Courtesy, New York State College of Agriculture and Life Sciences, Cornell University)

Improving a herd. Improvement of a swine herd requires careful selection of breeding stock over a period of years. Breeders who sell breeding stock will wish to maintain purebred herds. They will seek to bring their herds to a high level of type and producing ability, so that their breeding stock may be sold readily to other breeders.

Breeders of market hogs frequently mate good purebred boars with grade sows. Crossbreeding is being followed by many who produce market hogs. They frequently rotate boars from two or more breeds in successive years.

In improving a swine herd, breeders should select sows that produce a large number of fast-gaining pigs. Gilts for replacements and boars should be selected from litters with these qualities.

In recent years, some breeders have selected hogs that convert feeds into lean pork faster and with less feed. Performance tests at special swine-testing stations are used to select lines of breeding that rate high in these characteristics. In some cases, one or more hogs from each litter are slaughtered to determine the quality of the meat. By using similar methods over a long period of time, swine breeders in Denmark have greatly improved their efficiency in pork production.

Improving beef cattle

Breeders of beef cattle wish to produce fast-gaining animals of good type. To do this requires good parent stock.

Selecting heifers and cows. A beef heifer or cow should be large for her age. The body of a good heifer or cow is broad and deep. She has a straight topline and shows good development over the loin and in the rear quarters.

A mature cow should have a record of producing one calf per year. These calves should be rapid growers and uniformly good in type. Close relatives of a heifer or cow should show good type and growthiness.

Selecting a beef bull. A good beef bull is large for his age. A bull that has grown rapidly from birth, and especially from weaning to 12 months of age, will usually produce calves that gain rapidly. He should be a purebred.

Improving a herd. Breeders of purebred beef cattle pay considerable attention to type. Increasingly, in selecting breeding stock, they are emphasizing rate of gain as well as type.

A breeder of beef cattle for the market should secure good bulls of approved type with rapid gaining qualities. Cows retained in the herd should be selected on the basis of their type and on the type and rate of gain of their offspring. Heifers for replacements should be selected from the cows that have the qualities mentioned. Systems have been developed for performance testing, or production testing, in beef breeding herds. Accurate checks are made on the rate of gain of different individuals in the herd, and this is taken into account when breeders are selecting breeding stock.

In some beef herds, crossbreeding is used to produce the desired type of market animal. By some estimates, the beef going to market from crossbreeding has increased to nearly 50 per cent of the total. Included

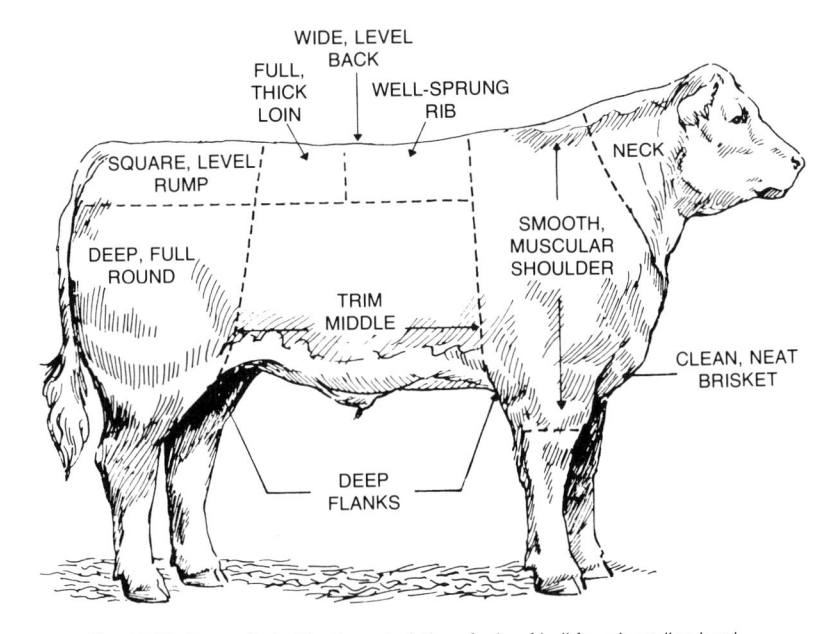

Fig. 10-17. Some desirable characteristics of a beef bull for a breeding herd.

among the crosses are the English breed crosses, European breed crosses, dairy breed crosses, Brahman breed crosses, and other combination crosses. Many of these crosses, including the Charolais, Simmental, Limousin, and Maine-Anjou, have become known as exotic crossbreeds.

The Agricultural Research Service Livestock and Range Research Station, Miles City, Montana, has one of the longest continuing beef cattle linebreeding programs in the United States. The Line 1 program was established in 1934, and the animals have been maintained since then as a closed herd with no introduction of bulls or cows from other herds. Weaning weight and yearly weight have been increasing by 22 pounds and 35 pounds, respectively, per generation (about every four years). Almost half of the registered Hereford breeders have cattle from or descended from the Line 1 herd.

Improving sheep

Sheep breeders seek to develop flocks that will produce large lamb crops and heavy, good-quality fleeces. Breeders of wool-type sheep stress wool-growing qualities, but they also wish to raise fast-growing lambs.

Selecting ewes. A good ewe lamb should be large for her age. She should weigh 75 to 90 pounds at six months of age, depending on the breed. Twin lambs weigh somewhat less as individuals at this age, but most

Fig. 10-18. Ultrasonic scanning techniques are used to photograph an interior cross section of a sheep. Fat thickness and muscle area can then be measured to aid in the selection of genetically superior animals. (Courtesy, *Agricultural Research,* U.S. Department of Agriculture)

Improving Herds and Flocks

breeders prefer twins because they result in more pounds of lamb per ewe. However, in range flocks, twins are not as desirable as in farm flocks because it is hard for ewes to care for twins on the range. A desirable mature ewe should have produced lambs regularly each year. Her lambs should be large and of approved type.

A ewe should be fairly wide and deep in the body. Her back should be straight and strong. Her rear quarters should be well developed. Her fleece should be compact, with moderately long fibers of good quality. Her udder should be sound. In selecting a ewe, the sheep raiser should inspect some of her close relatives to determine whether or not they are an approved type. Ewes for the farm flock and especially for the range flock should be free from wool around the eyes so that they can readily see to eat and move about.

Selecting a ram. In selecting a ram, the sheep raiser should consider all the desirable characteristics for a ewe. A purebred ram should be selected. If possible, he should have already sired good lambs.

Improving a flock. By breeding the best rams obtainable to the ewes in the flock, the sheep raiser may improve the flock. In some range flocks, ewes are carefully chosen on the basis of length of wool, weight of fleece, size of body, freedom from excessive wool on their faces, and uniformity of fleece grade. These ewes are bred to rams with similar qualities. Some sheep breeders are crossbreeding to produce high-quality market lambs.

For an effective breeding program, the sheep breeder should retain ewes that produce fast-gaining lambs of good type with heavy, good-quality fleeces. Ewe lambs retained for breeding should be selected from the

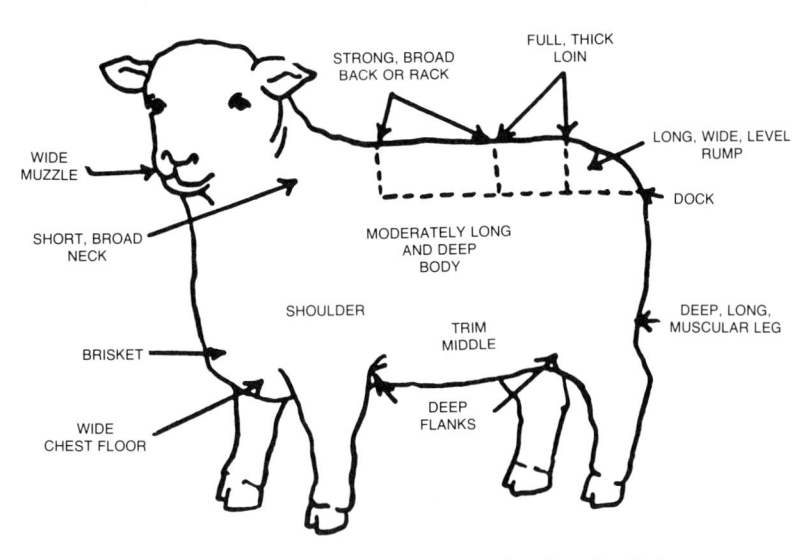

Fig. 10-19. Desirable features of a sheep for a breeding flock.

Fig. 10-20. A Rambouillet ram showing ruggedness and size, which are desirable for range flock improvement. Note freedom from wool around the eyes, which is desirable for good vision. (Courtesy, U.S. Department of Agriculture)

daughters of these kinds of ewes. Rams purchased for the flock should have similar qualities.

Improving horses

Horses are usually raised for pulling, riding, or racing.

Selecting draft horses. A good draft horse is large, with sound feet and legs. The old saying "no feet, no horse" indicates the importance of good feet and straight, strong legs. The pasterns should have some slope to provide for springiness when the horse moves about. The draft horse is the athlete of the farm. Hence, it should be muscular, with sloping shoulders that provide a good surface for the collar. The horse should have a good disposition. If possible, before purchasing a draft horse, the farmer should see it in action and note its pulling ability.

Selecting light horses. A horse for riding or racing purposes should have good feet and legs and a springy stride. It should be well trained and have a good disposition. Size is not as important with a light horse as with a

Improving Herds and Flocks 349

draft horse. A horse should be seen in action and consideration should be given to its performance along the desired line.

Improving a herd. As with other kinds of livestock, good parent stock must be used if horses are to be improved. Records in racing are considered as a basis for selecting breeding stock for producing future racing horses. The performance of the offspring is an important check on the value of parent stock for breeding purposes.

SUGGESTED ACTIVITIES

1. Describe any experiences you might have had in making a pet of some animal that exists naturally in the wild state. Mention some of its traits needed for survival in the wild state. What traits persisted the longest during the taming process?
2. From your own experiences, reading, and observation, show how heredity and environment influence the lives of people and animals.
3. With others in your class, visit a farmer who has high-producing livestock. What are some of the methods the farmer uses for selecting and improving livestock? What are the present levels of production obtained from them?
4. Attend a fair and observe an expert judge at work. As classes of livestock are judged, make your placings and check them with those announced by the judge. What are the shortcomings of the show-ring method of judging?
5. If you were to raise some kind of livestock, what methods might you use to secure good animals? How would you improve your herd or flock over a period of years?
6. Concepts for discussion:
   a. Farm animals are descended from animals that existed in the wild state at one time.
   b. Domesticated animals are those that people have tamed and changed to suit their purposes.
   c. The domestication of animals took place because people needed a reliable source of food, clothing, shelter, and other things animals could provide.
   d. Wild animals improve through the process of natural selection, or "survival of the fittest."
   e. Improvement of animals through controlled breeding is based on the desire to get more from a particular breed or kind of animal of whatever it is that is desired.
   f. Farm animals are bred for increased production, which leads to increased farm income per animal.
   g. Animal production is influenced by both heredity and environment.
   h. Heredity refers to characteristics an animal transmits to its offspring.
   i. The transmittal of characteristics from parents to offspring follows "rules of nature," which makes it possible to predict with some accuracy the best breeding program to follow.
   j. Both desirable and undesirable traits may be inherited.

k. Inheritance in animals is similar in many respects to that in plants.
l. Inherited characteristics determine the potential of an animal for production; the environment determines how closely the animal will come to reaching its potential.
m. New "breeds" of animals can be developed.
n. Inbreeding, the mating of closely related animals, increases the chances of enhancing both desirable and undesirable traits.
o. Crossbreeding, the mating of animals from different breeds, is used a great deal in livestock improvement.
p. Artificial insemination makes it possible to use a good male to breed many more females than would be possible by natural breeding.
q. Success in breeding livestock depends on selecting parent stock capable of transmitting desirable characteristics to their offspring.
r. In livestock selection, physical appearance is not as important as is actual production.
s. Offspring of high-producing animals are more likely to be high producers than are offspring of low-producing animals.
t. The best measure of a bull is the production of his offspring.
u. Animals for production and breeding should be obtained from breeding stock of high production.
v. Low-producing animals should be culled.
w. Performance testing, or production testing, is important in the long-time process of improving livestock through breeding.

REFERENCES

Ensminger, M. E. *Animal Science*. Interstate Publishers, Inc., Danville, IL 61832.

Ensminger, M. E. *Animal Science Digest*. Interstate Publishers, Inc., Danville, IL 61832.

Ensminger, M. E. *Dairy Cattle Science*. Interstate Publishers, Inc., Danville, IL 61832.

Ensminger, M. E. *Poultry Science*. Interstate Publishers, Inc., Danville, IL 61832.

Ensminger, M. E. *The Stockman's Handbook*. Interstate Publishers, Inc., Danville, IL 61832.

Ensminger, M. E. *Stockman's Handbook Digest*. Interstate Publishers, Inc., Danville, IL 61832.

Herman, H. A. and F. W. Madden. *The Artificial Insemination and Embryo Transfer of Dairy and Beef Cattle*. Interstate Publishers, Inc., Danville, IL 61832.

Lee, J. S., and D. L. Turner. *Introduction to World AgriScience and Technology*. Interstate Publishers, Inc., Danville, IL 61832.

Osborne, E. W. *Biological Science Applications in Agriculture*. Interstate Publishers, Inc., Danville, IL 61832.

U.S. Department of Agriculture, *Agricultural Research*, a monthly publication. Agricultural Research Service, U.S. Government Printing Office, Washington, DC 20402.

Chapter 11

Feeding Livestock

Farmers and ranchers who are most successful with their livestock consider factors such as: (1) improving their herds and flocks for the purposes for which they are kept, (2) feeding the animals properly, (3) giving them proper care and protection, and (4) maintaining them in good health. It should be emphasized that success is dependent upon no one factor alone but upon the proper combination of all of them. If any one factor is neglected, failure is likely to result.

In the preceding chapter, methods of selecting and improving livestock were discussed. In this chapter, methods of feeding livestock are given consideration.

1. Describe how animals used feed to carry on the life processes.
2. Discuss and explain the purposes of nutrients in feed as applied to the life processes.
3. Discuss feed selection and methods effective for various kinds of livestock.

Why Is Good Feeding of Livestock Important?

Feeds make up an important part of the costs of raising livestock. During one year, a single dairy cow in the Middle West, where she gets pasture part of the year, eats about 3 tons of silage, more than 1 ton of hay, nearly 1 ton of grain, and the grass on 1 acre of good pasture. An average hen requires about 90 pounds of feed annually. To produce a 225-pound hog for market requires about 1,000 pounds of feed, consisting largely of grains and special protein feeds. To develop a 450-pound steer to 1,000 pounds requires about 50 bushels of corn, 200 pounds of protein feed, nearly 1 ton of hay, and 1 acre of average pasture.

Proper feeding has an important bearing on whether livestock will show a profit or a loss to the owner. As a matter of fact, feed is the largest item in livestock production, even when feeding is properly done. In the production

Fig. 11-1. Dairy cattle feeding on forage. Calcium deficiencies rarely occur where cattle eat sufficient forage. (Courtesy, U.S. Department of Agriculture)

of meat animals, such as swine, beef cattle, and sheep, feed costs account for about 80 per cent or more of the total expenses. In egg and milk production, feed amounts to half or more of the total costs. The need for careful attention to feeding methods is evident. Maximum returns must be secured from each feed dollar. Underfeeding, overfeeding, poor combinations of feeds, abrupt changes in feeding, irregular feeding, and failure to use the most economical kinds of feeds are some of the mistakes that may prove costly to livestock raisers.

How Do Animals Use Their Feed?

Farm animals, like factories, convert raw materials into useful products. These raw materials in the form of feeds are changed by the animal body into meat, work, milk, eggs, wool, etc. Although it is possible to compare an animal body to a factory or a machine, an animal is far more complex than any mechanical contrivance to which it might be compared.

Uses of feeds in the animal body

A part of the power or energy of a factory is required to keep the machinery in motion before any products can be manufactured. In a similar manner, an animal requires energy to carry on the life processes before any products can be obtained. This part of the feed is used by an animal to maintain its body weight, keep its body warm, repair its tissues, furnish energy for the operation of organs such as the heart and lungs, and provide for

Feed Grains: Feed and Residual

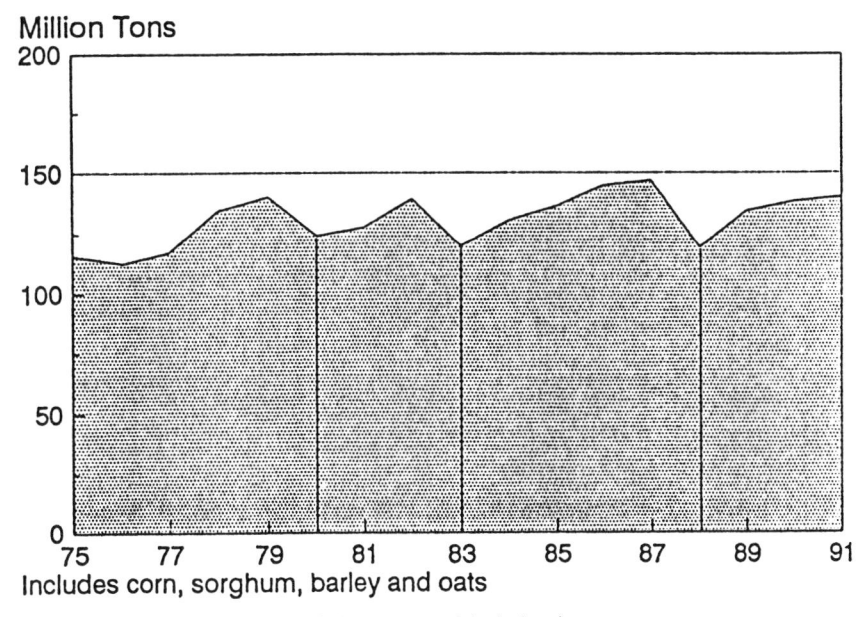

Million Tons

Includes corn, sorghum, barley and oats

Fig. 11-2. (Courtesy, U.S. Department of Agriculture)

slight movements of its body. The portion thus used is called the *maintenance ration*.

Before machines in a factory can manufacture useful products, these machines must be supplied with power that is greater than that required for keeping them in motion without a load. In the same way, an animal must be given more feed than that necessary for maintenance before it will produce in a satisfactory manner. The feed that is used in the manufacture of useful products by the animal body is called the *productive ration*. The useful products may be milk, eggs, wool, fat, work. One-half or more of the feed that a properly fed animal receives is used for maintenance. When improperly fed, an animal may require most of its feed for this purpose and thus have little left for other purposes.

In addition to requiring feed for maintenance and for the production of useful products, an animal may need feed for growth. This is, of course, true of young animals that are still growing. Pregnant animals require feed for the development of unborn young.

Nutrients in feeds

Feeds contain specific materials that the body uses for the various purposes mentioned in the preceding paragraphs. The materials that are neces-

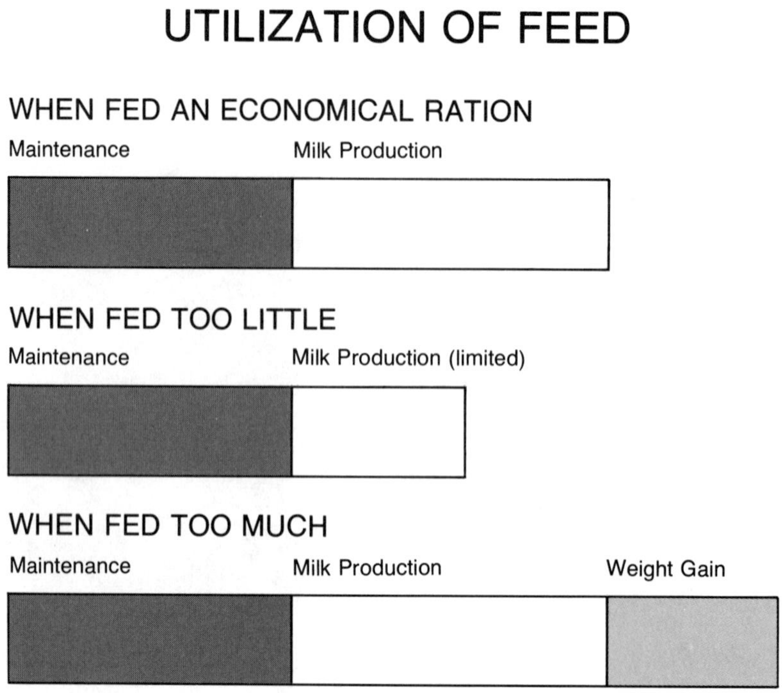

UTILIZATION OF FEED

WHEN FED AN ECONOMICAL RATION
Maintenance   Milk Production

WHEN FED TOO LITTLE
Maintenance   Milk Production (limited)

WHEN FED TOO MUCH
Maintenance   Milk Production   Weight Gain

Fig. 11-3. A mature dairy cow uses feed chiefly for maintenance and milk production. When fed an economical ration, as shown in the top portion of the above chart, a dairy cow uses about half of the nutrients for each purpose. When she is fed too little, what result is indicated? When fed too much? (Courtesy, Michigan State University)

sary for the proper nourishment of the body are called *nutrients*. The important nutrients are carbohydrates, fats, proteins, minerals, vitamins, and water. Each of these materials serves specific functions in the nourishment of the body.

Carbohydrates are used in the body for the production of heat, energy, and fat. They are also used in the production of milk and eggs. Carbohydrates include the starches and sugars and also the fibrous material known as cellulose.

Fats are used in the body for many of the same purposes as carbohydrates. For the most part, the feeds that are given to farm animals are low in fat, but the animal body has the ability to convert carbohydrates into fats for supplying fatty acids needed for growth, for storage in the body, and for use in yielding products such as milk and eggs.

Proteins are used for the repair of body cells, for growth, and for the development of unborn young. Proteins are also essential parts of milk, eggs, and wool. The amino acids that make up protein are often called the building blocks of blood proteins, body tissues, and enzymes.

Minerals are used primarily in bone development, but they are also

Feeding Livestock

used in the production of various animal products, including milk and eggs, in the formation of body tissues, and in the body processes. Blood and certain other fluids of the body contain small amounts of minerals. Phosphorus and calcium, in addition to the amounts in regular feed, are required by animals. Other minerals needed by animals are iodine, cobalt, iron, copper, manganese, magnesium, sodium, potassium, sulfur, chlorine, zinc, molybdenum, selenium, chromium, and fluorine. Most of these are needed only in trace amounts. Commercial mineral mixes are often used to provide the needed minerals. Calcium is of particular importance, being needed for blood coagulation, membrane permeability, and tissue functioning, as well as for the bone development noted above. Phosphorus also is used in bone development, providing strength. It is also needed for efficient use of feed, for growth and reproduction, and for the production and use of energy.

All animals should have common salt. Salt provides sodium and chlorine, both needed for animal life. Animals that live mostly on forage need more free salt than do swine or poultry. Salt also assists with many body functions, among them blood formation, saliva secretion, enzyme action, body acidity control, body cell maintenance, and body fluid formation. Salt also provides some of the trace minerals needed.

Fig. 11-4. The nutrient in least adequate supply becomes a limiting factor in the efficient use of other nutrients and in animal growth or production.

Water is the largest single component of nearly all living things, as well as of products such as milk and eggs. It is one nutrient from which there are apparently no ill effects from excessive consumption.

Water is used extensively in the body of an animal. It makes up over 90 per cent of the blood that acts as the carrier for transporting nutrients to the body cells and for removing from the body the wastes, or by-products, of body functions such as digestion, assimilation of food, metabolism, and respiration. Water also regulates the body temperatures and helps to disperse the heat generated by certain chemical reactions that take place in the body. Water helps lubricate the joints, protects the nervous system, contributes to good vision, and enhances hearing.

Water is available to animals from the water they drink, from the feed they eat, and from their own metabolism—the breaking down of carbohydrates, proteins, and fats. Life would not be possible without water.

Vitamins of several kinds are necessary to keep the body healthy and vigorous. A shortage of specific vitamins may cause serious disturbances in the animal, stunted growth, and lack of thrift in other ways. Up to now, scientists have found that 12 or more vitamins are of some importance to animals. Each of these serves one or more specific functions in the animal body.

Vitamin A promotes growth in young animals and seems to help them resist certain types of infections. A type of eye infection in chickens, for example, is likely to result if the ration is low in vitamin A.

Fig. 11-5. Beef cattle licking salt blocks. Proper distribution of salt in a pasture aids in distribution of stock and proper use of range. (Courtesy, U.S. Department of Agriculture)

Vitamin B is really a group of vitamins, each with a different purpose. In sheep and cattle, most of these vitamins are manufactured in the digestive tract. Vitamin B_{12} is necessary for the normal growth of young pigs and chickens and may be provided by feeds from animal sources.

Vitamin C, or ascorbic acid, is manufactured in the bodies of farm animals and, therefore, does not need to be given attention in feeding them under most conditions.

Vitamin D is provided in sufficient amounts in most cases. This vitamin is necessary for the use of phosphorus and calcium by the body. It is sometimes called the sunshine vitamin because animals that receive the direct rays of sun have the power to manufacture their own supply.

Under normal conditions and with good rations, it is likely that most farm animals secure enough vitamins. However, livestock raisers need to be alert to new discoveries in feeding and to conditions that merit the addition of vitamins to the rations of livestock.

Materials called *growth stimulants* are being added to the rations of some types of livestock with beneficial results. Some of these are called antibiotics. Some kinds of antibiotics in small quantities increase the growth rate of pigs and chickens as much as 25 per cent, decrease somewhat their feed requirements per pound of gain, and increase the thriftiness of some animals. Some antibiotics have shown promise in the feed of calves, cattle, and lambs that are being finished for market. A hormone called thyroprotein has increased the milk production of dairy cows. Growth stimulants are drugs and should not be considered to be nutrients. Because of the spectacular effects of some of these materials under some conditions, they have frequently been called wonder drugs. Just how these materials produce the effects indicated is not fully known. They must be used carefully and in accordance with recommendations from reliable sources. Some of these substances cannot be used because they may be harmful to humans.

Digestion of feeds

Before animals can utilize the nutrients in feeds, eating and digestion must take place.

Animals differ in the capacities of their stomachs and the other portions of their digestive tracts, as shown in Fig. 9-7. Cattle and sheep are ruminants. They can handle large quantities of bulky feeds known as *roughages*. As explained in Chapter 9, these animals are called ruminants because they chew much of their feed a second time in the form of cuds that are brought back to the mouth from the paunch, the first compartment of the stomach. This material in the form of cuds is chewed and swallowed again. This process, plus the action of large numbers of microorganisms in the stomach, aids in digesting coarse, fibrous feeds. Horses have comparatively small stomachs, but their capacity for digesting bulky feeds is increased by a large caecum located at the same place as the appendix in humans. Hogs and chickens have less capacity for bulky feeds and therefore require much of their feeds in forms that have high food value in proportion to their volume. Such feeds are called *concentrates*.

During the digestive processes, the various ingredients of feeds are changed into forms that are soluble in the fluids of the body. These soluble portions are absorbed through the walls of the intestines and carried by the blood to the various organs and tissues of the body, where they are used for maintenance, production, and growth.

Some of the materials in feeds are not digested by the animal body. Those portions of feeds that are capable of being digested are called the *digestible nutrients*. The remaining portions are eliminated from the body.

How Should Feeds Be Selected for Farm Animals?

It is important that farm animals be given rations that supply the necessary nutrients in sufficient quantities. A ration is the amount of feed that an animal receives during a period of 24 hours. A balanced ration contains the nutrients needed by the animal for body maintenance, growth, reproduction, and the production of animal products such as milk and eggs.

Every animal requires protein, carbohydrates, fats, minerals, vitamins, and water. Some types of livestock require these nutrients in proportions different from those necessary for other types of livestock. For example, a young pig that is growing and fattening requires more protein than a full-grown hog that is being fattened. A dairy cow that is producing milk needs more protein and minerals in its ration than a mature horse. Thus, each kind of animal must be fed according to its needs. In each case, feeds should be selected that will come nearest to fulfilling these needs, as well as being suitable in other ways.

In feeding their farm animals, livestock farmers should consider the costs of the various feeds. While it is important to select a combination of feeds that will furnish the needed nutrients, it is also important to choose feeds that are reasonable in cost. Because homegrown feeds are usually the cheapest, rations probably should consist largely of such feeds. However, livestock farmers frequently must purchase some feeds, especially feeds high in protein and minerals, for some kinds of livestock. Before they purchase these feeds, they should investigate prices to determine which feeds will be the most economical for the desired purposes.

Farm animals, like people, have *likes* and *dislikes* with respect to the feeds that are put before them. Dairy cows, for example, usually like corn silage better than dry cornstalks. When selecting feeds, good livestock farmers consider these likes and dislikes.

When selecting feeds, livestock farmers must also take into account the bulkiness of the ration. As previously explained, some types of livestock, particularly cattle and sheep, have digestive systems that are suited to bulky feeds. Hay, pasture crops, and silage are bulky feeds, or roughages. Other animals, such as hogs and chickens, require a less bulky ration. Grains, linseed meal, and tankage are examples of these feeds. Although all animals require some feeds from each group, it is important to provide these feeds in the proportions most favorable to the animals in question.

When selecting feeds, livestock farmers should also consider the effects of a feed upon the animals and their products. Some feeds, such as wild onions, will taint dairy cows' milk if they are eaten in sufficient quantities. The meat from hogs that have been fed large quantities of soybeans is often soft and inferior in quality. Soybean oil meal, from which the oil has been removed, however, does not produce this effect. Horses may have serious digestive disturbances if they eat spoiled feeds. Sheep and cattle may bloat if they consume large quantities of green clover or alfalfa to which they have not become accustomed.

Thus, it is important to select carefully the feeds that are used in rations for the various kinds of farm animals. Good-quality hay is important in the livestock ration. Unofficial standards for hay have been developed by the American Forage and Grassland Council. The address of the council is 121 Dantzler Court, Lexington, Kentucky 40503.

Good pasture is an important source of feed on farms and ranches where livestock are raised. Pasture may provide a major portion of the feed for cattle, sheep, and horses if it is available. Hogs and even chickens also benefit from good pasture crops. Good pasture crops supply proteins, minerals, vitamins, and carbohydrates and reduce the amounts of grain and other concentrates needed. In addition, livestock on pastures are furnished with sunshine, exercise, and sanitary conditions that are favorable to livestock health. Animals that graze on pasture harvest their own feed, so that human labor is saved. Land in pasture is less subject to erosion than land in cultivated crops. Newer methods of fertilization, improved varieties of pasture crops, and better methods of using pastures contribute to increased returns per acre of pasture. In some cases, it is profitable to cut the grass crop and haul it while green to the livestock for eating. By using these practices, a livestock farmer can increase the capacity of a pasture and the pounds of gain or milk per acre.

What Feeding Methods Are Effective for Each Kind of Livestock?

Each kind of livestock should be fed in accordance with practices shown to be effective. In most cases, it pays to utilize homegrown feeds as the principal part of rations. Purchasing some protein feeds, salt and other minerals, and perhaps other portions of the rations is usually necessary. Many scientific experiments on animal feeding are conducted by state colleges of agriculture and the U.S. Department of Agriculture.

Feeding dairy cattle

Probably the first people to keep cattle for milk production depended solely on pasture for the feed supply of these animals during the entire year.

If a dairy cow is given too little feed, she will provide for her own body needs first; and, as a result, her milk production will be reduced. Young dairy cows need feed for growth. Pregnant dairy cows require feed for the development of their unborn young. For these purposes, cows require some of each of the common nutrients.

Carbohydrates and some fats are supplied in feeds such as corn, oats, barley, and other grains. Hay and silage also contain fats and carbohydrates. Rations are often too low in protein, which can be provided in legume hay, linseed meal, soybean oil meal, bran, and, in limited amounts, the common grains. Minerals are also important in the rations for dairy cows. Almost all the required minerals, other than salt, can be supplied by good legume hay or good pasture. The concentrate mixture may also be supplemented with steamed bone meal.

Water is required in large quantities, because milk is about 87 per cent water. Between 6 and 7 gallons of water are required by a cow for each gallon of milk she produces. A continuous supply of water is important for highest milk production.

Some vitamins are combined in whole grains, legume hay, silage, and green feeds. Cows can manufacture vitamin C and some other vitamins. Usually, cows with a good assortment of feeds will be abundantly supplied with vitamins, with the possible exception of vitamins A and D. If cows are given good-quality hay or pasture crops, they will usually have ample amounts of vitamin A. Good-quality hay that is cured in the sun has considerable vitamin D. Cows benefit from direct sunlight; thus they should be given the opportunity to bask in the sun at intervals throughout the year, when weather permits.

Feeding producing cows. Dairy cows producing milk should be given as much good roughage as they will eat. Most cows on good pastures require little or no feed in addition to grass. Some of the best producing cows may benefit from the addition of grain. If pastures are poor, additional roughages and some grain should be fed to most of the producing cows.

During the winter, legume hay is a good roughage. Many dairy producers provide corn silage or grass silage in addition to hay. Cows that are capable of producing considerable milk need some concentrated feeds,

Feeding Livestock

largely grains, in addition to roughages. These concentrated feeds may consist mostly or entirely of ground homegrown grains. Corn, oats, barley, and sorghum grain are some of the grains available, depending on the section of the country. It may be desirable to include some high-protein feed, such as linseed meal, cottonseed meal, or soybean meal, in the grain mixture. If roughages of poor quality are fed, the proportion of high-protein feed in the grain mixture should be increased. The following are some suggested mixtures of concentrates for use with various roughages. In all cases, the grains should be ground.

1. On good pasture or with good-quality legume hay and some silage:
    a. Corn or corn and cob meal.
    b. Corn and oats, equal parts by weight.
    c. Corn, oats, barley, and wheat bran, equal parts by weight.
2. With medium-quality hay and silage:
    a. Fifty pounds corn, 30 oats, 20 soybean meal or cottonseed meal.
    b. Fifty pounds corn, 35 bran, 15 soybean meal or cottonseed meal.
3. With poor-quality hay and silage:
    a. Forty pounds corn, 35 oats, 25 soybean meal or cottonseed meal.
    b. Forty pounds corn, 40 bran, 20 soybean meal or cottonseed meal.

A good dairy producer usually makes a mixture of concentrates for the entire herd and varies the amount for each cow in accordance with her production of milk and the test of her milk. An approximate guide for a cow producing milk testing about 3.5 per cent butterfat is to feed 1 pound of concentrate mixture for each 4 or 5 pounds of milk produced daily. Thus, a cow that produces about 35 pounds of milk per day should receive about 7 to 9 pounds of concentrate mixture per day. A cow that produces milk that tests 4 to 5 per cent butterfat should have about 1 pound of concentrates for each 3 to 4 pounds of milk produced per day. An experienced dairy producer will watch each cow and decide whether to increase or decrease the amount of concentrates.

A dairy cow eats about 1 pound or more of hay and about 3 pounds of silage for each 100 pounds of her weight. Thus, a cow weighing 1,200 pounds eats at least 12 pounds of hay and 30 pounds of silage daily. If hay is the only roughage fed, she will usually eat 2 to 3 pounds daily per 100 pounds of liveweight.

In addition to grain, hay, and silage, a dairy cow should be given plenty of water. This should be made available in drinking bowls or in water tanks at all times of the day. A cow producing considerable milk will drink about 30 gallons or more daily.

Cows should be provided with salt, preferably in boxes available to them at all times. For providing phosphorus, a small amount of steamed bone meal may be mixed with the concentrates at the rate of about 1 per cent by weight, or it may be mixed with salt. In some areas, iodine may be needed. If so, an iodized form of salt may be purchased. In some areas, small amounts of cobalt may be needed.

Cows should be dry for a period of about six weeks previous to the date that they are expected to give birth to a calf. This means that they should produce no milk during this period of time. During this period, they should be fed sufficient amounts of grain and other feeds to get them in good condition for freshening. After freshening, the amount of feed should be increased gradually.

Feeding dairy heifers. When in the barn, dairy heifers should have plenty of good-quality hay, haylage, or silage. Corn silage is a good feed for heifers and may be used to replace some of the hay. Up to 20 pounds of silage per day may be fed. Grass silage is also a good feed. Enough grain should also be fed to keep the heifers in good body condition. This may require up to 6 pounds of grain per day. The amount of grain should be increased just before calving time.

Heifers may be turned out to pasture when they reach about 6 months of age. It is best to continue feeding some hay and grain to make sure that they grow well and stay in good condition.

The mineral supplements and salt used for the cows may be used to meet the needs of the heifers.

Feeding dairy calves. A calf should be permitted to nurse its mother for the first two or three days after it is born because the first milk from a fresh cow has a laxative effect on a young calf and contains large amounts of vitamin A and other beneficial materials. A calf should then be taught to drink. A special pail with a large nipple attached is best for this purpose. The milk allowance should be given in two or three feedings per day. A calf weighing about 60 pounds should be given approximately 6 pounds of milk per day. This amount may be increased as the calf gets older. Whole milk may be continued until the calf is about two weeks old. Skim milk may then be gradually substituted. Milk in some form should preferably be continued until a calf is about six months of age. The milk that is fed should always be at about the temperature it was when it was taken from a cow, and it should be fed in a clean pail in order to prevent digestive troubles. If skim milk is not available, other feeds may be substituted, such as calf meals or other milk replacers that may be purchased.

A calf will start to eat hay and grain when it is a week or two weeks of age. Good alfalfa or clover hay should be provided at this time. A grain mixture of coarsely ground corn and oats with a little bran or linseed meal is suitable. At first, a calf can safely be given all the grain it will eat. After a few weeks, the grain may be limited to about 3 pounds per day.

Feeding chickens

Experiments have shown that many hens are poor layers because they

Feeding Livestock

are fed improperly. As a basis for proper feeding, the needs of hens should be considered. Hens, like other animals, require feed for body maintenance. About 70 per cent of the 80 or 90 pounds of feed that a properly fed hen consumes each year is used for this purpose. About 10 per cent of the feed eaten by a growing pullet is used for growth, another 10 per cent for building a reserve of body fat, and 10 per cent for egg production, in addition to the 70 per cent for maintenance. Before a hen will produce eggs in keeping with her ability, all her body needs must be provided.

For her body needs, a hen requires feeds that contain proteins, carbohydrates, fats, minerals, vitamins, and water. An egg contains approximately 66 per cent water, 13 per cent protein, 10 per cent fat, and 11 per cent minerals. These nutrients are required by a laying hen in amounts additional to those needed by her body.

Parts of a good ration for a laying flock. A good ration for a laying flock usually contains the following parts: (1) a dry mash mixture, (2) a grain mixture, (3) minerals, (4) grit, and (5) water. The exact feeds for the mash and grain mixtures depend somewhat on the prices and availability of feeds in a given community. Therefore the following feeds should be considered as suggestions only.

A dry mash mixture is usually made from ground grains, feeds high in protein, small amounts of minerals, alfalfa meal, and vitamins A, D, and riboflavin. The grain mixture ordinarily contains whole or coarsely ground

Fig. 11-6. Automation in poultry feeding. Mr. Frye adjusts controls in the mill room of his food processing center so feed will be delivered to an auger wagon for transfer to the field. Pipes bringing unprepared feed to the mill (behind Mr. Frye) converge above it. Prepared feed drops at right into a bin. (Courtesy, U.S. Department of Agriculture)

grains. Where available, yellow corn is preferable to white because it contains considerable vitamin A.

Suitable commercially made mash mixtures may be purchased from reliable feed dealers. Home-mixed mashes may be mixed in accordance with recommendations in publications from state colleges of agriculture. A mash mixture should be fed dry in a self-feeder, and it should be available to the hens at all times.

Growing chickens need more protein than do mature chickens. Chickens about one month old will need a ration with 20 per cent protein. The percentage of protein should be reduced gradually until it reaches 15 per cent to 16 per cent when the chickens are 8 months of age. At this time, each chicken will eat 2½ to 3½ pounds of feed each day.

Chickens may also be turned out on pasture or fed chopped green plants to supplement the ration. Green, tender plants provide valuable feed for chickens. Of course, chickens should be put out on pasture only when the weather is suitable.

A grain mixture may consist of various combinations of whole grains or coarsely ground grains, such as two parts of shelled corn and one part of oats. Another mixture consists of equal parts, by weight, of shelled corn and wheat. The mixture is sometimes called a scratch mixture because it may be fed in the straw or other litter on the floor of the poultry house in order to promote exercise. The grain mixture may be placed in one feed hopper and the dry mash in another hopper, so that hens can choose the amounts they eat from each. In some cases, an all-mash ration is fed without a separate grain mixture.

The percentage of protein used in the mash should be sufficient to provide the proper percentage of protein for the ration. For example, if 12½ pounds of grain is fed to each 100 hens each day, the percentage of protein for the mash should be 20 per cent to 22 per cent. If 5 pounds of grain is fed to each 100 hens each day, the percentage of protein for the mash should be 15 per cent to 16 per cent. For feeding grain, it is best to feed no more than the chickens will eat in about 20 minutes.

Minerals are important in a poultry ration because they are needed for certain body structures of the hen. Certain minerals are also used in the production of eggs. Some minerals, such as small amounts of ground limestone, bone meal, and salt, are usually added to the mash mixture. Ground oyster shells, which are valuable because they supply large amounts of lime, needed especially for egg shells, may be placed in a container available to the hens at all times.

Grit, consisting of hard materials, such as small pieces of gravel, is helpful in the grinding of food that takes place in the gizzard. Ground granite is a desirable form of grit that can be purchased. Grit should be placed in a box available to the hens.

Water should be provided at all times. In the winter, it is important that water be kept from freezing. Electrically heated automatic waterers are available at reasonable prices.

Some poultry raisers are using lights to lengthen the "working day" of hens during the winter season. This method results in higher egg production during the winter season when egg prices are high.

Many chickens are raised for meat purposes. Some persons raise small numbers for home use. However, many commercial producers have large-scale operations and feed out thousands at a time. These are usually marketed as broilers at about 3 pounds in weight. In most cases, mixed feeds that contain proper amounts of the various nutrients are purchased. Antibiotics are included in most of these feed mixtures. If properly fed, broilers weigh 3 pounds at about eight or nine weeks of age.

Feeding hogs

We often hear the expression "greedy as a pig." As a matter of fact, the pig appears to use more common sense in its eating than do most other farm animals and many people.

Feeding hogs for market. In feeding hogs for market, hog producers usually find that getting them to market weight in the shortest possible time is most profitable. When properly fed, pigs may reach 200 to 225 pounds at six months of age or less.

Hogs being fattened for market can be put on *full-feed;* that is, they are allowed to eat as much as they wish. The method known as *self-feeding* has

Fig. 11-7. The self-feeder is filled with feed so the hogs can help themselves whenever they wish to eat.

been shown to be a profitable practice. Sufficient quantities of feeds that will last for several days or longer are placed in a self-feeder. Shelled corn in one compartment of the feeder and a suitable protein mixture in a second compartment are a widely used combination of feeds for the self-feeder. A hog apparently is able to choose fairly accurately the amount of each of these feeds that it needs and thus is able to balance its own ration.

Suitable protein mixtures may be purchased from reliable dealers. For growing fattening pigs these mixtures usually contain two or more feeds high in protein (such as tankage, soybean oil meal, and dried skim milk) and alfalfa meal. In addition, small amounts of minerals are included, such as salt, ground limestone, and steamed bone meal. Several kinds of vitamins and some antibiotics are also desirable in a mixture, especially for pigs up to 100 pounds in weight.

Many swine feeders prefer to mix ground corn or other grains in suitable proportions with the protein feeds, vitamin feeds, antibiotics, and minerals and thus provide the ration as one mixture. Suggested mixtures of appropriate kinds are available in publications from state colleges of agriculture.

Feeding too much protein to pigs may be wasteful. A 20 to 40 per cent protein diet by dry weight is sufficient up to 50 pounds. From 50 to 100 pounds, pigs need a 16 per cent protein diet; and from 100 to 200 pounds, a 13 per cent protein diet is satisfactory.

Water should be provided, preferably in a self-waterer of a type that will not freeze in winter.

Barley or sorghum grain may be used as a substitute for corn, but it should be ground coarsely for feeding to hogs.

Sometimes, hogs are turned into a field of corn and permitted to do their own harvesting. This practice is called "hogging-down" corn.

Good pastures, such as ladino clover, alfalfa, and lespedeza, should be provided for hogs during as much of the year as possible.

Feeding a sow and her litter of pigs. It is essential that pigs be given a good start so they will live and develop rapidly. A bred sow should be given a good ration so that she will farrow a litter of vigorous pigs.

A good ration for a bred sow consists of protein feeds in ample amounts and sufficient corn or other grains to keep her in thrifty condition. During gestation, a gilt needs only an 8½ per cent protein diet. At farrowing, she needs a 16 per cent protein ration. During seasons when good pasture is not available, she should be given alfalfa hay. This may be placed in a suitable rack where she can eat it when she chooses, or alfalfa meal may be placed in the protein mixture. Some minerals, such as salt and steamed bone meal, may be mixed in small amounts in the protein mixture, or a mineral mixture may be placed in a container available to her at all times. She should be given all the water she will drink. Some of the feed, such as ear corn, may

Fig. 11-8. A pig creep for feeding young pigs. Only the little ones can get in to the feed. (Courtesy, U.S. Department of Agriculture)

be fed at a considerable distance from her sleeping quarters so that she will be encouraged to get ample exercise.

Sometimes, pigs are born without hair. Such pigs are usually dead at birth or are so weak they do not live very long. This condition is due to a shortage of iodine in the ration of the mother previous to farrowing. Feeding a mineral mixture containing iodine or feeding iodized salt to the sows during pregnancy will supply the proper amount of iodine.

Sows suckling large litters of pigs should have about all the feed they will eat. For this purpose, a self-feeder with shelled corn in one compartment and a suitable protein mixture in another compartment may be used. As an alternative, the corn, and other grains, may be ground and mixed in suitable proportions with the protein feeds. This mixture may be placed in a self-feeder.

In addition to nursing the sow, little pigs should have grain and protein feeds. These are fed in a creep, which is designed to admit little pigs only. Rations for the pigs should include suitable amounts of minerals, vitamins, and antibiotic feeds to insure thriftiness and rapid growth.

Feeding beef cattle

Cows in a beef breeding herd can secure sufficient feed from good pasture during much of the year in many parts of the United States. In some parts of the country, pasture is not available during the winter months. In these cases, the cows can be wintered satisfactorily on hay alone or with silage or other roughages in addition. For body maintenance, it will take about 2 pounds of dry roughage for every 100 pounds of body weight. Silage may be substituted for hay on the basis of 3 pounds of silage to 1 pound of hay. Depending on the quality of the roughage, it may be necessary to provide a grain supplement. Good-quality legume hay is desirable for part of this ration. In the northern portions of the range areas, hay is frequently fed to the breeding herd during the winter. In the South and in some of the range states, cottonseed cake is frequently used to supplement the winter pasture. This is made from crushed cotton seeds from which the oil has been removed.

It does not pay to overfeed cows on winter pasture. If good pasture is available, a pound or a little more of cottonseed meal daily is usually sufficient. The exact amount of meal to supplement the pasture will vary somewhat from one area of the country to another. Overfed cows may wean calves that will weigh slightly more than the calves of cows fed a lighter ration, but the overfed cows will not be as long-lived and the percentage of calf losses may be twice that of the cows on the lighter ration.

Some farmers in various parts of the country raise their own calves for feeding. If these calves are fed for rapid growth, they are ready for market at an age of 14 to 18 months. In some cases, the calves are fed corn or other

Fig. 11-9. Beef cattle on good pasture. They will start eating again when the photographer is finished. (Courtesy, The American Hereford Association)

grain and some protein feed, such as cottonseed meal, while they are running with their mothers on pasture. These feeds are placed in creeps, which only calves can enter. In many cases, pastures provide the only feed for the cows and calves. In such cases, care must be taken not to overstock the pastures. Overstocking can result in serious losses in the weaning weights of the calves. Some research by USDA scientists has indicated that the weaning weights of calves reflect overstocking of pastures better than does the condition of the cows.

Many cattle raised on ranges are fattened in the feedlots of the Middle West and elsewhere. This is the final stage in the fattening of beef cattle for market. In the feedlot, cattle are fed some legume hay and perhaps silage, large amounts of corn or other grains, and some protein feed, such as linseed meal, cottonseed meal, or soybean meal. Sometimes small quantities of growth hormones are added to the feed to increase the rate of gain and to decrease feed costs. The use of hormones is being reduced because of possible danger to human health. For example, DES is now banned because it may cause cancer.

USDA researchers have found that steers apparently need no more fiber in their diet than is contained in ground corn. Three groups of good grade feeder steers (nine animals per group) were self-fed all they wanted for 200 days. Group A was fed a ration containing 68 per cent corn and cob meal, 20 per cent ground corn, and 500 units of vitamin A per pound of feed. Each steer averaged 2.3 pounds of gain daily. Group B was fed ground corn (88 per cent) and 500 units of vitamin A. Each animal gained 2.1 pounds a day. Group C averaged 2.3 pounds of gain a day on the 88 per cent ground corn diet and 2,500 units of vitamin A. About 12 per cent of all rations were soybean meal, bone meal, and mineralized salt. Groups B and C also converted feed into meat more efficiently than did Group A. Steers in Group C

Fig. 11-10. Hundreds of beef cattle in a large feeding establishment in a western state.

consumed only 6.6 pounds of feed per pound of gain; Group B animals ate 7.3 pounds of feed per pound of gain; and Group A animals ate 8.5 pounds of feed per pound of gain. All animals gained about the same total amount of weight. The increased amount of vitamin A in the ration for Group C was to determine if it would help avoid stress and fatigue in the summer months, and it apparently did. Other experiments indicate that the protein level in the feed must be maintained. On rations containing only half the recommended amount of protein, the daily feed consumption of the cattle dropped more than 37 per cent.

Because steers like to eat many times during the day, animals fed frequently outgain those fed only once or twice a day. Some cattle raisers recognize this by never letting feed bunks become empty. They feed to satisfy the animals' needs rather than when it is most convenient.

In feeding cattle for market, many cattle raisers use good pasture extensively. By this method, less grain is required and the costs are reduced.

Beef cattle should be provided with salt, preferably in a box available at all times, and with a good supply of water.

Feeding sheep

During the months when good pasture is available, sheep thrive with little or no additional feed. A flock of breeding ewes should be fed a suitable ration during the winter season. Until a month before lambing time, they may be fed almost entirely on a good legume hay, such as clover or alfalfa.

Fig. 11-11. A flock of sheep in a clean, dry lot. (Courtesy, New York State College of Agriculture and Life Sciences, Cornell University)

A little silage or corn fodder may also be included in the ration, but the amount should not exceed 2 pounds per head daily. About a month before lambing time, some grain should be added to the ration. For this purpose, about ½ pound daily of a mixture of equal parts of corn and oats or other grains may be used. After lambing time, more grain should be fed to stimulate milk production. Up to 1½ pounds of corn and oats, supplemented with a protein supplement such as soybean oil meal, should be fed daily as long as the ewes are feeding the lambs and the ewes are confined.

Young lambs soon learn to eat grain and hay. A recommended procedure is to supply these in a special creep, which allows the lambs to enter but excludes the mature sheep. The lambs can be given cracked corn and crushed oats until they are able to eat shelled corn. A good legume hay should also be placed in the creep.

Lambs from the western ranges are frequently purchased and fed by farmers in the Corn Belt and elsewhere. These lambs are fed fattening rations for a period of two to three months. Corn, a protein supplement, clover or alfalfa hay, and small amounts of silage are commonly used for this purpose. Some feeders make mixtures of feeds consisting of ground legume hay, ground grains, and other feeds. In some cases, the feed mixtures are compressed into small pellets. Overeating grain will kill sheep. Sheep raisers should keep roughages equal to about half the ration.

Sheep need salt, minerals, and a supply of water. A mixture of one-half steamed bone meal and one-half mineralized salt can be self-fed.

Feeding horses

In order to work most efficiently, horses must be properly fed. The digestive system of horses seems to be more sensitive than those of other farm animals. Therefore, considerable skill is necessary in feeding horses.

Horses have a small digestive system in proportion to their size and thus do not use as much forage as cattle. Too, horses use more of their feed for body maintenance than do meat animals.

Horses are creatures of habit and respond better when fed at regular times. Regular feeding helps to prevent digestive upsets and stable vices such as cribbing. Changes made in types of feed should be gradual to avoid causing colic, off-feed, or diarrhea.

Horses need well-balanced diets that include carbohydrates, proteins, vitamins, and minerals. Salt and water are also required. The average horse will drink up to 15 gallons of water per day. To prevent bloating, horses should be watered before being fed grain or dry forage.

Timothy hay and oats are the feeds that are commonly fed to horses. For mature horses, this combination has been shown to be a good one. Good-quality hay should be free-fed or fed at the rate of 1 to 1½ pounds per 100 pounds of body weight per day. Moldy or dusty hay may cause heaves, a respiratory ailment. Horses are also susceptible to moldy feed toxicosis. Mold is frequently found in silage as well as in hay. Good-quality hay should provide the needed minerals. Calcium and phosphorus may have to be provided as a supplement if hay is of poor quality.

Corn may make up part of the ration, and clover hay or mixed hay may be used instead of timothy hay. Idle horses should have little grain. Pastures may provide much of the ration for horses, especially during periods of idleness.

Horses should not be overfed. Fat horses will not perform well and may not live as long as will properly conditioned horses.

Temporary pastures, grown as a part of a rotation, are best because they usually have fewer parasites than permanent pastures. Since horses are not likely to bloat, legume pastures are good for them.

Horses in poor condition, working, or on poor pasture should receive concentrates in addition to pasture.

SUGGESTED ACTIVITIES

1. Describe the methods you use in feeding some of your pets or other animals for which you are responsible. After studying suggestions from various reliable sources, make improvements that you believe are desirable.
2. With your class, arrange a field trip to a farm of a successful livestock farmer. Ask the farmer to tell and show you how each kind of livestock is fed. Be prepared to ask questions.
3. Write to your state college of agriculture for a bulletin on feeding some kind of livestock in which you are interested. Write a report on the methods that are recommended.
4. Plan on raising some chickens or some other kind of livestock for yourself. Write to your state college of agriculture for literature and plan carefully the rations you should use.
5. Assume responsibility for feeding some kind of livestock, if you live on a farm. Study the recommended methods for feeding them, and discuss with your parents the desirability of making changes in the ration.
6. Concepts for discussion:
    a. Feed is the largest single cost factor in livestock production.
    b. One-half or more of the feed an animal receives when it is properly fed is used for maintenance.
    c. The nutrient in least adequate supply becomes a limiting factor in the efficient use of other nutrients and in animal growth or production.
    d. Feed contains many nutrients and feed elements needed by animals for maintenance, growth, and reproduction.
    e. Each nutrient and other feed elements serve specific functions in the body of the animal.
    f. Ruminants consume large amounts of roughages; nonruminants require a high proportion of concentrates.

g. Animals differ in their ability to use feed efficiently.
h. During the digestive process, feeds are changed to forms that can be absorbed into the bloodstream and carried to the various parts of the body.
i. The useable portions of feed are called digestible nutrients.
j. A balanced ration contains the proper proportion of nutrients needed by animals.
k. Different animals require nutrients in different proportions.
l. Farm animals have likes and dislikes with respect to feed.
m. Economical animal feeds that are used for a desired purpose should be chosen.
n. Feeds that will result in the desired quality of livestock products should be selected.
o. Standards for evaluating and grading some kinds of feed have been developed.
p. Animals receiving too little feed will provide for body maintenance first.
q. The nutrient requirements of animals vary with size, age, environment, state of productivity, and kind of digestive system.
r. Water is necessary for most body functions.
s. The method of feeding varies according to the kind of animal.
t. Feeds vary in nutrient content and energy value.

REFERENCES

Baker, J. K., and E. M. Juergenson. *Approved Practices in Swine Production*. Interstate Publishers, Inc., Danville, IL 61832.

Ensminger, M. E. *The Stockman's Handbook*. Interstate Publishers, Inc., Danville, IL 61832.

Haynes, N. B., S. W. Sabin, H. F. Hintz, and H. F. Schryver. *A Horse Owner's Guide*. Cooperative Extension Information Bulletin 153 New York State College of Agriculture and Life Sciences, Cornell University, Ithaca, NY 14853.

Juergenson, E. M. *Approved Practices in Beef Cattle Production*. Interstate Publishers, Inc., Danville, IL 61832.

Schryver, H. F., H. F. Hintz, and J. E. Lowe. *Feeding Horses*. Cooperative Extension Information Bulletin 94, New York State College of Agriculture and Life Sciences and College of Veterinary Medicine, Cornell University, Ithaca, NY 14853.

Van Riet, W. J. *Beef Production in California*. Cooperative Extension Leaflet 21184, University of California, Berkeley, California 94720.

Chapter 12

Caring for Livestock and Their Products

Good livestock producers are interested in giving their livestock good care. Many tasks are involved in caring for livestock and their products. Barns and other buildings must be provided and maintained to protect livestock from adverse weather conditions. Chore routines must be performed regularly. Young animals need special care and protection. Labor-saving methods are important to reduce costs of production. Eggs and milk must be handled in ways that will assure a high-quality product.

OBJECTIVES

1. Discuss the need for proper care, buildings, and equipment for livestock.
2. Describe the care necessary for dairy cattle and dairy products.
3. Describe the care necessary for poultry and eggs.
4. Describe the care necessary for hogs, sheep, beef cattle, and horses.

Livestock are creatures of habit, just as are most other animals, including humans. It is easy to see evidence of habits by observing animals carefully. For example, animals choose the same spots for resting, they start to move toward shelter or to where food will be provided about the same time each day, changes in what they eat often upset their digestive systems, animals return to their own stalls, and pets have favorite toys. The examples of activities of animals, called habits, are almost endless. Many people get a great deal of pleasure from animals because of their habits, which makes some animal behavior predictable. Good caretakers of livestock are aware of the habits of animals and make use of this knowledge in caring for livestock and their products. When the same routine is followed every day in caring for livestock, the animals become accustomed to the routine and look forward to it rather than being frightened by it. What are some of the habits of animals you have seen?

Why Should Proper Care Be Provided for Livestock and Their Products?

We all benefit greatly from the animals raised on farms and ranches. Consequently, it is our obligation to give them proper protection and care and to contribute to their well-being in other ways.

If livestock are injured by careless handling, those who raise them are the losers. Those individuals who seek to outsmart animals rather than to subdue them may use methods that may result in injury to the animals and perhaps to themselves. Good livestock caretakers are familiar with the temperaments and traits of their livestock and exercise patience and self-control when working with them. They succeed in gaining the animals' confidence by moving among them slowly and by talking to them kindly. Back of all their actions, good livestock caretakers have a liking for their livestock.

During the first few days after they are born, animals are quite weak and susceptible to injury. If given poor care, many will die. Every animal that dies represents a loss in feed and labor. For example, every young pig that dies soon after birth represents a loss of about 140 pounds of feed that had been fed to its mother prior to its birth. Much of this loss can be avoided if proper care is provided.

Labor ranks next to feed as an item of expense in livestock production. Labor is estimated to be about 20 per cent or more of the total costs involved in producing eggs and milk. Labor is estimated to be about 10 per cent of the total costs involved in raising beef cattle. Farmers who use inefficient methods in working with livestock have high labor costs and, consequently, make less profit than would be possible with improved methods. To reduce labor costs, they should (1) locate and arrange equipment for convenient use; (2) use power equipment, feed carts, and other devices that reduce hand labor; and (3) develop efficient chore schedules that will save steps and time. In addition to saving time, these methods eliminate much of the hard work of raising livestock.

What Buildings and Equipment Are Needed for Livestock?

Spacious barns are characteristic of farms in many parts of the United States. In pioneer days, the construction of a barn, called a barn raising, was frequently made a community and cooperative affair. Modern barns are a vast improvement over some of the dark, cold, improperly ventilated structures of the early days.

In addition to barns and other buildings, much equipment is needed for different kinds of livestock. This equipment includes silos and other feed storages, fences, feeders of various kinds, stanchions, stalls, watering devices, heat lamps, brooders, barn cleaners, and many other items.

Providing buildings

The principal building on many farms is a large barn. In former years, horses were kept in one part of this barn, but today most large barns are used primarily for beef cattle or dairy cattle and for storage of hay, since there are many farms on which no horses are kept. Usually, separate buildings are provided for every kind of livestock, such as hogs, sheep, and poultry.

Buildings for livestock should be planned to provide comfort and healthful conditions for the animals, convenience for the caretaker, economy of cost, economy of labor, and durability. The interior of a barn or other building can be arranged to conserve labor. If careful attention is given to location of feeding space and feed storage, much saving in labor is possible. Floors made of concrete provide for easy cleaning and sanitation.

The designs of barns have changed in recent years. One-story structures have been built to replace old barns that had a hay mow as a second story. In these modern barns, hay and straw are stored on the first floor, since this saves labor at harvest time and in feeding livestock. Special attention is usually given to providing ample window space for natural lighting and electric lights for artificial lighting.

The space in a barn should be planned carefully for the number of livestock to be kept. This is important to reduce cost and, in northern states, to make it possible for the heat from the bodies of the animals to keep the interior at a comfortable temperature in cold weather. Ventilation is needed for the removal of undesirable odors and excess moisture, for the admittance of fresh air, and for the maintenance of a comfortable temperature.

Dairy producers are concerned with the temperature-humidity index (THI). THI is a numerical expression of the discomfort people are likely to feel due to the combined effects of temperature and relative humidity. Average THIs are available from local weather stations. Researchers have found that the THI has an effect on milk production. The effects of temperature and humidity on milk production of 56 Holstein cows were studied so that data that could be used as a guide for determining cooling needs in dairy shelters could be developed. The study found that if the THI averages more than 71 during the summer, dairy producers can expect less milk than usual from their cows unless they make their dairy shelters cooler.

Livestock also require shade if they are to produce well. In experiments to determine the best kind of shade, researchers found that the most efficient shade was a 6-inch layer of hay. Painted steel and aluminum surfaces, painted or aluminum foil-covered fiberboard, plastic and plywood surfaces, and neoprene-coated nylon were also very effective.

Fig. 12-1. Barns come in many designs. A barn in Pennsylvania with an overhanging second story *(top)*. A round barn in Illinois *(bottom)*.

Fig. 12-2. A dairy cattle barn in Indiana. (Courtesy, Farm Security Administration)

Providing equipment for livestock

A convenient supply of water is important for livestock. This is best done by having water piped from a well or a storage tank. Automatic watering devices are desirable; in northern states, these should be provided with a source of heat to prevent freezing in cold weather.

Much has been done in recent years to provide improved equipment for feeding livestock. Feed racks near the supply of hay or silage save considerable labor in feeding dairy cattle or beef cattle. Silos may be equipped with mechanical devices to remove the silage. Bunker-type silos and pit silos are used on some farms. These are economical to construct and may be filled with a minimum of labor. Some of these are equipped with a movable panel, which allows the cattle to eat their way into the silage from one end of the structure.

Self-feeders are being used extensively for feeding hogs. If properly designed, these supply feed without waste. Much labor is saved, especially if the feeders have sufficient storage space for several days. Self-unloading trucks are used on some farms to save labor in filling feeders or feed bunks.

Many large feeding operations provide for the movement of feeds from the storage areas to the self-feeders and feed racks by means of pipelines or augers. Some feeding operations are automated to the point that the farm operator stores all components of concentrate feeds in huge bins, sets a series of dials to indicate the amount of each component desired in the feed mixture, and then presses an electric switch. The machine measures out each of the feed components, mixes the feed, and then moves the feed to the feeders. In these operations, the amount of labor required for feeding is reduced to a minimum.

Barnyards and concrete feeding floors keep the animals out of mud, prevent waste of feed, and save labor in cleaning. Manure-loading attach-

Fig. 12-3. An automatic feeding system for beef cattle. The silo unloader *(top)* takes the feed from the silo and dumps it on an auger that carries the feed down a long pipe *(bottom)*, spilling feed gradually from holes in the pipe into a feed bunk where the cattle can reach it.

What Care Should Be Provided for Dairy Cattle and Dairy Products?

Probably no farm animal is more responsive to good care than is the dairy cow. Dairy cows have been improved to such an extent that they have become highly specialized animals. They have a rather nervous temperament; consequently, they are very sensitive to improper treatment. Regularity and kindness result in more milk and greater profits. Good dairy producers take pride in their cows and are particular about the care they give them.

Carelessness in handling bulls has frequently resulted in serious injury to farmers and in some cases death. Having a special safety pen and leading the bull with a staff fastened to a ring in his nose can prevent most of these accidents. It is sound advice never to trust a bull, no matter how gentle he may appear to be.

All farm animals, especially dairy cows, are creatures of habit and are easily upset by any change in routine. Feeding, milking, and other daily chores should be done at regular times if fluctuations in milk flow are to be avoided.

Dairy cows, as well as other farm animals, should be provided with an

Fig. 12-4. A nose ring is being used in the handling of this prize dairy bull. This is one of the most important farm safety practices. (Courtesy, U.S. Department of Agriculture)

abundance of bedding materials, such as clean, dry straw, in the barns where they are kept. These materials aid in keeping the animals warm and clean. Cows should be allowed some exercise out-of-doors for short periods in the direct sunlight when weather permits. During summer months, cows may be kept on pasture day and night, except at milking times.

Milking properly

Dairy cows at all times should be treated in a quiet, gentle manner. This is particularly necessary during the milking process.

Cows have been milked for thousands of years. However, until recent years, the process of milk production in the cow has not been understood. Techniques of milking that reduce the time of milking and in some cases increase the milk produced have been developed. With larger herds of cows, milking machines have come into wide use. A method of milking known as managed milking includes the following steps:

1. Prepare the cow by wiping and massaging the teats and udder. Use a clean cloth wrung out of water at 130°F. This cleans the udder and stimulates the cow to let down her milk. Use of a suitable material, such as chlorine solution, in the water reduces bacteria.
2. Milk two or three streams of milk from each teat into a strip cup. This makes it possible to get rid of first-drawn milk, which is high in bacteria, and it provides a check on the condition of the udder.

Fig. 12-5. This Wisconsin dairy farmer is using a strip cup to inspect the milk before he attaches the milking machine. This is an important step in the detection of abnormal milk. (Courtesy, U.S. Department of Agriculture)

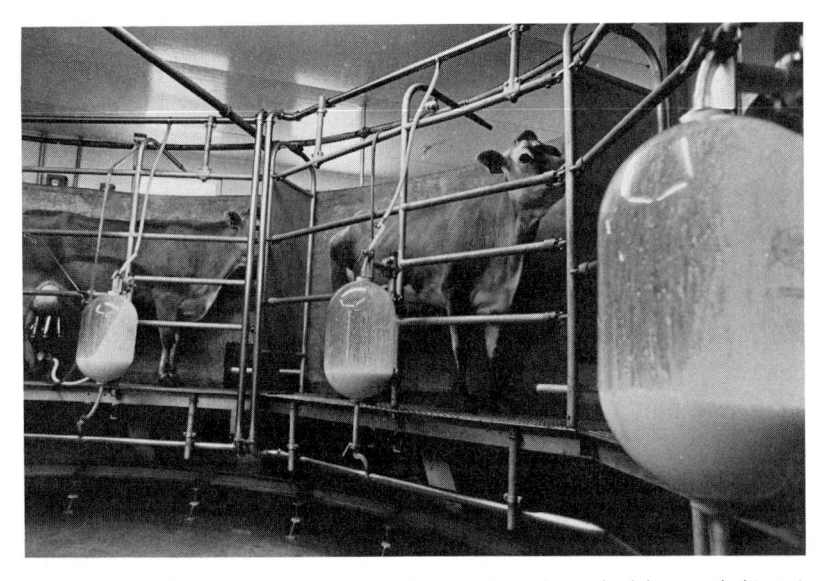

Fig. 12-6. This automated revolving milking parlor reduces the labor needed to put milk on the table. (Courtesy, U.S. Department of Agriculture)

3. Attach the milking machine about one minute after Step 1 and leave it on for about three to five minutes, depending on the individual cow.
4. Remove the milker as soon as the milk flow ceases.

Providing barns for dairy cattle

The building requirements for a dairy farm include a milking parlor, a feed processing and handling center, a feeding system, housing, and a manure handling system. Additional types of structures are needed for calves and young stock.

A dairy barn should be dry, well-ventilated, and light, and it should be maintained at a temperature comfortable for the cows and the caretakers. There should be a good floor, and the entire barn should be constructed so that it can be kept clean and sanitary without excessive labor. Flexible stanchions that permit the cows to turn their heads readily are preferred.

Some dairy farmers use pen-type barns that provide loose housing for dairy cattle. In this type of building, space is provided for a loafing room, in which the cows are free to move about freely and consume feed from racks and troughs. The cows are milked in a special room, called a milking parlor, where two or more cows are milked at a time. The cows are confined in stalls only while they are being milked, and at this time they are usually given a grain mixture. With this arrangement, less time is required for caring for the cows, and the barn itself is less expensive. Old buildings may frequently be remodelled economically into pen-type barns. Cows do well in these types of barns, even if the temperature becomes quite cold.

The type of housing built depends, in part, on the feeding system and hay and other roughage storage system. Good-quality hay is best stored under a roof, and good-quality silage should be stored in a silo.

Producing high-grade dairy products

Special precautions are needed to produce high-grade milk. Good-quality milk is clean, has a good flavor, and contains few bacteria, none of which are harmful. The production of high-quality milk does not necessarily require expensive equipment, but it does require great care and cleanliness.

Clean barns, clean cows, clean persons doing the milking, clean utensils, and prompt cooling of the milk are among the essentials for the production of high-quality dairy products. The dairy barn or milking parlor should be clean, and the air should be free from dust and odors. Cows should be free from disease. They should be clean and their udders should be wiped with a moist, clean cloth just prior to milking. The people who do the milking should wear clean clothing, and they should wash their hands before they begin milking. All milking equipment should be thoroughly sterilized at least once daily.

Silage should be fed at least three to four hours before milking time. If this is not possible, it should not be fed until milking has been completed. Feeding silage just before milking will result in the milk having the flavor of the feed eaten.

The milk should be cooled as quickly as possible after it is drawn from the cows. When milk is cooled to below 50°F, the development of bacteria is greatly retarded. The usual method for cooling is to place the cans of milk in a tank of cold water or to place the milk in a bulk-type tank with a cooling mechanism. The equipment for cooling the milk and for washing the utensils is usually kept in a special building or milk house near the barn. Specialized types of equipment for handling milk on dairy farms include pipelines from the milking machines to a bulk cooling tank where the milk can be held at about 38°F.

Spraying the walls and ceilings of dairy barns at intervals and spraying the cows with suitable chemicals to control flies will help make the cows comfortable and will aid in keeping the dairy barn sanitary. However, dairy producers must exercise great care when they are using these chemicals to avoid having the flavor of the milk affected. Some dairy producers believe that paints, fly sprays, and other odor-producing chemicals have no place in the dairy barn.

Practically all cities require that milk sold for drinking purposes be produced by healthy herds and under sanitary conditions. Herds and barns where this milk is produced are inspected periodically, and the milk from each herd is checked frequently. Milk is hauled from dairy farms in large insulated tanks mounted on trucks. At the processing plant, the milk is pasteurized, which means that it is heated at a specified temperature for a given period of time, to insure its safety for human consumption. Some milk is homogenized, which means that it is given a type of treatment that breaks up the fat particles so that the cream does not rise to the top of the containers in which it is distributed to customers.

Saving labor

Many people do not appreciate the amount of human labor involved in producing milk on farms. Actually, as an average, about 42 hours of human labor is required per dairy cow each year. The labor per cow is greatly decreased when large herds of high-producing cows are kept and when labor-saving methods and devices, such as milking machines, pipelines and bulk handling of milk, barn cleaners, efficient barn arrangement, automatic watering systems, and improved feeding methods, are used.

The best way to save labor is to take proper care of all the dairy animals so that the extra work and expense of treatment for problems can be avoided. The caretakers should be experienced in that work, should like working with livestock, should be willing to work the long hours required, and should know how to observe the animals to see when all is not well with them. Labor can also be saved, and the herd kept at a high level of production, when the most efficient routine for doing the work is worked out and that same routine is followed each day.

Caring for dairy calves

Many youth on farms raise calves of their own. It is important that these calves be given good care so that they will develop properly. Since many of these calves are shown at fairs, attention should be given to their appearance. Calves, to show at their best, should be properly fed, trained, and fitted. The training of a calf partly consists of breaking it to lead and to stand correctly. The hide and hair can be improved by practices such as washing, grooming, clipping, and blanketing.

Calf losses are one of the major costs to some dairy operations. Many of these losses occur because the same degree of care and attention that is given to the milking herd is not given to the calves. Some studies have shown that calf losses are much lower when the owner, and/or members of the family, take care of the calves. The owner and members of the family are more likely to observe the calves closely and thus can detect early signs indicating that a calf is off-feed or is developing some other problem.

Many farmers have good success housing calves in separate small shelters, or hutches, which are open on one side. The calves can be put in the hutches as soon as they are strong enough to be active and to run around. Since this type of housing exposes calves to the weather, the caretaker must observe them very carefully each day to make sure they are healthy. At the first sign that a calf may not be feeling well, the caretaker should move it inside for treatment and should keep it there until it is well and active again.

Fig. 12-7. Many dairy producers raise calves in individual small hutches open only on one side.

What Care Should Be Provided for Chickens and Eggs?

Good care is especially important for young chicks and for the laying flock.

Raising chicks

In raising chicks by modern methods, poultry raisers have replaced the mother hens on most farms with special brooding equipment. Houses about 10 by 12 feet are frequently used for young chicks. Such a house is suitable for about 200 chicks until they are a few weeks old.

For the first few weeks of their lives, young chicks must be provided with artificial heat. Electric brooders are commonly used. In one type, the heat is provided by electric infrared heat lamps. These are suspended about 18 inches from the floor. For the first few days, a temperature of 95°F is needed. This may be maintained by a thermostat that automatically turns part of the lights off and on. The chicks soon learn to find the zone most comfortable to them. As the chicks get older, the height of the heat lamps

Fig. 12-8. An electric hover for warming baby chicks. Note the food container on the left. (Courtesy, Rural Electrification Administration)

from the floor is gradually increased so that the amount of heat received by the chicks is decreased.

The floor of a brooder house should be covered with a suitable type of litter, such as ground corn cobs, wood shavings, peat moss, or chopped straw. This litter should be gradually increased and built up to a depth of several inches.

Feeds of the proper kinds should be provided in containers available to the chicks at all times. Water should be supplied in suitable types of fountains. As chicks get older, they should be allowed to run out-of-doors on grass or legume pasture that has not previously been used by chickens.

After chicks are a few weeks old, they should be given increased space in shelters. Roosts should be installed in these buildings.

Caring for the laying flock

Good housing is needed for a laying flock. The house should be large enough to provide 3 to 4 square feet of floor space per hen. A house should be constructed for warmth in areas where the weather gets cold. Ventilation should be provided, but this should be done in such a way that will prevent drafts. Special ventilating systems may be installed. There should be sufficient window space for good natural lighting. Electric lights are desirable, since they allow for a 14-hour working day of hens during the winter season, thus increasing egg production during the winter months.

Many poultry raisers find it best to confine the laying flock to a building throughout the year. Special litter should be used to cover the floor. Materials, such as straw and ground corn cobs, may be built up gradually to a depth of 6 to 8 inches. Such material, if frequently stirred, needs to be cleaned out only about once per year.

A well-equipped poultry house has self-feeders for dry mash mixtures and for other feeds and suitable containers for water. There should be separate hoppers for grit and oyster shells. For large flocks, automatic feeders and waterers are being used to reduce labor. Watering devices should allow about an inch of watering space, or a 10-gallon capacity, for every 10 hens. About 4 inches of feeder space per hen is needed.

Roosts should be provided for the birds to use during resting periods. These are usually placed at the rear of the house. Under the roosts, special pits, which keep hens from contact with the droppings, are frequently constructed. These need to be cleaned out only two or three times per year.

Layers should have approximately 1 square foot of nesting space for every four hens. Hens are less likely to crowd into certain nests if all nests look alike. A type of nest, called a community nest, is used in some poultry houses. This is a large compartment into which several hens may go at one time to lay eggs. This device results in less breakage of eggs, helps to keep

Fig. 12-9. Hens in individual cages. The hens eat from the trough in front. The eggs roll into the tray as they are laid. (Courtesy, New York State College of Agriculture and Life Sciences, Cornell University)

Fig. 12-10. Turkeys raised in confinement on a wire platform to provide sanitation. Note the feeders at the side and the covered roosting space in the rear.

eggs cleaner, and makes gathering the eggs easier. Sufficient nesting space will prevent crowding. Other improved types of nests may be used.

Some poultry raisers are using special cages or battery laying coops for confining each hen in a separate compartment. Each hen has adequate feed and water. The floor of each cage is sloped so that the eggs roll into a tray. These cages result in lower death loss of the hens, clean eggs, and accurate culling of hens, since production of each hen can be checked. The hens in these cages are thrifty and contented, even though they have very little space in which to move about. This type of system is seldom used for small flocks.

Producing high-quality eggs

In recent years, much emphasis has been placed on the production of high-quality eggs. Such eggs command top prices.

An egg of highest quality is clean and sound. The air cell at the blunt end is small. The yolk is well centered and only dimly visible when the egg is held before a candling device. The yolk has no defects or blemishes. The albumen (white) is firm and clean. When an egg of high quality is broken in a dish, the yolk remains in a rounded form, and the white is fairly thick and tends to remain on a relatively small space. According to poultry experts, the following practices will aid in producing high-quality eggs: (1) feed hens a uniform ration, (2) keep house and nests clean, (3) gather eggs frequently and put them in a cool place, (4) store eggs in a place with a temperature of 60°F or less and with a relative humidity of about 85 per cent, and (5) market eggs frequently, so they will be used within about two weeks after being laid.

Eggs should be candled and graded and then placed in attractive containers for marketing. These steps are usually performed after the eggs leave the farm. Eggs should be kept refrigerated so they will retain their high quality.

If eggs need cleaning, it is best to clean them with light sandpaper or a similar material. Special egg detergent-sanitizers are available for washing eggs. In no case should eggs be washed in cold water.

What Care Should Hogs Receive?

Hogs are not filthy animals by nature, though they are often forced to live in surroundings that are far from clean. It is just as important for hogs to be given a clean and comfortable environment as it is for other farm animals to have these conditions.

Providing shelters and equipment

Many farmers who raise hogs in large numbers have a central hog house for them. There are several types of these buildings, but it is essential that they be convenient, light, warm, dry, well-ventilated, and easy to keep clean. These factors are especially important in the farrowing houses (where the young pigs are born) for the prevention of disease, the control of parasites, and the prevention of accidents to the sow and litter.

Most of these hog houses are constructed with an alley through the middle for the caretaker to use in cleaning and feeding. The areas along each side of the alley are usually designed so that partitions can be inserted to provide farrowing pens, each with about 50 to 60 square feet of floor space. Electricity should be available for lighting and other uses. Ventilation may be provided through windows that are constructed so they can be tilted inward at the top. In some cases, electric fans of special design are used to provide ventilation in the hog house.

Many farmers use small, portable hog houses, either to supplement the central building or to replace it entirely. Some of these houses are of a size suitable for one sow and litter, while others are sufficiently large for two or more sows and their litters. These houses are desirable because they can be moved into fields that are free from contamination with certain diseases and parasites of hogs.

It is important to keep sows cool during breeding and gestation. This is a special problem in warmer climates. Water foggers, shallow concrete wallows, shade, and maximum exposure to breezes are some of the methods used to bring about cool conditions. At least 20 square feet of shade is required per sow for natural cooling.

Self-feeders, watering devices, and troughs should be provided for hogs.

Fig. 12-11. Shade is a necessity for pigs out on pasture.

Handling hogs

Young people on farms frequently own and raise sows and litters of pigs. It is important that such pigs be fed properly and that they be given good care. Pigs that are to be shown at fairs should be properly "fitted." In this fitting process, they should be thoroughly scrubbed with soap and water previous to the time of showing. Frequent brushing with a stiff brush also aids in getting the hair and skin in good condition. The feet should be carefully trimmed so that the pig will stand properly. Oils should be used to make the skin and hair glossy. In addition, the pig should be trained so that it will show to the best advantage in the show ring.

Considerable injury may result if hogs are beaten or otherwise bruised prior to marketing. Not only is this painful to animals, but the carcasses may be damaged seriously for use as food after the hogs have been slaughtered. In driving hogs or in separating them into groups, the caretaker will find a light hurdle made of lumber to be a useful device. Such a piece of equipment aids in heading them off and in guiding them to the place desired. A "slapper" made of several thicknesses of canvas or a piece of an old inner-tube from an automobile tire is a good device to provide "encouragement" in driving hogs without injuring them.

Caring for pigs

A sow and her litter must be given special care at the time the pigs are born and for several days thereafter. Small pigs frequently are quite weak at birth and may die unless they are given protection. Guard rails may be placed around the edges of the pen to prevent the sow from crushing the pigs. Special farrowing stalls or farrowing crates may be used to confine the sows and to prevent injury to pigs as they are born. Pig brooders equipped with heat lamps help to prevent chilling of pigs born in cold weather. These permit the pigs to stay in a warm, sheltered compartment when they are not nursing the sow.

Recent studies reveal that, of pigs born alive, nearly 25 per cent die before weaning, with most deaths coming in the first two or three days because of chilling. A temperature of 90°F is needed the first day after birth, and 80°F is needed from the second day to the end of the third week.

Additional practices often used in caring for baby pigs are clipping nee-

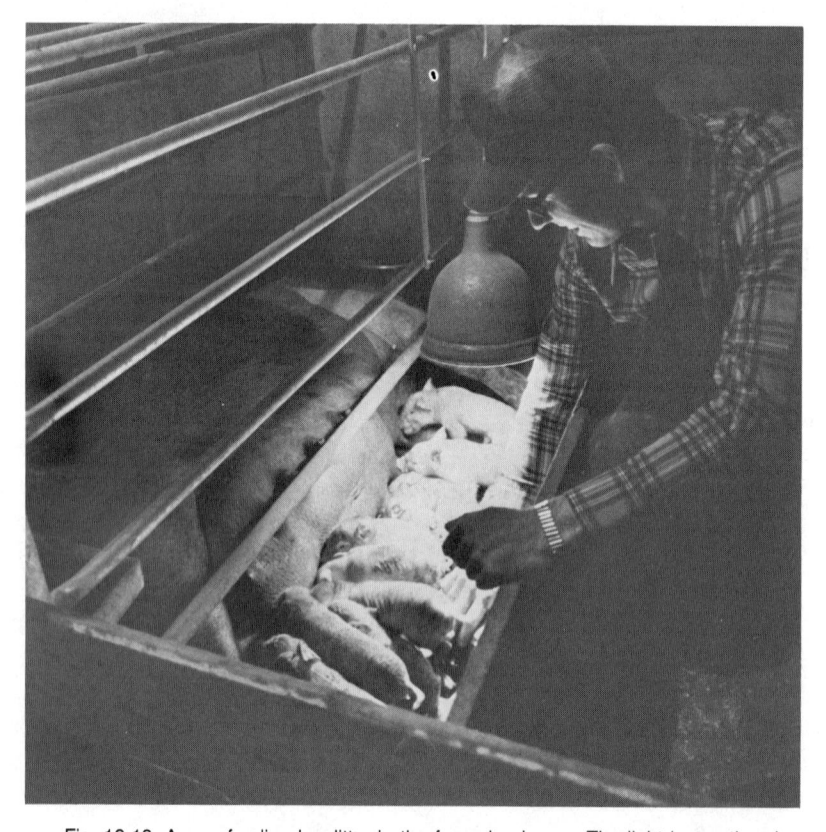

Fig. 12-12. A sow feeding her litter in the farrowing house. The light keeps the pigs warm; the farrowing crate prevents injury to the small pigs. (Courtesy, U.S. Department of Agriculture)

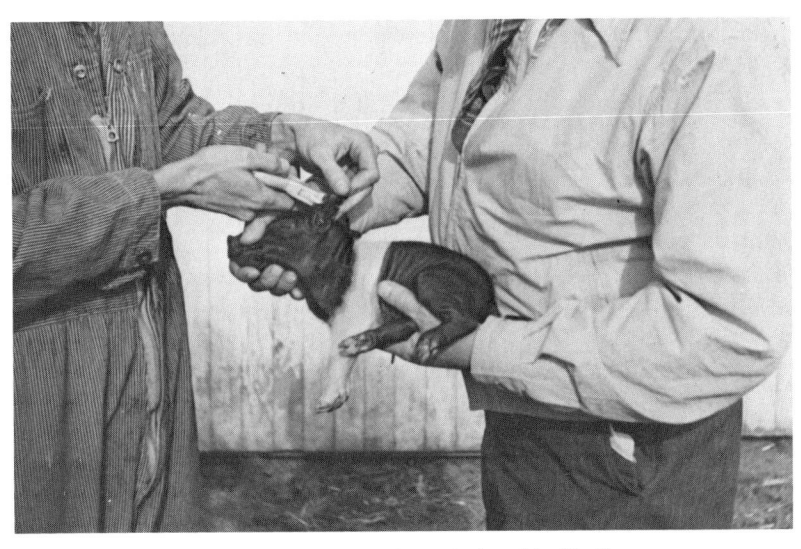

Fig. 12-13. Notching a pig's ear for later identification.

dle teeth, dipping the navel in iodine, giving iron or iron shots, castrating boar pigs, and vaccinating for cholera.

One development in the care of sows in the summer is the use of air conditioning. Since hogs are not equipped to adapt to high temperatures, experiments have been conducted to determine if a modified form of air conditioning will help. Thus far, the studies have indicated that air conditioning for a period before and after farrowing is beneficial. The sows with air conditioning farrowed a larger number of live pigs, and in addition, the pigs gained more during the first few weeks than did the pigs with normal temperatures in their building.

What Care Should Be Given to Sheep and Beef?

Sheep and beef cattle do not require as expensive shelters as do most other animals. Sheep have a natural protection in the form of wool, and beef cattle are protected by thick skins and a layer of fat beneath the skin. These animals should be given shelter, such as sheds open on one side, during periods of windy and stormy weather.

Sheep should be provided with yards and buildings separate from other farm animals in order to protect them from possible injury. Sheep require fairly warm quarters during the season that lambs are born, which is usually in early spring. Lamb brooders warmed by heat bulbs are desirable if the lambing season occurs during cold weather. Special care should be taken in raising sheep to keep the wool free from burs and other foreign materials, as the quality of wool may otherwise be lowered. The yards for sheep should not be damp and muddy, as their feet may become diseased under such conditions.

Herds and flocks may need special help during calving and lambing time. This is especially true for beef heifers calving for the first time. Daily checks should be made during calving and lambing time to locate heifers and ewes needing help and to locate newborn calves and lambs needing help. Unless ewes have been shorn of their wool, some clipping of wool around the udder is needed three of four weeks prior to lambing so that the new lamb will not suck wool when it is trying to feed. With purebred animals, it is also necessary to mark newborn animals for identification and to keep accurate records of birth dates and ancestry.

Sheep and beef cattle, like hogs, should be handled so as to avoid injuries, which will detract from the value of their carcasses when they are slaughtered for use as meat. Some people in handling sheep grab them by the wool. Not only is this painful to a sheep, but if the animal is slaughtered shortly thereafter, injuries will show on the carcass, which will decrease its value.

Care should be used to prevent crowding when sheep or cattle are driven through narrow doorways or gates. Otherwise, serious injuries may result, especially to the animals that are pregnant.

Fig. 12-14. Dipping lambs near Dubois, Idaho. (Courtesy, U.S. Department of Agriculture)

Caring for Livestock and Their Products 395

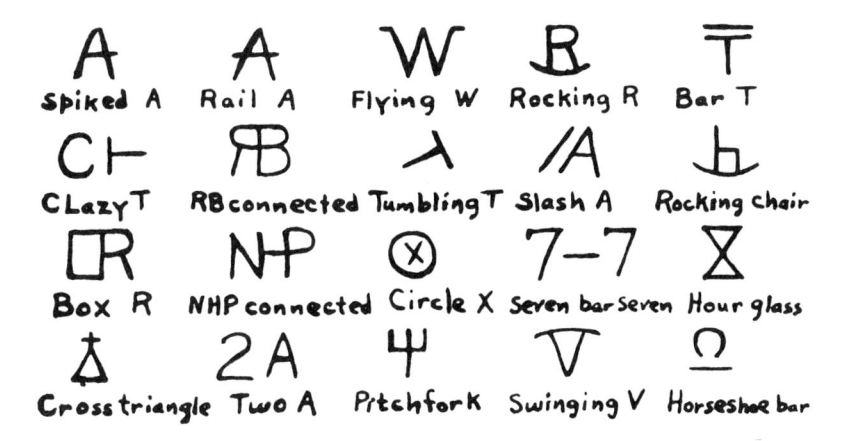

A	A	W	R	T
Spiked A	Rail A	Flying W	Rocking R	Bar T
C⊢	RB	⋌	/A	⊔
C Lazy T	RB connected	Tumbling T	Slash A	Rocking chair
⊏R	NP	⊗	7-7	X
Box R	NHP connected	Circle X	Seven bar seven	Hour glass
△	2A	ψ	∇	Ω
Cross triangle	Two A	Pitchfork	Swinging V	Horseshoe bar

Fig. 12-15. Many kinds of brands are used by western cattle ranches. Some are shown above.

Cattle kept on ranches are usually branded so that they may be identified if they stray away from the herd or if they are stolen. This practice dates back at least 4,000 years. Inscriptions and pictures on the walls of ancient Egyptian tombs indicate that cattle were branded as early as 2,000 B.C. Early Spanish settlers introduced the use of cattle brands to Mexico and parts of South America, and from these areas the practice spread to the United States. Branding of calves is usually done at roundup time. The calf is thrown on its side and held while a heated branding iron is pressed momentarily against the skin. The burn is not painful, but the resulting mark leaves a permanent scar. Each ranch has a special brand, which is registered along with official brands of other ranches in each state. There are thousands of brands registered in the United States.

Super-cold irons may be the future tool for branding cattle. Freeze branding works much the same way as fire branding. The cold kills the pigment-producing cells that color an animal's hair follicle, making the hair grow back white. If white cattle or spotted dairy breeds are being branded, the cold iron is left on long enough to make a bare-skin brand (where the hair follicles are completely destroyed). Freeze branding leaves no open wounds, is less painful, and keeps hide damage to a minimum.

Other ways for identifying cattle, used mainly by purebred breeders, are tattooing numbers in the ears, attaching metal number tags on neck chains, and fastening metal or plastic number tags to the ears. Ear tags and tattoos are used also to mark sheep for identification. Branding and tattooing are used to identify horses. Hogs are sometimes branded on the hip or loin with a hot iron.

Although many sheep and cattle on large ranches are confined by fences, many others are allowed to graze on open ranges for at least part of the year. Those on open range need to be watched to prevent straying, and

Fig. 12-16. Shearing sheep is a job calling for considerable skill. Note the care used to keep the fleece free from foreign material. (Courtesy, University of Minnesota)

Earnings Mount Up with Every Lamb Saved

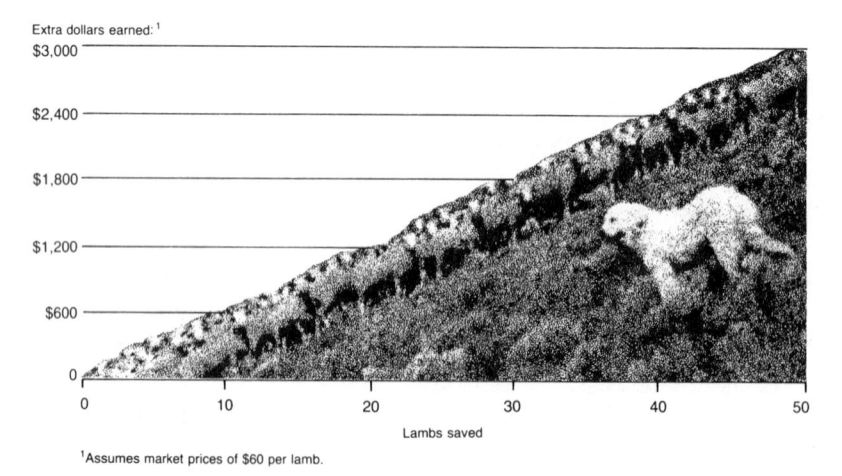

[1]Assumes market prices of $60 per lamb.

Fig. 12-17. As shown above, earnings mount up with every lamb saved from predators. In a single year, ranchers lose over $100 million of sheep, lamb, and calves to predators. Guard dogs may help to reduce these losses to manageable levels. (Courtesy, *Farmline*)

sheep need to be protected from coyotes and wolves. Coyotes, by killing sheep and lambs, are estimated to cost sheep raisers up to 20 per cent of their flocks in a single year. A coyote will bite sheep in the neck, while a domestic dog will slash at the sheep's hind legs and back; a cougar will bite the top of the head; and an eagle will leave talon marks on the back and head. Even animals confined by fences need to be checked often enough to see that all is well.

In recent years, sheep raisers have been turning to guard dogs to protect their flocks. Two breeds being used are the Komondor and the Great Pyrenees, both imported from Europe. Early findings indicate that the Komondor will work best in fenced pastures, and that the Great Pyrenees will work best on open rangeland. The New England Farm Center in Amherst, Massachusetts, is experimenting with using the Turkish Anatolian Shepherd, the Yugoslavian Shepherd, and the Italian Maremma. As a part of the research, several hundred guard dogs have been placed with flocks in 28 states.

Incidentally, the game of golf was originated by shepherds in Scotland about 600 years ago. While watching over their flocks, they played the game on golf links laid out on the green pastureland along the seacoast, with natural putting greens, fairways, rough areas, and sand traps.

What Care Should Be Given to Horses?

Horses are the athletes of farm animals, and their usefulness is dependent upon strength and endurance. Good care and attention is necessary to keep them in proper physical condition for doing their best work. Horses used for racing or pleasure purposes also need good care.

The stable for horses should be well-ventilated, free from drafts, and comfortable. Each horse should be kept in a separate stall, where it is tied to a manger, or in an individual box stall, which permits the horse to move about. The size of the box stall for the average horse should be 100 square feet or more, with a ceiling height of about 9 feet. Concrete floors are durable and provide for ease of cleaning. Sufficient dust-free bedding will prevent dampness and give the horse a cushion of material to lie on.

Caring for horses

Horses respond to kindness more than do many other animals. Thus, it is important that the person caring for a horse understand its traits.

Horses enjoy company and become anxious when separated from other horses. This trait makes it possible for horses and people to develop close bonds. People, however, must be sure to be the dominant figures in this relationship for safety's sake. Horses are also easily frightened and fear the unknown. They will struggle to get free when trapped, even if this results in harm to themselves. It is important that a horse have some animal companionship, since it may suffer from loneliness, which may lead to the development of various kinds of undesirable behavior.

The Arabians, who are noted for being good equestrians, say that "rest and fat are the greatest enemies of the horse." During periods when horses are not working, their feed should be reduced, and they should be permitted to exercise in yards or pastures. When horses have been idle for a considerable period of time, it is desirable to accustom them to work gradually.

The feet of a horse should be trimmed frequently enough to keep them level and even. If the trimming of the feet is neglected, permanent injury may result to the feet and legs due to improper strains on these parts. The feet should be examined regularly to see that no hard substances have become wedged in the soles. Shoeing is necessary if horses are driven on paved streets and hard-surfaced roads.

The teeth of a horse should be examined periodically, preferably by a veterinarian. Nature has provided horses with teeth that continue to grow to compensate for wear. Sometimes the teeth in the back part of the mouth wear irregularly, with the result being that sharp points or edges, which may injure the tongue and inner surfaces of the cheeks, are formed. Horses having teeth in this condition often become thin because eating for them is such a painful process that they do not eat sufficient amounts of feed or they fail to chew it properly. These irregular projections on the teeth can be removed by filing, which should usually be done by a veterinarian, who is trained to do it properly.

Careful grooming keeps the skin in good condition and improves the appearance of the coat. A brush and a curry comb are necessary for the grooming process. Horses that are given the freedom of a yard or a pasture have the opportunity to roll, which helps to keep the skin in good condition.

Fences are needed to confine horses when they are not in the stable. Younger horses require the strongest and most visible fences. A woven wire fence with a mesh opening that the horse's foot will not fit through is the easiest to use. A board fastened along the top will make the fence easy to see. A smooth electrified wire along the top will keep the horse from leaning against the fence and breaking it down. Wooden fences can be attractive and effective, but they are expensive to build and to maintain.

Training horses

The usefulness of a horse is dependent to a great extent upon its training. This "education" should begin early in life. It is important to win the confidence of a colt before proceeding with the training. Frightening and/or teasing a colt should not be permitted. A small halter should be placed on a colt when it is a few weeks old. A colt should then be taught to lead and to stand when tied. These habits can be formed quite readily when the colt is still with its mother. A colt should be accustomed to grooming early in life. Intensive "schooling" is provided when the animal reaches the age of two to three years. At this time, the colt should be harnessed and trained for driving, or saddled and broken for riding. These lessons should proceed regularly and slowly.

Much skill is required to train a horse properly for the purpose for which it will be used. It is important never to mistreat a colt. Gentleness, coupled with a certain degree of firmness, is important. If you have helped in breaking a colt or have watched someone do it, describe the process to the class.

SUGGESTED ACTIVITIES

1. With your class visit a good dairy farm. Notice the barns and equipment. What methods are used that provide for the comfort of the cows? For the production of high-quality milk?
2. Visit a farm on which hogs, poultry, beef cattle, or sheep are raised successfully. Note the barns and equipment. What special methods are used in the care of the livestock?
3. Develop plans for giving good care to some kind of animal in which you are interested. If possible, get some animals of your own and carry out the plans you have prepared.
4. If you have had experience on a livestock farm or ranch, describe some of your experiences to the members of your class.
5. Concepts for discussion:
    a. Proper care of livestock is essential for greatest production.
    b. Labor ranks next to feed in livestock production costs.
    c. Buildings can be planned for providing proper care for livestock with a minimum of labor.
    d. Temperature and humidity affect livestock productivity.
    e. Animals are creatures of habit and may be seriously upset by changes in their routines.
    f. Animals require exercise.
    g. Comfort and cleanliness are important in animal care.
    h. Improperly handled livestock products may soon deteriorate and be unfit for human use.
    i. Newborn animals require special care.
    j. Proper grooming and training make animals look better and handle easier.
    k. Slow movements are essential in working with livestock.
    l. Proper care of livestock is essential for preventing disease.
    m. The kind of housing needed varies with the kind of animal and the climate.
    n. Some way of identifying animals is necessary in certain farming operations.
    o. Careful and regular observation of animals is essential to good management.

REFERENCES

Albaugh, R. *Horse Behavior*. Cooperative Extension Leaflet 21002, Division of Agricultural Sciences, University of California, Berkeley, California 94720.

Ensminger, M. E. *Horses and Horsemanship*. Interstate Publishers, Inc., Danville, IL 61832.

Ensminger, M. E. *The Stockman's Handbook*. Interstate Publishers, Inc., Danville, IL 61832.

Lee, J. S., and M. E. Newman. *Aquaculture: An Introduction*. Interstate Publishers, Inc., Danville, IL 61832.

Romans, J. R, K. W. Jones, W. J. Costello, C. W. Carlson, and P. T. Ziegler. *The Meat We Eat*. Interstate Publishers, Inc., Danville, IL 61832.

Chapter 13

Keeping Animals Healthy

In preceding chapters, emphasis was placed on the importance of having farm animals that are well bred, well fed, and well cared for. Health is another important consideration in the successful production of livestock.

OBJECTIVES

1. Explain why it is important to keep farm animals healthy.
2. Discuss methods used to prevent disease in the food supply.
3. Identify causes of diseases in farm animals.
4. Describe how animal health can be maintained.

Why Is It Important to Protect the Health of Farm Animals?

The health of farm animals is of vital importance to everyone. Not only must the economic benefits to the nation and the livestock industry be considered, but the health of the public must be considered as well. Many diseases of humans are contracted from diseased livestock or animal products. Not only do farmers benefit from healthy herds and flocks, but all of us, who need the nutrients of meats, milk, and other foods of animal origin, also benefit.

Farmers need to be concerned about diseases among their farm animals both because their own personal health could be affected and because serious losses of money could be incurred. For example, it has been estimated that brucellosis cost the livestock industry $200 million or more a year. Federal and state funds are being used to wage control programs for brucellosis of cattle and hogs. The time is near when this disease may be wiped out. More than $2 billion is lost yearly because of infectious, noninfectious, and parasitic diseases.

Young animals are especially susceptible to various diseases and accidents that can cause death. Upwards of 20 to 30 per cent of all pigs farrowed and all chicks hatched die before they reach maturity. Probably 10 per cent or more of calves die. Most of these deaths are in the animals' first few weeks of life. Many farm animals grow slowly or produce at a low level because of ailments of various kinds. These losses, as well as deaths, constitute the toll that diseases and parasites take in the livestock industry.

Economic losses because of animal diseases are not limited to farmers. In addition to costs from diseases contracted from livestock, the entire population must pay more for the food purchased in the grocery store because of the increased production costs that are a direct result of livestock ailments.

As stated before, a major reason for public concern for the health of farm animals is that some kinds of animal diseases may endanger human health. Medical authorities recognize about 80 diseases and parasites of domestic animals that are transmissible to humans. About 20 of these are important to public health in the United States.

Health authorities have reported that hundreds of people in the United States contract undulant fever each year. Most of these people get it by direct contact with cows or pigs that have brucellosis. Many of the people who contract undulant fever are living on farms, or are packing-house workers or veterinarians. Some people get the disease by drinking unpasteurized milk produced in dairy herds in which some of the cows have brucellosis. To be safe, people should consume only milk that has been pasteurized. Most states and cities have strict regulations for all milk offered for sale. These regulations require healthy herds for the production of milk, sanitary methods in the production of milk, and pasteurization of the milk.

People may contract *rabies* from animals that have this disease. Any warm-blooded animal can have rabies. Dogs infected with rabies are especially dangerous to people. All dogs should be vaccinated regularly against the disease. A person bitten by a dog should see a physician at once. If possible, the dog should be located and captured alive so that it can be placed under the observation of a veterinarian. In the event the dog is diagnosed as having rabies, the persons bitten by it should be given treatment. Otherwise, death is almost certain.

People contract *septic sore throat* by drinking unpasteurized milk from cows with certain types of udder infections. People who are careless in treating diseased animals or who handle unsterilized products made from animals that had the disease may contract *anthrax*, a disease that affects horses, swine, cattle, and sheep.

Tularemia from rabbits, *psittacosis* from parrots, and *erysipelas* from hogs are other examples of diseases transmissible to people. *Psittacosis* is known as *ornithosis* in turkeys, an infectious disease caused by a virus-like organism. The virulent form can produce serious illness and sometimes can cause death in persons who handle sick turkeys.

Some ailments of people may be caused by parasites that are harbored by farm animals. The muscular tissues of some hogs harbor small roundworms known as *trichinae*. Persons who eat undercooked pork from these diseased animals may contract the disease called *trichinosis*. Hogs fed on garbage containing uncooked pork scraps are most likely to become infested with this parasite. By discontinuing the feeding of such materials to hogs and by cooking pork thoroughly, people can almost certainly eliminate any danger from this disease.

Some types of *tapeworms* may be transmitted to people who eat improperly cooked beef, pork, or fish. Tapeworms may also be transmitted from dogs. Several kinds of ticks may be transmitted from animals to people, and some carry organisms of diseases such as Rocky Mountain spotted fever.

People should not become so alarmed about these diseases that they discontinue eating animal products. The danger of getting these diseases is reduced greatly by proper educational measures regarding the use of the products, by campaigns for the control of diseases in livestock, and by government regulations for the inspection and distribution of food products. For example, the nationwide campaign for the control of tuberculosis in cattle has almost eliminated the form of tuberculosis in humans formerly contracted from cattle. This has been the largest undertaking of its kind in history, and it has been supported in part by funds from federal and local sources. Since all people, urban and rural, benefit, such use of public funds seems highly justifiable.

How Can Disease Be Prevented in the Food Supply?

The U.S. Department of Agriculture is responsible for guarding against disease in livestock and the food supply. The procedures it has set up are designed to assure that the meat and poultry products consumers buy in the stores are wholesome and fit to eat and that livestock on U.S. farms are protected from diseases from foreign countries and other sources.

The USDA *Food Safety and Inspection Service* is responsible for the inspection of all meat and poultry sold in interstate or foreign commerce, for the correctness of the labels, and for the wholesomeness of the meat. In addition, the Food Safety and Inspection Service must inspect meat and poultry sold within any state if that state is unable to provide an adequate inspection service of its own.

In 1981, federal inspectors examined more than 129 million meat animals and more than 4.4 billion birds. Of the birds and animals inspected, over 49 million were condemned as unwholesome. In addition, 459,000 meat animals and 37,000,000 pounds of processed products were condemned as unwholesome. Meat and poultry that are judged to be unwholesome, adulterated, or mislabeled cannot be sold to customers for food.

Each foreign plant that ships meat or poultry to the United States, as well as that foreign country's own inspection system, must be approved by the U.S. Department of Agriculture. Federal veterinarians visit the foreign plants to inspect them, and they inspect the meat and poultry at the ports where the shipments enter this country. Because of the many chemicals used in the production of meat and poultry, special attention is given to the detection of possible drug, pesticide, and chemical residues.

Under the *Egg Products Inspection Act,* the U.S. Department of Agriculture is also responsible for the healthfulness of egg products that reach the consumer. Continuous inspections are made of all plants processing liquid, dried, or frozen egg products. The act also controls the disposition of restricted shell eggs, those that might contain harmful bacteria that could cause illness. Egg products can be imported only if the foreign country's inspection system is as good as the U.S. inspection system. The Canadian inspection system is the only one approved, so Canada is the only country exporting egg products to this country.

Protecting the health of the nation's livestock, poultry, and other animals is the job of the *Veterinary Services* of the USDA *Animal and Plant Health Inspection Service*. The service began in 1884 when Congress created a special agency within the U.S. Department of Agriculture to combat *bovine pleuropneumonia,* a disease that was killing many cattle in the northeastern and midwestern states. Within eight years, the disease had been eradicated. Diseases that have been eradicated since then include foot-and-mouth disease, Texas cattle fever, fowl plague, Venezuelan equine encephalitis, sheep scabies, screwworms, exotic Newcastle disease, and hog cholera. Diseases currently being combated include brucellosis, cattle fever ticks, tuberculosis, scabies in cattle, and pseudorabies in swine.

Measures used to fight animal diseases include (1) quarantines to stop movement of infected or exposed animals, (2) testing and examination, (3) treatment to eliminate parasites, (4) vaccination, and (5) cleaning and disinfection of contaminated premises. Import regulations to keep out dangerous diseases such as foot-and-mouth disease, African swine fever, and rinderpest are also administered by the Veterinary Services. Under the Virus-Serum-Toxin Act of 1913, the Veterinary Services enforces regulations to assure that animal vaccines and similar materials are safe and effective. The laws governing humane treatment of animals also are a responsibility of Veterinary Services. These laws regulate the handling of livestock transported by railroad; care and treatment of animals used in research, in the wholesale pet trade, and in zoos and circuses. The Horse Protection Act of 1970 was amended in 1976 to prohibit "soring," the use of cruel and inhumane practices to enhance the gait of show horses.

The work of the Veterinary Services is carried out by a force of about

600 veterinarians, about 600 lay inspectors, and about 250 laboratory technicians.

Animal disease eradication workers identify and eliminate diseases on farms and ranches before such diseases can cause serious economic losses. They prevent the spread of animal diseases from state to state and insure humane treatment of livestock and other animals. They participate in nationwide cooperative state-federal programs and assist foreign governments in controlling and eradicating serious livestock and poultry diseases.

What Are the Causes of Diseases Among Farm Animals?

A *disease* is considered in a broad sense to be any disorder of the body. (When the word is analyzed, it really means "lack of ease," which is probably its original meaning.)

Improper feeding as a cause of diseases

Some diseases are caused entirely by improper feeding methods. The disease called *rickets,* in which the bones of an animal fail to develop properly, is due to improper amounts of certain minerals and vitamins. Young animals are sometimes born with enlargements in the throat region. This ailment, which is known as *goitre,* or "big neck," is caused by insufficient amounts of iodine in the ration of the mother animals during the period previous to the birth of the young. *Colic* in horses and *bloat* in cattle and sheep are usually caused by improper methods of feeding.

Animals may be poisoned by lead, which they can get through licking paint pails or freshly painted surfaces. They may also consume feeds that have a poisonous effect on their bodies. Poisonous weeds such as loco weed and white snakeroot are examples. The latter weed, if eaten by dairy cows, may injure not only the cows themselves but also the people who drink the cows' milk. Halogeton is a weed that has spread rapidly in semidesert areas; at times, it has caused the death of many sheep in some western states. Other plants that have a poisonous effect on livestock are alsike clover, when eaten wet; bouncing Bet, cocklebur, and corn cockle roadside plants; bracken fern, found in dry and abandoned fields; buttercup, when eaten green; larkspur; ergot; horse nettle; jimson weed; Johnson grass; nightshade; hemlock; milkweed; and wild cherry.

Ornamental plants toxic to goats include oleander, azalea, castorbean, buttercup, rhododendron, philodendron, yew, English ivy, chokeberry, laurel, daffodil, jonquil, and many members of the lily family. Even tomatoes, potatoes, rhubarb, and avocadoes contain toxic substances. Goats may also be harmed by toadstools, mushrooms, mistletoe, and milkweed.

Each year, additional plants are found to be toxic to animals. Most cases of livestock poisoning are reported during periods of time when there is insufficient growth of desirable forage. Animals then begin eating any available plants, including plants that are poisonous.

If poisoning of livestock is suspected, the animals should be removed from the area where the poisonous plants are located and a veterinarian should be called. Affected animals should be placed where they can be comfortable and quiet and where treatment can be given easily. Other farm management practices that will help prevent losses are:

1. Avoid placing animals in areas where poisonous plants can be found.
2. Provide sufficient amounts of good pasture so livestock will not be tempted to eat undesirable plants.
3. Eradicate poisonous plants, even those that can be reached by livestock through the fence.

Physical injury and overheating

Injuries and overheating may be the causes of ailments in animals. Cuts and other injuries should be given careful attention when they occur, since they may become serious. All animals on pasture should be provided with adequate shade on hot, sunny days. Plenty of water should be available at all times. Pieces of barbed wire and various sharp objects that may cause external injuries should be kept out of the places used for livestock.

Indirect causes of diseases

Animals that become weakened by improper feeding contract some diseases more readily than stronger and more rugged animals. In the same manner, animals that have been weakened by improper care and by exposure are more susceptible to some diseases. Some animals have inherited a weakness for or a lack of resistance to certain diseases. These factors are often referred to as *predisposing* or *indirect* causes of disease.

Parasites as causes of diseases

Many diseases of farm animals are caused by organisms. The term *parasites* may be applied to these organisms because they live *on* or *in* the bodies of animals and at the expense of these animals. Some of these parasites are called *germs,* or *bacteria,* which are really simple plants so small that a microscope is needed to see them. Some ailments are caused by microscopic animals called *protozoa.* Some ailments are caused by *fungi,* which are small plants with multiple cells.

Some parasites, such as lice, ticks, and worms of various sorts, are sufficiently large to be seen readily by the unaided eye. Commonly, the ailments that are caused by germs are called *germ diseases*, and those that are caused by the larger parasites are called *parasitic diseases*. For another group of diseases, no specified organisms have been isolated, but these diseases are transmissible to other animals, nevertheless. Some of these are called *virus* diseases.

Many ailments caused by parasites have increased because the barns, yards, and other parts of the environment in which farm animals have been forced to live for many years have often become badly infested with parasites in various stages. Pioneers who moved into a new region had fewer livestock losses from diseases of this type because there was not an accumulation of worm eggs and disease germs in the soil. Thus, the production of farm animals requires more skill today than formerly so that diseases can be effectively controlled.

For the most part, each particular parasite, whether large or small, does damage to only one species of farm animal. Thus, there is a specific germ that causes blackleg in cattle, but that does no harm to other species of farm animals. The swine roundworm does its greatest damage to hogs, while the sheep stomach worm does it greatest damage to sheep. Lice that affect one species usually do little damage to other species.

A few diseases, however, may be contracted by several species of farm animals. Anthrax commonly affects cattle, sheep, and horses, and in some cases, humans. Most animals may contract tuberculosis, but there seem to be differences in the bacteria that commonly cause this disease in each species of livestock. It is known, however, that hogs may contract the form that affects chickens and the form that affects cattle. On the other hand, cattle rarely contract the form that affects chickens.

Many of the germ and virus diseases that affect one animal in a herd or flock are readily contracted by other animals of the same species and in some cases by animals of different species. Such diseases are spoken of as being *contagious*. Most contagious diseases are apparently spread from one animal to another in one of three ways. One means of spreading is by direct contact between an infected animal and a healthy animal. Another is from food or drinking water that is contaminated with the germs or virus. A third method by which diseases are spread is by insects or other forms of animal life. Texas cattle fever, which used to be a serious disease in cattle in some of the southern states, is conveyed from one animal to another through the Texas fever cattle tick. Sleeping sickness in horses may be carried by mosquitoes and possibly by other insects. Erysipelas in hogs may be spread by flies. The tapeworm in chickens passes one stage in the bodies of stable flies, houseflies, snails, slugs, earthworms, and grasshoppers. Chickens become infected with tapeworms by eating some of these intermediate hosts that contain one stage of the parasite.

How Can the Health of Farm Animals Be Maintained?

The highest aim of all livestock farmers should be to prevent their animals from becoming diseased. There is an old saying, "An ounce of prevention is worth a pound of cure." What is the meaning of this statement in its application to diseases of farm animals?

In cases where diseases have not been prevented entirely, every effort should be put forth to control them. Since the health of farm animals concerns everyone, the layperson should know something about the prevention and control of animal diseases.

Preventing diseases

The farmer who is careful in using good methods of selection, feeding, and care of livestock is doing a great deal to prevent diseases among them. If the more vigorous animals are selected from each generation, and provisions are made for giving them the proper feeding and care, these animals will have a greater ability to ward off certain diseases with which they may come in contact.

Sanitation, the maintenance of clean surroundings, is one of the important ways of preventing diseases. Sanitary conditions can best be maintained if the buildings are constructed with concrete floors and smooth walls and ceilings that can be cleaned readily. Yards should be well drained and clean. Rotating pastures, for sheep and hogs especially, will aid in preventing parasites, such as intestinal roundworms in swine and stomach worms in sheep.

Disinfectants of various sorts help keep an environment sanitary by destroying disease germs that may be present. The interior of barns and other livestock buildings should be disinfected at intervals with solutions of cresol compounds or coal-tar preparations. This is particularly important if disease has broken out. All foreign materials should be thoroughly cleaned from the walls and floors before they are disinfected so that all germs can be reached. A sprayer that applies the solution with considerable force is a desirable piece of equipment for the process. Sunlight is nature's disinfectant. It kills many disease germs, but it must shine directly on them to be effective.

Proper feeding and care of farm animals aids in the prevention of diseases, as already mentioned. The failure to provide adequate amounts of some vitamins and minerals not only may lead to specific ailments but also may reduce the general vigor of the animals. Improper protection from severe weather may also lead to loss of vigor, which in turn makes animals more susceptible to diseases. Animals need considerable exercise, prefera-

Keeping Animals Healthy

bly out-of-doors when weather conditions permit. There should be ample space in buildings for sheep, hogs, and chickens, since overcrowding makes it difficult for them to exercise properly. Furthermore, buildings should be properly ventilated and lighted if healthful conditions are to prevail.

Animals may be protected against some germ and virus diseases by *immunization*, which makes an animal resistant to a specific disease. For example, a hog may be immunized against hog cholera by being injected with certain materials. In a similar manner, a calf may be immunized against blackleg. If these immunized animals later come in contact with the disease germs against which they are protected, they will not contract the

Fig. 13-1. A University of Illinois experiment in raising hogs in confinement. A rotating stream of water keeps the area clean.

Fig. 13-2. Both pigs received the same ration, but the ration for the pig on the left was deficient in phosphorus. (Courtesy, Purdue University)

disease. The explanation of this resistance is that the body, through immunization, or vaccination as it is commonly called, produces certain substances known as antibodies that destroy the germs or virus that may enter the body.

Care in purchasing animals is an important consideration in the prevention of diseases in the herd. Animals infected with certain diseases are sometimes brought into healthy herds. Some diseases, such as brucellosis in cattle and hogs, may be brought into herds by animals that do not show outward signs of infection. Fortunately, this disease and some others may be detected by the use of specific tests, which can be made by veterinarians. It is safest to purchase animals from healthy herds or flocks. All purchased animals should be isolated for a considerable period previous to being placed with other animals so that any possible symptoms of disease can be detected. Most states have strict regulations preventing the shipment of diseased animals from one state to another.

Controlling diseases

In case there is an outbreak of disease in a herd or flock, the first step in its control is to isolate all ailing animals. An alert caretaker usually notices any animals that do not appear normal. Following the appearance of a contagious disease, the premises should be thoroughly cleaned and disinfected. In some cases, such as outbreaks of cholera in hogs or blackleg in cattle, the immunization or vaccination of the unaffected animals may be helpful in checking the disease.

Some diseases may be detected from tests of the animals or their products. Examples are the tuberculin tests for detecting tuberculosis in cattle, blood tests for detecting brucellosis in cattle and hogs, chemical tests of milk for detecting mastitis in cattle, and laboratory tests for detecting rabies in dogs and other animals.

Federal and state agencies aid in preventing and controlling diseases of livestock. Under the U.S. Department of Agriculture, all animals shipped into this country are checked carefully before they are allowed to enter. By quick action of this department, some outbreaks of foot-and-mouth disease of cattle in the United States have been entirely eliminated. The United States aided in curbing an outbreak of this disease in Mexico and kept it from spreading to the United States. In the event of serious outbreaks of animal diseases in a state, state officials take an active part in controlling them. Specialists in colleges of agriculture and veterinary medicine are constantly searching for methods of preventing and controlling diseases.

Veterinarians can do much to help farmers maintain the health of their farm animals. In the diagnosis and control of livestock diseases, instead of resorting to "cure-all" remedies and quack treatments, livestock farmers should seek the expert advice and help of veterinarians. Some veterinarians

Fig. 13-3. This cow came from a Mexican dairy herd infected with foot-and-mouth disease. Salivation, caused by mouth ulcers, is one of the symptoms of the disease. (Courtesy, U.S. Department of Agriculture)

plan complete herd health programs for the herds in their care. Increased milk production and reduced loss of animals can pay for the services of a health care program many times over.

How Can the Principal Ailments of Cattle Be Prevented and Controlled?

Much superstition has been connected with the health of cattle and other farm animals. Cows were formerly treated for imaginary diseases, such as "hollow horn" and "loss of cud." Parts of all horns are hollow, and therefore no particular ailment can be associated with this condition. All cows stop chewing their cuds when they have serious digestive disturbances. The real cause of the sickness should be treated, rather than attempting treatments that are useless. Livestock owners have enough difficulty preventing and curing real diseases without spending time on purely imaginary ones.

Bloating is an ailment in cattle in which the compartment of the stomach known as the paunch becomes filled with a large amount of gas formed mainly by rapid bacterial action. Bloating is really a type of indigestion. Unless treatment is given, preferably by a veterinarian, the animal may suffocate because of the pressure on its lungs. Cows bloat most frequently on legume pastures, such as clover and alfalfa, especially if they graze on them

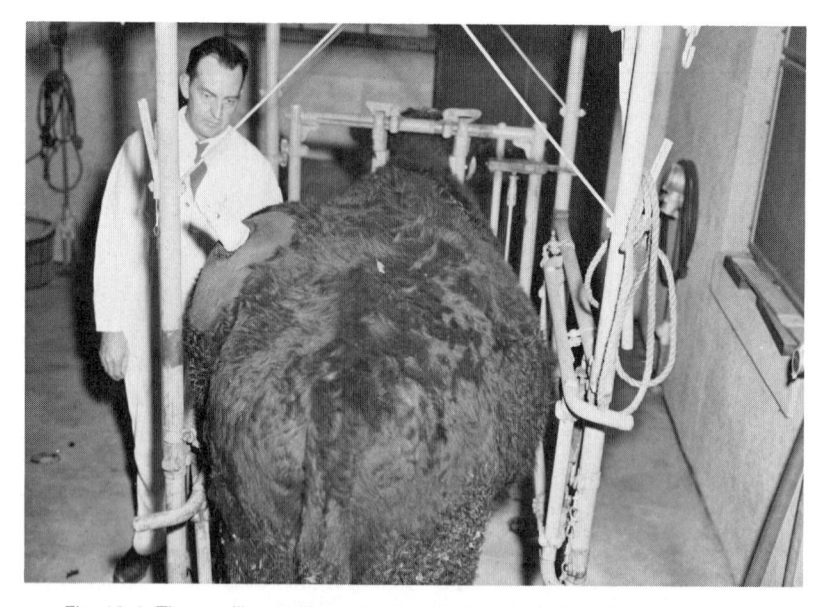

Fig. 13-4. The swelling on the animal's side is typical of bloat and reaches a peak within an hour after a feeding. Researchers study bloat through the fistula—the hollow, pipe-like bulge on the animal's left side—which opens directly into the rumen through a steel cannula. (Courtesy, U.S. Department of Agriculture)

when the crops are wet from dew or rain. Legumes mixed with grass for pasture crops are less likely to lead to bloat. Before cattle are placed on these pastures, the cows should be given a full feed of dry hay so they will not gorge themselves on the green legumes. Water and salt should be available at all times. An added safety practice is to provide hay in a rack near the resting place of cattle in a pasture where bloat is likely to occur.

Pinkeye is a common infectious disease of cattle, especially in young cattle with white faces or white pigment around the eyes. It is believed to be caused by a micro-organism accompanied by other organisms causing secondary infections. It is spread by contact, by means of dust, tail switching, and insects. In mild forms, the eyeball develops a pink color; in serious cases, the cornea becomes cloudy, ulcers develop, and blindness occurs. Treating with antibiotics, covering infected eyes with cloth patches, controlling flies, and confining infected animals in dark quarters are all recommended.

Milk fever is a disease of dairy cows, which may affect them soon after calving. It is caused by the removal of calcium from the blood faster than the calcium can be replenished. Cows with this ailment become paralyzed and lie in a position in which the head is turned toward the rear portion of the body. A veterinarian should be called as quickly as possible, because death will probably result unless relief is given. As a treatment, the veterinarian usually injects a solution of calcium gluconate into the affected animals.

Before calving, cows with past milk fever history may be fed heavy doses of vitamin D, which will usually prevent the disease.

Tuberculosis is a disease caused by a germ. Infected animals cannot be detected by outward appearances. The tuberculin test is the best method for detecting this disease in cows. As there is no known cure for infected animals, all of those tested positive are slaughtered under the supervision of federal veterinarians. The federal government partially repays the owners for their losses. Counties throughout the United States have carried on campaigns to eradicate tuberculosis in cattle.

Brucellosis, or Bang's disease, causes serious losses in many herds of beef and dairy cattle. Cows having this disease often abort their calves; that is, the calves are born before the normal time and usually are born dead or die soon after birth. Infected cows produce less milk, and breeding difficulties frequently occur. The ring test of milk is one method of detecting brucellosis in a milking herd. A sample of milk, taken from a mixture of all the milk produced by a herd in one day, is tested in a laboratory. If the test is positive, one or more cows in the producing herd probably have the disease. A special blood test by which infected animals can be detected is then made for each cow in the herd. Cows found to be reactors to the blood test are usually sent to market for slaughter. All calves raised in such a herd are usually vaccinated at four to eight months of age; the special vaccine immunizes them against the disease.

Fig. 13-5. A veterinarian vaccinating a calf against brucellosis, a disease that causes abortion and reduced milk production in cattle. (Courtesy, *Agricultural Research,* U.S. Department of Agriculture)

The Secretary of Agriculture is authorized to cooperate with the states in conducting brucellosis eradication programs. The *Brucellosis Eradication Methods and Rules* forms the basis for cooperation between the U.S. Department of Agriculture and the states. This set of rules has been developed by the U.S. Animal Health Association and approved by the U.S. Department of Agriculture.

Mastitis is an inflammation of the udder, which is typically caused by germs. Different types are caused by different germs. Injuries to the udder, improper practices of milking, and chilling of the udder may be contributing causes. Some cows may inherit a greater susceptibility to mastitis than others.

In the most serious form of this disease, the udder swells and becomes inflamed, and the cow may become seriously ill. However, healthy-looking cows may have mastitis. The disease reduces the milk flow and may result in permanent injury to one or more quarters of the udder. One symptom of the disease is the appearance of thick, flaky milk, although cows may carry the germs without showing this symptom. The livestock farmer should consult a veterinarian, who will recommend specific control measures. Good sanitation and careful milking are necessary in a control program.

Ox warbles, or cattle grubs, are serious pests of cattle. These insects cause millions of dollars of losses in hides from infested cattle. They also reduce the gains and milk production of infested animals. Ox warbles pass through an unusual life history, as shown in Fig. 13-6. These insects can best be controlled while they are located in the skin beneath the back. Rotenone powder may be rubbed into these openings, or it may be mixed with water and sprayed onto the backs. When these grubs are killed, damage to the hide is decreased and the next year's crop of heel flies is reduced.

The *Gulf Coast tick* can cause enormous losses to cattle owners. The ticks attach themselves to the ears of cattle, causing sores and swelling. Cattle injure themselves by rubbing their ears on trees and posts, thus causing lesions that become infested with the screwworm fly. Researchers believe that ear tags impregnated with the proper insecticide will control the tick.

Hardware disease of cattle is caused by pieces of metal, such as nails, wire, and other sharp objects, that are swallowed with feed. Because a cow swallows its feed rather quickly with little chewing, these pieces of metal are swallowed along with the feed, and they accumulate in the stomach. In time, the sharp metal objects may puncture the wall of the stomach, pierce the tissues surrounding the heart, or penetrate other vital organs, resulting in inflammation. Such cows usually die, since there is no very successful treatment. In some cases, operations for the removal of these objects have worked. Prevention consists of keeping metal objects out of the feed and away from places where cattle may pick them up in searching for feed.

One way to limit the danger from hardware disease is to use small magnets. Some farmers have found that the insertion of a small, flat magnet in a

Keeping Animals Healthy 415

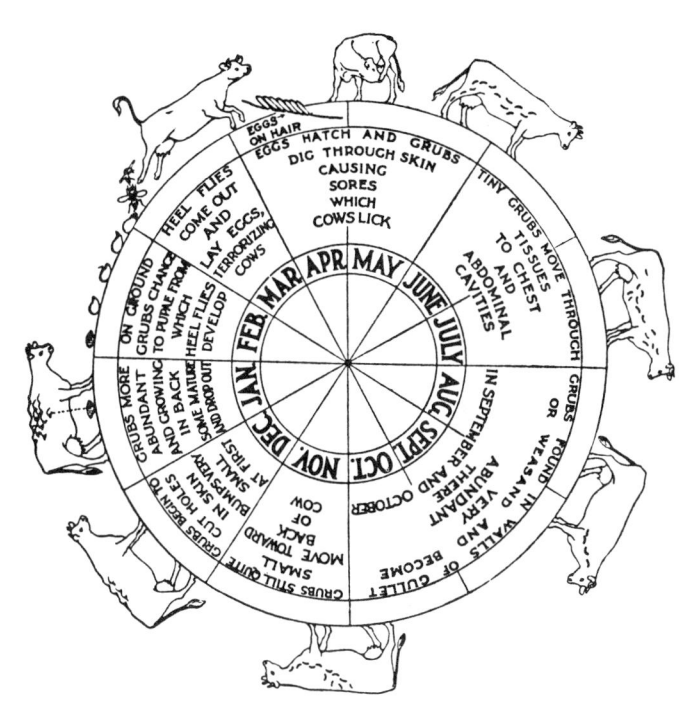

Fig. 13-6. The life cycle of ox warbles, or cattle grubs, showing various stages and the months each occurs in Kansas. The stages are reached earlier in states south of Kansas and later in states north of it. (Courtesy, U.S. Department of Agriculture)

Fig. 13-7. Back scratchers containing chemicals help protect cattle against insects.

cow's stomach can prevent damage by catching the nails, wire, and other pieces of metal and keeping them in one place.

Vibriosis is the name given to a disease of dairy and beef cattle that is responsible for 40 per cent of the infertility in cattle. The only signs of infection may be poor conception rates of heifers, which may require multiple services before settling. Cattle owners can develop vibriosis-free herds from infected ones by breeding heifers only to bulls proved free of the disease or by artificial insemination. Healthy animals and their offspring should be isolated from the rest of the herd.

Blackleg is a disease that attacks cattle in some areas of the country. If cattle are in an infected area, they should be vaccinated annually prior to the blackleg season.

Photosensitization is a disease characterized by sensitization of the skin to sunlight. It usually occurs when animals consume certain feeds, forages, and medicines. Faulty liver function may also be a cause. The disease may affect all animals and is common in cattle, sheep, and swine. The sign of the disease is severe sunburn. Preventing access to plants that may cause the disease is the main method of control.

Enterotoxemia is a bacterial disease characterized by internal bleeding. It may affect cattle, sheep, and swine. It is usually fatal.

Coccidiosis is a parasitic disease that may cause stunted growth in cattle, sheep, and swine. In severe cases, it may cause death. A veterinarian should be consulted for treatment.

Footrot is a painful foot ailment in sheep and cattle. A veterinarian should be consulted for treatment. Prevention includes providing pens with good drainage, using footbaths for animals to walk through, and trimming hoofs regularly.

Pneumonia is a serious respiratory disease affecting all animals. It is often the result of poor ventilation. Many agents can create the symptoms. Treatment should be given by a veterinarian.

Vesicular stomatitis is a rare virus disease spread by flies, mosquitoes, and eye gnats. The disease may affect cattle, horses, swine, dogs, and people. Blisters usually appear on the mouth, feet, and teats. Blisters on the tongue and lips are painful and may prevent the animals from eating. Animals need special help, including soft, highly nutritious feed.

How Can the Principal Ailments of Chickens Be Prevented and Controlled?

Chickens are subject to a large number of diseases. Every effort should be made to prevent diseases from getting started. Some of the specific precautions to take with chickens are the following:

1. Buy healthy chicks from an approved hatchery, put them in a clean

and disinfected brooder house, and provide them with heat at a suitable temperature.
2. Provide a clean range on which grasses and legumes are growing and on which no chickens have been kept for at least a year.
3. Keep chickens of different ages separated, as older birds may spread various diseases and parasites to younger birds.
4. Provide dry litter, which should gradually be built up to a depth of several inches.
5. When an outbreak of disease occurs, take appropriate steps for finding the causes and bringing the disease under control.
6. Avoid overcrowding of chickens in houses.
7. Provide a balanced diet and a comfortable environment.
8. Observe the flock daily for any signs of stress, such as changes in feed and water consumption or in egg production.
9. Remove and dispose of dead birds promptly.
10. Do not permit persons who have been to other poultry farms, or to livestock and poultry shows and fairs, to go near the poultry houses and yards unless they have changed clothes. Many poultry diseases can be carried from one place to another on clothes, especially on shoes.

Newcastle disease is a serious disease in chickens. The symptom is difficult breathing, followed by paralysis. The disease is caused by a virus, which is carried by infected chickens or by visitors. The disease spreads rapidly through a flock. There is no effective treatment for diseased birds. Care should be used to prevent the introduction of the disease into a flock. Vaccinations have been developed that make chickens immune to the disease.

Pullorum disease is one of the most serious and widespread diseases of baby chicks. Infected chicks stand about listlessly and frequently have white and pasty droppings. Death losses may be high. The disease is caused by bacteria, which may be transmitted by a carrier hen to chicks by means of her eggs. There is no satisfactory treatment once the disease starts. To prevent this disease, chicks should be secured from hatcheries that sell stock with a rating of "U.S. Pullorum passed." This means that the chicks are from eggs laid by flocks that have passed a clean test for the disease.

Tuberculosis is present in flocks in some communities. This disease is caused by bacteria and is rather slow-acting in infected individuals. Tubercular chickens lose weight and may become lame and extremely weak before they die. The internal organs of suspected chickens should be examined. The liver, spleen, and walls of the intestines of infected chickens usually contain many yellowish growths, called lesions, or tubercles. Some hogs sent to central markets are found to be infected with the fowl form of tuberculosis.

Badly infected flocks should be sold on the market, subject to inspection. The premises should then be thoroughly cleaned and disinfected. The young chicks on these farms should not come in contact with the diseased flocks. After the adult chickens (among which the disease is present) have been disposed of and after the premises have been thoroughly cleaned, the younger chickens may be safely brought to the permanent quarters. In flocks where tuberculosis is a problem, it is desirable to dispose of all hens after the first laying season.

Fowl pox is spread by insects and is difficult to guard against. It pays to vaccinate against this disease.

Fowl cholera is frequently a problem in commercial laying flocks and breeder replacement flocks. Flocks can be vaccinated to prevent the disease.

Many other diseases, such as coccidiosis and chronic respiratory disease (CRD), can cause severe losses in a poultry flock because of the rapidity with which they can spread among many birds confined in a relatively small area. Providing a good environment for the flock, in addition to other good management practices, is the best weapon against these diseases. Early detection and treatment with proper drugs can limit losses.

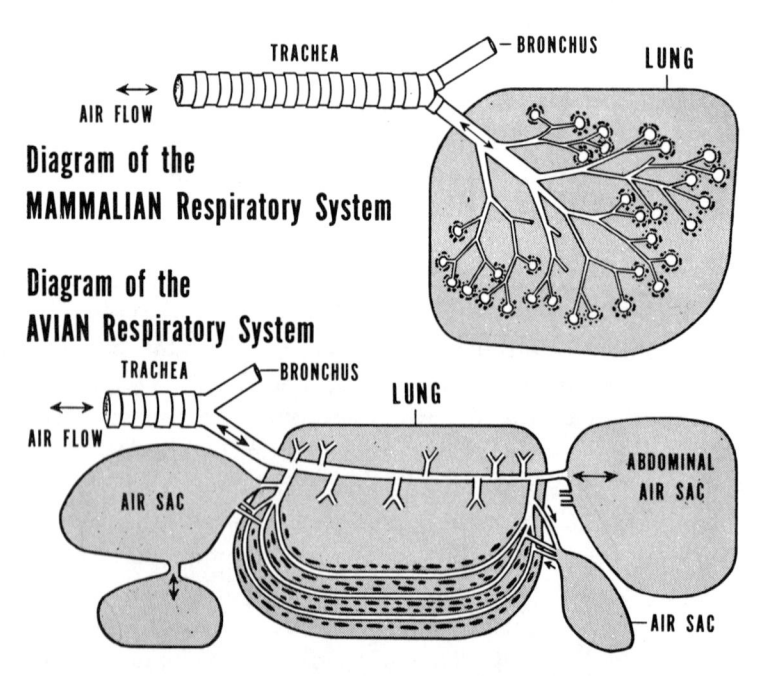

Fig. 13-8. One reason for chronic respiratory disease (CRD) in poultry is that the breathing system of a bird includes many air sacs. Air sacs in bones make the bird lighter. Other functions of air sacs may include storing air for use during flight, helping control body temperature, and helping evaporate water within the body. (Courtesy, *Agricultural Research*, U.S. Department of Agriculture)

Intestinal worms of various kinds cause severe losses in some poultry flocks. Unless the poultry lots are rotated and reseeded at intervals, the soil may become contaminated with certain stages of these parasites. Chicks should be raised on clean ground. By treating the flock with certain drugs, poultry raisers may control large roundworms and some other types of internal parasites.

Lice cause irritation and loss of vigor among chickens. They feed and live on the bodies and feathers of the chickens. Lice powders of various sorts may be used. One of the most practical methods for controlling lice is to put nicotine sulfate on the top surfaces of the roosts about one-half hour before the chickens roost for the night. The fumes from this material kill many of the lice on the birds. A second treatment in two weeks is desirable.

Mites of several types are parasites of chickens. The common type lives in the cracks of the roosts and walls during the day. At night, they crawl onto the roosting hens and suck blood from them. Thoroughly cleaning the poultry house and treating the roosts, nests, and nearby walls with creosote, carbolineum, lindane, or another preparation is the best control method.

How Can the Principal Ailments of Hogs Be Prevented and Controlled?

Hog cholera is probably the most destructive disease of hogs. Before an effective method of prevention was developed, the losses from this disease in certain years were tremendous. Even now, there are serious hog cholera outbreaks in some years because some farmers fail to take the necessary measures to prevent this disease.

Hog cholera is caused by a virus. Infected hogs become stiff and inactive, lose their appetite, become very weak, and may die quickly. Whenever cholera is suspected, a veterinarian should be called at once. In case of an outbreak in a herd, it is usually advisable to vaccinate those animals that have not as yet contracted the disease.

The main tools for the prevention and eradication of hog cholera are: (1) vaccination of every pig in the herd; (2) isolation of replacement pigs until they are proved free of cholera; (3) quarantine of swine quarters; (4) sanitation of premises, vehicles, equipment, and workers' footwear; and (5) cooking of raw garbage fed to pigs.

Influenza, or flu, is an ailment that may appear suddenly in a herd of hogs, but it seldom results in death. The affected hogs pant and cough and refuse to eat. Good feed and dry, draft-free sleeping quarters help to prevent this disease. If hogs get the flu, they should be fed lightly, kept warm and dry, and given antibiotics or other medications as recommended by a veterinarian. They will usually recover in a few days, but they lose considerable weight.

Fig. 13-9. These pigs are showing the symptoms of the advanced stage of hog cholera. (Courtesy, U.S. Department of Agriculture)

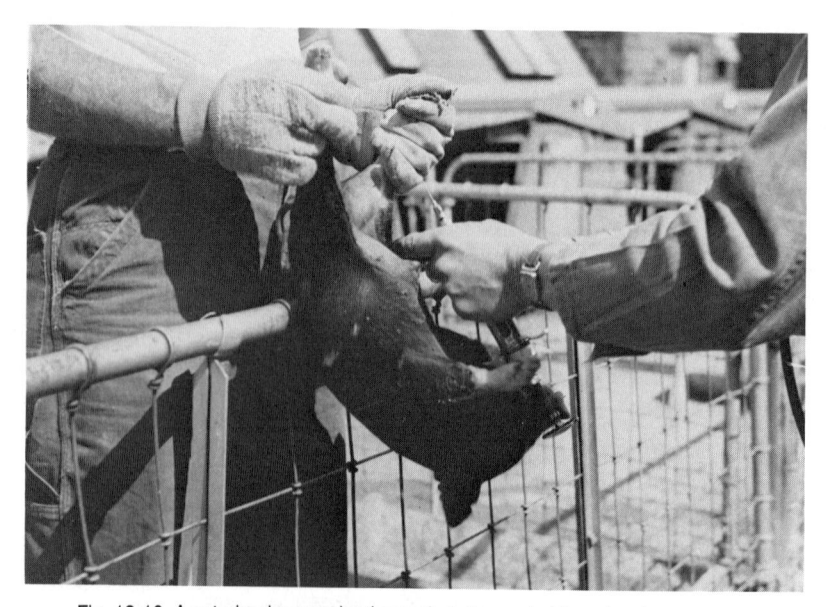

Fig. 13-10. A veterinarian vaccinating a pig to immunize it against hog cholera.

Rhinitis of various kinds affects hogs. Atrophic rhinitis and bull-nose are two diseases in this group. The former is more serious than the latter, but both affect the nose in some way, frequently causing it to be twisted and deformed. Common bull-nose can be effectively prevented by a sanitation program. Atrophic rhinitis is much more difficult to control. Breeding stock should be replaced from herds known to be free from the disease and placed on clean ground.

Brucellosis in swine is a disease comparable to brucellosis in cattle, but it is caused by a different germ. If brucellosis appears or is suspected in a herd of hogs, the sows kept for breeding stock should be blood tested, and the reactors should be sent to market. It is well to test the breeding stock at intervals. Any breeding stock purchased should come from brucellosis-free herds.

Erysipelas is a disease that has caused serious losses in swine. It is caused by a germ that may live a long time in the soil. The disease is difficult to diagnose because it has several different forms, some of which are similar to cholera or other diseases. In the acute form of the disease, pigs die suddenly. In the chronic form, various symptoms are present, such as enlarged joints, stiffness, and skin eruptions. The help of a veterinarian is needed. When the disease is present, the veterinarian will usually vaccinate the animals. Healthy hogs in the herds should be moved to clean quarters and pasture. The infected quarters should be thoroughly cleaned and disinfected. In purchasing hogs, hog raisers should secure them from herds free from this disease.

Gastroenteritis is usually a fatal ailment. It is very contagious and is probably caused by a virus. There is no known treatment for pigs with this disease. Sows that have not farrowed should be moved to clean quarters and kept away from the infected stock.

Lice and mange may cause considerable unthriftiness to hogs. One treatment for lice consists of applying crude oil to the skin of the hogs. Mange is caused by tiny parasites called *mange mites*. These burrow into the hide and cause severe itching. A thickened and scabby skin results, and hogs become unthrifty. For the control of mange, hogs should be sprayed with benzenehexachloride (BHC) or lindane. The same treatment will also control lice.

Intestinal roundworms cause hogs to be unthrifty. An effective method for preventing roundworms and some diseases was worked out by the U.S. Department of Agriculture in cooperation with hog breeders of McLean County, Illinois. The results have been remarkably satisfactory wherever this method (called the McLean County System of Swine Sanitation) has been correctly used. In this method, the central hog house is thoroughly cleaned, and the floor is scrubbed with boiling water and lye, or it is sterilized with steam. Previous to being placed in the farrowing pens in this building, the sows should be scrubbed with soap and warm water. About two weeks after the pigs have been born, the sows and pigs should be hauled to a pasture on which hogs have not been kept since the field was plowed. The pigs should be kept on this pasture, with portable hog houses for shelter, until they are ready for market or until they are at least four months of age.

A modification of the McLean County System is to let the sows farrow in clean, movable hog houses, placed on a clean pasture. The pigs should be kept on this pasture until they are at least four months old. Another modification is to use a central hog house with several adjacent fields on which crops can be rotated so there is a fresh pasture each year.

Infested hogs may be treated to eliminate the worms. One method is to mix sodium fluoride at the rate of 1 per cent in a day's ration. The hogs are allowed to eat this mixture for one day.

Anemia may be contracted by small pigs that do not have sufficient iron and copper. Pigs with this disease are often said to have the thumps, due to the irregular breathing that accompanies it. Thumps was once thought to be caused by lack of exercise, but this is not the case. The easiest method for preventing anemia is to place a piece of grass sod in the pen where the pigs are kept. They will nose around in this soil and eat enough to get sufficient iron and copper. It is important to provide soil that is not contaminated with roundworm eggs. Swabbing the sow's udder with an iron sulfate solution, feeding the pigs iron sulfate, and injecting the pigs with an iron compound are other ways of preventing anemia.

Swine kidneyworms are a parasite that costs hog producers millions of dollars each year. Swine kidneyworms were eradicated from heavily infested experimental pastures in three farrowing seasons. Only first-litter gilts were used for breeding; then the gilts were removed after they had weaned their pigs. Kidneyworms may require as long as a year to attain egg-laying maturity in swine. A gilt normally weans her pigs and can be disposed of before she starts passing kidneyworm eggs in her urine. After about two years, hog lots will be free of the parasite so that older sows can be used more profitably.

Pseudorabies is a disease of major economic importance. It may cause abortions, reproductive failures, and death. Cattle, dogs, cats, and wild animals also may contract the disease. In baby pigs, the symptoms are fever, dullness, loss of appetite, vomiting, weakness, lack of coordination, and convulsions. Mature pigs may have fever, dullness, loss of appetite, nasal discharge, sneezing, nose rubbing, coughing, difficulty in breathing, itching, trembling, lack of coordination, blindness, vomiting, and diarrhea or constipation. Prevention includes keeping people and other animals away from swine areas, wearing clean clothes and equipment, and using decontamination procedures. Several states require testing to certify pseudorabies antibody-free status before breeding animals may be introduced.

How Can the Principal Ailments of Sheep Be Prevented and Controlled?

Stomach worms are the most serious parasite of sheep. They often become very numerous in the fourth, or true, stomach of the sheep. The individual parasite is thread-like and an inch or less in length. When the parasites become exceedingly numerous in a sheep, they do considerable damage. Infected sheep become thin, weak, stunted, and they often have diarrhea. When these are suspected in a flock, one or two sheep should be killed and the contents of their stomachs inspected to determine if stomach worms are the cause.

A chemical called phenothiazine (PTZ) is effective in controlling stomach worms and parasites of the intestines known as nodular worms. Each sheep is given a dosage of this material in the fall and again in the spring prior to being put on pasture. In addition, while the sheep are on pasture, they should be provided with a mixture of 1 part of phenothiazine to 12 parts of salt.

Changing the pasture of sheep every two or three weeks helps to control stomach worms, as one stage of this parasite is passed on the ground and reinfests the sheep when it is eaten with the grass. Sheep on the range

Fig. 13-11. Yearling rams feeding at the breeding station of the Southwest Range and Sheep Breeding Laboratory in Fort Wingate, New Mexico. (Courtesy, U.S. Department of Agriculture)

are seldom infested with stomach worms because they graze over wide areas and are shifted at frequent intervals.

Sheep may become infested with *ticks,* which are parasites that live in the wool next to the skin. They injure the sheep by sucking the blood. Spraying or dipping all sheep shortly after shearing will control them. Rotenone mixed with water and used as a dip, or some other recommended material, should be used.

Lambs born in cold weather may become *chilled* and die unless cared for properly. Chilled lambs should be rubbed briskly with a coarse cloth, or they should be placed under a heat lamp.

Poisoning from the weed halogeton is a serious problem in some areas. Hungry animals are apt to eat lethal quantities of this weed. Some herders have lost nearly 20 per cent of their flocks from halogeton. Researchers have found that dicalcium phosphate, a mineral supplement that is commonly used in livestock feed, prevents the weed from killing sheep. Sheep owners have been advised to supplement the diets of sheep with pellets of alfalfa and 5 per cent dicalcium phosphate before moving sheep from the winter ranges.

Pregnancy disease commonly occurs a few weeks before lambing, primarily in ewes carrying twins or triplets. Ewes become less active, weak, walk slowly and stiffly, stand with their heads against an object, lie with their heads turned to one side, grind teeth, breathe rapidly, and have a characteristic sweetish odor to their breath. Early treatment under the direction of a veterinarian is essential. If the disease is not too far advanced, drenching or bottling with ½ pint of molasses or ½ pint of a 25 per cent to 50 per cent fructose solution twice daily will help the animals recover.

In warm weather, the *screwworm fly* may lay eggs on lamb navels and on ewes. If egg masses are seen, the animals should be treated immediately.

How Can the Principal Ailments of Horses Be Prevented and Controlled?

Venezuelan equine encephalomyelitis (VEE), a disease of horses and humans, first appeared in the United States in 1971. Other strains of the disease have been known for some time as *sleeping sickness* and *blind staggers.* It is caused by a virus carried by biting insects. The symptom is nervous disturbances, as it attacks the central nervous system. The mortality rate in horses is high—from 90 to 100 per cent. Vaccines have been developed for both horses and humans.

Influenza is a respiratory disease of horses, which can also be found in people. It is often spread at horse shows. A vaccine does combat some of the viruses causing it. Complete rest is needed for full recovery.

Thrush is a foot disease resulting in lameness. It is caused by poor sanitation. Other foot problems are *grease-heel,* *hoofcracks,* and *corns.* Professional help should be secured for treatment.

Worms and *horse bot* can also be serious problems. The bot fly eggs, laid on the hair of the horse, hatch into larvae that grow in the stomach and intestines. Both horse bot and worms require internal treatment, which should be given at the direction of a veterinarian.

Equine infectious anemia, or swamp fever, causes fever, depression, depressed appetite, and rapid weight loss. Spread by a virus, it is present in all parts of the world. Prevention includes blood tests to identify infected animals, sterilization of surgical instruments, control of biting insects, and euthanasia or complete isolation of infected animals. No vaccine is yet available.

Heaves, or broken wind, may be caused by repeated exposure to dust and mold spores in hay and bedding. Dry coughing, weight loss, and nasal discharges may develop. The best prevention is good management.

Vaccinations are recommended for *tetanus,* or lockjaw, and for *viral rhinopneumonitis,* or virus abortion.

Digestive tract disorders include *colic* and *diarrhea.*

A program of good management plus vaccination as directed by a veterinarian should be followed to keep all farm animals healthy.

SUGGESTED ACTIVITIES

1. Arrange to have a local veterinarian talk to your class about controlling and preventing diseases of farm animals and pets. Ask the veterinarian to bring specimens of parasites and diseased tissues, if available.
2. With other members of your class, make a survey of animal losses during the past year in your community. What ailments are most serious? What measures would you suggest for controlling them?
3. Prepare a report on the topic of "The Relation of Animal Health to Human Health."
4. With some other student, plan a demonstration on some phase of animal health and present the demonstration to your class.
5. With others in your class, make charts demonstrating some slogan, such as "Animal Health Means Wealth," or "Livestock Health Improves Human Health."
6. For some of your livestock or pets, plan a health program to prevent diseases and control parasites.
7. Concepts for discussion:
    a. Animal health is a major economic consideration in livestock production.
    b. Many diseases of humans may be contracted from diseased livestock and livestock products.
    c. Young animals are especially susceptible to diseases.
    d. Some human illnesses may be caused by parasites harbored by farm animals.

e. Federal inspections and controls over livestock and their products provide the United States with the safest animal products in the world.
f. Federal inspection of livestock movement into this country helps prevent the introduction and spread of diseases.
g. A disease is any disorder of the body.
h. Diseases have many causes, direct and indirect.
i. Most parasites do damage to only one species of animal.
j. Some diseases may be contracted by several species of animals.
k. Contagious diseases are those that spread from one animal to another.
l. The best control for diseases is prevention.
m. Animals that receive proper feed and care are more able to resist disease than are animals that are not well cared for.
n. Animals may be immunized against some diseases.
o. All purchased animals should be kept in isolation and tested until the owner is certain that they are free of disease.
p. Early detection of disease is essential to treatment.

REFERENCES

Baker, J. K., and W. J. Greer. *Animal Health: A Layperson's Guide to Disease Control.* Interstate Publishers, Inc., Danville, IL 61832.

Ensminger, M. E. *Horses and Horsemanship.* Interstate Publishers, Inc., Danville, IL 61832.

Haynes, N. B., S. W. Sabin, H. F. Hintz, and H. F. Schryver. *A Horse Owner's Guide.* Cooperative Extension, Information Bulletin 153, New York State College of Agriculture and Life Sciences, Cornell University, Ithaca, NY 14853.

Vocational Agriculture Service. *Horses and Horsemanship.* VAS 1047, Vocational Agriculture Service, College of Agriculture, University of Illinois, Urbana, IL 61801.

Vocational Agriculture Service. *Plants Poisonous to Livestock.* VAS 4020, Vocational Agriculture Service, College of Agriculture, University of Illinois, Urbana, IL 61801.

Chapter 14

Characteristics of Farm Crops

As we walk through fields of grain or view the various crops of the countryside from the highway or from the air, many of us fail to appreciate the importance of these crops to farmers and to humankind in general. We may realize in a general way that bread is a product of wheat and that meat, milk, and eggs come from farm animals that eat various crops. However, many of us have failed to analyze the full importance and varied uses of crops in our everyday living.

OBJECTIVES

1. Discuss plant photosynthesis and trace to animal life and our food supply.
2. Discuss the use of plants for food, fiber, and enjoyment.
3. Identify important crops produced in the United States.
4. Diagram the life cycle of a plant.
5. Identify parts of a flower.
6. Discuss the botanical classification of farm crops.

In the United States, we tend to take for granted our ample supplies of food. Some people even become greatly disturbed because there are surpluses of some grains and other food products in the United States. The opening quotation in this chapter, taken from the Bible, indicates a shortage of food in ancient times. The word corn was used for the common grains, probably wheat, barley, and rye. Corn, as we know it, originated in the United States, and was not grown in other parts of the world at that time.

Even in modern times, famine haunts many countries of the world. More than half of the world's population fails to get enough food to prevent the pangs of hunger. Only as we become aware of these facts will we begin to appreciate the importance of domesticated plants and feel thankful that we live where crops are produced in abundance. The shortage of foods in some countries is one of the basic causes of unrest in the world today. In helping to relieve this situation, we should try to find ways to distribute foods from areas of abundance to areas of shortage. We should also show people in developing countries how to produce more food on the land available to them.

Fig. 14-1. The sign and the storage buildings are indications of our abundant supplies of grain.

How Are Plants Useful?

Most forms of animal life would perish without plants. Our food supply comes directly or indirectly from green plants. Some animals live almost entirely on plants; some live on the flesh of other animals; and some (including most people) live on both. However, somewhere along the line, all foods trace back to plants. Plants produce carbohydrates and oxygen through the process of photosynthesis. The chemical equation for photosynthesis is:

$$CO_2 + H_2O \xrightarrow[\text{green plants}]{\text{light}} CH_2O + O_2$$

One molecule of carbon dioxide combines with one molecule of water in the presence of light within the pigment of green plants to produce carbohydrates and oxygen. The oxygen and organic matter produced through photosynthesis provide for the continuity of life.

Early tribes of people found it necessary to shift their herds and flocks from place to place where feed was available. A settled form of life became possible when they learned to cultivate crops to provide feed for their livestock and for themselves. This stage in civilization really marked the beginning of agriculture, which literally means "field tillage" or "land culture." To a large extent, civilization has progressed as people have recognized the values of an increasing number of plants and have learned to cultivate and improve these plants. The plants discussed here are the ones commonly grown on farms and ranches in the United States.

Plants as a source of food

At times people have been slow to recognize the value of certain plants for food. For example, potatoes were taken from America to Europe by early explorers. For a long time, people were reluctant to use them for food because they were considered unhealthful, and some people refused to eat them because they were not mentioned in the Bible.

Soybeans have been an important food crop in China for thousands of years. Only during the last 60 years have they become an important crop in the United States. Many industrial products have been developed from soybeans, and soybean meal is widely used as a livestock feed. Other than the oil, which is used in food products, soybeans are still of minor importance as a food crop in the United States.

Much of our food is secured from grain crops of various kinds. Bread is often called the "staff of life" because of its importance as a food. Most of the bread in the United States is made from wheat flour, so wheat is considered one of the important crops in the United States, as it is in many parts of the world. It is interesting to note, however, that rice is the principal food crop for about half of the people in the world, more particularly those in the Far East.

Long before recorded history, grains furnished much of the food for humans. At first, these were used without grinding or cooking. Bread had its origin over 5,000 years ago, and it was one of the first cooked foods. The ancient Egyptians made bread out of the wheat and rye that they grew along the Nile River. At first, they mixed flour with water to make a thick dough and baked it on stones with heat from the sun. Later, they baked bread on stones heated by fires.

Ancient peoples first made flour by placing the grain in a hollowed-out stone and using another stone to pound the grain into coarse meal. Later, they used flat millstones for grinding grain, by rotating a stone with a grooved surface on another similar stone held in a stationary position. Water power was frequently used for this operation. During the past century, iron rollers were developed for crushing grain into flour, and gradually more complicated types of flour mills were designed for making flour as it is known today. American Indians were the first to make bread from ground corn.

People consume small quantities of grains such as rice, barley, rye, and corn. Some of these are made into breakfast foods and processed in other ways. Other foods, which come directly from plants, include vegetables and fruits.

In addition to foods secured directly from plants, we eat crops such as

Fig. 14-2. A well-tended apple orchard. Note the young trees planted as replacements for some of the older trees. (Courtesy, U.S. Department of Agriculture)

grasses and grains *indirectly*—after they have been converted by livestock into meat, milk, and eggs. Actually, the largest possible number of people could be fed from an acre of cultivated land if plant products were used directly for food. This is necessary to a large degree in thickly populated countries, such as China and India. We should recognize, however, that about half of the land area in the United States is in grass crops and much of this portion is not suited for other crops.

Plants as a source of shelter and clothing

Our homes are constructed largely or partially of lumber, which is made from trees. Fiberboard and other materials used in construction are made in part from wood, cornstalks, and other fibrous materials of plant origin.

Much of our clothing is made in part from cotton fibers. Some synthetic fibers are made from wood and other plant materials.

Industrial uses of plants

Many industrial products are made from plants. Oils are extracted from crops such as soybeans, cottonseed, flaxseed, and corn. Some of these oils are used in food products, such as margarine and cooking oils. These vegetable oils are also used in many other ways.

In addition to being fed to livestock, corn has hundreds of other uses. Many uses have also been found for peanuts and soybeans.

The relation of plants to soil fertility

A considerable portion of the material in fertile soil is decayed plant material, or humus.

Plants called *legumes* add nitrogen to the soil when grown under favorable conditions. The nitrogen is taken from the air by special bacteria, which are present in nodules on the roots of these plants. Clover, alfalfa, and lespedeza are some of the common legumes that, if grown properly, benefit the soil in this way. Various crops, such as legumes and grasses, may be grown and plowed under to provide organic matter in the soil. These plants also help to prevent washing of the soil and are important in controlling erosion.

Many crops are fed to livestock and the animal manure is returned to the soil. This practice helps to maintain soil fertility, as much of the mineral matter originally taken from the soil by the plants is thus returned. In addition, manure provides organic matter, which improves the physical condition of the soil.

Plants as a source of enjoyment

As people have more leisure time, working with and enjoying plants become more important. In recent years, the recreational values of plants have received greater recognition than ever before.

Beautifying the home, as discussed in an earlier chapter, is one kind of value. Another kind of value is in the enjoyment travelers get from observing various kinds of vegetation.

One of nature's most spectacular sights is the fall color of leaves on trees. Sunny days and cool nights result in the production of sugars and tannins during the day, which are kept in the leaves by the cool nights and from which anthocyanins are produced. The anthocyanin is the pigment that provides the oranges, reds, and purples so much enjoyed by many people. The shade of orange, red, or purple depends on the acid or alkaline condition of the sap of the plant. If the sap is acid, the color will be orange or red. A neutral sap results in light purple, while an alkaline sap produces the dark purple or blue. Yellow is common to most leaves and is hidden by the green chlorophyll. When the chlorophyll disappears, the yellow becomes visible. If no yellow pigment is present, the leaves become brown. Each plant has its own shade of color, which is influenced by climate and soil conditions.

If the weather in fall is cloudy, sugar production is low. If the nights are warm, the sugars move to the branches for storage, and foliage color is quite dull.

Another source of enjoyment from plants is the cooling effect of large expanses of grass. One acre of grass will lose about 2,400 gallons of water in the air during one warm summer day. The release of this amount of water has the cooling effect of a 70-ton air conditioner.

Plants also help control noise. Green leaves have the ability to deflect sound waves, changing their direction and reducing the intensity of the sounds. Thus, buffer zones of plantings are often used on highways and near industrial plants.

What Are the Important Crops Produced in the United States?

The principal crops grown in the United States, in terms of farm value, were as follows:

Crop	Thousands of dollars
Corn, grain	10,671,890
Soybeans	9,125,434
Fruits, nuts, and berries	7,084,018
Nursery and greenhouse crops	5,774,391
Wheat	4,827,887
Vegetables, sweet corn, and melons	4,696,083
Cotton and cottonseed	4,207,891
Hay, silage, and field seeds	2,598,615
Tobacco	1,745,417
Grain sorghum	943,684
Barley	711,241
Oats	223,134
Other crops	4,482,146

Corn remains the top value crop grown in the United States.

Grain crops

Grain crops are grown primarily for the seeds they produce. The three most important grain crops grown in the United States are corn, soybeans, and wheat. Other grains include oats, barley, rye, and rice. All these grain crops, except soybeans, belong to the grass family, and they are often called cereal grains because they are grass crops grown primarily for the seeds, which are used for feed and food. Some grain crops not of the grass family are soybeans, field beans, peanuts, and buckwheat. All these grains are edible and, with the exception of soybeans, are used rather extensively for food. Other crops grown largely for the grain produced are grain sorghum and flax.

Characteristics of Farm Crops

Corn leads all grain crops in the United States in acreage and value. In fact, its total value usually exceeds the combined value of wheat, barley, rye, and buckwheat.

Corn was grown in this country by the Indians many years before the white settlers came. The early settlers at Jamestown and Plymouth Rock would have starved if the Indians had not given them corn and shown them how to prepare it for food and how to plant and raise it. From the Indians, the colonists learned how to fertilize the soil by placing a fish in each hill of corn. In order to prevent dogs from digging up the fish used as fertilizer, the colonists passed a law that all dogs should be "tied by ye leg" during the planting season. These early settlers paid their taxes in corn. Early settlers made the original "Johnny cake" from corn by grinding it into coarse meal, mixing it with water, and baking the mixture on hot stones.

The United States is the leading corn-producing country in the world. Almost half of the corn production in the world is in the United States. Other leading corn-producing countries are China, Brazil, France, Argentina, and Mexico. Yugoslavia, Rumania, South Africa, and the Soviet Union also raise considerable corn.

Corn is grown in nearly every state in the United States and on over half of all farms. The Corn Belt, located in the upper part of the Mississippi Valley, is the area most important in corn production. In 1980, over 83 per cent of the corn was produced in the states of Iowa, Illinois, Indiana, Ohio, Minnesota, Wisconsin, Nebraska, Texas, Kansas, and Michigan. Corn is also an important crop in many other states.

Over half of all corn raised in the United States is fed to livestock, of

Fig. 14-3. Corn leads all grain crops in the United States in production and value. This field is in Iowa, the leading corn state. (Courtesy, Wettach, Iowa)

which hogs are the major consumer. The other portion has hundreds of uses, including some for food in the form of cornmeal, hominy, salad oil, corn syrup, and corn sugar. Corn starch is also an important manufactured product.

Soybeans have become the second leading crop in value of production in the United States. Since 1909, the acreage of soybeans harvested in the United States has increased from 2,000 acres to over 70,000,000 acres. The United States produces two-thirds of the world's soybeans. Soybeans are one of the oldest cultivated crops of the world. The leading countries in producing soybeans in recent years have been the United States, Brazil, China, and Argentina.

In the United States, soybeans are grown most extensively in the Corn Belt. The leading states in production of beans in 1980 were Illinois, Iowa, Minnesota, Indiana, and Missouri.

Over 80 per cent of all soybeans harvested as grain are sold to processing plants, and oil is one of the chief products. Most of this oil is used for making edible products such as margarine, mayonnaise, and salad dressing. Most of the meal that remains after the oil is removed from the crushed beans is used as a valuable feed for livestock because of its high protein con-

Fig. 14-4. Some plants of the "miracle" crop—soybeans—growing on an Illinois farm.

tent. Soybeans have many industrial uses, including the manufacture of plastic products, soaps, paints, glues, and many edible products in addition to oil.

Wheat ranks third in importance as a grain crop in the United States. As a source of food in the world at large, wheat ranks second only to rice. The leading countries in the production of wheat in recent years have been the Soviet Union, the United States, China, India, France, Canada, Australia, Turkey, and Italy.

As the United States was settled, the chief areas of wheat production shifted westward. Today, wheat is grown in nearly every state in the United States and on nearly a million farms. The leading wheat-producing states in order of production in 1980 were Kansas, North Dakota, Texas, Washington, Oklahoma, Montana, Minnesota, California, Nebraska, and Missouri.

The United States is fortunate in having two great wheat-producing areas: the spring wheat area in the north and the winter wheat area farther south. Weather conditions are seldom unfavorable for wheat in both areas at the same time.

Nearly two-thirds of all wheat produced in the United States is used for human consumption. Most of this portion is made into flour, and small amounts are used for breakfast foods and other food products. Much of the flour used in making bread is manufactured from hard red spring wheat and hard red winter wheat, both of which are rich in gluten. Flour from soft winter wheat is especially well suited for baking fluffy cakes and pastries, but some of it is also used for making bread. Durum wheat is used chiefly for making macaroni.

Sorghum for grain is produced chiefly in areas of limited rainfall and fairly long seasons. Texas, Kansas, Nebraska, and Missouri are the leading states. The grain from sorghum plants is frequently called milo or milo maize. This grain is used chiefly as a livestock feed.

Two kinds of sorghums are grown principally for purposes other than the grain produced. One kind is the sweet sorghum, from which the sap is pressed and made into sorghum that is used for food purposes and for livestock feed. The other kind is called broomcorn, from which portions of the stalks are used in broom making.

Forage crops

Forage crops include hay and pasture plants from which most or all of the portions above ground are used for feed.

About half of the land area of the United States produces pasture crops that are grazed by livestock. Much of this area consists of land that is hilly and mountainous or is unsuited for cultivated crops because of low rainfall or other climatic reasons. Common pasture grasses are Kentucky bluegrass, timothy, Johnson grass, brome grass, fescue, and the various native grasses

Fig. 14-5. Sweeping cobwebs out of the sky! This is broomcorn, and the heads provide broomstraw for making brooms.

of the range states. Several legumes, including white clover, red clover, lespedeza, and kudzu, are used for pasture. The last two are especially adapted to southern parts of the United States. White clover and red clover are frequently combined with bluegrass for pastures and lawns. A variety of white clover, called ladino clover, has become an important pasture crop in some states.

Alfalfa, clover, and timothy are important hay crops. In some places, lespedeza, cowpeas, and vetches are also used for hay. Soybeans may be cut for hay. All these crops, except timothy, are legumes. Legumes are preferred for hay by most livestock raisers.

In recent years, increasing numbers of farmers have been using certain legumes and other crops for making silage to feed to livestock. Corn is one of the preferred silage plants because of the high tonnage secured per acre.

"Grassland" farming is being emphasized in many parts of the United States at the present time. Legumes, as well as grasses, are included in this program for increasing the acreage and yield of forage crops. This type of

Characteristics of Farm Crops 437

Fig. 14-6. Red clover is grown alone and in mixture with grasses for hay, pasture, and soil improvement. (Courtesy, U.S. Department of Agriculture)

farming is especially suited to some types of land where erosion is likely to be serious if cultivated crops are produced.

Roots and tubers

The white, or Irish, potato is an important crop in the United States. Potatoes are really tubers, or underground stems, as explained later in this chapter. The leading Irish potato-producing states in 1980 were Idaho, Washington, Maine, and Oregon.

Potatoes and bread are important items of food for most people. Potatoes are actually three-fourths water, and the remainder is primarily starch, which provides heat and energy for the human body. It is now known that potatoes are also a good source of certain vitamins and minerals. Potato chips are a favorite in the U.S. diet; about 10 per cent of the crop is used for this purpose.

Sweet potatoes are an important root crop used for food and many industrial purposes. North Carolina, Louisiana, California, and Texas are leading states in the production of this crop.

Many other root crops are grown in the United States. These include beets, carrots, rutabagas, and parsnips.

Fiber crops

Cotton is the fiber crop most commonly grown in the United States. Cotton ranks fifth among the principal crops. It is a very important crop in several southern states.

The United States produces about one-fifth of the world's cotton. The leading countries in cotton production are the Soviet Union, China, and the United States, in that order. Other leading producers of cotton are India, Pakistan, Brazil, Egypt, Turkey, and Mexico.

The Cotton Belt has been extended westward in recent years. The leading cotton-producing state is Texas. California, Mississippi, Arizona, Louisiana, Arkansas, Alabama, Georgia, Tennessee, Oklahoma, and Missouri are other important cotton-producing states.

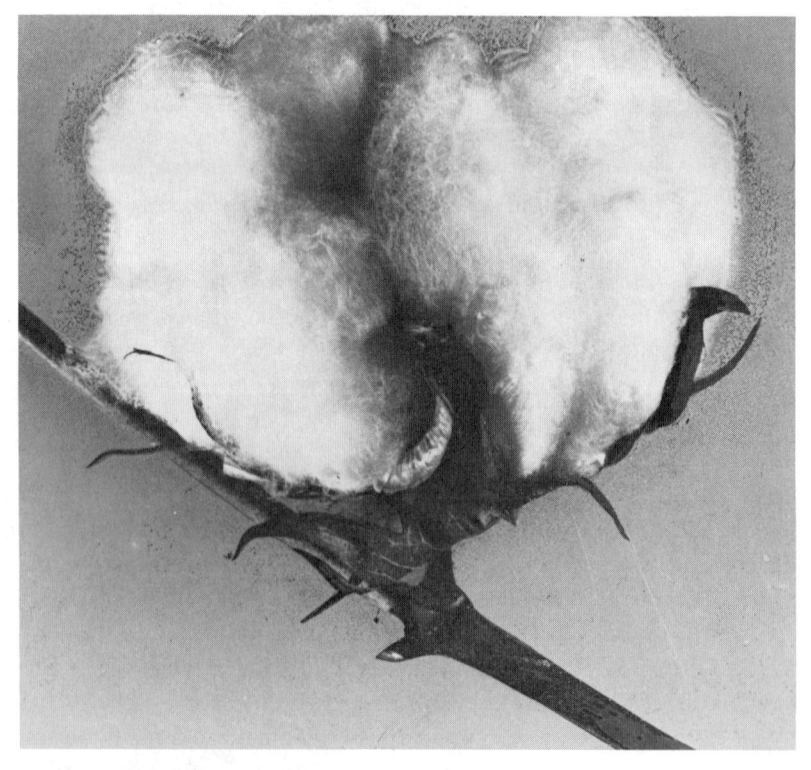

Fig. 14-7. A close-up of an open cotton boll. Cotton is a major fiber crop in the United States. (Courtesy, U.S. Department of Agriculture)

Characteristics of Farm Crops

Cotton is the only major crop that produces fiber, food, and feed. The fiber, or lint, is extensively used in making various kinds of cloth. Some of it is used in the manufacture of rayon, varnishes, cellophane, explosives, and many other products. From the seeds, oil is extracted to be used in making margarine and various salad and cooking oils. The crushed portions that remain are used for feeding livestock.

Flax and hemp are two other fiber crops produced in the United States, but they are of minor importance. Flax is used for making linen cloth and hemp is used for making rope. Linseed oil, which is commonly used in paints, is obtained from flaxseed.

Sugar crops

Sugarcane and sugar beets are the most important sugar crops in the United States. The growing of sugarcane for sugar is limited primarily to

Fig. 14-8. This cut-away shows how young sugarcane plants grow from the joints of an old cane that has been planted. (Courtesy, U.S. Department of Agriculture)

Florida, Louisiana, Hawaii, and Texas. The leading producers of sugar beets are California, Minnesota, and Idaho.

The early settlers in this country found the Indians making sugar and syrup from the sap of maple trees. The settlers used this as a source of much of their sugar. Maple sugar and maple syrup are still important products in some states, notably Vermont, New York, and other states in that region.

Yesterday's "sugaring off" involved carrying pails of sap over rough terrain and watching fires all night. The modern maple sugar producer uses power drills to bore holes, sanitizing pellets to prevent organisms from growing in the holes, plastic spouts, and miles of plastic tubing to take the sap to roadside tanks. Modern evaporators are pans fired by oil or gas and equipped with devices that automatically draw off finished syrup. All these developments make sap boiling a modern technology.

Oil crops

Soybeans and cottonseed are the two chief sources of vegetable oils in the United States. Other crops from which oils are secured are peanuts, flaxseed, coconuts, tung nuts, olives, and corn.

The chief cotton and soybean states were given earlier in this chapter. Peanuts for grain and oil are an important crop in Georgia, Alabama, North Carolina, Texas, and Virginia. Minnesota, North Dakota, South Dakota, and Texas raise considerable flax. Tung nuts, a tree crop, are produced in some southern states.

Wood crops

Much of the land included in farms in some states is wooded. In some of the southern states, as much as half of all land on farms consists of woodlands. Trees in farm woodlands are really a farm crop. These trees provide timber for lumber, paper products, posts and poles, railroad ties, and other products. In addition, many of these products are obtained from forestlands, which may be privately owned or government owned.

What Factors Affect the Kinds of Crops Raised?

From the preceding pages, we have seen that there is considerable variation in the kinds of crops grown in different parts of the United States. In general, the farmers in any area produce the crops that give them the highest returns for their labor and other costs. The returns from crops are affected by natural factors such as temperature, soil, and rainfall. Irrigation has made it possible to produce crops in some areas where the natural rainfall would not be sufficient as a source of water.

Some crops, including corn and wheat and many forage crops, can be grown under widely varying conditions. However, varieties of each crop, which are adapted to the conditions in specific regions, have been developed.

Corn needs a frost-free period of 125 to 150 days and at least 20 inches of rainfall annually, much of which occurs during the growing season. Temperatures of 80° to 90°F are needed for corn during much of the growing season. Oats and white potatoes grow well in a cool, moist climate, and they will mature in a fairly short season. Cotton requires a long growing season, with around 200 frost-free days and a summer temperature of about 80°F or more.

Some crops have specific preferences as to soils. For example, corn and soybeans do best in loamy, mellow soils. Potatoes grow best in loamy soils that are especially mellow in texture. The fertility of the soil is also a factor, as some crops, such as corn, are heavy "feeders" and require abundant supplies of nutrients.

Legumes and grasses of many kinds and varieties are grown in the United States. Some of these are adapted to areas where rainfall is plentiful and some to places where rainfall is light; some to cool climates and some to hot climates.

The topography of the land is a factor in determining the kinds of crops to raise. Cultivated, or row, crops, such as corn and cotton, are best suited to land that is fairly level, because erosion would be serious on hilly land. Grass crops can be grown on sloping land because the top growth and fibrous roots help to hold the water and thereby prevent serious erosion of the soil. Trees grow well on steep slopes and in swampy land not suited to most other crops. Many temperate zone woody plants require varying hours of exposure to low temperatures before they will break normally from their dormancy and resume growth. Fruits such as apples, peaches, and blueberries will not bloom and bear fruit properly if they have not been exposed to enough hours of low temperature (30° to 40°F).

How Do Plants Work for Us?

Altogether, throughout the world, there are over 180,000 different species of plants that have been identified. Some are beneficial, some are harmless or of no recognized value, and some, such as weeds and poisonous plants, are definitely harmful. Actually, use has been found for only a small percentage of plants in the environment. About 200 species are grown in commercial quantities in the United States, and many less than these account for most of the production on farms.

Plants differ markedly in many ways. For example, they vary in size from tiny bacteria, which cannot be seen with the naked eye, to the giant

First the seed—then the plant

To understand how plants work, we should be familiar with some of the life processes from seed to mature plant. Practically all farm crops are multiplied by seeds that have been planted. Each seed contains a tiny embryo plant in a resting stage, a supply of food, and a seed coat on the outside.

Four conditions are necessary for a seed to germinate, or sprout, a process in which the embryo develops. These conditions are (1) a live seed in which the embryo is capable of growth, (2) favorable temperature, (3) enough air to supply oxygen, and (4) sufficient moisture.

A dry seed takes up considerable moisture in the germination process. As growth takes place, the developing plant is at first dependent upon the food stored in the seed. As the plant continues to grow, it develops a top and

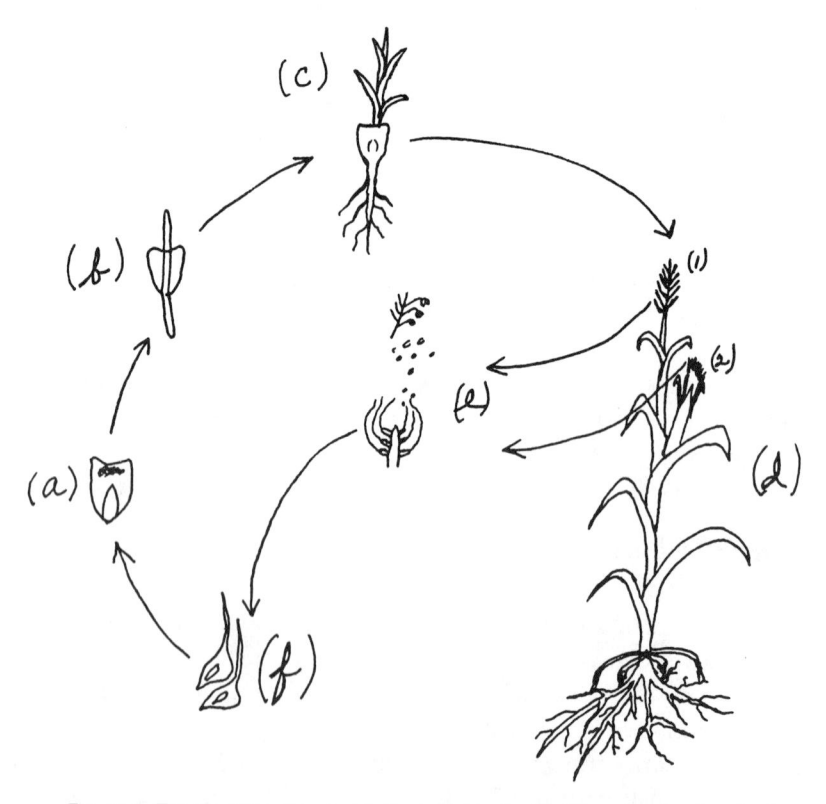

Fig. 14-9. The life cycle of a plant is illustrated by the sketches showing the development of corn: (a) mature kernel; (b) germination of seed; (c) seedling; (d) mature plant showing tassel (1) and silk on developing ear (2); (e) pollen falling from tassel to silk; (f) fertilization by pollen of portion of developing ear that becomes the kernel.

roots. Soon it is dependent on outside sources of nutrients and on light, by which it manufactures the food needed for further growth and development.

Green plants can be thought of as factories. In fact, green plants alone are equipped to utilize energy from the sun for manufacturing foods useful to people and other animals.

The soil provides the plant with important mineral elements and water. These mineral elements, dissolved in water, pass through the thin coverings on the tiny root hairs and from there are carried to various parts of the plant. Actually, little is known about the processes by which plants absorb nutrients through their roots and keep other materials from entering. A plant gets food materials from the air in the form of carbon dioxide. This is taken in through tiny openings in the leaves. The raw materials secured from the soil and air form compounds needed by the plant to grow and to produce fruit or seed.

The process of manufacturing simple sugars, which takes place in the leaves of green plants in the presence of light, is called *photosynthesis*. (*Photo* means "light," and *synthesis* means "putting together.") Thus, a plant captures and stores energy from the sun and makes this energy available as plant products.

The process of photosynthesis has long been a mystery. Just what goes on in the green coloring matter, *chlorophyll*, of the leaves has never been fully determined. It is known, however, that the chlorophyll in the plant cells is necessary for the plant to combine carbon dioxide and water in the presence of light to make carbohydrates. The chlorophyll itself is not used up in the process. Water from the soil and carbon dioxide from the air are combined chemically to form sugar, and oxygen is given off into the air in the process. The sugar may be changed by the plant into starch and other compounds for use by the plant for growth or for storage in the seeds, roots, stems, or leaves. It can also be changed into oils or fats or into proteins. In making protein compounds, the plant adds nitrogen and often phosphorus and sulfur to carbohydrates. Much of the material made by a plant is stored in the seeds, which are the most valuable part of grain plants.

Other compounds manufactured by different plants include vitamins, hormones, dyes (such as indigo), latex (used in rubber), nicotine, quinine, and many others.

At least 16 chemical elements are needed by plants for growth and development. Phosphorus, potassium, nitrogen, sulfur, calcium, magnesium, iron, boron, manganese, zinc, copper, molybdenum, and chlorine are supplied directly by the soil. Carbon, oxygen, and hydrogen are obtained principally from air and water. Actually, exclusive of water, only about 2 to 5 per cent of a green plant's weight comes from the soil. This portion from the soil is represented by the ash that remains if a plant is burned as completely as possible. About 95 per cent of a plant's substance is due to carbons, hydrogen, and oxygen, which come from the air and water.

Plants require large amounts of water. Most of this water is taken into the plants through the roots and much of it is given off in the leaves. Plants require more water if the soil fertility is low than if the fertility is high. For example, an experiment in Missouri showed that every bushel of corn produced on fertile soil required 5,600 gallons of water, while on a field of low fertility, every bushel needed 21,000 gallons. As an average, it has been estimated that for every pound of corn that is harvested, ½ ton of water must pass through the plant.

Some plants grow very rapidly if the conditions are highly favorable for growth. For example, a stalk of corn may grow several inches in a 24-hour period. Actually, the corn plant makes a noise when it grows! This was proved by a scientist who put a sensitive microphone in a corn field.

Scientists have found that flowering habits of plants are affected by the length of darkness in a 24-hour day. Some plants are favored by short nights, some by long nights. Poinsettias and chrysanthemums, for example, are favored by long nights, while shasta daisies and wheat require short nights to flower. Varieties of the same kind of plant may differ somewhat as to conditions that favor flowering. For example, varieties of soybeans that are suited to some southern states but don't do well 300 miles farther north have been developed. Recommended varieties of most grains differ considerably among states.

Plants have a variety of root systems. The carrot has a taproot that grows almost directly downward to considerable depths and has many small, lateral branch roots. The roots of the onion are a fibrous, extensive, shallow network of small roots. Some plants, such as the apple and other tree fruits, have several large roots for anchorage, with many smaller branch roots for taking up water and nutrients.

How Do Plants Reproduce?

All plants develop from either seeds or buds. Every plant that produces seeds passes through a flowering stage. Even oat plants and wheat plants produce flowers, although the blossoms are very inconspicuous.

A typical flower has green leaf-like structures near its base. These are called *sepals*, and they protect the flower in the bud stage. Next to the sepals is a row of *petals*, which are usually colored. The shape and the color of the petals are different in each kind of plant. Inside the petals are slender, erect, thread-like structures with knobs at their tips. These are called *stamens*, and the fine powder on the knobs (or *anthers*) is called *pollen*. At the center of a flower is an erect structure known as the *pistil*. The enlarged tip of the pistil is called the *stigma* and the enlarged base is the *ovary*. The ovary contains the young seeds called *ovules* (see Fig. 14-10).

Each flower on some kinds of plants has stamens and a pistil. Some

Characteristics of Farm Crops

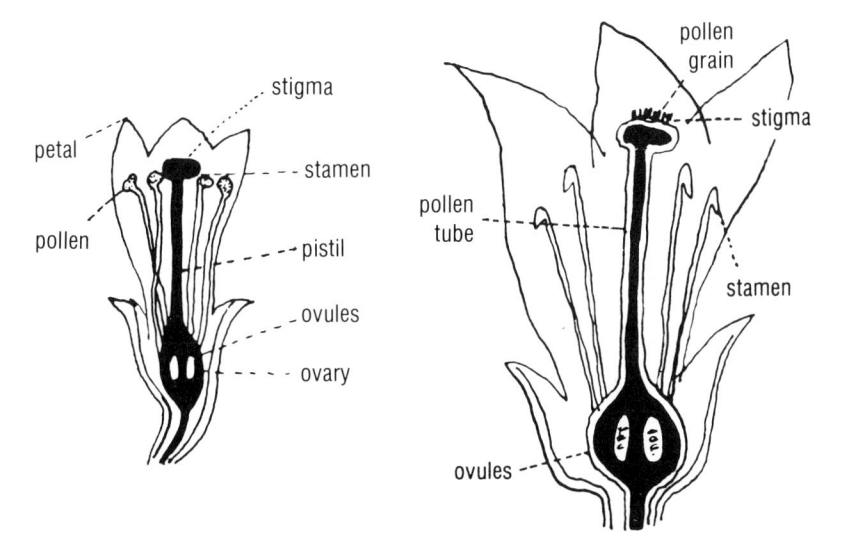

Fig. 14-10. Parts of a complete flower are shown at left. The process of fertilization is indicated at right. (Courtesy, Swift and Company)

other plants have these two parts in separate kinds of flowers. For example, the tassel of a corn plant is a cluster of many flowers, each of which bears stamens only. The young ear of corn is also a cluster of many flowers, each of which has a pistil consisting of an embryo seed, or ovule, and a silky thread, or stigma. Still, other kinds of plants have pistils on some plants and stamens on other plants.

The flowers of some plants are self-pollinated; this is true of wheat, barley, soybeans, and oats. In some plants, the pollen from one flower must be carried to the stigma of another flower. This is done by wind, by gravity, or by bees and other insects. Such plants are said to be cross-pollinated; these include corn, clover, and rye.

After the pollen falls on the stigma, *fertilization* takes place if conditions are favorable. In this process, a single microscopic pollen grain sends a tube down through the stigma until it reaches the ovary and connects with an ovule, or immature seed. The ovule then develops and forms a seed. In the case of corn, each ovule is fertilized through a silk on which a pollen grain has fallen.

New plants may be reproduced from buds. Frequently called *vegetative reproduction*, this may be done in various ways. For example, a *cutting*, or *slip*, may be taken from a geranium and placed in moist sand. It produces roots in the sand, develops a top growth of leaves and stems, and finally becomes a plant. Strawberry plants are produced from *runners*, which develop from a "mother" plant. These take root and produce new plants. White clover plants multiply in much the same way. Grafting of apple trees is done by transferring a bud or portion of a stem from one tree to another in such a way that the two become joined and develop new branches from the grafted portion. Potatoes produce plants from the eyes, or buds, on the tubers.

Seeds of plants differ in the length of time they may remain dormant and still retain the ability to sprout or grow. Under favorable conditions, some seeds may remain dormant for many years and under suitable conditions will grow. This is true of many weed seeds. Seeds of most common farm crops can be kept for only a few years. Some scientists in Ohio sealed grains of corn in a tube, refrigerated them at about 25°F for 20 years, and found that 95 per cent of the kernels developed into normal plants. However, under usual conditions, corn kept for three or four years will not germinate well. Wheat seeds may retain their ability to grow after a period of about 40 years, but under normal conditions, it is only for a few years.

In a few cases, wheat grains and other seeds have been found in the tombs of ancient kings, some of whom were buried several thousand years ago. According to some reports, a few of these seeds have sprouted and produced plants. Most authorities doubt the truth of this. One unusual instance has been verified in which seeds of ancient lotus plants germinated and produced plants. These seeds were recovered in deposits in southern Manchuria and were made available to scientists in the U.S. Department of the Interior. These seeds had hard seed coats and had been buried under conditions favorable to the seeds so that the tiny germs of some of the seeds remained alive for a period estimated to be in excess of 1,000 years, and possibly as long as 50,000 years.[1]

It is surprising how rapidly plants can multiply from seeds planted under favorable conditions. For example, a small religious group in Michigan heard a sermon based on the story of a grain of wheat that fell on fertile soil and brought forth "fruit." As a result of this sermon, this group formed an organization known as "Dynamic Kernels." They made plans for planting approximately 360 wheat seeds, which had a total volume of only 1 cubic inch. These seeds were planted in a plot 4 by 8 feet. The wheat harvested from this represented a fifty-fold increase. The church was given its tithe of 10 per cent, and the remainder was planted the following year. By the third year, enough seeds were produced, after deducting 10 per cent, to plant an acre. By the fifth year, 230 acres had been planted, which yielded 4,468 bushels. For the first five years, the late Henry Ford supplied the land. In the sixth and final year, 230 farmers planted 2,500 acres, which yielded 60,000 bushels valued at $100,000. If they had continued the project for a total of 13 years, it is estimated that all the land in the world would have been required to plant the wheat seeds, which would have been raised from the original 360 seeds, after 10 per cent had been deducted each year.

---

[1]Confirmed by a letter to G. P. Deyoe from Horace V. Wester, plant pathologist, U.S. Department of the Interior, August 10, 1951.

How Are Common Farm Crops Classified?

Farm crops may be classified according to the chief products obtained from them. Thus, there are grain crops, forage crops, roots and tubers, fiber crops, sugar crops, oil crops, and wood crops, as discussed earlier in this chapter.

Length of life

On the basis of life habits, plants may be classified into annuals, biennials, and perennials.

Annuals are plants that start their growth in the spring or summer, produce flowers and seeds, and die within one year. Many of the important farm crops, including corn, wheat, barley, oats, beans, cotton, and peanuts, are annuals. Many weeds also belong to this group. Some plants are called *winter annuals* because they can be planted in summer or fall, start their growth, become dormant during winter, resume growth the following year, produce seeds, and die. Winter wheat, winter oats, and winter rye are examples of these crops.

Biennials require two years to complete their growth. Seeds are borne the second year and the plants then die. The sugar beet is a crop that produces fleshy roots the first year but will not produce seeds unless the roots are left a second year. Medium red clover is typically a biennial, although some varieties will live longer under favorable conditions.

Perennials are plants that live more than two years. Most hay and pasture crops, including alfalfa, Kentucky bluegrass, bromegrass, and timothy, are perennials. Some of the worst weeds are perennials.

Botanical classification of farm crops

Farm crops are members of the *Plant Kingdom*. Nearly all farm crops belong to Phylum IV, Spermatophyta, the seed-bearing plants. The other phyla in the Plant Kingdom are Phylum I, Thallophyta, which includes algae and fungi; Phylum II, Bryophyta, which includes the liverworts and mosses; and Phylum III, Pteridophyta, which includes the horsetails, club mosses, and ferns.

Phylum IV is further divided into Subphylum A, Gymnospermae, which includes the evergreens; and Subphylum B, Angiospermae, which includes the grains, grasses, trees, shrubs, and flowering plants. The Angiospermae is still further divided into the classes Monocotyledoneae and Dicotyledoneae, and then into orders, families, genera, and species.

Most farm crops belong to one of two families, the grass family (Gramineae) or the legume family (Legumiosae). Many forage crops and all the ce-

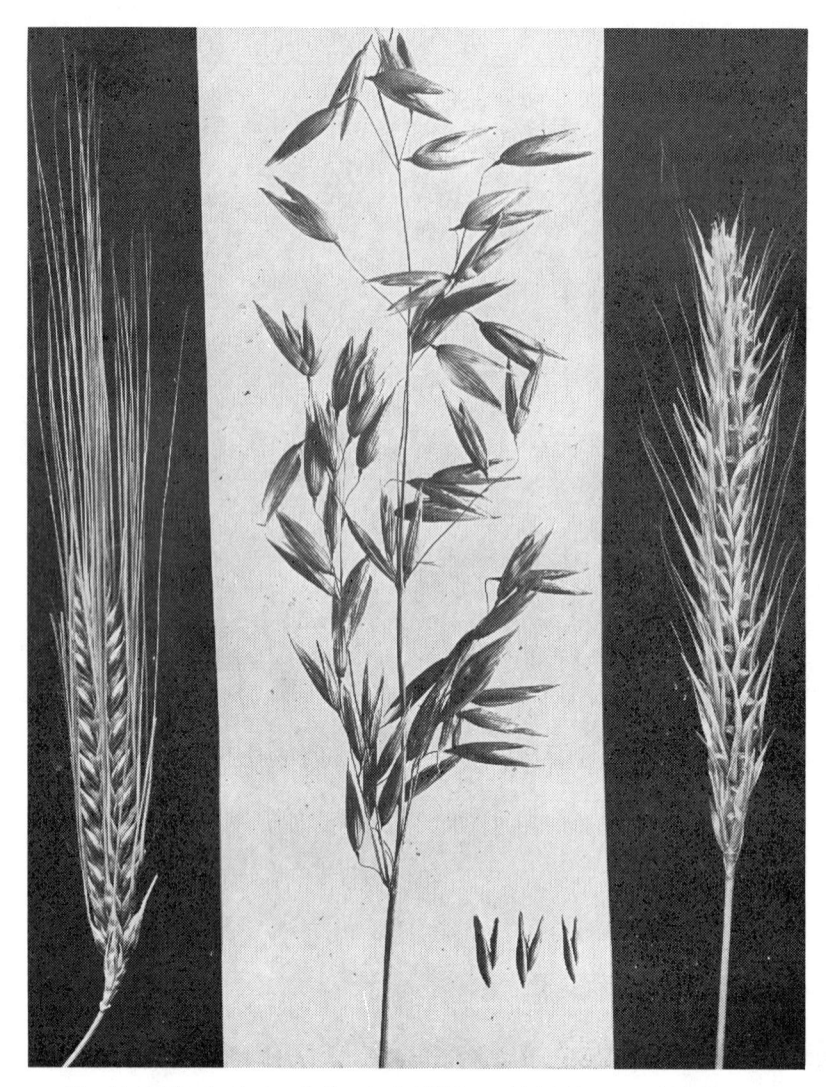

Fig. 14-11. Head of barley *(left)*, oats *(middle)*, rye *(right)*. (Courtesy, U.S. Department of Agriculture)

real grains belong to the grass family. These crops have plants with cylindrical stems that have cross partitions at joints, or nodes; long, narrow leaves; and fibrous roots. The legume family includes alfalfa, clovers, beans, and peas. The roots of legumes have enlargements called nodules, caused by beneficial bacteria that take nitrogen from the air and make it available to the plants. Legume plants have compound leaves and bear their seeds in pods. A third family, the nightshade family, includes the potato, the tomato, and tobacco.

The botanical name for white clover is *Trifolium repens*. The first word is the genus and the second is the species. In this case, the first word means

Characteristics of Farm Crops

three-part leaves, and the second means creeping, as white clover has a creeping habit of growth. The botanical name of common corn is *Zea mays*.

As a further breakdown in plant names, there are varieties to which farmers often refer. For example, there are many varieties of oats, such as Clinton, Benton, Mindo, and Vicland.

What Are Some Interesting Features of Some Common Farm Plants?

Most people can very likely name at sight many of the crops raised on the farms in their vicinity. Even so, there are many features of interest about farm plants that individuals may not have had occasion to learn about.

The corn plant

The features of a corn plant can be studied best through close observation of corn plants growing in a field. However, if this is not convenient, a corn plant can be brought into the classroom. This plant is best studied after it has produced a tassel and ears of corn. A plant should be secured with some of the roots attached. The main parts of a mature stalk of corn are roots, stem, leaves, tassel, and ear.

The *roots* of a corn plant are fibrous and hence are many branched and finely divided. The rather coarse roots that appear above the fibrous roots and grow from the stem into the ground are called *brace roots*. The main root system of a corn plant under favorable conditions penetrates the ground to a depth of several feet and spreads over a diameter of several feet.

The *stem* of a corn plant is cylindrical and is separated into sections. These sections are connected by joints from which the leaves and ears arise. The stalk is filled with pith in which stringy materials are found.

Each *leaf* of a corn plant shows three quite distinct parts. These are (1) the *sheath,* which encloses a portion of the stem; (2) the *blade,* which is the flat or spreading portion; and (3) the *rain guard,* which is the portion of the upper surface where the sheath and blade meet. The rain guard actually collects some of the rain and drops of dew and channels the moisture down to the ground near the base of a stalk.

The *ear* is attached to the stalk by a shank and is covered with husks. The husks are really modified leaves. An ear includes kernels, a cob, and silks. Actually, an ear develops from the female portion of the flower. The other portion of the flower is the tassel, which produces the pollen. The tassel may be thought of as the male flower and the tiny ear as the female flower. Normally, corn cross-pollinates; that is, it receives pollen from some corn plant other than its own.

Fig. 14-12. Tobacco, an annual of the nightshade family, is ready for harvest on a farm near Waldorf, Maryland. (Courtesy, U.S. Department of Agriculture)

The plants of wheat and other small grains

Plants of wheat and other small grains can be secured for study in late spring or early summer.

The *roots* of a wheat plant are fibrous, as is the case with corn. A wheat plant develops from a single seed, and several stems usually arise on one plant. This process is called *tillering,* or *stooling.*

The stem of a wheat plant is jointed, similar to a stalk of corn. However, the interior of the stem is hollow. Leaves similar to those of corn are produced at the joints, or nodes. The *head,* or *spike,* of wheat is produced at the tip of a stem. A spike of wheat produces two rows of kernels arranged alternately on a central axis. Chaff-like structures enclose groups of kernels and each kernel separately. In most varieties of wheat, *awns,* or *beards,* are produced on some of the chaff-like structures.

The roots, stems, and leaves of oat, barley, and rye plants are much like those of the wheat plant. These plants differ from each other and from wheat principally in the seed-bearing portions.

All small grains pass through a flowering stage, although the flowers are inconspicuous and partially covered by chaff. Flowers of wheat, oats, and barley are self-pollinating. In the case of rye, some cross-fertilization takes place.

The plants of red clover, white clover, and alfalfa

When these plants are studied, the root systems should be as nearly intact as possible. If these root systems are observed closely, enlargements known as nodules are evident. These nodules contain bacteria that are too small to be seen with the unaided eye. These bacteria have the power to take nitrogen from the air and store it in these nodules in the form of nitrogen compounds, as previously explained. This nitrogen becomes available to the legume plants. When the clover plants decay, the nitrogen compounds become a part of the soil and the nitrogen may be used by other plants, such as corn and small grains, that are not able to provide their own nitrogen.

The root system of a red clover plant usually consists of a main root, or *taproot,* with attached branch roots. This taproot is particularly evident in the root of an alfalfa plant. Alfalfa roots may extend to great depths in the soil if the conditions are suitable. Cases are on record of alfalfa roots that have penetrated to depths of over 20 feet. White clover is a low, creeping plant with stems that take root along the surface of the ground.

The *leaflets* of clover and alfalfa plants are produced in groups of three. Each collection of leaflets comprises a *compound leaf.* Occasionally, leaves are formed with four or more leaflets.

The flowers of clover and alfalfa plants are produced in clusters, each of which is called a *head.* A head usually consists of a group of *flowerlets. Pods* that enclose the seeds are formed. In the case of alfalfa, the pods are coiled in spirals similar to a ram's horn or a curved sickle.

The plants of common grasses

Two of the grasses common to many parts of the United States are timothy and Kentucky bluegrass.

Timothy is one of the best known of all grasses and is not likely to be confused with any other. A single plant usually produces several erect stems that attain a height of about 2 feet, but in some instances they may grow to a height of 4 or 5 feet. The roots are fibrous. A timothy plant has few leaves compared with some other grasses, and the leaves are relatively fine. The upper ends of the stems produce spikes, or heads, which are usually 3 to 4 inches in length, and occasionally much longer. If the plants are permitted to mature, the heads produce seeds. When timothy is threshed for seed, small bits of chaff usually remain on the seed, thus giving it a grayish color.

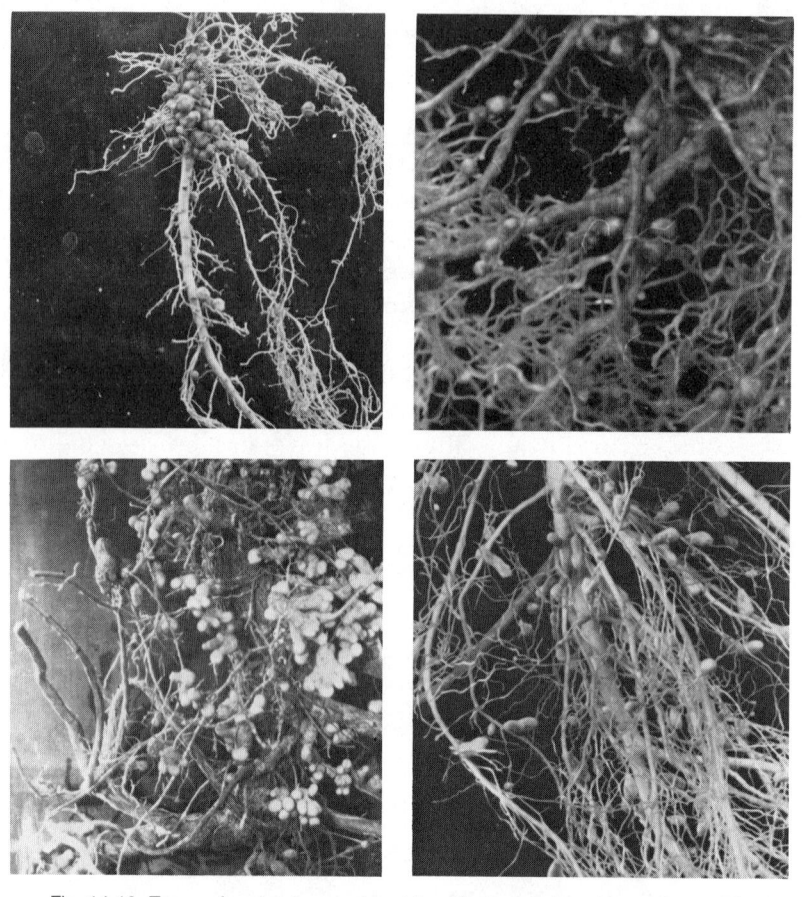

Fig. 14-13. Types of nodulation produced by effective strains of bacteria on different legumes. *(Top left)* lespedeza, *(top right)* trefoil, *(bottom left)* black locust, and *(bottom right)* crimson clover. (Courtesy, U.S. Department of Agriculture)

Kentucky bluegrass is well known for the beautiful lawns that it makes, as well as for the excellent pasture that it furnishes. Kentucky bluegrass has creeping rootstocks, each bearing a tuft of leaves at the tip. These later develop stems, which may grow in height from a few inches to 20 inches or more. The stems produce seed-bearing portions, the heads, or *panicles*, which are erect and spreading. A panicle as it matures assumes a brownish, purplish tint. Bluegrass can be recognized by the leaves, which are V-shaped in cross section, and by the peculiar leaf tip, which resembles the bow of a boat. Both timothy and bluegrass can be identified by the heads.

Characteristics of Farm Crops 453

Fig. 14-14. Alfalfa roots grow large and deep, as this photo of a single alfalfa plant demonstrates. (Courtesy, U.S. Department of Agriculture)

The potato plant

The potato plant at maturity has roots, stems, and leaves. Occasionally, plants are found that bear seeds in seed balls at the tops of the stems.

The portion of a potato plant of special interest is the *tuber*. The tuber is not part of the root system but of the stem system; to be exact, it is an underground stem. It is not a seed because it is not produced by a flower. (Fruits and seeds are always the products of flowers.) The real "fruit" in the potato is a spherical body, called a seed ball, which sometimes is borne on the top parts of the branches. Occasionally, farmers plant the seeds in these seed balls to try to get new and superior varieties of potatoes.

If a potato tuber can be imagined much lengthened, its likeness to a stem, such as the twig of a tree, is evident. The covering of the potato tuber is the skin. The skin shows specks (not those that may be due to disease injury). These specks are the pores through which the tuber exchanges gases from within with those on the outside. The most noticeable structures on the surface of the tuber are the eyes, which correspond to the buds on stems. If a potato tuber is planted, stems and roots grow from these eyes. While it is true that the potato tuber is not a seed, still it is customary to speak of it as the seed and of the entire tuber as a seed potato.

Fruits, Vegetables and Melons, and Nuts; Horticultural Specialty Crops; and Forest Products

Although fruits, vegetables and melons, and nuts; horticultural specialty crops; and forest products are produced in many states, the major portions of these crops are found in only a few states.

Fruit production

The leading states in fruit production are Washington, New York, California, South Carolina, Oregon, Michigan, Texas, and Florida.

The varieties of apples with the highest volume of production are Red Delicious, Golden Delicious, McIntosh, Rome Beauty, Jonathan, York Imperial, Stayman, Winesap, Cortland, and Yellow Newton. The Bartlett continues as the most popular pear produced.

Apples lead in fresh fruit consumption, followed by grapefruit, oranges, peaches, pears, tangerines, plums and fresh prunes, and lemons.

Vegetable and melon production

The major states in vegetable and melon production in 1980 were California, Florida, Texas, New York, Oregon, Minnesota, Washington, Michigan, and Arizona.

As judged by the value of total production, the most popular vegetables and melons, in order, are potatoes, tomatoes, lettuce, onions, sweet corn, snapbeans, cucumbers, cabbage, celery, carrots, cantaloupe, watermelons, broccoli, sweet potatoes, and peppers.

The *cucurbit* crops include the cucumber, gourd, muskmelon, watermelon, pumpkin, and squash. These plants consist of sprawling vines with many runners. Each vine has two kinds of flowers, a female with no anthers and a male with no pistils. Some commercial muskmelons have perfect flowers. The botanical name of the cucurbit fruit is *Pepo*. Some squashes weigh up to 350 pounds and some gourds almost as much.

Asparagus has deep roots called rhizomes. These store nutrients for producing the spring crop of spears, which are eaten. The male and female flowers are grown on separate asparagus plants.

Fig. 14-15. Growing pineapples in Puerto Rico. (Courtesy, Extension Service, Puerto Rico)

Rhubarb is grown for its thick stalks. The leaves, if eaten, can cause sickness and even death because of their oxalic acid content. Rhubarb is one of the most acidic of all vegetables, with a pH of 3.1 to 3.2. The green or red stalks are 94 per cent water.

Vegetable plants have some very interesting characteristics. Broccoli has small buds in clusters on thick stems as the edible portion. The center shoot is very large, ranging in size from 5 to 10 inches across.

Tomatoes, peppers, and eggplants are all members of the same family. Tomatoes range in size from the small cherry to the large beefsteak varieties. Colors may be red, yellow, orange, or pink. The tomato originated in the Andes Mountains of South America and was first grown in the United States as an ornamental. The pepper also had its origins in South America. Peppers may be red, yellow, or green. In size they range from quite small to 3 inches wide and 4 inches long. Eggplants also vary widely in shape and color, with the large purple fruit being common.

Celery is native to many countries—among them, the Scandinavian countries, Algeria, Egypt, and India. The two most familiar kinds are the green and the golden, or self-blanching. With celery, the stalks are the edible part. It is one of the most difficult plants for the home gardener to raise because it requires a long growing season and a great deal of moisture. The seeds are slow to germinate, and the seedlings are delicate.

Endive is native to the eastern Mediterranean area. It can have curly, loose leaves or broad, thick leaves forming a loose head. Outer leaves are apt to be bitter, so they are not eaten.

Onions are a popular vegetable domesticated in Asia and in the Middle East. Onion sets can be purchased in white, red, or brown. Round onion sets produce flat onions, while elongated, or tapered, sets mature into round onions.

Chicory produces edible shoots. It is native to Europe and Asia. Some chicory is grown for the roots, which are dried, ground, and used in coffeemaking. Chicory is often used in salads.

Garlic, shallots, chives, and leeks are all members of the onion family. Garlic produces a group of cloves, or sections, rather than a single bulb. Shallots also have multiple-section bulbs and are grown for the bulbs as well as for the young, green shoots. Chives have bulbs but are grown for the fresh, young leaves. Leeks are grown for the stems.

Root crops come in a wide range of sizes, colors, and shapes. The carrot is a bright orange taproot; the beet is deep red and bulb shaped; a radish can be small with a red or white skin and white flesh, or it can be a beet-shaped 50-pound giant.

Greens include a great many different plants, all grown for their tender and succulent leaves and stems. They are among the easiest plants to grow.

Beans and peas are favorite legumes of the home gardener. Both are grown for the immature pods and for the immature (shelled) seeds. The mature bean seeds are called dry beans. Beans also grow as either low-growing bush plants or as vines on poles or trellises.

Nut production

English walnuts received their name because they were brought to this country on ships from England. Probably originating in Persia (now Iran), they are also called Persian walnuts. The trees sometimes reach heights of 90 feet, with a limb spread of 60 feet. Roots go as deep as 15 feet. About 1,000 hours of temperature below 45°F are needed to meet the winter chilling requirement for viable blossoms.

Pecans are native to North America and require a deep soil and a warm growing season. About 30 trees can be planted on an acre of land, with a yield of 20 or more pounds of nuts per tree.

Chinese chestnuts have replaced American chestnuts, which were destroyed by the chestnut blight fungus. Many people find Chinese chestnuts to be an attractive addition to the home landscape, as well as a source of nuts. The trees grow well in temperate climates.

Horticultural specialty crops

Horticultural specialty crops include the following:

1. *Nursery crops:* trees, shrubs, vines, and ornamentals.
2. *Products grown largely (but not entirely) under glass:* cut flowers, potted plants, florist greens, and bedding plants.
3. *Other:* vegetables grown under glass, flower seeds, vegetable seeds and plants, bulbs, and mushrooms.

Lumber production

Over half of the saw timber is found in the western part of the country, with about 40 per cent in the Pacific Northwest and 10 per cent in the Pacific Southwest.

In the West, the softwood species of trees providing the greatest volume of lumber are the Douglas fir, western hemlock, true firs, ponderosa pine, and Jeffrey pine. The hardwoods are the red alder, cottonwood, aspen, and various oaks.

In the Northeast, the most productive softwoods are white pine, red pine, spruce, balsam fir, and eastern hemlock. The hardwoods are red oak, hard maple, white oak, soft maple, cottonwood, aspen, beech, and hickory.

In the Southeast, the greatest softwood production is from loblolly pine, shortleaf pine, longleaf pine, slash pine, and cypress. The hardwoods are red oak, white oak, sweet gum, hickory, tupelo, black gum, and yellow poplar.

Forest products are produced in many states and consist of firewood and fuelwood, sawlogs and veneer logs, standing timber, pulpwood, fence posts, Christmas trees, and maple syrup.

SUGGESTED ACTIVITIES

1. What foods have you eaten during the past few days that came directly from plants of various kinds? What steps were taken in the preparation of these foods prior to the time they were placed on the family table?
2. What crops common to your state are used for human food? What crops in your state are commonly fed to livestock? What other uses are made of the crops in your community? What products from crops not grown in your state do you use for food, clothing, and other purposes?
3. What articles of clothing are you wearing that were made from plant products?

4. Make a special study of some grain crop in which you are interested. Secure information including the principal states in which it is grown and its principal uses.

5. What are some uncommon or unusual crops grown in your state or community? What uses are made of them?

6. Place a piece of blotting paper in each of two plates and moisten. Put some corn and bean seeds on one of the blotters. Invert the other plate with blotting paper over the seeds. Keep the materials at room temperature and moisten the blotters as needed. From day to day, note the ways in which the seeds germinate and the appearances of the roots and stems.

7. Obtain flowers from some plant or plants and identify the parts.

8. Obtain plants of some of the common small grains grown in your community. Study the seed-bearing portions and note the similarities and differences among the different small grains.

9. Obtain plants of some of the common legumes, such as red clover, alfalfa, or soybeans. Note the roots, leaves, and seed-bearing portions.

10. Obtain plants of some grasses common to your community. Study the roots, stems, and seed-bearing portions. What are the similarities and differences between those studied?

11. Get a potato and study its parts.

12. With others in your class, make a collection of the seeds of most of the crops grown in your community. Note the distinguishing features of each and learn to identify each sample when you see it. With the rest of your class, plan an identification contest.

13. Concepts for discussion:
    a. Much of the world's population does not have enough food.
    b. All foods can be traced back to plant life.
    c. The oxygen and organic matter produced through photosynthesis provide for the continuity of life.
    d. Grains are the principal food crop for over half the people of the world.
    e. More people could be fed if plant products were used directly rather than indirectly when plant products are converted to livestock products.
    f. Plants provide food, clothing, shelter, and other materials for human use.
    g. Decayed plant material, called humus, helps keep soils fertile.
    h. Plants called legumes support bacteria, which take nitrogen from the air and thus help put nitrogen into the soil.
    i. Plants help conserve soil and water.
    j. Plants are a source of pleasure.
    k. Plants help cool the environment.
    l. A few crops account for most of the value of crop production.
    m. Grain crops are grown primarily for the seeds produced.
    n. Most of the corn produced is fed to livestock.
    o. Most of the wheat produced is used for human consumption.
    p. Forage crops include hay and pasture plants.
    q. Various root crops are important sources of food for people.

Characteristics of Farm Crops

r. Various crops are identified with the regions best adapted to their production.
s. Climate, soil, and markets are all important in determining crops grown in a particular area.
t. Plants have characteristics by which they can be identified.
u. Plants have parts that perform specific functions.
v. Plants need water, light, nutrients, and proper temperatures for growth.
w. Flowering habits of plants are affected by the length of the day.
x. Different kinds of plants have different growth requirements.
y. All plants develop from either seeds or buds.
z. The dormant period of seeds varies according to the kind of plant.
aa. Farm crops are classified according to the product obtained from them.
ab. Some plants live for one year, some for two years, and some live for an indefinite number of years.

REFERENCES

Boone, L. V. *Producing Farm Crops*. Interstate Publishers, Inc., Danville, IL 61832.

Lee, J. S., and D. L. Turner. *Introduction to World AgriScience and Technology*. Interstate Publishers, Inc., Danville, IL 61832.

Swaider, J. M., G. W. Ware, and J. P. McCollum. *Producing Vegetable Crops*. Interstate Publishers, Inc., Danville, IL 61832.

U.S. Department of Agriculture. *New Crops, New Uses, New Markets*. Yearbook of Agriculture: 1992, U.S. Government Printing Office, Washington, DC 20402.

U.S. Department of Agriculture. *Agricultural Statistics, 1992*. U.S. Government Printing Office, Washington, DC 20402.

Chapter 15

Improving Farm Crops—Biotechnology

Nature was the first breeder and improver of plants. The "laws" of nature, apply to plants in the following ways: (1) nearly every species of plants tends to reproduce in large numbers through seeds and, in some cases, by other means; (2) variations within a species occur in each generation; and (3) competition develops between plants of the same species and between species, resulting in the death of the weakest and the survival of the fittest. Because of these laws around us, most plants in the wild state are very hardy; consequently, they are able to survive under rather severe conditions.

OBJECTIVES

1. Discuss why crop improvement is important.
2. Trace the origin of important farm crops.
3. Apply the biology law of variation to crop improvement.
4. Discuss how plant breeders develop improved varieties.

Probably all cultivated, or domesticated, plants originated from wild species. For a long time (perhaps thousands of years) primitive peoples used the plants that nature provided and made little or no attempt to cultivate or improve them. In the course of time, they began to gather seeds from some plants of value to them and to plant these seeds to produce more of the kinds of plants they wanted. As they worked with these crops, they observed differences between the plants and saved seeds from the plants that suited them best. These methods resulted in some improvements beyond those that nature alone had made. For example, ancient tribes of Indians in North and South America brought about improvements in corn that were quite remarkable. Until recently, with the development of hybrid corn, very little increase had been made in the yielding ability of corn over that grown

Fig. 15-1. "Whatsoever a man soweth . . ." A stone plaque on Beaumont Tower, Michigan State University, portrays the well-known parable of the sower.

by some tribes of Indians. Many other plants, such as wheat and the common Irish potato, have been greatly improved since they were domesticated from forms that grew in the wild state.

Why Is Crop Improvement Important?

Everyone benefits from improvements in crops. Most of these improvements have resulted in increased yields per acre. Improved yields make it possible for farmers to produce crops at lower costs per bushel or per pound and thereby increase their profits per acre. As farmers become more prosperous, they purchase more products, and this in turn benefits industry and many people employed in various occupations. Furthermore, some of the

Improving Farm Crops

lowered costs of production are passed on to the consumers as lowered prices for the purchased goods that are made from these crops.

In some countries, increased yields are greatly needed to provide more food and fiber from the land available. In China, India, and Japan, this problem is especially acute. Recently, scientists in the United States and elsewhere have shown that yields can be increased greatly in some countries.

Increased yields of crops result from a combination of favorable factors, one of which is improved varieties of crops. Other factors are improved soil fertility, improved culture, and improved methods of controlling plant enemies (weeds, insects, and diseases).

Average yields of crops in the United States are still low when compared to the possibilities. This is especially evident if average yields are compared to yields obtained by some of the best farmers. For example, in 1980, for the United States as a whole, corn averaged 91.0 bushels per acre, wheat averaged 33.4 bushels, and soybeans averaged only 26.8 bushels. Until about 1930, very little increase had been made in the average yields of most crops in the United States. Since that time, the general trend in yields of most crops has been upward. Some of these increases have been due to favorable weather conditions, but the greatest influences have been the use of improved varieties and improved methods of production.

Under favorable conditions, remarkable yields have been obtained. For example, some farmers have produced more than 200 bushels of corn per

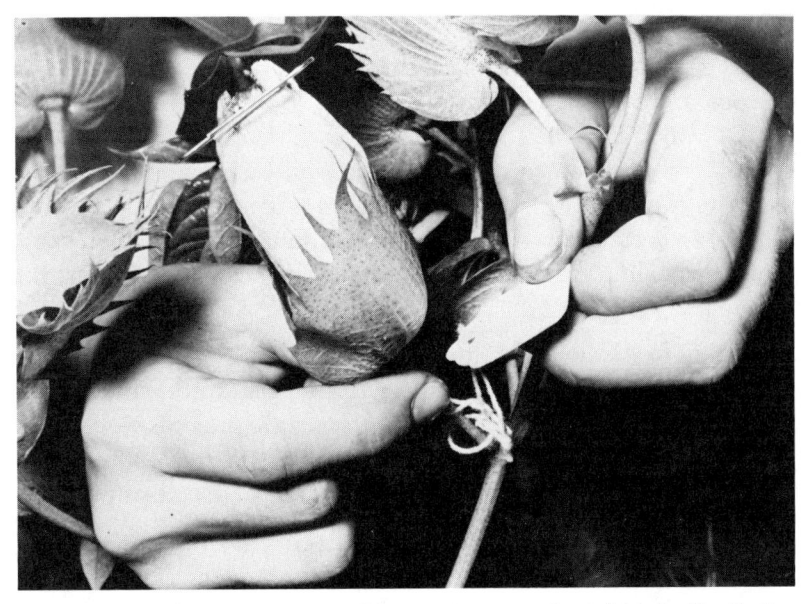

Fig. 15-2. This is a close-up photo showing the paper clip method of self-pollinating cotton flowers. The flowers are of the Sea Island × Egyptian × Tanguis cotton plant. (Courtesy, U.S. Department of Agriculture)

acre, and yields of 100 bushels are fairly common. An Illinois farmer secured 265 bushels on 1 acre; an Indiana farmer raised an average of 241 bushels per acre on 5 acres; and an Iowa farmer secured an average of 224 bushels per acre on 10 acres. In 1955, a teenage farm boy in Mississippi raised 304 bushels of corn on an acre; he was the first to reach the 300-bushel mark. Yields of 60 or more bushels of wheat are fairly common, and one wheat grower produced 102 bushels on an acre. Breeding research has increased red winter wheat production about 1 million bushels since 1946. Some of the more popular varieties developed are Knox, Monon, Red Coat, Arthur, Arthur 71, Abe, Oasis, and Sullivan. In each new variety, improvements have been made in pest resistance and other qualities. Yields of 50 bushels or more of soybeans per acre have been obtained by several farmers, and one Indiana farmer got 63.6 bushels from an acre. Of course, all farmers cannot produce such high yields, but there is much room for improvement on most farms. On the other hand, the top yields just cited can be expected to be surpassed.

Besides increasing yields, crop improvements are being made in several other ways. Among these are the following: (1) better quality, (2) resistance to disease, (3) resistance to insects, (4) adaptations to climatic conditions, and (5) adaptations to crop rotations.

Better quality. Plant breeders have developed varieties of corn and other crops that are uniformly good in quality. New sweet corn varieties are sweeter than some of the older varieties, and new popcorn varieties pop into larger and more tender kernels. Flower varieties are more beautiful and bloom for longer periods. Plant breeders have also developed varieties of wheat from which superior baking flour is made. Also available are varieties of fruits and vegetables that have been improved in flavor, in texture, and in other ways. For example, Lemhi Russet is a potato variety higher in vitamin C, more resistant to malformation, and better yielding than Russet Burbank, the top potato variety in the United States. This new variety is also resistant to certain diseases, including net necrosis, common scab, early blight, and verticillium wilt. It is, however, more susceptible to bruising and to hollow heart.

Resistance to disease. Plant breeders have developed strains of hybrid corn that are resistant to diseases such as leaf blight and smut. Some varieties of small grains are more resistant to certain rust diseases than older varieties. Tobacco breeders have produced strains resistant to wildfire, a serious disease of that crop. Varieties of alfalfa that are resistant to wilt diseases are available. Some varieties of potatoes, such as the Lemhi Russet, are resistant to scab.

Resistance to insects. Plant breeders are trying to find strains and varieties of wheat that are resistant to the Hessian fly, a serious insect pest. Corn breeders are perfecting strains of corn that are resistant to European

Improving Farm Crops

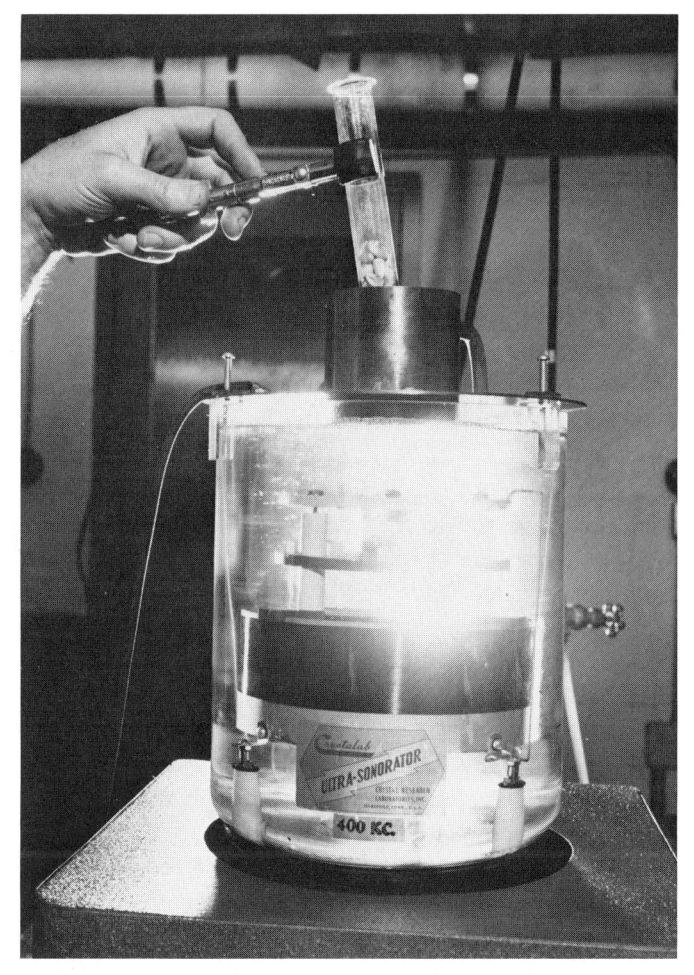

Fig. 15-3. The seed corn in the test tube is being exposed to ultrasonic (high frequency sound) radiation to determine its effect on corn growth and yield. (Courtesy, U.S. Department of Agriculture)

corn borers and corn ear worms. Plant breeders are working on a way to eliminate damage to wheat from the *greenbug*. Sap-sucking greenbugs, members of the aphid family, cause millions of dollars of losses to wheat growers each year. The *Afghanistan goatgrass,* a wild relative of wheat, is resistant to greenbug damage. This resistance can be transferred to wheat with a gene from the goatgrass.

Adaptations to climatic conditions. The Corn Belt has been extended northward, as strains and varieties of corn that are adapted to cool weather and short growing seasons have been developed. Hardy alfalfa varieties, which can survive the cold winters of northern states, are available. Drought-resistant varieties of sorghum and other grains and grasses have been developed for areas of low rainfall. Citrus fruits that are able to withstand colder temperatures than current varieties are being developed. An evergreen citrus relative from Australia and a cold-hardy tree from China that loses its leaves in winter are being crossed with established citrus varieties to produce new cold-resistant varieties.

A low-input feed barley, which requires only one irrigation treatment and no nitrogen fertilizer while producing 4,000 pounds per acre, has been developed. When planted near Tucson, Arizona, in December, after only one 6-inch irrigation treatment, the barley matured about May 1. Present irrigated barley varieties require 24 inches of irrigation water, plus nitrogen fertilizer. The new variety was developed from multiple crosses of drought-resistant plants in about seven years.

Adaptations to crop rotations. By the late 1990's farmers should have an annual alfalfa variety for their use. Alfalfa is one of the most effective plants in capturing nitrogen from the air. As an annual, alfalfa could fit into crop rotations to improve soil fertility and could be planted on land now left fallow for a year as a means of preventing wind and water soil erosion.

In addition to the above ways in which crops are being improved, plant breeders are developing varieties that have strong stalks and root systems. Losses are thereby reduced, as these varieties are less susceptible to wind damage. In these, and in many other ways, crop breeders are improving farm crops.

The costs of these improved crops are remarkably small when compared to the returns from such work. For example, the development of one variety of white navy beans by plant breeders at Michigan State University brought that state added income that exceeded the total cost for agricultural experimental work in Michigan for a period of 70 years. During a period of 30 years, the U.S. Department of Agriculture and the state colleges of agriculture spent about $10 million on the development and improvement of hybrid corn. In any one year, the increased income to farmers from the use of hybrid corn amounts to many times this figure.

As crop breeders continue their work in the years ahead, continued progress in the improvement of crops of various kinds can be expected.

How Did Important Farm Crops Originate?

Most farm crops of importance today have been raised for thousands of years. Many of these crops are of foreign origin, but some were native to the Americas and had been grown by Indians in North and South America long before white explorers came to these continents. In fact, various tribes of American Indians developed about 2 dozen crops of present-day importance in the United States. These include corn, cotton, tobacco, peanuts, white potatoes, sweet potatoes, pumpkins, beans, squash, tomatoes, peppers, pineapples, grapes, blueberries, blackberries, and cranberries. However, of

78 leading crops in the United States in a recent year, only 10 were native to the area now covered by the United States. Some came from Mexico, Central America, and South America, and many from Europe and Asia.

Some of the plants imported to the United States came by long and roundabout routes. Even the common white, or Irish, potato, which was developed from wild plants by ancient tribes of Inca Indians in South America, is believed to have been brought to North America for the first time by the early colonists from Europe. Previously, it had been taken to Europe from South America by Sir Francis Drake. Some historical reports indicate that Sir Walter Raleigh introduced the potato, which he obtained in North America, to England. However, other historians believe that it was another crop similar to the potato that the Indians in North America were growing at that time. Because Ireland was one of the countries in the Old World where the potato was grown extensively after it had been introduced into Europe from the Western Hemisphere, the potato began to be called the Irish potato. Furthermore, a colony from Ireland that settled in New England in 1719 brought potatoes and raised them extensively in that area. Wild forms of potatoes still grow today in South America. In fact, some wild species have been used in crosses to develop varieties resistant to some forms of a disease called late blight.

Alfalfa is another example of a farm crop that reached the United States in a roundabout way. Alfalfa is believed to have first grown wild in Asia near the Caucasus Mountains. The Moors took seeds to North Africa. Later, the Romans, Arabs, and Moors took alfalfa seeds to Spain. Early Spanish explorers took seeds to South America and started the crop in the region of Chile. Some of the forty-niners who sailed around the horn of South America to reach California in the gold rush days are believed to have stopped in Chile where they saw alfalfa growing. They took seeds to California and planted them there. Gradually, seeds were carried eastward in the United States and planted. Probably, too, seeds from Europe were taken by early colonists directly to the Atlantic Coast. During this period of expanded use in the United States, strains adapted to various regions were developed (see Fig. 15-4).

The navel orange industry in California was started from some trees imported from Brazil, where seedless oranges were reported by a missionary. In 1873, trees with grafted buds were shipped to California, where they produced seedless fruits of excellent quality. From these trees, grafting material was secured to produce many more trees. The navel orange industry may have been of more value to California than its famous gold mines.

Another interesting aspect of fruit production in California is the development of the prune industry. A Frenchman went to California in the gold rush of 1848. While working in the mines, he noted the high prices of fruit and decided that fruit production was a good way to find "gold in them thar hills." He and a brother went to France and secured twigs from plum trees

Fig. 15-4. Alfalfa reached the United States in a roundabout way. From Asia to Morocco to Spain to Chile, it was then taken to California by the forty-niners. (From Circular 375, The University of Wisconsin)

of a good variety. These were stuck into potatoes, to keep them moist, packed in sawdust in two trunks, and shipped to California. The twigs were grafted onto plum trees then grown in California. The plums produced were dried, and delicious prunes resulted. Thus began an industry that is today one of the most important in California.

Many fascinating stories could be written about how common U.S. farm crops were developed from wild species. The wild relatives of some of these plants are still found in different parts of the world, but in most cases they differ considerably from the corresponding domesticated crops. Although scientists and historians do not have all the facts regarding the origin and early development of many of the present-day crops, some interesting features are known. Brief glimpses of these developments are given in the following paragraphs.

The origin of corn is a real mystery. No wild relative has been discovered, and corn is so highly developed that it cannot survive without human help. Scientists are quite certain that corn originated in South America, Central America, or Mexico. The ancient tribes of Inca Indians in South America, Mayan Indians in Central America, and Aztec Indians in Old Mexico all had highly developed civilizations. All are known to have grown corn

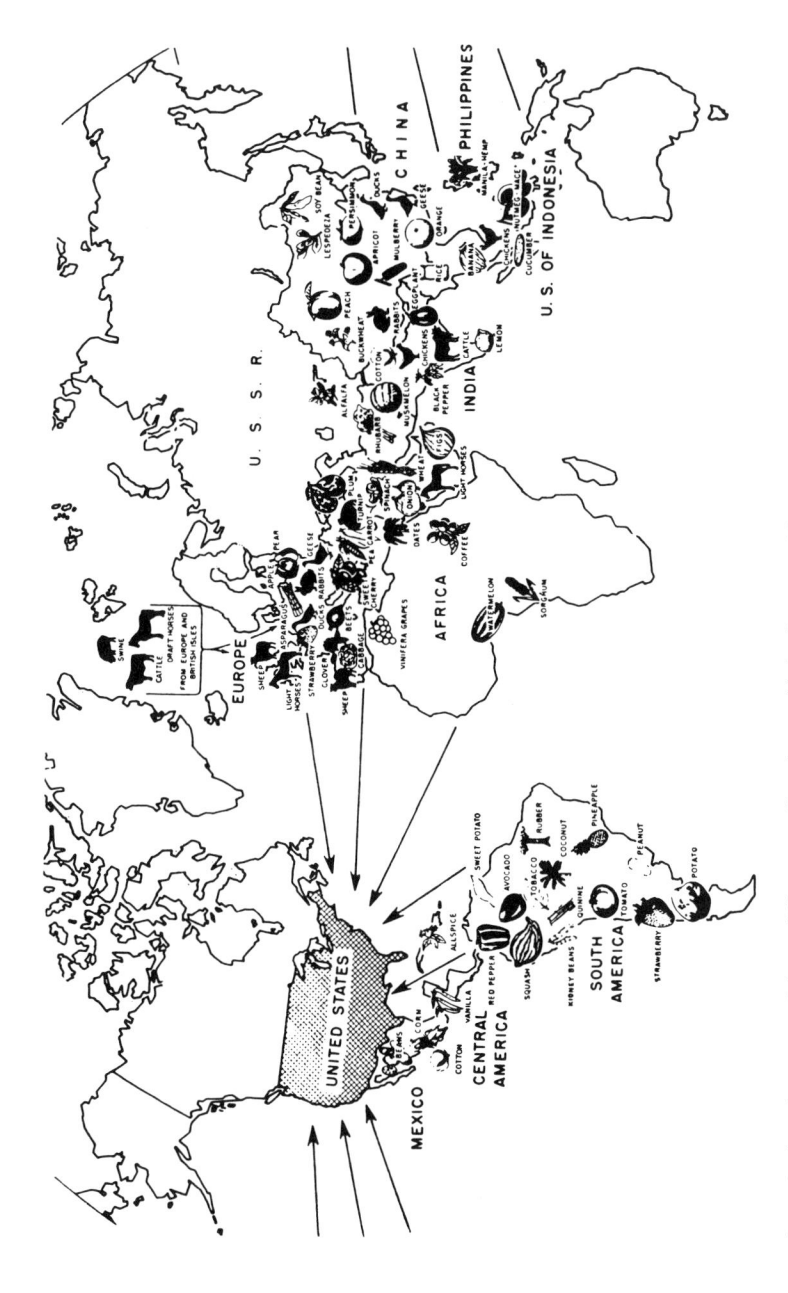

Fig. 15-5. Countries in various parts of the world have contributed many kinds of crops and livestock now grown in the United States. (Courtesy, Technical Collaboration Branch, OFAR, U.S. Government)

and to have used it extensively for food. This has been verified by the discovery of portions of ears of corn among the ruins of ancient buildings, pieces of old pottery in which replicas of ears of corn appear, and old drawings and paintings made by these ancient tribes. Some scientists have suggested that the ancestor of corn may have been a cross between a wild grass called teosinte and an unknown plant. However, some fossilized corn pollen grains estimated to be more than 60,000 years old have been unearthed in Mexico. Teosinte was not in existence at that time, so the ancestry of corn is still a mystery. However, corn was domesticated thousands of years ago. Corn was carried into North America by Indians 1,000 years or more ago, and gradually it was taken eastward by various Indian tribes. Columbus took corn to Spain in 1493. From there, it has moved around the world, and it is now grown so widely that every week of the year, corn is planted, cultivated, and harvested in various places in the world.

Wheat was probably one of the first plants to be taken from its wild state and cultivated. Wheat is believed by some scientists to have originated in Asia. Wild types of wheat are found today in parts of the Soviet Union and parts of Asia, although domesticated wheat will not grow as a wild plant. Some wheat grains believed to be more than 6,000 years old have been found in Egyptian tombs and in other remains of ancient civilizations. Wheat has been grown in China for at least 5,000 years. It was brought to North America (to areas now in the United States) about 1602.

Barley is also one of the most ancient of cultivated plants, and wild forms have been found growing in Western Asia. Samples of barley have been found in the tombs of Egyptians and among the early Lake Dwellers of Switzerland.

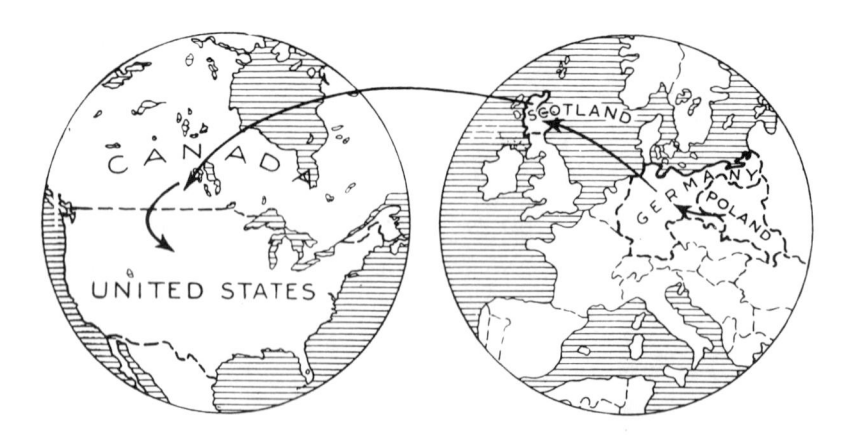

Fig. 15-6. A variety of wheat that founded the hard spring wheat industry in the United States was a world traveler. It came to the United States by way of Poland, Germany, Scotland, and Canada, where a farmer saved one superior plant from an entire field and from it came the Red Fife variety. Until recently, this variety was widely grown. (Courtesy, U.S. Department of Agriculture)

Oats, a much more recent plant than either wheat or barley, was cultivated by the ancient Greeks and Romans. A wild oat that grows in Eastern Asia is considered to be the ancestor of the tame oat.

Rye is a much more recent grain even than oats. Evidently it was not grown by any of the ancient peoples, as it was not mentioned in their writings, nor has it been found in their tombs or ruins. It probably originated in Southwest Asia. Rye maintains itself easily as a wild plant, and many species are growing wild today in Western Asia and Southwestern Europe.

Soybeans have been called the modern "miracle" crop in the United States. As recently as 1925, the acreage of soybeans in the United States was small, even though some soybeans were brought here a century ago. In recent years soybeans have become a major crop in the United States. Actually, soybeans are one of the oldest cultivated crops, having been grown in China for more than 5,000 years. In that country, soybeans have been an important food crop for a long time. In the United States, soybeans are the leading oil crop. Most of this oil is used for food, but some is used for paints and other non-food purposes.

Rice was probably domesticated in India or China several thousand years ago, but it was also an old crop in Southeast Asia. Some species of wild rice are native to the United States; the Indians used rice for food. However, cultivated varieties were imported, some as early as 1685.

Timothy, Kentucky bluegrass, and some other grasses are only semi-domesticated; that is, they are still much like the wild plants from which they originated. Timothy is a native of Europe. The name is believed by some to have been taken from the name of Timothy Hanson, who carried some of the seeds to Maryland from New England where the grass was grown for the first time in this country.

Kentucky bluegrass, with a name acquired in the United States, is a native of Northern Europe and the cooler parts of Asia. Bluegrass was probably brought to this country as seeds mixed with grain and was seeded with it by early colonists. It is likely that seeds of bluegrass were carried westward in hay used for livestock. As pioneers moved across the country, they scattered some of these seeds in the process of feeding the hay.

Red clover grows wild in Europe, Asia, and Africa. Early settlers brought seeds to the United States from England. At first, most of the clover plants grown from these seeds were killed by a small insect not found in Europe. Most of the plants that lived had hairy stems that protected the plants from these insects. From the seeds saved from these hairy-stemmed plants, red clover that prevails today was developed.

Ladino clover was brought to the United States about 1890 from the Po Valley in Italy. First grown in the western United States, it has become an important crop in many states.

Tobacco is a native American plant, which probably originated in South America. Columbus saw the Indians use tobacco. The Spaniards took the tobacco plant to England, where it is believed Sir Walter Raleigh introduced the use of tobacco.

Cotton was grown in India at least 5,000 years ago. India may be the original home of cotton in the Eastern Hemisphere. Fragments of cotton cloth believed to date back to 3,000 B.C. have been unearthed in that country. Egypt has also grown cotton for thousands of years. Cotton was grown in South America and Mexico by ancient Indian tribes, as evidenced by cotton cloth found in ancient burial grounds. Columbus found cotton growing in the Bahama Islands in 1492. However, the cotton grown by early settlers in Virginia was started from seeds they brought with them.

Hemp is probably the oldest cultivated fiber plant. It was grown about 5,000 years ago in China, and it was also grown at an early date in the Philippines. Flax is a fiber crop that has been grown for a long time. It is mentioned in the Bible. The ancient Egyptians grew flax and used the fibers for making linen. Before placing mummies in tombs, the ancient Egyptians wrapped them in linen cloth.

The tomato was "discovered" by Spanish explorers in South America. Its original home may have been Peru or Mexico. Seeds were taken to Spain and later to other parts of Europe where tomatoes became known as "love apples." Believed to be poisonous, tomatoes were grown as ornamental plants for over 300 years. Colonists brought seeds to the United States, where they also grew tomatoes as ornamental plants. Only since about 1830 has the fruit been used for food. Today, tomatoes are widely recognized as a highly nutritious food.

The watermelon is a native of tropical Africa. The cantaloupe came originally from Persia, although it probably got its present name from the Castle of Cantalupo in Italy.

Peanuts are a native of the tropics of South America. They were carried from there to Africa by early explorers. Peanuts were introduced to the United States by the early colonists.

Apples are believed to have originated in Asia, where they descended from wild crab apples. They were taken to Europe, and early colonists brought seeds to this country and planted them. The story of Johnny Appleseed, whose real name was Jonathan Chapman, is partly legend and partly fact. He planted apple seeds and sold young apple trees as he travelled on foot in Pennsylvania, Ohio, Indiana, and other areas for about 50 years, ending a century or more ago.

How Do Plant Breeders Improve Crops?

Plant breeders are interested in improving the inherited makeup of plants of various kinds. These persons strive to improve plants in ways that will make them more suitable for the purposes for which they are grown.

Improving Farm Crops

One of the "laws" of biology related to crop improvement is the *law of variation*. Although "like produces like," this is true only within broad limits. Variations occur from time to time. Some are minor and some are marked. Some are transmitted to the generations that follow, and some are not. Some are desirable and some are undesirable. Plant breeders take advantage of variations of desirable kinds that occur. They also have developed techniques that increase the number of variations and make desirable kinds more likely to occur.

An example of how changes have occurred over a long period of time is the cabbage group of plants. Ordinary cabbage, cauliflower, kohlrabi, Brussels sprouts, and Savoy cabbage are plants quite unlike in appearance, with different parts of the plants used for human food. It is believed that all of these had a common ancestor and that people played an important part in selecting variations as they occurred. Even today, plant breeders are seeking to improve each of these and other kinds of vegetables by developing varieties that are more flavorful, more resistant to diseases, and in other ways better than those now in existence.

One of the scientists who made important contributions to the science of plant improvement was an Austrian monk named Gregor Mendel. He experimented with garden peas and made many crosses of the flowers. Some of these crosses were between tall and short plants, red and white flowering plants, and smooth-seeded and wrinkled-seeded strains. He found that some of these characters were hidden in the first generation after a

Fig. 15-7. Dr. F. W. Poos, USDA entomologist, tests the effects of boron on the injury "alfalfa yellows." Used for the tests are plants grown in pure sand, to which known quantities of boron have been added. (Courtesy, U.S. Department of Agriculture)

cross was made, but they reappeared in the second generation in regular proportions. For example, the seeds produced from crosses of red and white flowers grew into plants that produced all red flowers. When seeds from these flowers were planted, about one-fourth of the plants produced white flowers. Thus, he found that the red character is *dominant* to white and that the latter is carried in a hidden, or *recessive,* fashion in the seeds of some plants that bear red flowers.

Mendel published the results of his experiments in 1866, but no further use was made of them until 1900, when his publications were rediscovered. Today, plant breeders are able to proceed most effectively in developing improved varieties of plants if they know which traits are dominant and which are recessive. However, many forms of inheritance in plants are much more complicated than those studied by Mendel.

The work of Mendel and other scientists shows that for certain plants, looks are deceiving. This is particularly true of a naturally cross-pollinated plant, such as corn. Seeds from good-looking ears of corn from fields that have been cross-pollinated usually do not produce uniformly good ears in the next generation, because some of the kernels may have been fertilized by pollen that carried undesirable traits. These undesirable characteristics may not show until later generations.

Three general methods used by plant breeders for improving plants are these: (1) introduction of new crops and varieties from foreign countries, (2) selection of improved strains from old varieties, and (3) crossbreeding, or hybridization, and the selection in later generations of desirable strains.

Introduction of new crops and varieties from foreign countries. As already noted, many of the kinds of farm plants were brought to the United States from foreign countries. The search for new kinds of varieties of plants still goes on. The U.S. Department of Agriculture sends people all over the world for this purpose and maintains a "bank" for about 200,000 of the world's plants. These are brought to the United States in the form of seeds, plants, cuttings, bulbs, and tubers. Many thousands of different seeds and plants have been secured for trial in this country. Over 10,000 varieties of soybeans alone have been brought to the United States. Only part of these have been found to be of value for direct use or for use in crosses with native varieties. Though only a few prove to be valuable, the search is worthwhile. Years of careful trials are usually needed to determine the value of any plants from these sources. The federal government is also cooperating with many countries by supplying them with varieties of crops developed here.

Nations throughout the world look to the USDA's collection of cereals, for example, which contains virtually unlimited germ plasm in its 16,000 wheats, 8,000 barleys, 4,000 oats, and 3,200 varieties and selections of rice. Germ plasm for all these many varieties of cereals, as well as for many other kinds of plants, is maintained in a form that can be used by crop breeders everywhere.

Many examples might be given of introductions of varieties, in addition to those indicated earlier in the chapter. Korean lespedeza was started from a few seeds imported from Korea. The Bond variety of oats was introduced from Australia in 1929 and the Victoria variety from Uruguay in 1927. Both these varieties have been used in crosses with native varieties, which has resulted in several improved varieties now being grown extensively in the United States. Other examples of importations of value are Balbo rye, Hungarian vetch, dallisgrass, and New Zealand white clover.

Selection of improved strains from old varieties. Selection of plants has long been a method used for the improvement of crops. By planting and comparing the varieties developed in this country and those brought from foreign countries, breeders can select the best ones for further use. Selection is also important when breeders choose the improved varieties resulting from crossbreeding.

Selecting by outward appearances is especially effective in securing improved varieties of naturally self-pollinating plants, such as wheat, oats, barley, and soybeans. This method is also effective with vegetatively produced plants, such as trees and shrubs, that may be multiplied by cuttings and grafts. Potatoes are increased vegetatively from the tubers.

For many years, plant breeders developed improved varieties of wheat and oats by selecting off-type plants that showed up by chance in plots or fields. Seeds from each plant were sown separately, and the amount of seed was further increased from those that proved superior.

Some alert farmers brought about new varieties, largely by selecting superior plants and saving seeds from them. For example, many years ago, David Fife, a Canadian farmer, started Red Fife wheat from a superior plant found in a field of wheat. The Fultz variety of wheat was developed in a similar manner. In 1862, Abraham Fultz, a Pennsylvanian farmer, noticed three beardless spikes of wheat growing in a field of Lancaster wheat, a bearded variety. He noted that the seeds were plumper and more numerous than in the heads of the Lancaster variety. He planted the seeds separately and gradually increased the amount of seed. Later, this variety was grown on nearly 5 million acres in the United States, although improved varieties have since largely displaced it.

In 1858, Wendelin Grimm immigrated to Minnesota from Germany. He brought with him a few pounds of alfalfa seed, which he planted. Due to the severe winters, most of the plants died. From the few that survived, the first winter-hardy variety emerged. This variety was named after Grimm.

Most of the plant breeding work in recent years has been done by specialists. Luther Burbank was a noted plant breeder for nearly 50 years prior to his death in 1926. He developed many new varieties of crops, including

Fig. 15-8. The first step in developing a new wheat variety. A plant breeder transfers pollen from the head of one variety to the head of another. (Courtesy, U.S. Department of Agriculture)

varieties of potatoes, tomatoes, squash, peas, asparagus, and sweet corn. His spineless cactus is useful as a feed for cattle in dry regions. Much of Burbank's work was done by skillful selection of variations among plants and by multiplying those that proved to be of special value. He also did considerable crossbreeding prior to selection. In all, he developed over 200 varieties of vegetables, fruits, and flowers.

Crossbreeding, or hybridization, and the selection in later generations of desirable strains. These methods of plant breeding are used extensively today. By crossing varieties of wheat, for example, plant breeders produce hybrids. If these are followed by careful selection in successive generations, it is thus frequently possible to combine the good qualities of two or more varieties. Crossing is done by a delicate process of removing the pollen-producing portions of flowers on one wheat plant, followed by transferring pollen to those flowers from the flowers on another plant.

Today, plant breeders are using various combinations of the three methods described on the preceding pages. They secure varieties from all possible sources, including many imported from foreign countries as well as those in the United States. They grow these under various conditions to determine their good and bad characteristics. Plants with desired characteristics are crossed. The offspring from the crosses are saved for further testing and selection.

One of the problems in improving small grains is to get varieties resistant to various forms, or "races," of rusts and smuts. New races of these diseases appear, and varieties that are resistant to old forms may be severely damaged by these new forms. Thus, plant breeders must continue to develop new varieties of small grains, such as oats and wheat.

Rapidan is a variety of barley resistant to mildew, resistant to some races of scald, and tolerant to leaf rust.

Hybrid corn is an outstanding example of changed methods for improving corn. In developing hybrid strains that the farmer can use, plant breeders first produce suitable inbred strains of corn. Inbreeding is done by putting pollen from a plant on the silks of the same plant. The resulting seeds are planted, and the process is repeated for four or more generations. This strain becomes more and more uniform in each succeeding generation, but the yield is considerably reduced. From many inbred lines, a few that have desired inherited qualities are selected, and these are crossed to produce hybrids. Usually, four inbreds are combined to yield the seed that is planted by farmers for commercial corn production. In producing hybrid corn, plant breeders strive to develop strains that are most desirable in yield, maturity, resistance to disease, quality of grain, strength of stalk, and other features.

Hybrid corn of superior strains has resulted in marked increases in yields. Experiments in breeding hybrid corn were started by scientists in the early part of the present century, and methods of producing seed were gradually developed. In a recent year, almost 100 per cent of the corn grown in the Corn Belt and around 90 per cent of all corn grown in the United States were grown from hybrid seed. Hybrid corn is truly one of the miracles of modern agriculture.

Varieties of soybeans have been developed similarly to those of small grains such as oats and wheat. New hybrids that are high yielding, resistant to certain diseases, and adapted to specific climatic conditions are being researched. Varieties imported from the Far East have been crossed with U.S. varieties to produce improved hybrids. In fact, superior varieties have been developed so rapidly that most varieties popular 15 years ago have been displaced.

Four soybean varieties resistant to phytophthora rot, a fungus infection, are Hawkeye 63, Clark 63, Harosoy 63, and Lindarin 63. In each case, the variety was developed from a cross of the old variety with a rot-resistant line. This was followed by four to seven generations of backcrossing—crossing each generation progeny with the original variety and selecting those plants that showed resistance to the rot. These were the first soybean varieties developed in the United States by this backcrossing procedure.

Potatoes normally multiply from the tubers. However, in seeking improved varieties of potatoes, the breeder crosses the blossoms and plants the resulting seeds. From thousands of such seedlings, the potatoes from a few may prove superior. In the development of new varieties of potatoes, yield, quality, and disease resistance are important. A period of several years is needed for crossing, growing seedlings, selection, and trials. Some years ago, wild potato specimens immune to some races of late blight were brought to the United States from South America. These were crossed with commercial varieties, and hybrids that were resistant to this disease were produced. Varieties that are resistant to scab have also been developed.

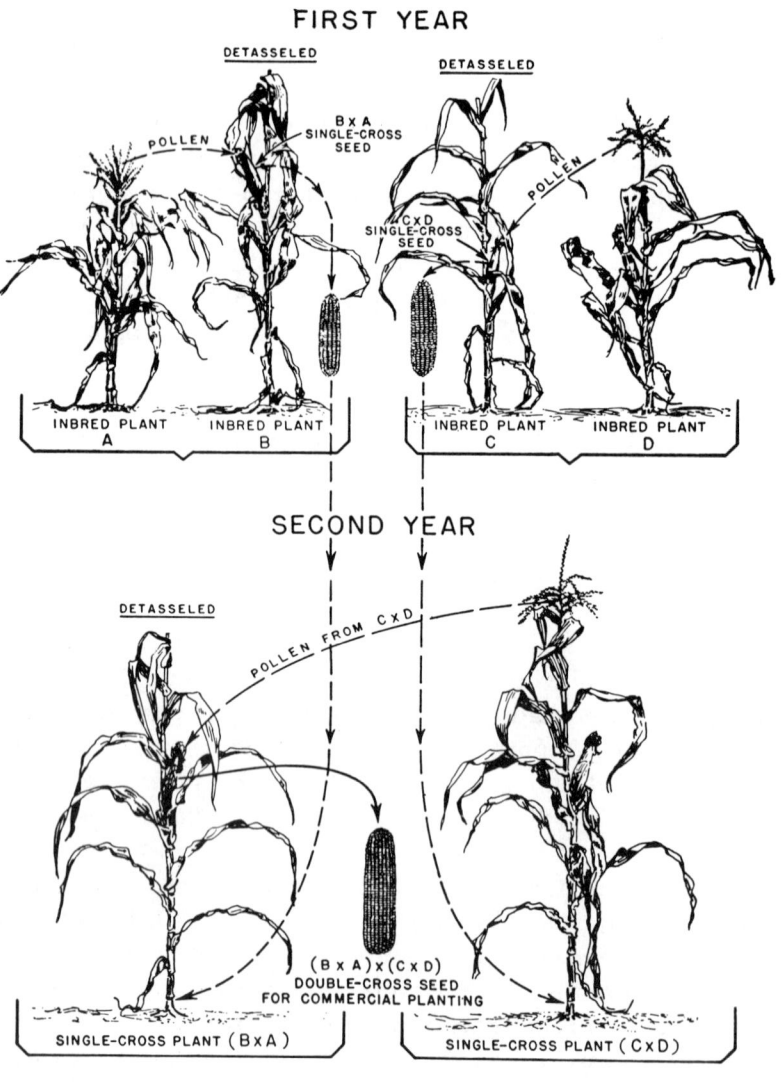

Fig. 15-9. The method of crossing inbred plants and the resulting single crosses to produce double-cross hybrid seed. (Courtesy, U.S. Department of Agriculture)

Improving Farm Crops

Further techniques are being used to obtain high-quality seed potatoes free from tuber-transmitted virus diseases, such as mosaic, curly dwarf, leaf roll, and blackleg, which cannot be controlled by spraying or dusting. One method is called "roguing," in which the diseased plants are dug up completely and carried from the field in which potatoes are being produced for seed. Another is tuber-unit planting, in which all the seed pieces cut from each single tuber are planted together. All the plants from the same tuber are removed if any of them are weak or show symptoms of a virus disease. Tuber indexing, in which a single eye from each tuber is planted in a greenhouse in winter, is also used. Only those tubers whose seed pieces have produced healthy plants are saved for increase in field plots the following spring.

Tomatoes are being improved through crossing and selection. Short-season varieties adapted to conditions in northern sections have been developed. Varieties resistant to some diseases are available. There have been improved varieties of many vegetables in recent years, largely from crossing and selection. Dr. A. F. Yeager at the University of New Hampshire has developed many improved varieties, including a midget watermelon, a honeydew melon for the North, a buttercup squash, a tomato high in vitamin C, and several others. In developing some of these varieties, he used varieties secured from foreign countries.

Fig. 15-10. New Hampshire midget watermelon was developed by Dr. A. F. Yeager, a famous plant breeder. Of excellent quality, it was produced through crossing and selection. (Courtesy, University of New Hampshire)

Apples, pears, and other tree fruits have been improved from chance trees that proved superior. Crossing techniques have been used extensively in recent years. The Elberta peach and the McIntosh and Delicious apple varieties each originated from a single tree that produced superior fruits. The original Delicious apple tree grew in southern Iowa, where it was planted in 1872. It died in 1940. In each of these cases, the varieties were multiplied by buds or small branches taken from the original tree and grafted onto other trees. The growth from grafted portions produced fruits exactly like the parent tree. Further grafting stock can be taken from these grafted trees as well as from the parent tree. In the development of some improved varieties of apples, bud "sports" that occasionally appear on a tree have been used. These produce fruit different from the rest of the tree, and some of the fruit is superior to that on the remainder of the parent tree. By various methods, plant breeders have established many varieties of hardy fruits for northern states.

In 1843, Ephriam Bull of Concord, Massachusetts, planted some seeds from wild grapes. From these, one seedling grew into a vine that produced superior grapes; thus, a new variety, which Bull named Concord, was started. From the parent vine, Bull took cuttings to produce many additional plants, and from these, further plants were produced. This variety was distributed widely. It is still important in the vineyards in eastern states.

Many varieties of strawberries, raspberries, blueberries, and other small fruits have been developed from crossbreeding and selection. These varieties show improvement in features such as color, taste, yield ability, and disease resistance.

Many varieties of grasses adapted to the soil and climatic conditions in various parts of the United States have been developed. Some of these have been developed from superior varieties of native grasses. Crossbreeding is also being used extensively for new varieties. One of these varieties is ryegrass meadow fescue. In making the cross, plant breeders doubled the chromosomes (the carriers of inheritance) by treating the seeds with a certain drug. The new hybrid excelled both parents in vigor, leafiness, and growth during periods of drought. Another hybrid variety is Coastal Bermuda grass, which is suited to many of the southern states. One of the varieties used in the initial cross was brought from South Africa. Coastal Bermuda grass is good for pasture and hay and is resistant to drought and to some diseases.

Better plant breeding has also resulted in the rapid improvement in flowers through the development of superior hybrid varieties. Increased beauty, hardiness, and disease resistance have resulted. Lilies, irises, peonies, roses, and many others have been improved in this manner. One example is the begonia. This plant traces back to the West Indies, where a French botanist became interested in it in 1690. At that time, it was a rather plain-appearing flower; and it continued to be so for over 200 years. About

Improving Farm Crops

1920, flower growers in the United States started to improve it. Through crossbreeding, hundreds of spectacular varieties have been developed.

Orchids are considered by many people to be the most beautiful of all flowers. In one flower show in New York City, over 1,000 varieties were exhibited. These were sent to the show by air express from 20 states and 10 foreign countries. The ancestors of these orchids came originally from widely separated parts of the world, including the high Andes of Colombia and the jungles of Burma. Present-day varieties differ widely in shapes and colors. Many of these were developed through crossbreeding.

The possibilities for getting improved plants through crossing to produce hybrids are almost unlimited. Hybrid poplar trees grow much faster and are more resistant to insects and diseases than the parent trees. Hybrid pines, black locusts, and chestnuts are other examples of improved trees. By making thousands of crosses, in some cases using stock from foreign countries, plant breeders have produced a few superior trees. These can be increased from cuttings and by grafting methods.

Improved varieties of almost every crop are available. These include cotton varieties with increased length and strength of fiber and resistance to certain diseases, tobacco varieties of improved leaf quality and resistance to some diseases, nuts of improved flavor and hardiness, improved sweet

Fig. 15-11. A plant physiologist examines tree limbs of the Page orange *(left)* and a hybrid *(right)* one month after a freeze. Note that the hybrid did not lose its leaves. (Courtesy, *Agricultural Research*, U.S. Department of Agriculture)

potatoes, sorghums resistant to drought, and sugar beets with increased sugar content and more resistant to some diseases.

Some Current Research on Plant Improvement

The agricultural research team of the U.S. Department of Agriculture, with all the many cooperating universities across the country, is engaged in many plant improvement projects that will eventually lead to more and better products for everyone. A look at some of the on-going research shows the extent of this work.

One of the major developments has been the discovery of genetic material in wheat that restores fertility in the male-sterile breeding lines. This genetic material, which scientists call restorer genes, is the second of two breeding tools that have been needed to start development of commercial hybrid wheat varieties. The other tool—cytoplasmic male sterility—has been known about for some time. Now that both male fertility and male sterility can be controlled, it is no longer necessary to cross wheat lines by the slow process of hand pollination.

In a few years grain sorghums that repel birds may be grown. Sharp awns or spines growing from the glumes (seed coverings) of some of the test varieties have been found to keep the birds from feeding. Since the bird-feeding problem makes it almost impossible to grow sorghums in some areas, the development of varieties with built-in protection against birds is of great economic importance.

The problem of developing plants with a high degree of tolerance to salty soil is of great importance, especially in irrigation areas. Research indicates that safflower, an oilseed crop, can be grown successfully in salty soils usable for only the most salt-tolerant crops. A finding that may have much significance in the future is that soybeans dwarfed by growth-retarding chemicals are tolerant to salty soil.

Wind and rain damage has always been a problem for cotton growers. Researchers are now working with several storm-resistant strains of cotton that can be harvested with a spindle-type picking machine. The combination of storm resistance and characteristics that will permit easy picking with a spindle-type harvester is of prime importance to cotton growers.

Improved raspberry plants are also available. Virus-free plants that produce four times as much fruit and three times as many sturdy canes as commercial planting stocks of the same variety have been developed. Red raspberry stocks free of mosaic viruses include Canby, Cuthbert, Durham, Indian Summer, Milton, Newburgh, Rideau, September, Taylor, and Willamette. Mosaic-free black stocks include Bristol, Cumberland, Dundee, Morrison, and New Logan.

Recent studies indicate that the atom is going to play an expanding role

Fig. 5-12. Growth regulators may lead to improved plants. Gibberellic acid, a growth regulator, was used to treat the plant on the right. As can be seen, the treatments heightened stems, lengthened flower stalks, and speeded flowering. The control plant on the left shows typical growth and stands about 12 inches high. (Courtesy, U.S. Department of Agriculture)

in agriculture. The studies have been with *dallisgrass,* a plant that produces seed without fertilization. As a result, the plant cannot be improved by conventional methods of crossbreeding. Attempts at developing superior lines of dallisgrass from planted seed of individual plants that appeared superior have been unsuccessful. Irradiation of common dallisgrass seed brought about tremendous changes in growth habits and in the structure of the plants, and these changes were passed on uniformly to progeny. The results indicate that radiation can produce a multitude of mutations that would not otherwise be available for study.

Scientists are trying to learn how alfalfa can capture so much nitrogen from the air. Studies are being done on the bacteria that cause the nitrogen-fixing nodules to form on the alfalfa roots. Scientists now believe that the bacteria are needed only for a short time and that they transmit the genetic code that causes nodules to form on the alfalfa. Thus, these studies suggest that it may be possible to use genetic engineering so that other crop species can fix nitrogen.

Scientists are discovering some new roles of the chemicals that plants produce. One group of chemicals functions as plant growth regulators, toxins, insect attractants, and feeding deterrents. One kind of chemical causes premature metamorphosis of insect larvae, creating miniature adults that die quickly and that cannot reproduce.

Insects obtain from plants the chemicals that enable them to grow and reproduce. This knowledge may make it possible to develop new kinds of safe pesticides.

Certain chemicals are responsible for the smell of fruits, vegetables, and other plants. In developing scents that will attract insects so they can be caught in traps, researchers may find this knowledge useful.

Some of the compounds produced by weeds, which may be toxic to animals, are being tested for possible use in fighting cancer in humans.

A new breeding tool was created when a national repository in Corvallis, Oregon, was constructed to collect, preserve, and distribute germ plasm of plants that are cloned or vegetatively propagated. Plant breeders use the germ plasm contained in seed to pass specific genetic characteristics on to new plant varieties. The U.S. Department of Agriculture maintains seed repositories for this work. Many perennial crops, such as fruits and nuts, have inheritance patterns too complex to be duplicated from seed in research work. These crops must be propagated through clonal or vegetative methods, and their germ plasm must be preserved in living plants.

The Oregon repository is only 1 of 12 that are to be built. It has germ plasm collections of small fruits, including strawberries, raspberries, and blackberries. It also maintains collections for pears, filberts, hops, and mints.

New collections come from all over the world. Information on disease and insect resistance is recorded, and the collections are then grafted and grown to verify resistance.

How Can Farmers and Others Improve the Crops They Raise?

Today, the development of improved varieties of crops is largely the job of specialists in plant breeding. However, a practical farmer should be alert to secure new varieties that may increase yields, produce crops of better quality, or result in other improvements over the varieties now grown. Likewise, other persons who grow plants should obtain new varieties of promise.

Seeds for any crop will not yield more than their inherited possibilities. Even if soil is improved in fertility, good cultural methods are used, various

Improving Farm Crops

plant enemies are controlled, and the weather is favorable, inferior varieties will yield poorly. Good seed is no harder to plant than poor seed, and the increased cost of good seed is small compared to the benefits that may result.

Individuals should choose varieties of crops that are recommended for their particular conditions. The college of agriculture in each state provides information of this kind for farmers and others. This information is available in bulletins, farm journals, and other publications. Reliable seed dealers, county agents, and teachers of vocational agriculture also assist persons in selecting suitable varieties.

Farmers and others who grow crops should purchase seeds from reliable sources. They should not be misled by offers of bargain seeds or by the extravagant claims sometimes made by salespersons.

In most cases, farmers should purchase certified seed, or they should secure supplies produced from seed known to have originated from certified sources. Most states and many crop improvement associations have strict regulations for the certification of seed. In the production of certified seed, authorized persons make and check samples of the harvested seed for purity of variety and for freedom from seeds of serious weeds. Samples are tested for germination. Certified seeds and other seeds from reliable sources carry a label on which information is given as to purity, weed seeds, inert matter, other crop seeds, and percentage of germination. Federal laws and regulations govern the sale of seeds between states.

Farmers should purchase new hybrid seed corn each year because seed saved from a regular field of hybrid corn will lead to reduced yields. For crops such as small grains and soybeans, they may save seed from their own crops after they get a start from certified sources. This homegrown seed should be cleaned with special seed-cleaning equipment that removes most of the weed seeds and other foreign matter.

Usually, farmers purchase seeds for crops such as alfalfa, clover, and grasses. Certified seed is desirable. Hardy varieties should be purchased for northern areas. Drought-resistant varieties are needed where rainfall is limited.

Individuals should make sure that the crop seeds they use rate high in the following:

1. A variety that is adapted to the area.
2. A variety that is high yielding.
3. Seeds that have a low percentage of mixture with other varieties and other crops.
4. Seeds that come from disease-free sources and that are resistant to diseases.
5. Seeds that have a very low percentage of weed seeds and that are entirely free from noxious weed seeds.
6. Seeds that are high in germination.

The Plant Variety Protection Act, administered by the U.S. Department of Agriculture, may result in more private plant breeders and companies increasing plant breeding efforts. Specifically, it covers varieties of plants that are reproduced sexually, that is, through seeds. In this group are virtually all garden vegetables, crops such as wheat and soybeans, all flowers, and various trees and shrubs. Not covered are okra, celery, peppers, tomatoes, carrots, cucumbers, and hybrids of any kind. Protected under the 1930 U.S. Patent Act are breeders of novel plants that are reproduced nonsexually, by buds or grafts. These plants include roses and fruit trees.

The Plant Variety Protection Act lists three qualifications for a novel variety. These are:

1. The variety must be distinct and must differ from all known prior varieties by one or more characteristics—for example, shape, color, resistance to disease.
2. The variety must be uniform. Any variations that appear in the plants must be describable and predictable.
3. The variety must be stable from generation to generation, remaining unchanged in its essential and distinctive characteristics.

Once a certificate of plant variety protection has been issued, the federal courts must protect the rights of the holder of that certificate. Infringement of the certificate holder's rights will occur, for example, if a second party, without authority, sells the novel variety, exports it out of the country, or multiplies the variety sexually as a step in selling it. The protection is valid for 17 years.

Magical results should not be expected from new varieties of crops. For example, though a variety of hybrid corn that is adapted to a given region is selected, it will not give good results with careless cultural practices or on rundown soil. Good seed and good cultural practices provide a combination that should prove profitable.

SUGGESTED ACTIVITIES

1. Study the growth of plants on a neglected piece of ground given up to weeds or on a wooded area left to itself. What evidences do you find that some plants are able to reproduce in large numbers? What features of some kinds of plants help them to survive?
2. From the information in this chapter and from other sources, indicate the native homes of the important crops grown in your state. For one of these, make a special study of its early development.
3. Make a special study of new varieties developed for some kind of plant in which you are interested. Study seed catalogs, books, bulletins, and magazines that provide this kind of information.
4. Look up materials on the life and work of some noted plant breeder, such as Luther Burbank.

5. If some farmer near your school specializes in the production of hybrid corn for seed, arrange to visit this farm and study the methods of production. Note methods of crossing, harvesting, drying, grading, packaging, and marketing.

6. Visit the crop improvement plots at the college of agriculture in your state and note the methods used. If possible, arrange to have some member of the staff of the college explain the experiments under way.

7. Visit a seed dealer who handles large quantities of farm seeds. Note the methods of labeling and the other ways in which information is provided about the various seeds. Note the tags that show certification and the information on the tags. What kinds and varieties are kept for sale?

8. Concepts for discussion:
   a. Plants reproduce sexually by seeds and nonsexually by buds or grafts.
   b. Competition between plants results in survival of the fittest, just as it does with animals.
   c. Through selection, variations that occur in plants of the same species are used in the development of desired plants.
   d. The best seed selections and cuttings were used in the domestication of wild plants.
   e. Improvement of plants is important to everyone.
   f. Many factors affect crop yields.
   g. Plants may be improved in both yield and quality.
   h. Plants can be developed for resistance to diseases and insects.
   i. Plants can be bred for resistance to adverse weather conditions.
   j. Most crops grown in the United States were not native to the United States.
   k. Plant breeders are able to proceed most effectively in developing improved varieties when they know which traits are dominant and which are recessive.
   l. Plant breeders improve plants by importing new kinds and varieties from foreign countries, by selecting improved strains from old varieties, and by crossbreeding.

REFERENCES

Boone, L. V. *Producing Farm Crops.* Interstate Publishers, Inc., Danville, IL 61832.

Lee, J. S., and D. L. Turner. *Introduction to World AgriScience and Technology.* Interstate Publishers, Inc., Danville, IL 61832.

Osborne, E. W. *Biological Science Applications in Agriculture.* Interstate Publishers, Inc., Danville, IL 61832.

Swaider, J. M., G. W. Ware, and J. P. McCollum. *Producing Vegetable Crops.* Interstate Publishers, Inc., Danville, IL 61832.

U.S. Department of Agriculture. *New Crops, New Uses, New Markets.* Yearbook of

Agriculture: 1992, U.S. Government Printing Office, Washington, DC 20402.

Chapter 16

Providing Fertile Soils

Soil is the most important natural resource. From soil come all of the necessities of life—food, clothing, and shelter.

The United States is blessed with large areas of highly productive soils. Because of these fertile soils, plus good cultural methods, farmers are able to produce abundant supplies of food, fiber, and other agricultural products. We often take these benefits for granted and fail to recognize that our nation's strength is tied closely to the soil. Nations that neglect their soil do not remain great or strong.

Soil should be considered a heritage to be passed from generation to generation in as good or better condition than it was received.

OBJECTIVES

1. Describe how soil is formed.
2. Identify basic types of soil and describe their composition.
3. Discuss soil fertility related to plant growth.
4. Determine the type and nutrient content of a bag of fertilizer.
5. Understand the purposes of tilling the soil.

Why Should We Be Concerned About Soil Fertility?

The parable of the sower in the Bible is really a parable of the soil. The seeds that fell on good soil yielded a hundred-fold, while those that fell on barren soil failed to produce a crop.

Farmers who are most successful produce high yields of crops. A fertile soil is the most important factor in securing good yields. Improved seeds, good culture, and control of plant enemies are important, but these will not bring high yields on soil that is low in fertility.

Traditionally, we have been wasteful of our greatest natural resource, the soil. The loss of soil fertility is a serious matter because it is a threat to our security as a nation. In no small way, our fertile fields contributed to victories for our country and its allies in the First and Second World Wars and thus made it possible for us to keep our democratic way of life. If we are to continue as a strong nation, we must give increasing attention to our soil.

Under proper methods of treatment, soils will remain highly productive or even improve in productivity. In some parts of Europe, land that has been farmed carefully for hundreds of years continues to produce high yields of crops. It should be recognized, however, that large amounts of human labor are responsible in part for these high yields. The Morrow Plots, a series of experimental plots at the University of Illinois, have been farmed in a variety of ways for more than three-fourths of a century. During this period, careful records have been kept of the yields from these plots. On one plot, where corn was grown year after year for over 75 years, with no fertilizer added, the yields gradually declined. Over one period of 10 years, the average yield of corn on this plot was only 27 bushels per acre. On a plot where corn was grown in a rotation with oats and clover, and fertilizers were added over a period of nearly 50 years, the yields increased. The average yields for the various crops in this rotation, during another 10-year-period, were 93 bushels of corn per acre, 75 bushels of oats per acre, and 2.9 tons of clover hay per acre. Thus, the corn raised in one year on the plot given good soil treatment was more than the total corn for three years on the plot given poor soil treatment. In another year, a portion of the plot on which corn had been grown continuously for 79 years, with no fertilizers added, was fertilized heavily and made to produce 86 bushels of corn per acre the first year of treatment.

Fig. 16-1. A profile of Houston Black Clay, with about 4 feet of rich top soil still in place. (Courtesy, U.S. Department of Agriculture)

Fig. 16-2. "Corn yields on these plots range from 19 to 113 bushels per acre according to the way the land has been handled."

Individual farmers who improve the fertility of soils on their farms benefit in several ways. The relation of soil improvement to better living is illustrated by an incident reported by a staff member of a college of agriculture in the Middle West. A farmer observed the increased yield that resulted from a field on which the soil had been improved. With considerable feeling, he said, "Why couldn't I have known this a long time ago? I could have given my children a better education and made life much easier for my family."

Individuals who raise gardens or other crops on a small scale, as well as farmers, are interested in improving the soil.

What Is Soil and How Did It Get Here?

"It may be soil to you, but it's still dirt to me." This is an idea common to many people. They think of the material covering the earth's surface, which is called soil, as a disagreeable substance that clings to their clothing and makes their houses dirty. As a matter of fact, directly or indirectly, all forms of animal life depend upon the soil for their existence.

As we examine soil, we are usually aware only of the solid particles that we can feel and see. These solid portions consist of mineral matter and organic matter. The mineral particles in soil are fine pieces of rock. The larger particles found in some soils are called gravel and sand, and the smaller particles are called silt and clay. The organic matter consists mostly of plant materials, with some animal materials, in various stages of decay. Decayed organic matter is called *humus,* which is an important part of fertile oil. Although the ordinary productive soil will contain approximately 2 to 5 per cent of organic matter, some soils known as *muck* contain over 75 per cent organic matter. Peat soils are composed almost entirely of organic matter materials.

Soil also contains water and air. A garden or lawn soil is about half solid material and half pore spaces between the soil particles. The pore spaces are filled with water and air in approximately equal proportions.

These physical features of soil are only a small inkling of what soil really is. The soil is in reality a remarkable storehouse from which plants secure many necessary food materials. Moreover, it is a chemical factory in which materials undergo chemical changes necessary before plants can use them. Aiding in these chemical changes are countless numbers of tiny plants and animals called *microorganisms,* or *microbes.* These include single-celled plants called *bacteria* and single-celled animals called *protozoa.* Other types of small plants known as *fungi* and *molds* are also in the soil. Millions of these tiny forms of plants and animals live in each ounce of fertile soil. Because of their large numbers and their great importance in plant growth, soil is a living substance, far different from dirt, which most people call it.

The idea that soil is "alive" is new to most people. Antibiotics are produced from certain molds obtained originally from soil. These drugs, such as

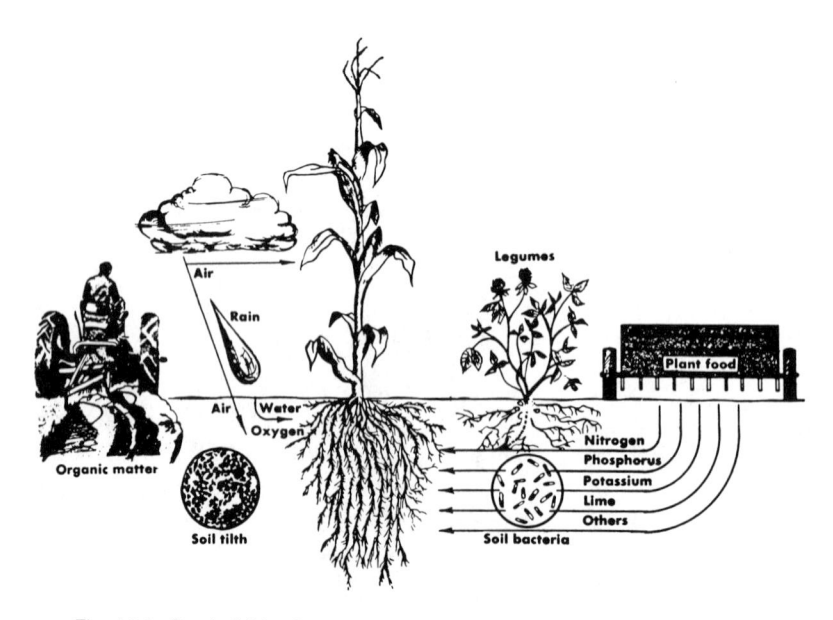

Fig. 16-3. Good yields of crops are the result of many favorable factors, some of which are shown. (From Circular 724, University of Illinois)

Fig. 16-4. These workers are harvesting radishes in the U.S. Army hydroponics garden at Chofu, Japan. (Courtesy, U.S. Department of Agriculture)

penicillin and streptomycin, are useful in checking certain harmful bacteria in people and other animals. Some antibiotics promote growth when they are added in appropriate amounts to the rations of some kinds of livestock.

Many of the cropping and tillage practices used by good farmers are important because they promote the growth of helpful organisms in the soil. These organisms need oxygen, which is provided if soil is properly aerated through tillage and drainage. A good supply of organic matter is desirable because it aids in aeration and in moisture retention. Furthermore, many of the beneficial organisms feed on organic matter and in the process change important plant food elements to forms that can be used by plants.

The major portion of soil seen with the naked eye has no direct part in plant growth. However, even this portion serves as a place in which roots grow and thereby anchor the plant in an upright position.

Raising plants without soil

It is actually possible to grow plants without soil. This science of soilless farming is called *hydroponics*. Hydroponics is widely used in all countries by plant scientists in plant nutrition studies. Basically, it consists of feeding plants by growing the plant roots in water in which fertilizer salts have been dissolved.

A variety of hydroponic systems have been developed. One method is to place water in a shallow tank and dissolve in it the chemicals needed by plants. A wire netting is placed over the surface, and seeds are planted in excelsior or peat moss placed on top of this netting. As the roots develop, they grow downward until they reach the surface of the solution containing food materials. Another method is to plant seeds in sand to which nutrients are supplied in water solutions added to the sand. During World War II, the armed forces on some remote islands covered with sand and volcanic ash were able to produce vegetables by this method. Other systems use pea gravel, volcanic rock, or other similar materials, which provide no nutrients for plant growth. These materials do provide support for the plant roots. The materials also hold enough water and dissolved plant nutrients so that applications of the nutrient solution throughout the day will provide the needed plant food, moisture, and air for the roots. If the plant roots grow in a solution placed in a container with no circulation of the solution, air must be bubbled through the solution to provide the roots with oxygen.

The simplest and most economical non-soil system for growing greenhouse crops is wood chips or an artificial soil such as peat-lite as the growing medium. The wood chips or artificial soil materials are put in plastic buckets, plastic pots, or black polyethylene bags with pre-punched drainage holes. A 3- to 5-gallon pot is good for tomatoes and cucumbers. The nutrient solution is applied several times each day in an amount that keeps the plants growing well. If there have been no disease problems, the growth medium and containers can be flushed, to clean out the accumulated salts, for use again. Pre-mixed complete nutrient solution salts in various formulations can be purchased. In general, however, soilless farming is a costly process and requires a great deal of skill. Some people carry on soilless culture on a small scale as a novelty or a hobby.

Formation of soil

Everyone should be interested in learning how something as important as soil, and with as many unusual features, "got here." Scientists, called *geologists*, who have made a thorough study of the structure of the earth say that at one time, probably millions of years ago, there was nothing but barren rock on all parts of the earth's surface not covered by water. In the long period since then, layers of soil of varying depths in different regions accumulated on the earth's surface.

The rock, or rock material of the earth, is the major parent material from which soil is formed. Many agencies of nature play a part in the formation of soil from this rock. If soil is observed carefully, many of the agencies responsible for soil formation can be detected. Many of these agencies work very slowly; but if we think in terms of hundreds, thousands, and even millions of years, we can begin to appreciate the length of time in which soil formation has been taking place. We should not get the idea, however, that all the agencies are in operation at one place at any one time, or that most of the processes responsible for soil formation can be detected by a casual inspection of our surroundings.

The early beginnings of soil formation consisted of processes that broke the rocks into smaller and smaller fragments. Then, as now, *rapid changes in temperature* probably were responsible for much of this breakage process in somewhat the same way that crockery and glass dishes crack when they are suddenly brought in contact with extreme heat or cold. Water that froze in the cracks and pores of rocks caused some breakage to take place.

Flowing water has a part in soil formation, due to the wearing process on rocks over which it passes. Water in motion also rubs rocks together, causing them to wear away. In these wearing processes, small particles of rock, or sand, are formed, and these are often carried for considerable distances before they are deposited. Most of the pebbles and even large rocks found in streams have smooth rather than sharp edges. This is in part due to the processes that have been mentioned, although in some cases glaciers may have been to some degree responsible.

Water has also helped to transport soil and soil materials from one place to another. Small ditches and gullies often form in the soil on a hillside or a slope during a rainstorm. In this case, *water* has carried soil to the bottom of the slope where the movement of the water becomes slower so that part of the soil is dropped. Some of the soil may be carried for considerable distances until it enters streams and rivers, where its presence is indicated by the muddy condition of the water. In many instances, these streams overflow their banks and leave much of the transported soil on the surface of the ground. Some of the richest farmland in the world has been created by the flooding of land by rivers and streams. These areas are called flood plains.

The ditches and gullies in fields are actually signs of soil erosion and provide vivid examples of the movement of soil by water and the effect of this soil movement on soil formation, both in the places from which the soil is taken and in the places where the soil is finally deposited.

Glaciers had considerable influence on soil formation in some regions in the past. Some can still be found in limited areas. A glacier is a huge mass of ice and other materials, usually accumulated over a long period of time, which moves slowly over the earth's surface. Because of its immense size and weight, a glacier grinds and crushes rock materials into smaller particles. As the glaciers advanced in the cold years and melted back in the warm years, the rock material was either pushed forward or deposited. Sometimes the rock material was piled up in hills, sometimes it was spread quite evenly over large areas, and sometimes it was left in deposits where the glacial streams and rivers flowed into glacial lakes, leaving the heavier pieces of material as soon as they entered the lakes and carrying the finer materials further out into the lakes. Our sand and gravel pits were formed in this way.

Glaciers, in addition to being soil-forming agencies, transport soil materials, often for considerable distances. Much of the soil in northern Iowa, Minnesota, northern Wisconsin, and parts of Illinois, as well as in many other areas, was influenced by the glaciers of the past. The carrying of soil materials by glaciers also resulted in the formation of soil that differs from the rock directly beneath the soil. Rather, the rock material in the soil is like the rock much further north.

Some glaciers exist at the present time, for example, in the Glacier National Park and at Mt. Rainier in the United States, in Greenland, and in other countries of the world. Most of Canada and the northern parts of the United States were at various times in the distant past covered by glaciers.

Wind has been very important in soil formation. In some places, rocks have actually been worn away by winds. Sand is blown by the wind, thereby aiding in this wearing process on the large rocks as well as on the particles of sand themselves. The small particles of rock that result become a part of the soil.

Strong winds can carry some soil particles for considerable distances. The long-time effects of this process can often be noted by the drifts of soil along fence rows and in other places where the force of the wind was decreased sufficiently to cause the soil to drop in a manner similar to the formation of a snowbank. On a much larger scale, most of the top soil found in some sections of the country was transported by the wind over a period of many years. This wind-laid soil is characterized by an absence of stones and boulders, which, of course, the wind could not carry. Sand dunes are examples of a kind of wind-blown soil deposit of no value for farming. Unless they are stabilized in some way, sand dunes can be moved along by the wind so that they cover up good soil.

Plants and *animals* have had an important part in soil formation. As time went on and soil formation took place, plants came into existence and their roots growing in the rock crevices served to pry the rock segments apart. Small plants called lichens dissolved some of the parts of the rocks with which they came in contact. Most prairie soils are high in organic matter for considerable depths, due to the accumulation of the tops and roots of grasses. On the other hand, most soils originally covered with trees have only a thin layer that is high in organic matter. Earthworms, gophers, and other burrowing animals aid in building soil by opening up channels permitting air to enter the soil and by mixing the soil constituents.

There are other soil-forming agencies, many of which cannot be noticed as readily as those that have been mentioned. For example, tiny plants, such as bacteria and molds, aid in the decomposition of plant and animal matter and thereby make it possible for these materials to become a part of the soil. Other processes, largely *chemical* in nature, also are responsible for important changes in soil-forming materials.

Climate has a direct bearing on soil formation beyond the breaking of rocks by temperature changes as already mentioned. High temperatures speed up chemical change and plant decay. If the temperature is high all year, then certain of the soil-forming processes continue to operate throughout the entire year.

The amount of rainfall determines the extent to which water erosion can take place and the availability of water for chemical reactions. Both rainfall and temperature help to determine the kind of plant and animal life that can exist and thus help to determine the kind of soil that will develop from the growth and decay of this life.

Topography also influences soil formation. It affects the amount and the rate of water run-off. It influences the amount of soil erosion from both water and wind. The effects of topography on soil formation are especially noticeable when the soil from one field is compared to that of a neighboring field.

Thus, the soil in any region has been influenced by many agencies, some of which were responsible for its formation and some of which carried it from other regions.

Before materials really become soil, decayed plant and animal matter, along with water and air in varying proportions, mixes with the fine rock particles. The formation of soil is a slow process; nature under favorable conditions requires 500 to 1,000 years to produce a layer of soil 1 inch deep. Of course, in restricted areas, soil may be laid down much more rapidly as it is transported by wind or water; but in these cases, it was formed elsewhere by the slow process just indicated.

Types of soil

Some soils high in clay are hard to cultivate. Sandy and loamy soils are easily cultivated. Between soils that are extremely sandy and those that are high in clay are various types, such as sandy loams, loams, silt loams, and clay loams, listed in the order of the fineness of the rock particles. By rubbing small amounts between the thumb and forefinger, a person can determine to some extent the type of soil. If it feels gritty, this is an indication of a fairly high content of sand. The clay loams have the least sand and feel smooth.

Sandy soils have low water-holding capacity, while clay soils have high water-holding capacity. Clay soils are sometimes referred to as cold soils because they take longer to warm up. Sandy soils usually benefit from the addition of manure and other forms of organic matter. Clay soils are frequently improved in the same ways; however, they may also need drainage. Clay soils must be worked at just the right time, or they will become sticky or lumpy and difficult to manage.

Loamy soils are frequently the most fertile. Such soils in good condition for plant growth are, by volume, about ¼ water, ¼ air, and ¹⁄₁₀ organic matter; the remainder consists of materials largely derived from rocks.

The soils of the United States have been classified into soil groups according to their many characteristics of parent material, origin, color, drainage, texture, depth, mineral content, organic matter content, and other

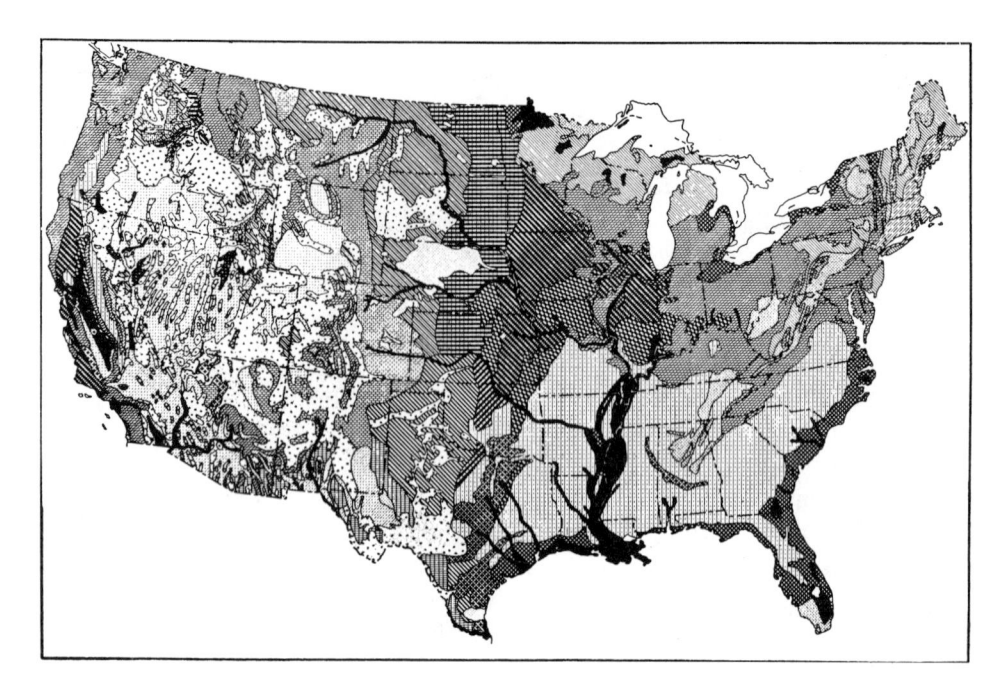

Fig. 16-5. General pattern of great soil groups in the United States. Legend is shown below and continues on the following pages. (Courtesy, U.S. Department of Agriculture)

LEGEND

The areas of each great soil group (zonal, intrazonal, and azonal) shown on the map include areas of other groups too small to be shown separately. Especially are there small areas of the intrazonal and azonal groups included in the areas of zonal groups.

ZONAL: Great groups of soil with well-developed characteristics, reflecting the dominating influence of climate and vegetation.

PODZOL SOILS

Light-colored leached soils of cool, humid, forested regions.

BROWN PODZOLIC SOILS

Brown leached soils of cool and temperate, humid, forested regions.

GRAY-BROWN PODZOLIC SOILS

Gray-brown leached soils of temperate, humid, forested regions.

RED AND YELLOW PODZOLIC SOILS

Red or yellow leached soils of warm and temperate, humid, forested regions.

PRAIRIE SOILS

Very dark brown soils of cool and temperate, relatively humid grasslands.

REDDISH PRAIRIE SOILS

Dark reddish-brown soils of warm and temperate, relatively humid grasslands.

Providing Fertile Soils 499

CHERNOZEM SOILS

Dark brown to nearly black soils of cool and temperate, subhumid grasslands.

CHESTNUT SOILS

Dark brown soils of cool and temperate, subhumid to semi-arid grasslands.

REDDISH CHESTNUT SOILS

Dark reddish-brown soils of warm and temperate, semi-arid regions, under mixed shrub and grass vegetation.

BROWN SOILS

Brown soils of cool and temperate, semi-arid grasslands.

REDDISH-BROWN SOILS

Reddish-brown soils of warm and temperate to hot, semi-arid to arid regions, under mixed shrub and grass vegetation.

NONCALCIC BROWN SOILS

Brown or light reddish-brown soils of warm and temperate, wet-dry, semi-arid regions, under mixed forest, shrub, and grass vegetation.

SIEROZEM OR GRAY DESERT SOILS

Gray soils of cool and temperate, arid regions, under shrub or grass vegetation.

RED DESERT SOILS

Light reddish-brown soils of warm and temperate to hot, arid regions, under shrub vegetation.

INTRAZONAL: Great soil groups with more or less well-developed characteristics reflecting the dominating influence of some local factor of relief, parent material, or age over the normal effects of climate and vegetation.

PLANOSOLS

Soils with strongly leached surface horizons over claypans on nearly flat land in cool to warm, humid to subhumid regions, under grass or forest vegetation.

RENDZINA SOILS

Dark grayish-brown to black soils developed from soft, limy materials in cool to warm, humid to subhumid regions, mostly under grass vegetation.

SOLONCHAK (1) AND SOLONETZ (2) SOILS

(1) Light-colored soils with high concentration of soluble salts, in sub-humid to arid regions, under salt-loving plants.
(2) Dark-colored soils with hard prismatic subsoils, usually strongly alkaline, in subhumid or semi-arid regions, under grass or shrub vegetation.

WIESENBÖDEN (1), GROUND WATER PODZOL (2), AND HALF-BOG SOILS (3)

(1) Dark brown to black soils developed with poor drainage, under grasses in humid and subhumid regions.
(2) Gray sandy soils with brown, cemented sandy subsoils developed un-

der forests from nearly level imperfectly drained sand in humid regions. (3) Poorly drained, shallow, dark peaty or mucky soils underlaid by gray mineral soil in humid regions, under swamp-forests.

BOG SOILS

Poorly drained dark peat or muck soils underlaid by peat, mostly in humid regions, under swamp or marsh types of vegetation.

AZONAL: Soils without well-developed soil characteristics.

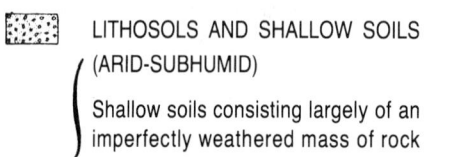

LITHOSOLS AND SHALLOW SOILS (ARID-SUBHUMID)

Shallow soils consisting largely of an imperfectly weathered mass of rock fragments, largely but not exclusively on steep slopes.

(HUMID)

SANDS (DRY)

Very sandy soils.

ALLUVIAL SOILS

Soils developing from recently deposited alluvium that have had little or no modification by processes of soil formation.

factors. Of perhaps more immediate interest from the standpoint of land use is the land capability classification used by the Soil Conservation Service of the U.S. Department of Agriculture in farm-planning work.

Land capability classification is designed to show the best use of the land, plus the soil management problems that exist and how to deal with them. Both surface soil and subsoil are considered. Some of the soil factors

Fig. 16-6. Land capability classes. Land Classes I, II, III, and IV are suitable for cultivation. Classes V, VI, VII, and VIII should be used for pasture, hay, woodland, and wildlife. (Courtesy, U.S. Department of Agriculture)

used to classify land according to capability are texture, color, structure, stoniness, slope, drainage, degree of erosion, and depth. Eight classes of land capability have been established. Land Class I is land that is level, well drained, free from erosion, and capable of growing cultivated crops with no special soil conservation practices being employed. Land Classes II, III, and IV all can be used for crop production, but they are progressively less capable in that respect because they are steeper, less well drained, and more eroded and more subject to erosion. These three land classes require the application of special soil conservation practices, with Land Class IV requiring the most specialized treatment. Land Class V is level land, but it is useful only for pasture, hay, or woodland because of poor drainage, stones, or other factors. Land Class VI is similar to Land Class V, but it requires the use of conservation practices because it is sloping. Land Class VII is eroded, rough, steep, or shallow land that with very careful management can be used for pasture or woodland. Land Class VIII is useful only for woodland, wildlife, and recreation.

How Is Soil Fertility Related to Plant Growth?

The soil has often been called the home of plants. Soil serves three important functions in plant growth. First, soil is a storehouse for chemical elements required by plants. These are technically called *plant nutrients* or *plant food materials*. Second, soil is a place where plant food materials are changed into forms usable by plants. This involves chemical changes that are aided by bacteria and other tiny organisms, as described previously. Third, soil serves as a place for plants to anchor themselves by their roots in a position so that they can obtain favorable surroundings with respect to air, heat, sunlight, and moisture.

Green plants contain upwards of 80 per cent water, most of which is obtained from the soil. In addition to water, plants contain many other materials, part of which are secured from the air and part from the soil. At least 16 chemical elements present in fertile soil are needed by plants, as indicated in Chapter 14.

In order for plants to grow well, the soil must contain the necessary chemical elements in ample quantities and forms usable by the plants. Furthermore, the soil must be in good physical condition to provide moisture and air and to permit the roots to grow and thus make use of the plant food materials that are there. Such a soil is *fertile*, because it is favorable to plant growth.

One of the important jobs of anyone who raises plants is to maintain a good balance of plant foods in the soil. When one element needed by plants is in short supply, even if it is a minor element, it limits plant growth. No matter how much of other elements is added, the one in short supply largely determines the yield.

Fortunately, only a few of the elements needed by plants are likely to be in short supply in most soils. In many soils, one or more of the elements nitrogen, phosphorus, and potassium are likely to be deficient. In some soils, calcium may be in short supply. In some cases, a shortage of one or more other elements may occur.

A farmer's relationship to soil fertility can perhaps be better understood by the following example. A child has a toy bank for pennies, nickels, dimes, etc. Unless the child places coins in the bank, it is obvious that none can be taken out. Unless all elements are present in the soil in a usable form, the growth of the plants will be checked. How are these two situations similar? Also, if the child puts only a few coins in the bank and then continues to take them out without putting more in, there will finally be an empty bank. In what ways is this similar to a soil that gradually loses its fertility?

A plant gets food materials from the soil through its root system. These materials in the soil must be changed to a form that is soluble before the roots can absorb them. Unless the phosphorus, potassium, and other elements in the soil are in a soluble form, the plant cannot use them, even though there may be large quantities in the soil. In other words, certain plant food materials may be locked up in much the same way that someone might lock up the child's bank and keep the key. The only remedy is for the farmer to try to manage the soil in such a way that the plant food materials will be "unlocked." For example, adding lime to high acid soils will unlock some of the phosphorus by special chemical reactions. Increasing the supply of organic matter in the soil may also aid in making certain food materials available by chemical and bacterial action. Drainage and good cultural practices increase the air in soil. Air aids soil organisms and chemical action and makes it possible for roots to take up soil nutrients. If the farmer fails to make available the food materials already in the soil, these will have to be added through commercial fertilizers or other means.

Some of the chemical elements needed by plants may be provided by *leaf feeding*, or foliar feeding. Nitrogen, boron, and iron may be dissolved in water and sprayed onto the surfaces of the leaves. Spectacular results may sometimes be obtained by doing this, but as yet this method of feeding plants has rather limited use.

The fertility of the soil, in some cases, affects the composition of plants grown on it. If certain plant food elements in the soil are in short supply, plants may still grow, but the protein and mineral content may be reduced. In most cases, the nutritive value of these plants to people and other animals does not seem to be affected seriously. However, the yields of crops under such conditions are likely to be seriously reduced.

Hunger signs in crops

Some elements have very specific effects on the growth and appearance of plants. A serious shortage of certain plant food elements in the soil is indicated by starvation or hunger signs in the crops grown. Of course, if the soil is deficient in several elements, the plants will not grow well, and it may be difficult to determine what elements are lacking.

In the case of corn and some other plants, *nitrogen deficiency* should be suspected if the leaves turn yellow or brown. The leaves start to turn yellow at the tips, and the yellowing follows the midribs of the leaves. *Phosphorus deficiency* is indicated if the leaves on young corn plants turn purplish or reddish, and the stalks are spindly and stunted. *Potassium deficiency* is indicated in corn if the edges of the leaves become yellow and then turn brown, and the leaves appear scorched along the edges. *Iron deficiency* is indicated by a loss of the green color in the young leaves at the top of the plants, with the leaves turning golden yellow. *Magnesium deficiency* is indicated if the leaves show yellow streaks or turn yellow while the veins remain green. A magnesium deficiency may be masked by other nutrient deficiencies.

These and other hunger signs of corn can be recognized by persons who are trained to know what these signs mean. Plants other than corn may also show evidence of starvation, but considerable skill is required to interpret these signs to decide what may be lacking.

How Is Soil Fertility Lost?

Anyone who grows plants on some scale (from the backyard garden on up) realizes that soil may lose much of its fertility. Soil loses its fertility principally in three ways as follows:

1. *Soil fertility is lost when plants are grown and removed from the land.* This happens even if these crops are fed to livestock and the manure is returned to the fields where the crops were grown. This is because some minerals, which the plants secure from the soil, are used by animals for body growth, and in some cases for the production of eggs, wool, or milk. Furthermore, the portions of the minerals that the animals eliminate in the form of manure undergo some loss before the manure is returned to the fields. Farmers who sell crops directly from the farm reduce the fertility more rapidly than livestock farmers, unless they take steps to offset the losses.
2. *Some plant food materials in the soil are lost when they are dissolved in water that falls on the surface and seeps away* to places beyond the reach of roots of plants. This is called *leaching*. In some soils, losses

through leaching may be more serious than the amounts removed by crops.

3. *A large amount of fertility is carried away by water that flows over the surface of the land and carries soil with it.* This is known as *water erosion.* In some areas, soil is carried away by *wind erosion.* These types of losses are described in Chapter 7. For the United States as a whole, the greatest loss in fertility comes about through water erosion. In fact, in some regions, the losses through this source are several times as much as from all the other causes combined.

The disturbing thing about erosion is that soils may be lost so gradually that people do not realize what is happening. This is particularly true of the effects of sheet erosion, which consists of a gradual removal of soil from the surface of the land. Only when gullies begin to develop, or when soil accumulates in large quantities on portions of fields, do some farmers realize what is happening. Another indication of erosion is a change in color of the soil on hillsides, due to the washing away of the darker top soil, thereby exposing some of the lighter-colored subsoil.

How Can Soil Fertility Be Maintained and Improved?

Everyone who grows plants is interested in improving the fertility of the soil. In recent years, scientists have made many advances in ways to maintain and improve soil fertility. The effective use of practices that improve the fertility of the soil increase crop yields. In some cases, through fertilization and other practices, land that was once practically worthless is now producing crops on a profitable basis. It is not unusual to find land on which the yields of various crops have doubled.

Some persons who grow plants are not using practices for improving the fertility of the soil, even though it would be profitable for them to do so. In one midwestern state where many samples of soil from numerous farms have been tested, it has been found that more than two-thirds of the soil is deficient in one or more essential elements. If farmers throughout that state were to apply suitable combinations of fertilizers, they probably could profitably increase their yields at least 30 per cent.

Crops require large quantities of plant nutrients secured from the soil, as discussed previously, to produce high yields. Unless these nutrients are available in the soil in ample amounts and in forms that can be used by the plant, the yields will be reduced. Since the yields are limited by the nutrients that are deficient, it is important to determine what the soil needs.

Providing Fertile Soils

Determining soil needs

Soil needs can be determined by a combination of several methods. These include (1) examining the physical characteristics of the soil, (2) studying yields and appearances of the plants grown on the soil, (3) making chemical tests of the soil, (4) making chemical tests of tissues of growing plants, and (5) using test plots for fertilizers and other soil treatments.

Examining the physical characteristics of the soil. A person who is familiar with soils can determine some of the shortcomings of a particular soil by noting its appearance and the way it "handles." If soil remains too wet and plants frequently "drown out," drainage is probably needed. If soil remains too dry most of the time, irrigation in some form may be helpful. If the soil is compact and forms hard lumps when dry, it probably needs additional organic matter. If it is extremely loose and water soaks away rapidly, organic matter is probably needed in this case also, as is usually true of sandy soils.

Studying the yields and appearances of the plants. If yields on a plot of ground are low year after year, the soil is probably low in fertility, provided other conditions are favorable. The specific needs can be determined

Fig. 16-7. Crimson clover root system. This is one of the deep-rooted annual legumes that requires deep, well-drained soil for best growth. (Courtesy, U.S. Department of Agriculture)

by chemical tests of the soil. Sometimes, if the deficiency of certain plant nutrients is particularly serious, the plants themselves show signs of starvation or hunger, as explained earlier.

Making chemical tests of the soil. One of the most helpful methods for determining what the soil needs is to make chemical tests of samples of the soil. One type of test, if properly made, is particularly useful for determining the degree of acidity and the amount of lime that may be needed to correct this acidity. Other types of tests help to pinpoint the amounts of phosphorus and potassium present in the soil, thereby making it possible for individuals to know how much fertilizer to add. Soil tests may be made by county agricultural agents, vocational agriculture teachers in high schools, and special testing laboratories in some counties. Soil samples for testing must be taken carefully and at appropriate places in a field so that they will be representative of the soil to be tested. A county agricultural agent or vocational agriculture teacher can tell how this should be done. Even a person who raises a garden or maintains a lawn should have the soil tested to determine what fertilizer may be needed.

Making chemical tests of tissues of growing plants. Some indication of soil needs in a field can be ascertained from chemical tests of the tissues of green plants grown on that field. This helps individuals to check symptoms noted from the appearances of plants, thus aiding in determining whether or not there are sufficient nitrogen, phosphorus, and potassium in forms available to the plants. The techniques for making these tests require considerable skill and are usually done by individuals with special training for making and interpreting them.

Using test plots for fertilizers and other soil treatments. If the various chemical tests and checkups do not seem to indicate specific soil needs, or if individuals are skeptical about the possible benefits from fertilizers, they may want to try out various treatments on portions of a field. By applying fertilizers on small test plots and comparing the resulting yields with the yields in untreated portions, they will know whether or not it would pay to treat other portions of the field.

Developing a soil fertility program for a farm or a plot of ground

Everyone who raises plants should develop a soil fertility program that is suited to the conditions on his/her farm or piece of land. Each farm or plot of ground has distinct problems because of the nature of the soil, the topography of the land, the interests of the people who till it, and other factors.

In developing a soil fertility program for his/her farm or plot of ground, the owner should make a careful study of the land. The soil should be tested, crop yields should be analyzed, physical conditions of the soil should be studied, and evidences of erosion should be noted. The steps described in the following paragraphs should be followed. In planning a soil fertility program, a person may secure advice from various persons, including the county agricultural agent, vocational agriculture teachers, and soil conservation specialists.

Planning suitable cropping programs. In establishing a fertility program for his/her farm or piece of land, the owner should put each portion to its best use. Some of the land may not be suitable for cultivation because of the steepness of slopes and the stony condition of the soil. Grass crops or trees may represent the best use of such land. Some land is suitable for several kinds of crops, and the kinds to grow are usually the ones that bring the highest returns.

Land suitable for tillage should be planted to crops that will adapt to the soil conditions and provide favorable returns to the person who raises them. If livestock are raised, the cropping programs should supply part or most of the feeds needed. Legume crops should be included in most cropping programs to add nitrogen and organic matter to the soil.

Land subject to serious erosion should be planted less frequently to row crops than land on which erosion is not a serious problem, and the cropping program should be planned accordingly. For example, a suitable rotation of crops for farmland not subject to serious erosion might consist of a four-year rotation of corn, corn, small grain, and legumes. On some level prairieland, a highly profitable rotation is corn, corn, soybeans, and wheat, with clover as a "catch" crop to be plowed under. On some land, row crops may be grown year after year if the proper methods are used to maintain soil fertility. On land subject to moderate erosion, a rotation of corn (or another cultivated crop), small grain, and legumes might be most appropriate. On land subject to considerable erosion, a suitable rotation might consist of a row crop one year and a small grain the next year, followed by three or more years of perennial hay or pasture crops. Many different combinations of crops are possible, depending on the section of the country, the condition of the soil, and other factors. Some land subject to serious erosion should be kept permanently in grass or trees.

Anyone planning to have a garden or other crops on a small plot of ground should consider the suitability of the land in deciding what to raise. Land that is nearly level and suitable in other ways is most desirable for garden crops.

Providing organic matter. In order to maintain a fertile soil, individuals should be concerned about the supply of organic matter, or humus. Organic matter benefits the soil in several ways, such as (1) providing conditions favorable to bacterial and chemical action necessary for making food materials available; (2) improving the physical condition of the soil, thereby making it easier to till, more absorptive of water, and less likely to erode;

Fig. 16-8. This is one way in which livestock farming helps to provide fertile soil for crops. (Courtesy, John Deere)

and (3) adding to the supply of nitrogen and certain other food materials as the organic matter decomposes.

One of the best ways to provide plant food materials for the crops is to apply animal manure. These waste products from animals contain a large part of the materials that the crops removed from the soil. If animal manure can be obtained, it is good material to use on garden plots.

Another way to add organic matter to the soil is to plow under crops grown for this purpose. Such crops are frequently called *green manure crops*. Rye and clover are two examples of crops used for this purpose. Some crops, such as rye, are also used as *cover crops,* which means they are grown to cover the soil and thus prevent erosion. Such crops also utilize plant food materials that might otherwise be lost by leaching. Persons who raise crops should try to keep the soil covered by crops of some kind during most of the year.

In producing some crops, such as corn and small grains, farmers frequently leave the stems and leaves on the field when the grains are harvested. These are called *crop residues,* and they provide organic matter when mixed with the soil by plowing or other tillage methods. The term *trashy cultivation* is used to describe the process of mixing these materials with the top soil in such a way as to check wind and water erosion. The practice of burning the top growth that remains after crops have been harvested is wasteful.

In recent years, chemical materials known as soil conditioners have been manufactured. These add no plant foods or organic matter, but they do help to improve the soil structure so that it is more easily tilled. As yet, these materials are too expensive to be used extensively.

Providing Fertile Soils

Applying lime if needed. Lime is added to soil for two principal reasons. The first and most important reason is to decrease soil acidity, or "sweeten" the soil. If soil is decidedly acid, most plants will not grow well. In such soils, conditions are less favorable for beneficial microbes, which help to change plant food materials into forms useful to plants. Furthermore, if soil is extremely acid, much of the phosphorus and some other elements in the soil may be locked up in a form unavailable to plants, and the tilth of the soil may not be as good as it would be if the acidity were reduced. Most crops yield best if the soil is slightly acid, although some crops, such as alfalfa, do best when the soil is about neutral in reaction.

A second reason for adding lime to soil is to supply calcium as a plant food material. Calcium is one of the food materials needed by plants.

Before lime is added to the soil, samples of the soil should be tested for acidity, as indicated earlier. Only enough lime should be added to reduce the acidity to the desired amount. In some cases, no lime is necessary. Excess lime added to soil may lock up some of the plant food nutrients so the plants cannot use them.

The degree of soil acidity, or of alkalinity if the soil is alkaline rather than acid, is measured in terms of pH. The term *pH* represents the concentration of hydrogen ions in a solution. Acid soils have a pH of less than 7, while alkaline soils have a pH greater than 7. A pH of exactly 7 means the soil is neutral, neither acid nor alkaline. Most crops grow best in a pH range slightly below 7. The soil microorganisms grow at their best for releasing

Fig. 16-9. These 12 lysimeter plots are used to measure the percolation of water through soil and to determine the soil constituents removed in the drainage. (Courtesy, U.S. Department of Agriculture)

nitrogen and other plant nutrients as they decompose soil organic matter at a pH of 6.5.

The material that is commonly used to correct soil acidity is agricultural ground limestone. In some cases, it may be hauled in a truck especially equipped to broadcast it on the fields. Lime may be added at any place in a rotation, although it is best added a few months before alfalfa or sweet clover is seeded so that the acidity will be corrected by the time these crops are planted. Lime may be distributed on plowed soil and worked into the surface soil. If limestone is applied in the amounts indicated by tests, it should last several years in most soils.

Using commercial fertilizers. Most soils benefit from suitable kinds and amounts of commercial fertilizers. Regardless of the other methods used to maintain and improve soil, some commercial fertilizer is usually desirable. For many soils in the United States, phosphorus is likely to be one of the elements in short amount. In some soils, potassium also may be needed. To obtain high yields of grain crops, forage crops, and vegetable crops, farmers will probably need to add nitrogen to the nitrogen already supplied by the growing of legumes.

Because of the resulting yield increases, each dollar spent for fertilizer may return several dollars in profits. However, anyone who raises crops should have the soil tested to determine the needed kinds of fertilizers and then follow the recommendations of reliable sources.

To summarize, the major materials added to the soil and the major benefits and functions of each are as follows:

Fig. 16-10. Because the drill stopped up, two rows of corn received little or no fertilizer. The height of the stalk in front of the hat is 7 inches. The average height of the stalks that received fertilizer is 40 inches. (Courtesy, U.S. Department of Agriculture)

Organic matter
1. Supports soil organisms such as bacteria and fungi.
2. Helps bring soil minerals into solution, thus making them available for plant growth.
3. Improves the physical condition of the soil.
4. Increases the water-holding capacity of the soil.
5. Improves the aeration of the soil.
6. Helps regulate soil temperature.
7. Provides certain plant nutrients.

Lime
1. Reduces soil acidity to desirable levels.
2. Provides calcium and magnesium, both of which are needed for plant growth.
3. Aids in the decay of organic matter.
4. Improves the physical condition of the soil.
5. Aids in making phosphorus available.
6. Increases crop yields.
7. Stimulates soil organism growth and activity.

Nitrogen
1. Stimulates rapid and luxuriant plant growth.
2. Increases yield of leafy crops.
3. Increases protein content of crops.

Phosphorus
1. Stimulates early growth of plants.
2. Stimulates root development.
3. Increases disease resistance of plants.
4. Hastens maturity of crops.
5. Gives winter hardiness to hay and grain crops.
6. Stimulates flowering and seed development.
7. Aids in the use of nitrogen by plants.

Potash
1. Stimulates good growth of plants.
2. Is necessary for production of carbohydrates and protein by plants.
3. Aids in development of stiff stems in plants.
4. Increases resistance to disease.
5. Increases resistance to dry weather and to cold temperatures.

The minor plant nutrients are becoming more important because of the intensive cropping program that is gradually bringing about shortages of them. These nutrients are important for many of the same reasons that apply to the major nutrients. For example, the minor nutrients promote growth of plants, development of roots, resistance to disease, use of other nutrients, development of seeds, and the various processes of growth. Yield increases as a result of their application point up their importance and indicate the need for including them in a sound fertility program.

Buying fertilizers

Fertilizers may be purchased in many forms. Where only one of the major nutrients (nitrogen, phosphorus, potash) is needed, it can be purchased and applied by itself. If more than one major nutrient is needed, it may be more convenient to purchase a commercially mixed fertilizer that has the correct amounts and proportions of each of the needed nutrients.

The amount of each plant nutrient contained in a particular fertilizer mixture is shown on the fertilizer bag; or if bulk fertilizer is purchased, the information can be obtained from the fertilizer dealer. The fertilizer analysis gives the amounts, expressed as percentages, of each of the major nutrients in the mixture. For example, a 5-20-10 fertilizer would contain 5 per cent nitrogen (N), 20 per cent phosphoric acid (P_2O_5), and 10 per cent water-soluble potassium oxide, or potash (K_2O). The percentages are always given in the same order, with nitrogen first, phosphoric acid second, and potash last. An 80-pound bag of 5-20-10 would contain:

$$\begin{array}{ll} \text{Total N} & 5\% \\ \text{Available } P_2O_5 & 20\% \\ \text{Water-soluble } K_2O & 10\% \end{array}$$

A 10-10-10 fertilizer would contain equal percentages (10 per cent each) of total nitrogen, available phosphoric acid, and water-soluble potassium oxide. There are many other fertilizer analyses available to fit the various combinations of soil needs. Some examples are 3-12-12; 5-10-5; and 5-10-10. One all-phosphorus fertilizer has an analysis of 0-20-0. One fertilizer containing phosphorus and potassium, but no nitrogen, has an analysis of 0-20-20.

Sometimes the fertilizer analysis is confused with the fertilizer ratio. The ratio refers to the amount of each major nutrient in relation to the other major nutrients in the fertilizer mixture. For example, the ratio of a 10-10-10 fertilizer is 1-1-1, indicating that there are equal percentages of nitrogen, phosphoric acid, and potash in the mixture. A 5-10-5 fertilizer has a fertilizer ratio of 1-2-1. An 8-16-16 fertilizer has a 1-2-2 ratio. In deciding on the mixed fertilizer to buy, a farmer should first translate the soil needs into the correct

Providing Fertile Soils

ratio or relative proportions of each of the needed nutrients. Once the ratio has been determined, the decision regarding the analysis to buy can be made.

If desired, fertilizer may be ordered mixed especially for a particular farmer's land. This is not usually practical if only small amounts are to be purchased.

In some areas, fertilizer can be ordered custom blended in terms of the number of pounds of each element to be applied per acre. For example, a farmer might order a 100-100-100 blend, with the numbers indicating that 100 pounds each of nitrogen, phosphoric acid, and potash will be put on each acre of land for which the fertilizer is purchased. A 60-40-20 blend would indicate that 60 pounds of nitrogen, 40 pounds of phosphoric acid, and 20 pounds of potash are to be applied to each acre of land. The blend, or mix, of fertilizer would be based on the soil tests and the needs of the crop to be grown.

Nitrogen fertilizers are available in several forms. Most of the present supply is produced by special factories equipped to convert nitrogen from the air into forms useful for fertilizer. Some of the forms in which nitrogen can be purchased for use as a fertilizer are anhydrous ammonia, ammonium nitrate, ammonium sulfate, urea, calcium cyanide, and nitrate of soda.

Phosphorus is secured primarily from phosphate rock, which is found abundantly in Florida, Tennessee, and the Rocky Mountain states. This material is mined, washed, ground finely, and in many cases treated with sulfuric or phosphoric acid. Phosphorus fertilizer can be purchased in the form of rock phosphate, nitric phosphate, ammonium phosphate, and calcium metaphosphate.

Potash also is mined from the earth. Although some is imported, most of it is secured from mines in New Mexico, California, and Arizona. The huge potash deposits in Saskatchewan, Canada, can meet world potash needs for years to come, even centuries. The major forms in which potassium can be purchased for use as a fertilizer are muriate of potash, sulfate of potash, and sulfate of potash-magnesia.

There are many ways of applying fertilizer to soil. Each method has its advantages, and the method to be used depends on many factors, such as the kind of crop, soil, climate, time and rate of application, and kind of fertilizer. Farmers and others can secure advice on methods of application from their state agricultural colleges, their county agricultural agents, vocational agriculture teachers, and other sources. In some cases, a combination of two or more methods is most desirable. The following are some common methods of applying fertilizer:

1. *Broadcasting.* This method consists of scattering fertilizer uniformly over the soil surface before or after plowing and then working it into

the soil while preparing the seedbed. Various kinds of equipment may be used.

2. *Banding along the rows.* This method concentrates the fertilizer near the plants and makes effective use of small amounts of fertilizer. The fertilizer is usually placed along the sides of the row and somewhat deeper than the seeds. Special fertilizer attachments are used on drills and planters to distribute the fertilizer in this manner.
3. *Sidedressing.* Some row crops may benefit from fertilizer after the crops have been started. Fertilizer is scattered between the rows. Special attachments on cultivators are frequently used for this purpose.
4. *Plowsole, or deep furrow.* Some fertilizer may be applied at the bottom of the plow furrow by special attachments on the plow. This method is beneficial under some conditions because it stimulates deep root growth.
5. *Top dressing.* For non-cultivated crops, such as pastures, orchards, small grains, and lawns, the fertilizer is frequently broadcast over the surface.
6. *Solutions.* Some fertilizers may be dissolved in water and distributed as solutions by special machines. Sometimes, fertilizers are placed in irrigation water and distributed to plants through this means. Usually, fertilizers in solution are called liquid fertilizers. In recent years many liquid fertilizer companies have been established for putting plant nutrients into solution for sale to farmers. Some farmers purchase or rent equipment for storing and applying liquid fertilizer, while other farmers order the liquid fertilizer and have the fertilizer company apply it to the fields in the amounts desired.
7. *Gas.* Certain forms of nitrogen may be applied to the soil in gaseous form. Special equipment forces the gas below the surface of the soil where it is absorbed by plants.
8. *Foliage, or leaf, feeding.* Some fertilizers may be applied as dusts or mists to the foliage of growing plants. This material is absorbed through the leaves. This method is not used extensively, but it may become more important in future years.

Recent experiments in some states seem to support a program of applying abundant quantities of the needed fertilizers so as to build up a reservoir of plant food nutrients in the soil. Under this program, heavy applications of fertilizers are made once every few years, usually in the fall or spring on plowed ground. The amounts and kinds to apply are determined by soil tests.

Properly fertilized plants produce abundant growth of tops and roots. This material, in turn, contributes to the amount of organic matter supplied to the soil.

Providing Fertile Soils

Controlling moisture supply

In some soils, too much moisture is present to permit the most favorable conditions for plants to grow. In this case, various methods of drainage may be used. Ditches may be dug to carry away the excess water. Frequently, cylindrical tiles are placed beneath the surface at appropriate depths so that surplus water may be carried away.

In some parts of the United States, the rainfall is too limited for growth of desirable types of plants. In cases where water may be secured from lakes, wells, or streams, irrigation is practiced. The water is distributed over the land by gravity, through small ditches, or by overhead sprinklers attached to pipes through which the water is forced under pressure.

In many states, irrigation has brought about what is known as round farming, or farming in circles. Each circle is about 130 acres and is irrigated by a center pivot irrigator, which consists of a pipe almost ¼ mile long mounted on wheeled A-frames. The end of the pipe in the center of the field is connected to a pump that forces water along the pipe into hydraulically driven cylinders that power the wheels to drive the rigs around the fields like the hands of a giant clock.

This system provides for good control of water use, makes leveling and grading unnecessary, is movable from field to field, and can apply fertilizer as well as water.

Even in areas where annual rainfall is considerable, periods of drought, which seriously affect the crops, may occur. Under some conditions, it is

Fig. 16-11. Tile for draining wet land must be properly laid to insure proper drainage. (Courtesy, Soil Conservation Service, U.S. Department of Agriculture)

Fig. 16-12. Tubes are used to siphon water from this irrigation ditch into the sugar beet field. (Courtesy, Kansas State Board of Agriculture)

Fig. 16-13. Light, portable pipe is often used for irrigation. Water is pumped into the pipes under high pressure and distributed through rotating nozzles. (Courtesy, Reynolds Metal Co.)

Fig. 16-14. Laser technology is used to keep this traveling trickle irrigation system level so it will work properly. (Courtesy, *Agricultural Research*, U.S. Department of Agriculture)

profitable to irrigate during such periods. This is called supplemental irrigation. In some areas, yields of certain crops have been greatly increased by this method.

Scientists have found that plants grown on fertile soils are least affected by periods of drought. This is largely because the root growth is much more extensive under such conditions.

Controlling erosion

Everyone who raises crops should take the steps necessary to prevent or reduce losses of the fertile top soil by wind or water erosion, as discussed in Chapter 7. By using good cropping methods, increasing the organic matter in the soil, and supplying fertilizers that stimulate the maximum growth of plants, farmers can reduce these losses.

On some soils, additional practices are helpful. One of these is grass waterways that allow surplus water to flow away after heavy rains, without damage to the soil. Rows planted on the contour of slopes help to prevent washing. Under some conditions, strips of row crops are alternated with uncultivated crops, thus checking erosion on sloping fields. This practice is called *strip cropping*. On some fields with rather steep or long slopes, terraces may be constructed around the contours to control the flow of water.

In areas subject to wind erosion, a modified form of strip cropping is also used. Trashy cultivation, in which plant materials are mixed with the

SOIL LOSSES THROUGH CULTIVATION

Fig. 16-15. Soil erosion is likely to be severe if cultivation is practiced on sloping fields. Loss of soil varied from practically none to 60 tons per acre on soil kept cultivated with no crop (fallow). (Courtesy, U.S. Department of Agriculture)

surface soil, also helps to control wind erosion. Further details on the control of water and wind erosion are given in Chapter 7.

Tilling the soil

One of the most important ways in which a fertile soil can be provided is through proper tilling of the soil. Tillage improves the condition of the soil in many ways so that high crop yields can be obtained. Some of the benefits of good tillage are:

1. It loosens the soil so that the seeds can come into contact with the soil moisture for good germination.
2. It loosens the soil so that the small plants can emerge easily and uniformly.
3. It helps to kill weeds.
4. It loosens the soil for good root development.
5. It helps to mix nutrients and pesticides in the soil.
6. It helps to open the soil so oxygen will be available to plant roots.
7. It helps to warm the soil by loosening it and exposing more of it to the sun.
8. It helps to maintain the best moisture conditions for plant growth by speeding removal of excess water and by holding moisture when it is in limited supply.

Plowing the soil with a moldboard plow is usually the first step in preparation for planting. It is important to have properly adjusted equipment to

Providing Fertile Soils

secure good coverage of trash and vegetation and to have a field with uniform furrows. The depth of plowing depends on the type of soil and the need to avoid the development of a compacted layer at plow depth (hardpan). For some conditions, such as fields that are rocky and fields where plow furrows are undesirable, a disk or rotary tiller may be used.

After the soil has been plowed, the next step is to go over the field with a disk or spring-tooth harrow and a cultipacker to break up the large clods of soil into small particles and to level the soil. If a good job of plowing has been done, the seedbed can be prepared for planting with only this one trip over the plowed field. Power-driven rotary tillers are especially effective in pulverizing and smoothing the soil surface. To reduce the number of times a field must be tilled, some techniques have been developed for tilling only the narrow strip of soil where the seeds are to be planted.

While some soil and climatic conditions require fall plowing, most plowing is done in the spring not too far in advance of planting. This prevents weed growth, which would require additional tillage for control, and soil compaction from the long wait until planting time.

At the present time, several new tillage systems are being tried. One of the most important of these new systems is *conservation tillage*. Actually, conservation tillage is a combination of several tillage systems, all of which leave residues on the surface of the soil to reduce or stop erosion. Some of the other systems are: (1) *no tillage*; (2) *eco-fallow*, in which a row crop follows a grain crop; (3) *sod planting*, which refers to forage crop farming; (4) *strip tillage*; (5) *rotary tillage*, in which strips are opened through surface residue with a rotary tiller (for vegetable crops); (6) *ridge planting*, in which narrow strips are made with a pair of coulter disks that create a ridge; (7) *disk planting*; and (8) *slit planting*, in which the compacted soil layer is penetrated with a narrow slit to create an opening for the roots.

The advantages of the new forms of tillage are (1) a savings of soil through reduced erosion, (2) a reduction in water pollution from soil and chemicals that are carried by the water run-off, (3) a savings of fuel by the reduction in the number of trips over a field, and (4) a savings in time. The major disadvantages of the new tillage systems appear to be the increased dependence on the use of herbicides and the large investment needed for new planting equipment.

SUGGESTED ACTIVITIES

1. Describe the manner in which the food habits of the following make each dependent upon the soil for existence: humans, dogs, robins, pigs, turkeys, cats, hawks, and squirrels. Which of these are directly dependent on plants for food? Indirectly?

2. Observe the soil in an open field, in a wooded area, along a stream, and in other places to determine some of the ways in which the soil has been formed and has accumulated. What are some of these ways?

3. Study soil at different depths, such as in a roadside cut, a basement excavation, or a dug hole. What are the chief differences between top soil and subsoil? Where do you find most of the roots of plants? What plants send roots to the greatest depths?
4. If possible, visit some experimental plots at a state college of agriculture or elsewhere where different fertilizers have been applied to the soil. What differences do you note in the crops where various fertilizers have been used?
5. Organize a contest in writing rhymes on some phase of soil fertility or soil erosion.

"A hillside field,
A surface bare,
A pouring rain,
And the soil ain't there."

"Where little care is given ground
Lower yields will be found."

6. Plan and organize a mock trial in which some member of your class is tried for "robbing" the soil. Have a judge, lawyers, witnesses, a jury, etc., consisting of members of your class.
7. Make exhibits with tables of sand or pans of soil to demonstrate the effects of different cropping practices on soil erosion—the use of strip cropping, terraces, gully control, etc.
8. Prepare a short talk on what would happen to your community if the soil became only half as productive as it is at present.
9. Concepts for discussion:
    a. Soil is our most important natural resource.
    b. A strong agricultural production industry is essential for a strong country.
    c. If proper practices are used, soil will remain productive even with intensive use.
    d. Proper practices can return depleted soils to a productive state.
    e. All forms of animal life depend on the soil for their existence.
    f. Soil is about half solid material and half pore spaces filled with water and air.
    g. Soil contains millions of tiny plant and animal forms.
    h. Soil particles vary in size from fine clay to coarse sand.
    i. Organic matter in the soil is needed by soil organisms, which change soil elements to forms plants can use.
    j. Plants can be grown without soil.
    k. Soil formation is a slow process influenced by many factors.
    l. Soils are classified according to their various characteristics.
    m. Land capability classification is designed to show the best use of land.
    n. Soil provides plants with water, nutrients, and physical support.
    o. Green plants contain upwards of 80 per cent water.
    p. A fertile soil is one in good physical condition with an ample supply of nutrients and adequate moisture.
    q. The plant nutrient in short supply in the soil is a limiting factor in plant growth.
    r. Some chemical elements can be provided plants by leaf feeding.
    s. Certain deficiencies in soil nutrients can be identified by hunger signs in crops.

Providing Fertile Soils

t. Soil fertility is lost through removal of crops, leaching by water, and erosion of soil.

u. Soil needs can be determined by observation, crop yields, soil testing, plant tissue testing, and test plot usage.

v. A soil fertility program should be built around utilization of a sound cropping system, addition of organic matter, and application of fertilizer.

w. Each of the materials and elements applied to the soil performs identifiable functions.

x. Fertilizers may be purchased in many forms.

y. Moisture control through irrigation or drainage may be needed for best plant growth.

z. Erosion control practices help maintain soil fertility and improve plant growth.

REFERENCES

Donahue, R. L., R. H. Follett, and R. W. Tulloch. *Our Soils and Their Management*. Interstate Publishers, Inc., Danville, IL 61832.

Johnson, H., Jr. *Hydroponics: A Guide to Soilless Culture Systems*. Cooperative Extension Leaflet 2947, Division of Agricultural Sciences, University of California, Berkeley, CA 94720.

Lee, J. S., and D. L. Turner. *Introduction to World AgriScience and Technology*. Interstate Publishers, Inc., Danville, IL 61832.

Osborne, E. W. *Biological Science Applications in Agriculture*. Interstate Publishers, Inc., Danville, IL 61832.

U.S. Department of Agriculture. *Agricultural Research*, a monthly publication. U.S. Government Printing Office, Washington, DC 20402.

U.S. Department of Agriculture. *Our American Land*. Yearbook of Agriculture: 1987, U.S. Government Printing Office, Washington, DC 20402.

Vocational Agriculture Service. *Hunger Signs in Crops*. VAS 4011a, Vocational Agriculture Service, College of Agriculture, University of Illinois, Urbana, IL 61801.

Vocational Agriculture Service. *pH Test for Soil Acidity*. VAS 4002a, Vocational Agriculture Service, College of Agriculture, University of Illinois, Urbana, IL 61801.

Wildman, W. E. *What on Earth Is Soil?* Cooperative Extension Leaflet 2637, Division of Agricultural Sciences, University of California, Berkeley, CA 94720.

Wilson, H. M., and C. S. Winkelblech. *Tillage: Basic Principles and Techniques*. Extension Bulletin 1176, New York State College of Agriculture and Life Sciences, Cornell University, Ithaca, NY 14850.

Chapter 17

Planting and Cultivating Farm Crops

In considering the culture of farm crops, we are concerned with the methods that people may use to secure high yields of good-quality products. While it is important to provide a fertile soil, there are many additional things that must be done, as suggested in the statement quoted above.

OBJECTIVES

1. Identify methods important in producing crops successfully.
2. Demonstrate cultural practices for specific crops.
3. Discuss how weather conditions affect crop production.

What Methods Are Important in Producing Crops Successfully?

In order to secure high yields of good-quality crops, farmers must use scientific methods. For most crops, these include:

1. Providing a fertile soil.
2. Providing good seed of suitable varieties.
3. Preparing a good seedbed.
4. Planting properly.
5. Cultivating effectively.
6. Controlling plant enemies (weeds, insects, and diseases).
7. Using efficient methods of harvesting and storing.

The specific methods for producing crops differ considerably among the various kinds of crops. Furthermore, methods for raising any one crop, such as corn, differ somewhat in various parts of the United States. Some of the methods and practices generally recommended for each of several important crops are described in this chapter. Methods for improving soils and for controlling weeds, harmful insects, and plant diseases are described in other chapters.

Weather conditions affect greatly the production of crops in any locality. In choosing the crops to raise, farmers must take into account some of these conditions. However, in any locality, variations that may affect the commonly raised crops occur in rainfall and temperature. To some degree, farmers are able to reduce the undesirable effects of too much rainfall by drainage and other practices, and of too little rainfall by irrigation, use of drought-resistant crops, cultural methods, and other methods. Fortunately, most of the important crops, such as wheat and corn, are grown widely, and adverse weather conditions usually affect only portions of the total acreages grown in the entire country.

Weather forecasts, made daily and for several days ahead by persons called *meteorologists,* can help farmers. In some cases, farmers can take steps to protect their crops, although frequently it is impossible to prevent the damaging effects of weather, even though advanced information is given.

Fig. 17-1. Large grain drills help make sure that the crops will be planted while weather and soil conditions are favorable. (Courtesy, John Deere)

One of the most recent developments in crop culture has been the compaction of the soil because so much heavy machinery is being used on it. Small-wheeled, heavily loaded fertilizer trucks are especially damaging in this respect. The small wheels concentrate all the weight in a small area, thus causing deep compaction. Heavier tractors and other power machinery are also contributing to this problem. The effects of soil compaction include the closing of the soil pores and the creation of layers of compacted soil through which plant roots are unable to penetrate. The closing of the soil pores reduces soil water intake, which increases erosion through greater surface run-off, and it reduces soil aeration, which impairs root growth. All these factors result in lowered crop yields.

What Are Desirable Cultural Practices for Corn?

Farmers who use good cultural practices in raising corn are able to secure high yields of good-quality corn. In recent years, marked increases in yields of corn have resulted from the use of hybrid seed combined with good cultural practices.

Providing a fertile soil

Corn is a high-profit crop in the Corn Belt and many other parts of the United States. For this reason, farmers who grow corn give it a favored place in the cropping program. If clover or other legumes are grown, corn frequently follows such crops in the rotation. The corn plants thus benefit from the nitrogen added to soil by the legumes.

Some rotations used in the Corn Belt are (1) corn, small grain, clover; (2) corn, corn, small grain, clover; (3) corn, soybeans, small grain, clover; (4) corn, corn, small grain, alfalfa (usually left for two or more years); and (5) corn, corn, soybeans, and wheat, with a legume crop grown the same year for plowing under. The last rotation is suited to level land and is one of the most profitable combinations of crops. By using large amounts of fertilizers on level land not subject to erosion, some farmers are able to grow corn successfully without legumes in the rotation.

Because corn is a "heavy feeder" of elements from the soil, commercial fertilizers are used extensively to secure high yields. Some fertilizer is usually broadcast and plowed under. In addition, some corn growers use a "starter" fertilizer that is placed in a band about 2 inches away from the row and slightly lower than the level of the seed by an attachment on the corn planter. Additional fertilizer, usually nitrogen in some form, may be applied as side dressing in the rows during the growing season if the supply of nitrogen in the soil is insufficient. Livestock manure is a valuable fertilizer. It is usually applied before the soil is plowed.

The plant nutrients most critical to good corn yields are nitrogen, phosphorus, potassium, sulfur, and zinc. In some areas, iron deficiency may be a problem.

Using good seed

Hybrid seed corn is used by most corn growers. This seed should be purchased each year from a reliable dealer. A variety adapted to the soil and climate should be secured. The college of agriculture in each state usually provides a list of recommended varieties for various conditions in the state. Seed corn from reliable firms is sold in bags that carry a label showing the variety name or number and the percentage of germination. Frequently, the corn has been treated with chemicals for protection against some corn diseases and insects.

There are many different hybrids available. The single-cross hybrid is the most expensive, the three-way cross the next most expensive, and the double-cross hybrid the least expensive. When selecting hybrids, farmers should consider factors such as standability, tolerance for high plant populations, breakage of stalk, yield of grain and silage, length of maturity, adaptability to climate, and the ability to fit in with other farm operations. The hybrid that contributes the most to the profitability of the farm operation is the best one to buy.

Preparing the seedbed

Plowing is usually the first step in preparing the seedbed for corn. The purpose of plowing is to cover trash and break up the soil. On land not subject to erosion, plowing may be done in the fall. Otherwise, plowing should be delayed until spring. A satisfactory depth for plowing most soils is 5 to 8 inches. If cornstalks are plowed under, they should first be chopped up by a disk or a special stalk cutter and then thoroughly covered. This helps in controlling European corn borers that are harbored in the old stalks of corn.

Several kinds of implements may be used to work up the soil, after it has been plowed, before the corn is planted. The purpose of working the soil is to get it into a fairly firm, crumbly condition. The disk is commonly used to cut up and stir the surface soil. This is frequently followed by a spike-tooth harrow. Other implements used in some areas are the spring-tooth harrow and the cultipacker. If the different operations in preparing the seedbed are performed at intervals of a few days, many weeds may be killed, thus reducing the labor and costs for weed control during the growing season.

The use of trucks to fertilize land for corn and the use of other heavy machinery in growing and harvesting corn contribute to soil compaction. Recently, for some types of soils, methods have been developed for prepar-

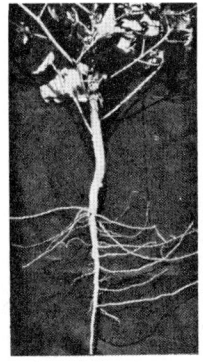

Fig. 17-2. The effects of soil compaction. *(Left)* Pigweed root system failed to grow through compacted soil layer. *(Middle)* In chisel row, cotton roots penetrated. *(Right)* Between chisel points, cotton roots fanned out. (Courtesy, *Agricultural Research*, U.S. Department of Agriculture)

ing the seedbed by "once-over" tillage, in which a special harrow or other implement is pulled behind the plow to pulverize the soil. Other forms of once-over, or minimum tillage, are planting the corn in the wheel tracks of the tractor that pulls the planter so that the soil between the rows of corn does not have to be leveled and worked to remove lumps and attaching both plow and planter to the same tractor so that plowing and planting can be done at the same time. With the plow-plant method, the corn is planted in the center of the turned-over slice of soil, and the area of soil in between the rows of corn is left unpacked. All minimum tillage methods are designed to reduce to a minimum the number of times tractors and machines go over a field to prepare the seedbed and plant the crop. Not only does this procedure reduce compaction of the soil, but it also reduces costs.

At the present time, experimentation is being conducted on zero tillage of corn. In some cases, minimum or zero tillage is known as conservation tillage. With this method, the corn is planted directly in sod that has been killed by herbicides a few weeks before planting time. The yields from the use of zero tillage appear to be about as good as from other methods, and there can be great savings in costs and in the prevention of soil erosion. In one Nebraska experiment, corn grown in sod produced 178 bushels per acre. Careful management is required when this method is used because of the danger from insects and rodents, from herbicide run-off, and to germination of seed from cool, wet spring weather. Since soil run-off is reduced to a small fraction of a ton per acre, however, the risks appear to be worthwhile. A record corn yield of more than 300 bushels per acre was established with conservation tillage by an Illinois farmer in 1975.

With no-til or conservation tillage, some changes are needed in the application of plant nutrients because the nutrients will not be mixed into the soil as they would be with conventional tillage. For example, soil samples for testing should be taken closer to the surface rather than at plow depth. Also, the soil remains cooler because of the mulch cover, which affects fertilization and seed germination.

Planting corn

Corn should be planted after the soil is warm. A soil temperature of at least 55 degrees at planting time is generally recommended. Seeds will not germinate well at lower temperatures and will rot in cold, damp soil.

The recommended time for planting is about 10 days after the average date of the last killing frost in a particular locality. Most corn in the Corn Belt is planted during the first two weeks of May. Progressively earlier dates are suitable for the areas south of the Corn Belt.

Most corn is planted with tractor-drawn planters of two-row to eight-row types. The kernels may be dropped in checkerboard fashion, or hills, so the corn can be cultivated both lengthwise and crosswise. In this case, three or more seeds are usually planted in each hill. The kernels may also be distributed along the rows or two or more seeds may be placed together along the row without regard to cross rowing. Rows are usually spaced 38 to 42 inches apart. Some farmers use wide-row planting, with rows 60 to 80 inches apart, and this practice may increase. In this case, legumes are planted between the rows after the corn reaches about 18 inches in height. The corn yields may be lowered a few bushels by this method, but corn may be grown successfully year after year on some types of soil.

Corn should be planted at a rate that in normal years will produce ears averaging about ½ pound each at harvest. Corn experts emphasize that the plant population should be matched to the fertility of the soil. Under favorable conditions, from 14,000 to 24,000 plants per acre are desirable; but for yields of 80 to 90 bushels per acre, 12,000 to 16,000 plants per acre are usually sufficient. To secure the desired number of plants, the farmer must plant about 10 to 15 per cent more seeds than the expected plant population. In highly fertile soils, an increased rate of planting is desirable; in soil of low fertility, it is less. One bushel of shelled corn will plant about 4 to 8 acres, depending on the rate of planting.

Corn is usually planted 1 to 2 inches deep in most soils. Deep planting should be avoided. The seedbed should be deep and loose, but it should not be overworked. In planting corn, the farmer should also be careful to:

1. Obtain the correct planter plates. Planter plates are different for the various seed sizes and shapes.
2. Check regularly for seed and fertilizer tube stoppage. Stopped tubes can cause big gaps in corn rows and poorly fertilized corn.
3. Prevent the planter from placing fertilizer in contact with seed. Seed placed in direct contact with fertilizer will not grow.
4. Maintain accurate row spacing to facilitate cultivation.

Planting and Cultivating Farm Crops

Table 17-1. Desired number of corn plants, kernels per acre, and distance between kernels in rows

Desired plant population[1]	Kernels per acre needed	Inches between kernels in rows			
		(30-in. rows)	(36-in. rows)	(38-in. rows)	(40-in. rows)
13,000	15,294	—	11.4	10.8	10.3
14,000	16,471	—	10.8	10.0	9.5
15,000	17,647	—	9.9	9.4	8.9
16,000	18,824	—	9.3	8.8	8.3
17,000	20,000	—	8.7	8.3	7.8
18,000	21,176	9.9	8.2	7.8	7.4
20,000	23,529	8.9	7.4	7.0	6.7
22,000	25,882	8.1	6.7	6.4	6.1
24,000	28,235	7.4	6.2	5.9	5.6

[1] Assuming 85 per cent of kernels produced plants.

If a field is likely to be quite weedy, it is a good practice to apply a pre-emergence spray. Pre-emergence spraying for weeds is also good insurance in case cultivation is delayed by wet weather.

Cultivating corn

The chief reason for cultivating is to kill weeds. Some of the newer chemical methods for controlling weeds are being used by some farmers to reduce the number of cultivations required. These methods are discussed in Chapter 18.

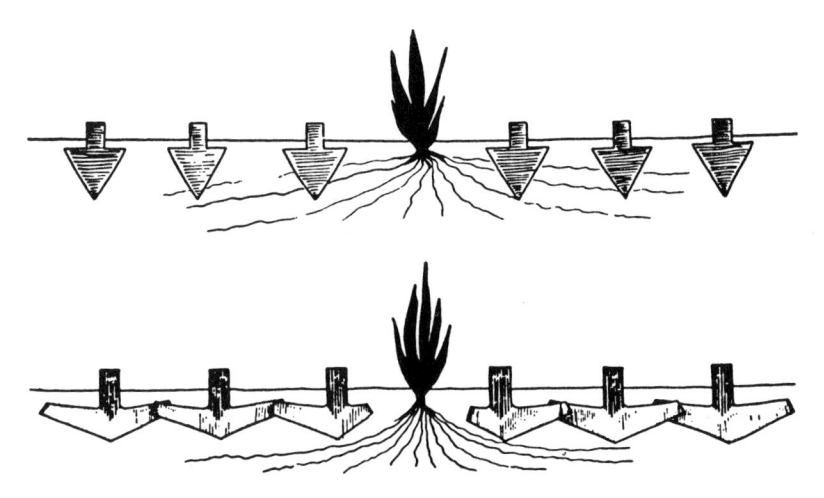

Fig. 17-3. Deep cultivation *(top)* injures corn roots. Shallow cultivation *(bottom)* is best. (Drawings from Zimmerman, Illinois)

A spike-tooth harrow is frequently used to destroy small weeds before the corn plants have emerged. After the corn is several inches tall, a rotary hoe is effective for killing many kinds of weeds. For later cultivations, two-row to eight-row types of tractor-drawn cultivators are frequently used. These are most commonly equipped with small shovels that stir the soil and kill weeds. The first cultivation may be fairly deep and close to the corn plants. Later cultivations should be shallow to avoid injury to corn roots. After the corn plants are about 30 inches tall, no further cultivation is necessary.

Irrigating corn

In some parts of the country, irrigation is needed for corn production. The amount of water needed depends on the location, type of soil, and method of irrigation. Up to 4 acre-feet of water may be required in some places, with the corn being provided water once every 7 or 10 days.

The times when water is most critical to plant growth are (1) when the plant roots are developing, (2) two or three weeks after the plants emerge from the ground when the tassel and ear start to develop, (3) during the pollination period when tassels and silks appear, and (4) during the period of time when the kernels of corn are developing.

Moisture will be needed until the corn is mature. Various signs of maturity, such as the kernels being well dented, should be observed.

Controlling insects and diseases of corn

Some insects and disease of corn are reduced when resistant varieties are planted. Various chemical and cultural methods help to control some insects and diseases of corn, as described in Chapters 19 and 20.

Harvesting and storing corn

Where corn is grown extensively, the most common method of harvesting is the use of mechanical pickers to husk the ears from the stalks in the field. In recent years, some farmers have been using picker-shellers, or corn combines, which pick and shell the ears in one operation. For safe storage in ear form, the moisture content should be below 25 per cent and preferably about 20 per cent. For shelled corn, the moisture content should be 15 per cent or less. Special drying equipment makes it possible to harvest corn with higher moisture content.

To prevent serious losses of corn in the field at harvest time, pickers must be adjusted properly. Operators of corn pickers should practice safety precautions, as serious accidents result from carelessness.

Fig. 17-4. Leaving the residue of previous crops in the surface soil helps reduce soil erosion and improves soil fertility. (Courtesy, John Deere)

Corn in ear or shelled form should be stored in well-ventilated, rodent-proof bins.

In areas where dairy cattle or beef cattle are raised, considerable corn is harvested for silage. In this case, the stalks and ears are best harvested when the kernels are well dented but the stalks are still green. The entire plant is chopped into small pieces and placed in an airtight structure called a silo. The common type is a round structure 12 to 16 feet in diameter and 30 or more feet in height. Trench silos and bunker-type silos are being used in many places. Silage properly made undergoes a partial fermentation process that preserves it for a long time. Moisture control is one of the most critical factors in making high-quality silage. If the moisture level is above 70 per cent, there is apt to be a loss of nutrients from poor fermentation and excessive drainage.

In some areas where small amounts of corn are grown, the stalks are harvested, with the ears attached, and placed in shocks in the fields. After it has dried, this material is called *fodder*. In some cases, the ears are husked out, and the stalks and leaves are shredded by special machines called shredders.

Livestock may also be used to harvest the corn. If livestock are used, the animals are turned into the fields to eat the corn directly from the stalks. When hogs are used, the process is called hogging down corn.

What Are Desirable Cultural Practices for Small Grains?

Wheat is the most widely grown of the small grains. Oats, barley, and rye are important in some areas. Rice, flax, and buckwheat are also classed as small grains.

Providing a fertile soil

Commercial fertilizers are being used rather widely in producing wheat and some other small grains. Fertilizers may be applied at the time of planting or before. For winter wheat, top dressing in early spring with fertilizers high in nitrogen has increased the yields on many soils. In some states, a successful practice is to apply all the fertilizer at the time the winter wheat is planted; a combination grain and fertilizer drill may be used. Under some conditions, nitrogen and other fertilizers also are beneficial if applied to spring wheat and oats at the time of planting. Nitrogen and phosphorus are the nutrients most needed for small grains. In potassium-deficient soils, potash will be needed to increase yields and to improve the ability of the plant to stand. Potash is also useful for forage legumes when these follow the small grain. In some areas, sulfur may also be needed. The college of agriculture in each state provides information on methods of fertilizing small grains.

Using good seed

Farmers in each state should use varieties of small grains that are recommended by the college of agriculture. Many new varieties of small grains that are resistant to some rusts and smut diseases are being developed. Certified seed is desirable, but after farmers get a start with a good variety, they may save their own seed. Seed should be high in germination. Small grain seed should be cleaned with a special fanning mill or seed cleaner to remove light seeds, weed seeds, and trash. Seeds should be treated with chemicals to kill the spores of diseases carried on the grains. Ceresan-M or panogen may be used for this purpose. Further details are given in Chapter 20.

Preparing the seedbed

Practices in preparing the seedbed for small grains vary greatly. In some sections, the land is plowed and worked into condition by disking and harrowing. If a small grain follows a cultivated crop such as corn or soybeans, a common practice is to prepare the soil by disking and harrowing without plowing.

In "dryland" areas, where rainfall is low, wheat is planted only one year out of every two. The practice of summer fallowing is used in alternate years. This consists of periodically working the soil during the summer that no grain is planted. By this means, weeds are destroyed and moisture is conserved for the wheat crop the following year.

Planting and Cultivating Farm Crops 533

Recent minimum tillage practices developed for corn may be adapted to growing winter wheat if additional research supports preliminary findings of the U.S. Department of Agriculture. In the experiment, winter wheat was drilled in cloddy soil. The fields had no seedbed preparation except the plowing of the dry soil in July. The wheat was seeded in the cloddy soil in October. Only slight reductions in yields were found. If this practice proves successful, two to four tillage operations per year can be saved, and the cloddy surface may contribute to efforts to control soil erosion.

Weeds can be troublesome in small grains; however, they can usually be controlled by good farming practices. Destroying the weeds by proper tillage just prior to planting, planting the best varieties, using the proper rate of seeding, and applying fertilizers and appropriate herbicides are good practices to follow.

Planting small grains

The time of planting small grains varies with the crop and the section of the country. In the northern areas, all small grains are planted in the spring of the year. Most of the oats in all areas are planted in the spring, although winter varieties planted in the fall are being used in areas where winters are fairly mild. In the northern Wheat Belt, spring planting of wheat prevails; but in Kansas and many of the surrounding areas that comprise the winter-wheat section, fall planting is the common practice.

Fig. 17-5. Wheat and summer fallow strips are placed alternately against the prevailing winds to prevent erosion in the famous "breadbasket" wheat-growing area in Montana. (Courtesy, Soil Conservation Service, U.S. Department of Agriculture)

Spring grains should be planted as early as possible. Late planted grain is likely to be damaged by hot weather at the time the seeds are forming. Planting can be done about as early as the soil is in condition for working. Small grains are not damaged by moderate frosts.

Fall planting of small grains, such as winter wheat, should be delayed until Hessian flies are least likely to be harmful. The safe dates for planting are determined by the college of agriculture in each state.

Small grains may be broadcast by special types of seeders and covered by a harrow. Usually, however, a grain drill is preferable, especially for wheat, because less seed is needed and the seeds are placed at a uniform depth.

The amount of seed to plant per acre varies with the kind of grain, the fertility of the soil, and the amount of rainfall. Under most conditions, about 2 to 3 bushels of oats are recommended. The amounts of wheat to plant vary from about ¾ of a bushel in the dry areas to 2 bushels in the more humid areas. The rates for rye, flax, and buckwheat vary from about ½ to 1 bushel or more, depending on the area.

For oats, the rate per acre varies from 2 bushels to 3½ bushels, depending on the method of planting.

From 125 to 150 pounds of rice is used per acre. On some farms, rice is seeded by airplanes, which makes it possible to plant fields covered with water.

A good rule to follow in determining depth of planting small grains is to place the seed just deep enough to be in moist soil. The usual depth is about 1 inch or less.

Controlling insects and diseases of small grains

Insects and disease are controlled by careful selection of resistant varieties and by the use of good cultural practices to establish vigorous growth. Treatment of seed prior to planting is a common practice. Chemical control of insects is sometimes effective when heavy infestations occur early in the growing season.

Harvesting and storing small grains

Small grains are ready to harvest when the kernels are in the hard-dough stage and the grain is dry enough to store without spoiling. A moisture content of 14 per cent or less is generally considered a safe level for storing grains. For harvesting small grains, wide use is made of combines, which cut and thresh the grain in one operation. In some cases, the plants are cut and placed in windrows to dry. Following this, combines are used to pick up the cut grain and thresh it. In some parts of the country, grain binders are still used. These machines cut the grain and bind it into bundles,

Fig. 17-6. The grain flows from the combines into the waiting trucks. (Courtesy, Division of Extension, Kansas State University)

which are placed in shocks. Later the grain is removed from the straw by a threshing machine. Rice fields are drained prior to harvest, and combines are commonly used for harvesting the grain.

What Are Desirable Cultural Practices for Soybeans?

Soybeans are an important crop in many states of the Middle West and in some other states.

Providing a fertile soil

Soybeans are grown in various rotations with other crops. A fairly common four-year rotation is corn, soybeans, small grain, and grass-legume. On very fertile, level soil, corn and soybeans may be grown in alternate years or in a rotation of corn, corn, soybeans, and wheat seeded to a legume.

Soybeans do best in a fertile soil. In most cases, fertilizers are applied to other crops in the rotation. However, on some soils, soybeans respond to fertilizers added at the time of planting. One method is to add the fertilizer in a band along each row.

Using good seed

Farmers should choose varieties of soybeans recommended for their area. They should secure seed from reliable dealers. Preferably, if they are adopting a new variety, they should purchase certified seed. After that, if they exercise care, they may use seed grown on their own farms. Before seed is planted, it should be inoculated, thus insuring that the proper bacteria will be present to supply nitrogen to the plants. These bacteria establish themselves in nodules on the roots of the plants. Commercial cultures for inoculating soybeans can be purchased; these should be applied to the seed in accordance with directions on the package.

Preparing the seedbed

The seedbed for soybeans should be prepared in about the same way as that for corn. Soybeans are also being grown using zero or conservation tillage. In one study, soil losses were reduced from over 13 tons per acre to less than $2/10$ of a ton per acre when the soybeans were planted directly in sod that had been killed by a herbicide.

Planting soybeans

Usually soybeans are planted in rows 10 to 36 inches apart. About 40 to 60 pounds of seed is recommended per acre if seeded in rows. A grain drill, corn planter, or special soybean planter may be used for planting. A grain drill may be used by closing some of the openings to space the rows at desired widths. Soybeans may be planted at about the same time as corn, or a little later.

Some farmers are using a planting practice called solid seeding. This type of planting involves planting with a grain drill and not cultivating the growing crop. It is best to select fields that are relatively weed free, in addition to using herbicides, for effective weed control. The use of adaptable varieties and the use of good cultural practices are necessary for profitable yields with this practice.

Cultivating soybeans

When bean plants are small, but after leaves have formed, weeds may be controlled with a rotary hoe, special weeder, or spike-tooth harrow. This is usually followed later by one or more cultivations with cultivators of the same type as used for corn.

Because soybeans can be more easily injured by chemical weed killers than most other crops, the use of chemicals for weed control in soybeans is subject to quite rigid directions. The best control of weeds in soybeans is

Planting and Cultivating Farm Crops 537

Fig. 17-7. For the best growth of soybeans, the seeds should be inoculated with a pure culture of a special kind of beneficial bacteria, as shown above. These bacteria develop on the roots of the plants and remove nitrogen from the air, which is used by the soybean plants. (Courtesy, U.S. Department of Agriculture)

early preparation of the seedbed to permit weed seeds to germinate so the weeds can be destroyed when the soil is worked prior to planting time. Often, two or more weed crops can be destroyed by this method before planting.

Controlling insects and diseases of soybeans

Farmers control insects and disease largely by using approved cultural practices and by selecting resistant varieties. Insecticides can be beneficial when infestations become very severe.

Harvesting and storing soybeans

Most soybeans are grown for the grain portions. A grain combine is commonly used for harvesting. At the time of harvest, the beans should have a moisture content of 12 per cent or less for safe storage. Usually, most of the leaves have fallen by the time the crop is ready for harvesting. In harvesting, farmers should try to prevent excessive shattering of the beans. The beans should be stored in well-ventilated, rodent-proof bins.

In order to hasten harvest, farmers sometimes apply chemicals called *defoliants* to fields of soybeans a short time before harvest. These materials kill weeds and cause the leaves of the soybean plants to drop off, thus making harvest easier.

Some soybeans are cut for hay. Usually, for this purpose, the beans are planted thickly. The plants are cut and allowed to cure like hay crops.

What Are Desirable Cultural Practices for Forage Crops?

Many crops grown on farms and ranches are commonly called forage crops. These are crops in which most of the portions of the plants above ground are used for feeding livestock. These include pasture crops, hay crops, and silage crops. Various grasses and legumes are the chief forage crops grown on farms and ranches in the United States.

Forage crops are widely grown in the United States. The grasslands, haylands, and forested rangelands of the United States cover more than a billion acres. This is over 50 per cent of the total land area. More than half of the farms and ranches in the United States depend primarily on grasslands for feed for livestock.

Selecting kinds and varieties of forage crops

Many kinds and varieties of grasses and legumes are available for the various purposes and for the differing conditions of climate and soil. The kinds and varieties to grow should be based on the recommendations of the college of agriculture in each state.

Alfalfa is one of the most widely grown legumes in the United States. While some alfalfa is grown in every state, it is more commonly found in the central and north central states and irrigated areas of the western states. Alfalfa is used primarily for hay and to some extent for pasture. For the northern states, winter-hardy varieties should be secured. Varieties that are hardy and resistant to bacterial wilt are Ranger, Buffalo, and Vernal.

Many kinds of clovers are grown in various parts of the United States. Red clover is grown widely in the north central and northeastern states.

Planting and Cultivating Farm Crops

Kenland, a variety that is resistant to anthracnose, is adapted to the southern portions of the area in which red clover is grown. Alsike, ladino, and white clovers are grown in many states. In the southern states, crimson clover is commonly grown. Other legumes grown for hay and pasture are sweet clover, bird's-foot trefoil, and lespedeza. Kudzu and sericea lespedeza are grown in some of the southern states.

Timothy is one of the widely grown grasses for hay. Kentucky bluegrass is a widely used pasture crop in the cool, humid areas of the United States. Other grasses for these areas include bromegrass, fescue, and red top. Dallisgrass, Coastal Bermuda grass, and carpetgrass are among the kinds used in the southern states.

In the grazing areas of the Great Plains, blue grama grass and buffalograss are widely distributed. Crested wheatgrass is frequently found in the northern areas and bluestem grasses in the southern portions.

Mixtures of legumes and grasses are commonly used for hay and pasture. These vary widely according to whether a short-term or a long-term crop is desired and according to the climate. Kentucky bluegrass and white clover are widely used in combination for pasture. Another pasture combination for some areas is ladino clover and orchardgrass or tall fescue. Alfalfa and bromegrass are frequently seeded together. Red clover, ladino clover, alsike clover, and timothy are frequently grown together for hay and pasture crops. Many other combinations are used.

Many new varieties of grasses and legumes are being developed by plant breeders. Grass and legume seed should be secured from reliable

Fig. 17-8. Mower-conditioners have helped to improve the quality of hay for feeding livestock. (Courtesy, Sperry New Holland)

dealers, and the bags should carry labels that show purity, percentage of weed seed, percentage of germination, and other information. It is particularly important to secure a supply of seed free from seeds of serious weeds. Certified seed is preferable even though it costs more.

Before they are planted, seeds of alfalfa, sweet clover, and some other legumes should be inoculated with beneficial bacteria that aid the plants in obtaining nitrogen from the air. It is important to secure the correct kind of bacteria for each legume. Bacteria in commercial cultures should be purchased, and these should be applied to the seeds in accordance with directions on the packages in which the cultures are purchased.

Providing a fertile soil

A fertile soil is needed for high yields of forage crops. In many farming areas, legumes and grasses are rotated with corn and small grain crops to which commercial fertilizers are frequently applied. These fertilizers help the legumes and grasses, as well as the other crops grown in the rotation. Fields kept in pasture or hay crops for several years may improve because of additional fertilizer applied on the surface. This is commonly called top dressing. Grasses frequently benefit from fertilizers high in nitrogen.

Some legumes, such as alfalfa and sweet clover, need a soil that is free from acidity. Soils that are to be planted to these crops should be tested and lime should be added in the amounts needed for correcting the acidity of the soil.

Preparing the seedbed and planting forage crops

The seedbed for legumes and grasses should be firm, but the soil should be fine and crumbly on top. In many areas, legumes and grasses are planted with a small-grain crop. This crop shades the tiny seedlings of the grasses and legumes and checks the growth of weeds. Because of this protection, such a crop is frequently called a *nurse crop*. The small grain is harvested and thus provides a crop the year the forage crops are getting started.

The seeds of most legumes and grasses used for hay and pasture are small and hence should be planted shallow. Usually, a depth of ½ inch or less is most satisfactory.

A special type of seeder or a special attachment on a grain drill is commonly used for planting legumes and grasses. In some cases, especially on the range, airplanes are used for scattering these seeds.

Harvesting and storing forage crops

Large amounts of forage crops are cut for hay in some areas. The crops used for hay should be cut at a time when the feed will be most nutritious

Planting and Cultivating Farm Crops

Fig. 17-9. A Prince Georges County, Maryland, farmer topping tobacco. (Courtesy, U.S. Department of Agriculture)

for livestock. Red clover should be cut at about the half-bloom stage, and most of the grasses when the heads appear but before they bloom.

The first step in making hay is to mow the crop. After the crop has been mowed, various methods are used to cure the hay so that it will keep properly in storage. A common method is to rake the hay into windrows after it has partially dried. After it has dried sufficiently, the hay may be loaded onto hay racks and hauled to a barn where it is elevated into the haymow. In recent years, an increasing amount of hay has been baled in the field and stored in this form. In the western states, much of the hay is placed in large stacks in the fields where it is grown. Modern machinery has reduced much of the hard labor formerly required in handling hay.

Hay dryers for drying hay artificially are used by some farmers. One type has flues installed in the haymow beneath the hay. Air is forced through the flues and surrounding hay by large fans operated by motors.

Some hay is stored in chopped form. In this case, the hay is chopped after it has dried, and it is then placed into the storage mows by blowers or elevators.

In order to produce good-quality hay, the crops must be handled in ways that will retain the leaves and prevent severe damage from weather. One of the bugbears of haying is rainy weather; hence the old proverb, "Make hay while the sun shines."

Some grasses and legumes are cut green and made into silage. Some are also cut green and hauled to livestock to eat. When silage is to be made, it is especially important that the crop be harvested when the plants will provide the greatest amount of nutrients.

Using pastures effectively

Many fields throughout the United States and most range areas are kept permanently in pasture crops. This is done because the land is hilly and subject to erosion if planted to cultivated crops, or the rainfall is too limited for other crops, or for other reasons pasture crops represent the best use of the land.

Good pasture will frequently provide more feed nutrients per acre and at a lower cost than will harvested crops. In some studies, farms that had one-half of the land in pasture and one-half in crops made more profit than did farms with one-fourth the land in pasture and three-fourths in crops.

Pasturelands should be handled in ways that will provide the most feed for livestock. Much damage to pastures results if they are overgrazed. The number of livestock should be limited to the amount of feed produced. In some of the more fertile areas where rainfall is abundant, an average of one or more cows may be grazed per acre, while some of the range areas with low rainfall will support only one cow for as many as 100 acres or more. In some areas, pastures are grazed alternately by livestock moved from field to field to permit the grazed areas to recover. On some farms, pasture crops are cut green and hauled to cattle or other livestock. This method requires considerable labor, but larger amounts of feed are produced than if the pastures were grazed by livestock.

USDA researchers have found that small grain pastures can be used to fatten steers. They discovered that the small grain pastures can best be utilized by combined grazing and drylot feeding. This resulted in lower feed costs than did feedlot-only fattening. Animals fattened on pastures and in the feedlot gained as fast or faster, and their finish grades equalled those of animals fattened only in the feedlot. Oats planted only for grazing should be seeded about twice as heavily and should be fertilized more liberally than oats planted for grain.

On some farms and ranches in the United States, a series of fields planted to different pasture crops is used in order to lengthen the pasture season. In some areas where water is available from wells, streams, or other sources, irrigation is used to supplement the moisture provided by rains.

Permanent pastures may be improved by reseeding and other practices. This process is frequently called renovation. The steps commonly

Planting and Cultivating Farm Crops

Fig. 17-10. A hay crop is being harvested on a grass waterway that prevents erosion when storm water runs off the land. (Courtesy, Soil Conservation Service, U.S. Department of Agriculture)

Fig. 17-11. Combines make quick work of harvesting grass seed. (Courtesy, Sperry New Holland)

include (1) working up the surface soil, (2) applying lime and fertilizer in accordance with needs as shown by soil tests, and (3) reseeding with appropriate varieties. Fertilizers will increase yields on most established pastures. In many cases, clipping established pastures with mowers is desirable to destroy weeds before they go to seed. In areas where sagebrush and other shrubs interfere, spraying with suitable chemicals may destroy them.

It is not considered good practice to seed land to be used as permanent pasture to one kind of plant. A mixture of several kinds of grasses and legumes will usually provide a uniform stand of plants, fairly uniform production through the growing season, and high production for existing conditions.

By using good methods of handling pastures, farmers may secure increased returns in terms of meat, milk, or wool produced by the livestock that feed on the pasture.

What Are Desirable Cultural Practices for Potatoes?

White potatoes are grown in many parts of the United States. In some states large acreages are grown for commercial purposes. In addition, many farmers and other people produce potatoes for home use.

Providing a fertile soil

The soil for potatoes should be high in fertility. Commercial fertilizers are used extensively to supply plant nutrients needed for high yields. Some fertilizer is usually plowed under and some placed in bands along the rows at the time the potatoes are planted.

In order for a mellow soil to be maintained, considerable organic matter should be added to soil for potatoes. Plowing under green manure crops will accomplish this.

Selecting and preparing potatoes for planting

The tubers of potatoes are planted, and these are usually called "seed" potatoes. Certified seed potatoes are preferable. Since these are produced under the close inspection of experts, they are true to variety and free from diseases likely to be carried by the tubers. Certified seed potatoes are usually purchased in bags with labels that indicate the certification and variety. Varieties recommended for the area should be secured. Some varieties commonly grown in various parts of the United States are Katahdin, Chippewa, Irish Cobbler, Triumph, Green Mountain, Russet Rural, Sebago, Pontiac, and Warba. Some of these varieties are resistant to scab.

Seed potatoes should be treated with a chemical if there is evidence of scab and black scurf. The proper materials may be purchased, and they should be applied in accordance with directions on their containers.

Before potatoes are planted, they should be cut into pieces about the size of a hen's egg, or larger. The pieces should be blocky in shape, with two or more eyes in each piece. Potatoes may be cut into pieces with a knife or with a special machine that cuts them rapidly.

Preparing the seedbed

The seedbed for potatoes should be deep and mellow. The first step is deep plowing. Before the tubers are planted, the soil should be worked several times to kill as many weeds as possible and to provide a moderately fine and firm seedbed.

Planting potatoes

The time for planting potatoes varies according to the section of the country. In the northern states, potatoes are planted during the spring and early summer, and in the southern states, during the summer months.

Machine planters are used where potatoes are grown on a large scale. From 10 to 20 bushels, or more in some cases, are planted per acre, depending on the fertility of the soil and moisture conditions. Commonly, potatoes are placed in rows 33 to 36 inches apart and spaced at intervals of a foot or more in the rows.

Cultivating potatoes

After the potatoes have been planted, a spike-tooth harrow or a special weeder may be used to destroy small weeds. This method of cultivation may be utilized before the plants are up and until they are about 4 inches tall.

After the plants reach a height of a few inches, the rows are cultivated. The first cultivation may be deep and close to the plants, but later cultivations should be shallow to prevent damage to roots. Cultivation should be done frequently enough to control weeds. In some cases, ridge cultivation is practiced. This means the cultivators are adjusted to move the soil toward the rows with a ridging or hilling effect so that the plants are higher than the spaces between the rows.

Irrigating potatoes

Potatoes require a large amount of moisture to produce high yields. In some areas, practically all the moisture is supplied by irrigation. In some other areas where potatoes are produced commercially, irrigation is used to supplement the water provided by rains.

Controlling diseases and insects of potatoes

Potato plants must be sprayed several times to control diseases and insect pests. Bordeaux or copper materials are commonly used for diseases, and malathion for insects.

Harvesting and storing potatoes

Potatoes are harvested some place in the United States during every month of the year. The major harvest season in the northern states is in the fall months. Mechanical potato diggers save labor where potatoes are grown on a commercial scale. The potatoes are ready for harvest when the vines wither and the peelings on the tubers are not easily rubbed off.

In handling potatoes, individuals should try to prevent bruising them. Potatoes should not be in sunlight for very long, as this causes sunburn, which is undesirable. Potatoes are usually graded into fairly uniform sizes before they are placed on the market. Here again, mechanical equipment has greatly reduced the labor involved. Potatoes should be stored in dark bins at about 40°F. The surrounding air should be quite moist to prevent shrinkage of the potatoes.

Fig. 17-12. An agricultural engineer programs a weather station computer for solar radiation, wind, air temperature, and humidity readout. The computer processes the information in order to calculate and deliver the right amount of irrigation water to a field. (Courtesy, *Agricultural Research,* U.S. Department of Agriculture)

What Are Desirable Cultural Practices for Sorghums?

Grain sorghums will grow under limited rainfall and high summer temperatures. With recent improvements in varieties, many farmers are deciding to grow sorghums, especially on their less productive fields and as a second crop.

Selecting varieties of sorghums

It is important that the best adapted varieties be selected for the particular conditions existing in each locality. Factors to consider are length of growing season required, resistance to bird damage, color, type of head, distance from the flag leaf to the base of the head, and resistance to disease. Hybrid varieties are the most popular.

Providing a fertile soil

Sorghum fertilizer requirements are similar to those for corn, although sorghums will yield better than corn on less fertile soils. Fertilizer applications should be placed so that there is no direct contact between the fertilizer and the seeds or roots, since sorghums are particularly sensitive to damage from such contact. Nitrogen, phosphorus, and potassium are likely to be needed, while iron, zinc, sulfur, and other elements may occasionally be needed.

Preparing the seedbed and planting sorghums

Sorghums require a fairly fine seedbed for adequate contact between the seed and the soil. Seeds should be planted about an inch deep. The soil temperature should be about 65 degrees at planting time. The most common row spacings are 20- and 30-inch rows for a population of 60,000 to 120,000 plants per acre.

Irrigating sorghums

Sorghums are heavy users of water. Where irrigation is needed, application rates for water will range from about 20 to over 50 acre-inches per acre. For maximum yields, an adequate supply of moisture is needed throughout the entire growing season. The leaves show signs of a lack of moisture by wilting, rolling up, twisting, and turning gray. The leaves may also become more erect and may develop brown edges.

Controlling weeds in sorghums

Good seedbed preparation is necessary for weed control. A rotary hoe can be used with some success when the weeds are small and not yet established. Some herbicides may be applied before and after the plants emerge from the ground. It is not recommended that grain sorghums be planted in fields where Johnson-grass infestations are heavy.

Controlling insects and diseases of sorghums

Insects can be a serious problem. They are controlled by good cultural practices and by insecticides. Diseases are controlled by selection of resistant varieties, by treatment of the seed before planting, and by use of rotation controls.

Harvesting and storing sorghums

The stalks and leaves of many hybrids may still be green when the grain is ready for harvest. In some states, sorghum plants remain alive until killed by frost, thus creating moisture problems at harvest time. A combine is used for harvesting. For storage, the moisture content should be 14 per cent or less.

SUGGESTED ACTIVITIES

1. For a period of several days, study the weather forecasts given in newspapers or over the radio and on television. In what ways might these forecasts be used by farmers who raise crops?
2. Describe the methods used by a person in your community who has been especially successful in raising some crop. What methods are used for improving the soil, securing seeds, preparing the seedbed, planting, and harvesting?
3. From bulletins published by the college of agriculture in your state, and from other reliable sources, secure the names of the recommended varieties of the kinds of crops commonly grown in your community.
4. Test some seeds from clover, small grains, or grasses for germination. Count off 100 seeds, taking them as they come. Place them between two pieces of blotting paper and then moisten. Put a plate on each side of the blotting paper. Keep in a warm place. After a few days, inspect the germinator to see if more moisture is needed. At the end of 7 to 10 days, count the number of seeds that have sprouted and calculate the percentage of germination.
5. What experiences have you had in raising crops? How successful were you? Since reading this chapter, what are some ways in which your methods could have been improved?
6. In agricultural reference books, study the methods recommended for raising some crop commonly produced in your community.

7. Concepts for discussion:
   a. The culture of farm crops consists of the practices used to grow them.
   b. Soil compaction is a problem in plant growth.
   c. Weather conditions affect crop production.
   d. High yields are made possible through crop rotations.
   e. Good seed is essential to high yields.
   f. Different crops require different fertilization practices.
   g. High yields of some crops can be obtained with zero tillage, as well as with well-prepared seedbeds.
   h. Proper soil temperature is needed for seed germination.
   i. Different plants require different planting times.
   j. Fertilizer placement affects plant growth.
   k. Depth of planting the seed affects plant growth.
   l. Amount of seed used per acre affects yields per acre.
   m. The chief reason for cultivating is to kill weeds.
   n. The timing of harvesting is important for each crop.
   o. Crops may be harvested and stored in many ways.
   p. Disease and insect damage may be controlled to a degree by proper cultural practices.
   q. Summer fallowing is a means of preserving moisture.
   r. For best results, legume seeds should be inoculated with the proper bacteria before they are planted.
   s. Weeds may be controlled with chemicals.

REFERENCES

Boone, L. V. *Producing Farm Crops*. Interstate Publishers, Inc., Danville, IL 61832.

Kearney, T. E., et al. *Field Corn Production in California*. Cooperative Extension Leaflet 21163, Division of Agricultural Sciences University of California, Berkeley, CA 94720.

Swaider, J. M., G. W. Ware, and J. P. McCollum. *Producing Vegetable Crops*. Interstate Publishers, Inc., Danville, IL 61832.

U.S. Department of Agriculture. *New Crops, New Uses, New Markets*. Yearbook of

Agriculture: 1992, U.S. Government Printing Office, Washington, DC 20402.

Vocational Agriculture Service. *Growing Grain Sorghums*. VAS 4055, Vocational Agriculture Service, College of Agriculture, University of Illinois, Urbana, IL 61801.

Vocational Agriculture Service. *Growing Oats*. VAS 4023b, Vocational Agriculture Service, College of Agriculture, University of Illinois, Urbana, IL 61801.

Vocational Agriculture Service. *Improving Permanent Pastures*. VAS 4014, Vocational Agriculture Service, College of Agriculture, University of Illinois, Urbana, IL 61801.

Vocational Agriculture Service. *Soybean Production*. VAS 4033a, Vocational Agriculture Service, College of Agriculture, University of Illinois, Urbana, IL 61801.

Chapter 18

Marketing Farm Products

The marketing of farm products is an extremely important part of a farmer's total business operation. The profit in farming can disappear quickly if a product that is ready for sale is not sold or is sold at too low a price. It is very expensive to keep feeding and caring for animals beyond the time when they are ready for sale. The animals may actually become less desirable when fed for the additional time and, in addition, feeding and caring for them uses time and facilities that would be used to better advantage for new animals. For products that must be stored, failure to sell at the right time adds to storage costs and may make it necessary for farmers to borrow money at high interest rates to pay for continuing operations. When the commodities to be marketed are perishable, as are fruits and vegetables, it is even more important to have good plans for marketing to avoid losses of the product or sales at low prices. Thus, farmers must be as efficient in marketing crops and livestock products as they are in producing them. Consumers, too, are interested in the marketing and distribution of farm products and particularly in the prices they pay for them.

OBJECTIVES

1. Define marketing farm products.
2. Discuss what happens to agricultural products after they leave the farm or ranch.
3. Discuss some of the reasons prices for products go up and down.
4. Identify ways consumers are protected in the marketing and distribution of farm products.

What Is Meant by Marketing Farm Products?

Farmers frequently speak of "marketing their wheat" when they take it to a local elevator and get paid for it. Likewise, homemakers refer to doing

Fig. 21-1. The final step in the marketing of food products takes place when the various processed foods are sold to the consumer in the grocery store. (Courtesy, U.S. Department of Agriculture)

marketing when they buy bread and other articles of food at the grocery store. In reality, these activities are only small parts of the total marketing process. Wheat harvested by Farmer Brown must travel a long route and go through several changes before it appears as a loaf of bread which a consumer purchases from the grocery store. Wool shorn from sheep on a farm or ranch passes through many processes before it is purchased in an article of woolen clothing.

Marketing of farm products includes all the processes from the time these commodities leave the farms until they reach the consumers. Thus, marketing is a complicated process for most products raised on farms.

Why Is Marketing Important to Farmers?

Marketing is important to farmers because they want to sell their products at the best possible prices. Basically, by decreasing the cost of production, by marketing more effectively, by producing a better product, and by doing some processing, or by combining efficient production and effective marketing, farmers may increase their net profits from given amounts of products.

Most farmers in early pioneer days were not much concerned about marketing the products they raised. At that time, most of the people in the

Marketing Farm Products

United States lived on farms and produced primarily for their own families, with very little to sell. Much of the processing of farm products was done on the farms where these products were raised or by nearby processing plants, such as small water-powered flour mills and woolen mills. Farmers traded or bartered some products at local stores for essentials that could not be produced on their own farms. Some pioneer farmers were able to produce and sell for cash products such as wheat, wool, butter, maple sugar, cotton, and meat. In most cases, farmers bartered or sold their products directly to local consumers, merchants, and small industries.

Present-day farming is quite different from pioneer farming. Only a small percentage of the total population is engaged in farming. Farmers today produce large quantities of products for sale and only small amounts for their own use.

Farmers are concerned about marketing their products because the prices they receive fluctuate constantly and often widely. Farmers have little control over the prices they receive. Even though surpluses occur in some farm products, it is often impractical for farmers to decrease production because they have many fixed expenses to meet. Farmers cannot save money by discharging vast numbers of workers, as industry can, because farm workers consist largely of the farmers themselves and members of their families. Furthermore, farmers find it difficult to switch to the production of other kinds of crops or livestock, because their farms are not suited to them or are not equipped to produce them.

Components of the Farm-Food Marketing Bill

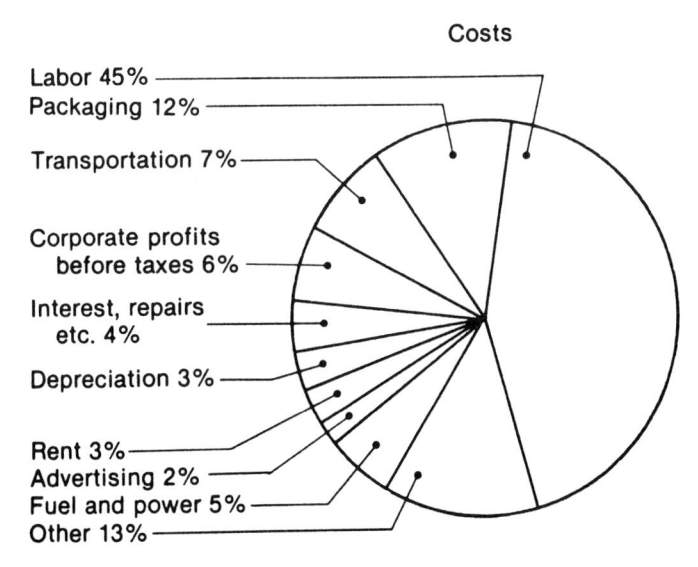

Costs

- Labor 45%
- Packaging 12%
- Transportation 7%
- Corporate profits before taxes 6%
- Interest, repairs etc. 4%
- Depreciation 3%
- Rent 3%
- Advertising 2%
- Fuel and power 5%
- Other 13%

Fig. 21-2. Labor is the largest single item of cost in the marketing of farm products. (Courtesy, U.S. Department of Agriculture)

Farm Share of Retail Food Prices

	Percent
Eggs	66
Poultry	52
Dairy products	51
Meat products	49
Average for market basket of farm foods	35
Fresh fruits and vegetables	29
Fats and oils	27
Processed fruits and vegetables	19
Bakery and cereal products	14

1981 data. Based on the payment to farmers for the farm products equivalent to foods in the market basket and the retail price.

Components of Retail Food Price

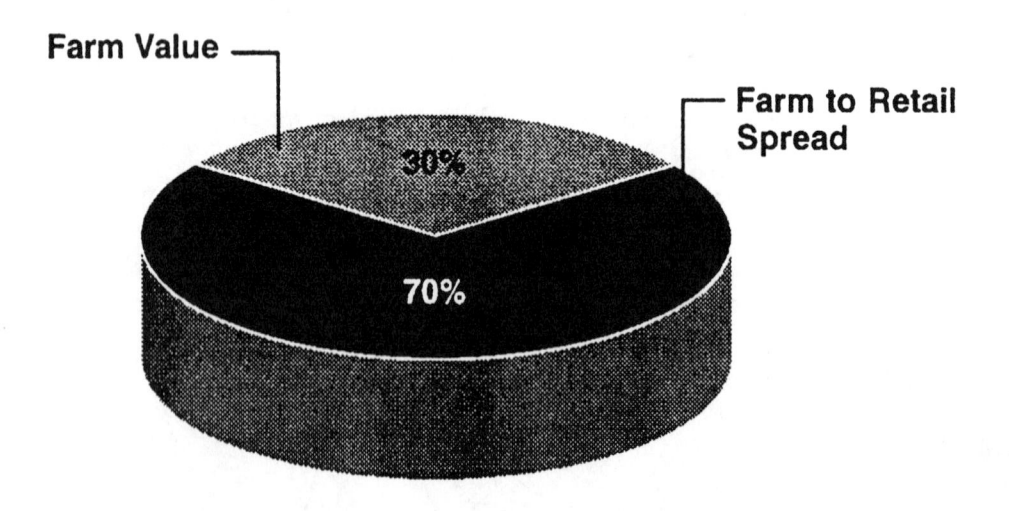

Farm Value — 30%

Farm to Retail Spread — 70%

Fig. 21-3. The farm share of retail food prices averaged 30 cents in 1992. (Courtesy, U.S. Department of Agriculture)

If farmers understand marketing, they may use this knowledge to some extent in planning their farming operations. For example, they may plan the production of hogs so as to place them on the market during seasons in which prices are likely to be most favorable. Farmers may make some reductions in the production of certain crops and livestock if prices are likely to be low, and they may increase somewhat the production of other crops and livestock for which prices are likely to be favorable. However, prices cannot be predicted accurately. Another way in which farmers may improve the marketing of their products is to change their methods of marketing, as discussed later in this chapter.

The prosperity of farmers and the prosperity of urban people tend to rise and fall together, although at times one group or the other may have an advantage. When farmers receive a fair return for the products they raise, this helps the United States to continue to be a strong nation.

Why Is the Marketing of Farm Products Important to Consumers?

The final step in marketing farm products occurs when consumers buy foods and other articles made from the raw materials produced on farms. Consumers are interested in getting the kinds of products they want at prices they consider reasonable. Many consumers wonder why these products cost so much and why the prices often show little change when farmers' prices go down.

Consumers need to understand the various steps in the marketing process and how these affect the prices they pay for food and other items made from agricultural products. Furthermore, they should learn how to choose products wisely and thus get the most for the money they spend.

Obviously, farmers want to obtain fair prices for the products they sell. Farmers are also interested in consumer prices that will encourage consumers to buy ample quantities of farm products. Farmers, too, are consumers of many agricultural products; hence, they have an interest in buying them at reasonable retail prices.

Consumers of bread do not buy wheat. They usually buy bread in sliced form in cellophane wrappers. Hence, they are buying the services of millers, bakers, distributors, etc., and paying for other marketing expenses, such as the cost of transportation at various stages from the farm to the grocery counter. Likewise, consumers do not buy cattle and hogs. They buy a part of an animal after it has been transported, slaughtered, processed, cut into pieces, and delivered to a retail store. Thus, consumers of agricultural products buy many services. The trend is toward more and more of these services, largely because consumers are willing to pay the additional prices in order to secure products in desired forms.

Population and Food Consumption

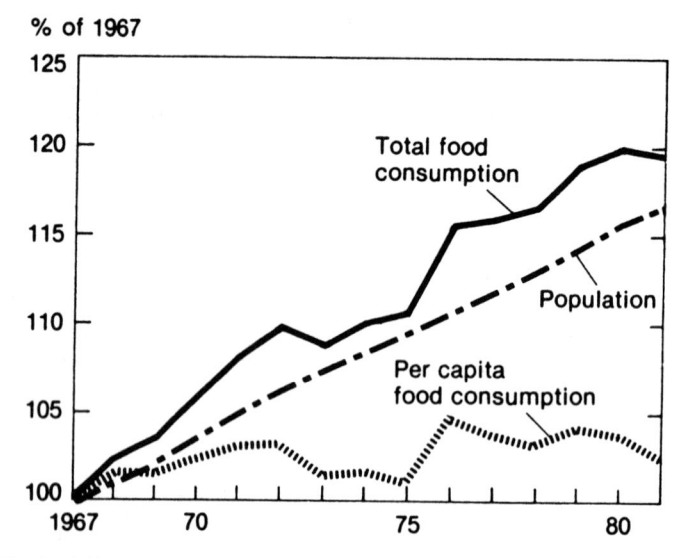

Fig. 21-4. Total food consumption rose with population increases, while per capita food consumption changed only slightly. (Courtesy, U.S. Department of Agriculture)

Marketing services add considerably to the prices of products after they leave the farms. An indication of this is the amount that farmers receive from each dollar spent by wage-earning consumers for a typical market basket of food. During 1955, the farmer's share of this food dollar was 41 cents, and marketing charges accounted for the remaining 59 cents. In 1962, the farmer's share of the food dollar was 38 cents, and the marketing charges were 62 cents. From 1977 to 1991, the farmer's share of the food dollar ranged from 27 to 37 cents, and the marketing charges ranged from 63 to 73 cents. In 1991, the farmer's share was 27 cents, and the marketing charges were 73 cents.

The farm share of retail food prices varied, depending on the product. In 1991 the share ranged from a high of 42 per cent for meat to a low of 7 per cent for bakery and cereal products. People in the United States spend only about 14.4 per cent of their personal income on food, while in Great Britain the percentage is 27.5; in India, 62.5; and in Russia, over 50.0.

Many people want foods that are ready to be used or that may be prepared with very little work in their homes. Examples are breakfast cereals, cake mixtures, precooked and packaged meats, frozen dinners that require only heating before they are eaten, bakery products, and canned foods of many kinds. These services are probably desirable, but consumers should recognize what each dollar they spend is buying. As the amount of processing done to foods after they leave the farm increases, the farmer's share of the dollar spent for the food decreases.

Marketing Farm Products

The cost of meat products is partly determined by the quantity and quality of edible meat that comes from each animal. For example, only about 60 per cent of a choice, 1,000-pound steer is edible meat, and only a small portion of the carcass consists of the highest-priced steaks. Much of the meat from such a steer is sold at considerably less per pound than the steaks that many consumers prefer. The actual weights of some of the cuts of retail meat from a 1,000-pound steer are about as follows:

   150 pounds of hamburger
   90 pounds of chuck roast
   70 pounds of round steak
   50 pounds of sirloin steak
   40 pounds of rib roast
   30 pounds of T-bone, club, and porterhouse steak
   20 pounds of rump roast

By most measures used in making judgments regarding food costs, food is a "good buy." The reason for this is that an hour of factory labor will buy so much more than it used to buy. For a comparable basket of food, the average wage earner in 1930 worked 118 minutes; in 1940, 107 minutes; in 1950, 94 minutes; in 1960, 73 minutes; in 1970, 58 minutes; and in 1980, 61 minutes. Thus, the average wage earner in 1930 worked almost twice as long as did the average wage earner in 1980.

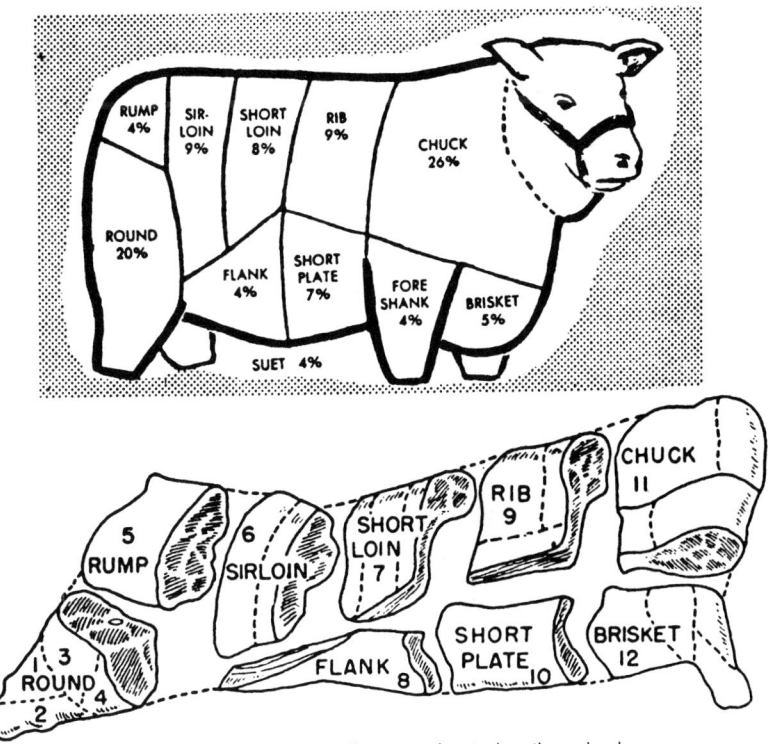

Fig. 21-5. Where the retail cuts are located on the animal.

What Happens to Agricultural Products After They Leave the Farms and Ranches?

It is a long way from a cattle ranch near Ten Sleep, Wyoming, to a butcher shop in New York City or from a hog farm near Grundy Center, Iowa, to a supermarket meat counter in Boston. Most of the meats, vegetables, fruits, and other foods consumed by urban families were transported many miles, processed in various ways, and handled by many people. These services make it possible for people everywhere to obtain crisp lettuce and other fresh vegetables and many kinds of fresh fruits throughout the year. These services are important for farmers and ranchers who produce foods and other raw materials, as well as for the persons who consume them. Without them, farmers and ranchers would find it difficult to market their products, and consumers would have difficulty in getting the products they want when they want them.

Fig. 21-6. The District of Columbia Open-Air Farmers' Market provides farmers with a direct market outlet to consumers. (Courtesy, *Agricultural Research*, U.S. Department of Agriculture)

Sometimes food products may be purchased directly from the persons who produced them. Some farm families operate roadside markets where they sell their fruits, vegetables, honey, maple syrup, and other food items. In some cities, there are farmers markets where farmers take their products and sell them directly to consumers. These are simple methods of marketing that are not very common today.

The marketing of farm products involves many establishments and many people. In a recent year, according to the U.S. Department of Agriculture, over 1 million firms in the United States handled food and other agricultural products that were raised by about 3.4 million farmers and ranchers. Included were firms that processed foods and beverages; manufacturers that made textiles, apparel, and leather products; firms engaged as assemblers, wholesalers, brokers and jobbers; retail food firms; eating establishments; and retail stores that handled clothing and shoes. The firms ranged from corner grocery stores and other small firms to supermarkets, large meatpacking companies, huge flour mills, and other giant establishments. Approximately 20 million people are employed to store, transport, process, and merchandise the output of the nation's farms.

The principal steps in marketing many of the products raised on farms include assembling, transporting, grading, processing, packaging, storing, buying and selling, financing, and carrying risks. Each one of these steps has a place in making products available in forms useful to consumers.

Assembling

As a first step in marketing, farm products are usually assembled in quantities larger than those produced by an individual farmer or rancher. Assembling, such as collecting grain by a local elevator before shipping it to a central market, may start locally. Livestock may be purchased by a local buyer who sends them to a central market, or by a direct buyer who assembles them to ship to a meatpacking firm in a large city. Milk from individual farmers is collected in large trucks and taken to receiving stations or directly to milk plants in cities where it is prepared for sale to consumers. In some places, cooperative establishments owned by farmers assemble products from individual producers.

Large terminal markets have been developed for assembling agricultural products of various kinds. These are centers for receiving, unloading, storing, and reshipping farm products. At these central markets, various other services, such as grading and selling, for some products are also performed.

Some of the main terminal markets for livestock are Peoria, Illinois; Kansas City, Missouri; East St. Louis, Illinois; St. Paul, Minnesota; St. Joseph, Missouri; Denver, Colorado; Fort Worth, Texas; Sioux City, Iowa; Indianapolis, Indiana; Wichita, Kansas; and Milwaukee, Wisconsin. Livestock are shipped by rail or truck to these centers and then placed in stockyards for feeding and watering prior to being sold. Thus, these stockyards serve as "hotels" for livestock sent by individual farmers, cooperative shipping associations, local buyers, and others. Less than 25 per cent of the cattle purchased by packers is obtained at these terminal markets. More and more, cattle are being marketed by direct sales from the feeder to the packer. The closing of the Chicago Union Stockyards in 1970 was just one sign of this changing marketing pattern.

In 1982, the three largest U.S. grain centers in terms of grain receipts were located at Chicago, Illinois; Toledo, Ohio; and Kansas City, Missouri. Fruits and vegetables are assembled in large markets in many cities, the largest of which is in New York City. Boston has a large central market for wool. Cotton markets are located in Memphis, Tennessee; Dallas, Houston, and Lubbock, Texas; Greenville, South Carolina; Augusta, Georgia; Montgomery, Alabama; Greenwood, Mississippi; and Fresno, California. Tobacco is graded and assembled in many large markets, where it is sold at auction.

Transporting

Most farm products are transported at several stages before they reach consumers. Trucks are used in most cases to haul the products from farms and ranches to an assembling point or directly to processors. Trucks also transport many foods and other agricultural products after they have been processed. Many agricultural products are transported by trains, and some by ships. Even airplanes are being used for transporting some high-value agricultural products. The development of refrigerator cars for railroads about 100 years ago was an important milestone in shipping meats and other perishable products. These cars were cooled chiefly by ice. Today, however, some railroad cars and many trucks are equipped with mechanical refrigeration units. Joint truck-rail service is sometimes used. Loaded truck trailers are placed on railroad flatcars and hauled piggy-back. Upon arrival at their destinations, the trailers are rehooked to truck tractors and hauled to warehouses, plants, and stores. This method shortens delivery time and saves labor and other expenses in marketing.

The importance of transportation is shown in the marketing of meat animals and meat products. In the early days, meat animals were driven on foot to central markets, often with serious losses in weight and even death losses. Today, about two-thirds of all cattle, hogs, and lambs are raised west of the Mississippi River, but about two-thirds of the meat is consumed east of that river. Much of this meat is consumed in the metropolitan centers of the East. On the average, the meat animals and meat are transported about 1,000 miles before the meats reach consumers.

Marketing Farm Products

A recent development in transporting milk has been the use of sanitary tank trucks, which are really huge vacuum bottles on wheels. These trucks stop at farms where the milk is pumped into the trucks from large bulk tanks in which the farmers store and cool the milk prior to arrival of the trucks. This method is replacing the old method of handling and hauling milk in 10-gallon cans.

Fig. 21-7. Inland waterways provide low-cost shipping for grain produced in the Corn Belt and elsewhere.

Grading

Livestock, eggs, grains, vegetables, fruits, nuts, wool, cotton, and most other farm products are sorted and graded so that quantities uniform in size and quality can be obtained. Much of this grading of products is done according to federal standards and, in some cases, is supervised by federal and state inspectors.

Meat animals sent to central markets usually are graded "on the hoof" before they are sold to meatpacking establishments. The carcasses from animals slaughtered in large packing houses are usually graded and stamped with appropriate grade names under the supervision of the U.S. Department of Agriculture. Eggs, milk, and some other products are graded, and the grade labels are carried on the containers. The consumers who purchase these products can depend on the quality; hence, the demand for these graded products has increased.

Processing

Most agricultural products are processed in various ways after they leave the farms. In making wheat into flour, flour millers use a complicated milling process. Additional processing is done by bakeries that make bread and other products.

Most meat animals are butchered in central packing plants. Considerable processing is done in preparing some meat products. For example, portions of hog carcasses are cured to produce bacon and hams. Wieners and hamburgers are other types of processed meats. The term *packing house* originated during the early days of marketing hogs and beef cattle when pork and beef were salted and packed into barrels to preserve the meat.

Wool is washed, carded, and made into yarns of various kinds. Cloth is woven from wool, which is made into articles of clothing sold at retail stores. Cotton is hauled to gins where the seeds and trash are separated from the fibers. Several additional processes are involved in making cotton cloth and articles of clothing.

Foods are processed in many ways in various kinds of plants and factories. Some foods are prepared and placed in tin cans. Freezing is a common practice for some kinds of fruits, juices, vegetables, and other food products.

Food factories are required to maintain sanitary conditions for the processing of foods. These conditions are enforced by various federal, state, and local regulations.

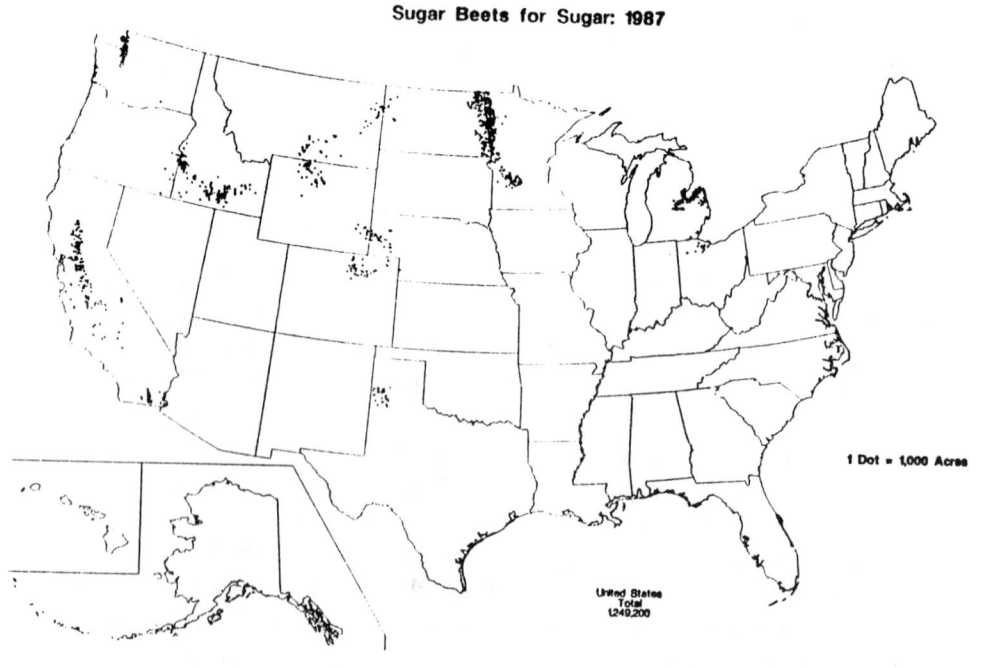

Fig. 21-8. The agricultural production on the grocery shelf may be imported, or it may be produced in the United States. (Courtesy, U.S. Department of Agriculture)

Packaging

Most food products that consumers buy have been placed in packages or wrappings of some kind, for convenience and sanitation. Milk is placed in jugs or cartons; margarine and many other foods in containers, boxes, or special wrappings; and some foods in cans. Apples and some other fruits are shipped in baskets or boxes, while some are individually wrapped. Cotton is baled for shipment and storage. As sheep are shorn, the individual fleeces are tied with special cord, and these fleeces are packed into huge sacks for shipment and storage. Processed foods of many kinds are packaged in various ways before they reach the consumers.

Storing

Many agricultural products are stored in their original forms or after various steps in processing. Some products, such as grains, wool, and cotton, may be stored quite easily. Perishable products, such as butter, meats, eggs, and fruits, are placed in cold storage. Some products are stored in frozen form.

Large warehouses of various kinds are used for storing agricultural products. Many of these are located at central markets. Some types of storages for grains have been provided by the federal government through the Commodity Credit Corporation. Farmers may store grain in these places and secure loans for the grain that they store. Considerable grain is also stored on farms and in privately or cooperatively owned elevators.

Buying and selling

Most farm products are bought and sold several times before they reach consumers. Firms that sell processed products in large quantities to retailers are called *wholesalers*. *Retailers* represent the end of the marketing process. They sell directly to consumers. In recent years, in the retailing of foods, large supermarkets and chain stores have been developed in addition to independent retail stores.

Various persons, such as buyers, brokers, and jobbers, are involved in buying and selling products. Each product may be bought and sold in several ways. For example, beef cattle fattened by an Iowa farmer may be sold to buyers for meatpacking firms and shipped directly to these firms, or they may be sold to private buyers who ship them to a terminal market. A farmer may ship cattle to a terminal market where they are sold by a commission firm to a buyer for packing firms.

Restaurants and other types of eating establishments have an important place in preparing foods and selling them to individuals who eat away from

Fig. 21-9. The cover has been removed from this egg cleaner to show eggs moving under hot water nozzles and abrasive bristle brushes. The cleaner processes 20 cases of eggs an hour. (Courtesy, U.S. Department of Agriculture)

home. In this case, the price spread is increased considerably, due to added costs in preparing and serving the foods.

Advertising in its various forms is widely used to promote the sales of agricultural products. Newspapers and other publications, signs, radio, and television are familiar media for advertising foods and other products.

Electronic marketing

Many agricultural products can now be purchased electronically. A buyer may sit at a computer terminal anywhere in the United States and participate in an auction for the products. A marketing association, the Na-

Fig. 21-10. Cattle are assembled in stockyards for sale to meatpackers and other buyers. (Courtesy, The American Hereford Association)

tional Electronic Marketing Association, Inc.,[1] was formed through a project funded by the USDA Agricultural Marketing Service. The association provides the buyer with easy access to livestock without the cost of attending live auctions or of searching the country for livestock ready for market. Products sold and bought include market lambs, feeder cattle, slaughter cattle, feeder pigs, slaughter hogs, and other agricultural products. Buyers receive a printed copy of the sale order before the sale, as well as a summary of their purchases immediately following the sale.

The National Electronic Marketing Association has no buyers of its own and does no buying. It provides services to marketing agencies to conduct computerized auctions. The seller works with a marketing agency in the area. Each marketing agency determines and solicits the buyers to participate in an auction. All livestock and money transfer takes place among the marketing agency, the buyer, and the seller. The marketing agency represents the livestock to the buyer, handles the transfer of livestock from the seller to the buyer, collects from the buyer, and pays the seller.

---

[1]National Electronic Marketing Association, Inc., Box 722, Christiansburg, Virginia 24073.

The following summarizes what takes place during an auction.

> ***Explanation of auction participation***
> ***(by computer terminal)***
>
> (1) Buyer gains access to auction program by typing *Auction Participation.*
> (2) Message displays sale sponsor.
> (3) Computer prints "WAITING FOR LOT" message periodically until auction starts.
> (4) Buyer receives message from control terminal operator.
> (5) Buyer reads brief description of starting lot.
> (6) Buyer notes additional comment on first lot's quality.
> (7) Buyer is instructed to bid by pressing Escape key.
> (8) First asking price is displayed once every 5 seconds.
> (9) ^ (uphat) symbol tells buyer this is an asking price.
> (10) 12 seconds remain until price will drop if no bid is received.
> (11) Preset time decrement drops with each display.
> (12) No bid was received in time limit so asking price drops preset 50¢.
> (13) Still no bid received and price drops to $61.00.
> (14) @ is printed on bidding terminal when bid key is pressed.
> (15) First bid is registered.
> (16) * tells buyer that he/she has the bid (* is displayed only on bidding terminal).
> (17) Time limit is increased for other buyers to bid.
> (18) Buyer notes * meaning he/she still has the bid; 15 seconds remain.
> (19) Space between price and seconds is blank; another bidder has bid $61.50.
> (20) Another bidder raised price by preset 25 cents; space between price and seconds is blank.
> (21) @ appears again when buyer bids $62.00.
> (22) * is present to tell buyer he/she has the bid; no more bids received in time limit.
> (23) Computer prints that auction of present lot is over.
> (24) Computer prints on *only* the successful bidder's terminal that he/she has purchased the lot and confirms the price. (Had he/she not been the successful bidder, would have printed "LOT SOLD TO ANOTHER BIDDER, PRICE $62.00.")
> (25) Computer informs buyer that today's auction is complete; thanks buyer for participating.
> (26) Computer summarizes each lot bought, giving lot #, lot size, average weight, location, phone number of sale management, sale price, and estimated cost.

Marketing Farm Products

The following summarizes what the buyer sees on the computer terminal.

---

(1) OPTION? *AUCTION PARTICIPATION.*

(2) SALE SPONSORED BY EQUITY COOPERATIVE LIVESTOCK SALES ASSN.

(3) WAITING FOR AUCTION OF LOT #544.
WAITING FOR AUCTION OF LOT #544.

(4) BE PATIENT—WAITING FOR A FEW MORE BUYERS TO LOG ON.

(5) SL—AUCTION #1   LOT #544   BRITT, IOWA   400 HEAD

(6) THIS IS AN EXCELLENT GROUP OF SPRING LAMBS!

(7) USE ESCAPE KEY TO INCREASE PRICE BY 25¢.

(8)(9)(10)	(11)		(12)	
62.00 ∧ 0:12	62.00 ∧ 0:07	62.00 ∧ 0:02	61.50 ∧ 0:12	61.50 ∧ 0:07
	(13)	(14)		
61.50 ∧ 0:02	61.00 ∧ 0:12	@		
(15)(16)(17)	(18)	(19)	(20)	
61.25 * 0:25	61.25 * 0:20	61.50 0:25	61.75 0:25	61.75 0:20
	(21)			
61.75 0:15	@			
(22)				
62.00 * 0:25	62.00 * 0:20	62.00 * 0:15	62.00 * 0:10	62.00 * 0:05

(23) AUCTION OVER.

(24) YOU HAVE PURCHASED LOT #544.   PRICE $62.00.

(25) THE SLAUGHTER LAMB AUCTION FOR TODAY IS COMPLETE.
THANK YOU FOR YOUR PARTICIPATION.
YOUR INDIVIDUAL SUMMARY FOLLOWS.

	LOT #	LOT SIZE	AVG WGHT	LOCATION	PHONE	PRICE	EST COST
(26)	544	400	110	BRITT, IOWA	(608) 356-8311	$62.00	$27,280.00

OPTION?

The Pacific Northwest Livestock Producers Marketing Cooperative, formed in 1974, is an example of a telephone auction system that brings many buyers and sellers together in a telephone-conference type arrangement. Feeder cattle also can be marketed by means of a video auction, where pictures of the cattle being auctioned are shown on a television screen. As another alternative, the buyers and the auctioneer can also tour the ranches that have cattle ready to sell.

In order to assure that each buyer can bid properly, most auction systems include information for buyers on points such as:

1. Use classification of cattle to be sold.
2. Size and uniformity of the lots of cattle to be sold.
3. Weight ranges and quality grade of the animals.
4. Time, place, and method of weighing and consideration of shrinkage.
5. Time, place, and method of delivery.
6. Responsibility for transportation and insurance.
7. Inspections and health guarantees.
8. Time and form of payment.

Financing and carrying risks

Huge investments are represented in establishments that assemble, process, and sell agricultural products. Large sums of money are needed for operating these businesses during any one year. Some of this money is loaned by various financing agencies. Firms that assemble and store farm products are taking risks, such as the possibility of lowered prices. By dealing in futures markets and by carrying insurance, firms can reduce some of these risks.

Marketing costs made up more than two-thirds of the $462 billion consumers spent for domestic farm foods. The estimated bill for marketing these foods was $361 billion in 1991. This amount included all charges for transporting, processing, and distributing foods that originated on U.S. farms. The remaining 101 billion made up the gross return that farmers received for producing the food.

The cost of labor is the biggest part of the total food marketing bill. Labor used by assemblers, manufacturers, wholesalers, retailers, and eating establishments cost $88 billion in 1981. This was 11 per cent more than in 1980 and 155 per cent more than in 1971. The total number of food marketing workers in 1981 was 7.7 million, compared with 5.4 million a decade ago. This growth in employment was confined largely to public eating places.

As we study the various steps and services in marketing farm products, it is easier to understand why prices for the processed products are increased over the prices paid to farmers for the raw products. The marketing of these products is highly competitive so that the profits from processing and marketing are usually not excessive.

Why Do Prices for Farm Products Go Up and Down?

Many factors affect the prices of farm products and cause them to go up and down. Some of these factors influence prices over periods of several years, some cause seasonal changes within each year, and some bring about changes for periods as short as one week or one day. Successful farmers study these changes and try to adjust to them in various ways.

The long-time fluctuations are due in part to general economic conditions in the country as a whole and in the world at large. Factors such as earning power of the workers, tariffs, depressions, government policies, and wars are involved. For example, during the depression of the 1930's, price levels of all products went down, but prices received by farmers fell faster and farther than most prices paid by consumers. During World War II and for a few years after it, prices of farm products rose more rapidly than for many products purchased by consumers. Some of these increases appeared to be greater than they really were because the value of the dollar was decreasing at the same time. Farmers and other people were handling more dollars, but the purchasing power of each dollar decreased to less than half of what it had been prior to World War II. This kind of situation is frequently called a period of price inflation.

Supply and demand

Two important factors that influence the prices received by farmers are *supply* and *demand*. At times, the supply of some farm products increases more rapidly than the demand for them. In general, this has been the situation in recent years. When prices of farm products are high, farmers tend to increase production. If the demand for these products does not increase in proportion to the supply and all the production is placed on the market, prices may fall to seriously low levels. Unless prices of the items farmers buy fall accordingly, farmers are at a disadvantage.

Weather conditions affect the supply of crops. Favorable weather in recent years in most parts of the United States has helped to increase the yields of crops. Furthermore, farmers are using improved production methods that also increase yields. These factors have caused the total production of many crops to increase even though the cultivated acreages have been reduced. Drought and other severe weather conditions, such as those experienced by many farmers during the summer of 1983, decrease production.

The demand for farm products is affected by the income or purchasing power of customers in the United States and by the demands from abroad. The foreign demands have fluctuated widely in recent years. Except for a few years following World War II and again during the Korean conflict, the exports of farm products decreased. This was a primary factor in the accumulation of large surpluses of wheat and cotton.

Prices of many farm products change yearly or seasonally. Prices paid for live hogs are normally highest in the summer months when the numbers going to market are lowest. The prices usually break during the fall and early winter when the numbers going to market are increased. Egg prices are usually highest in the fall and lowest in the spring, for similar reasons. Likewise, corn prices are normally lowest in November and highest in late summer, prior to the new harvest. These and other seasonal fluctuations are explained by the law of supply and demand, with changes in supply being the most important factor.

Fluctuations in prices occur from week to week and from day to day, depending on the supplies going to market, the supplies in prospect, and the demand for them.

Prices for some farm products tend to move in fairly uniform cycles over a period of years. This is illustrated by hogs. When prices for hogs are low in proportion to the prices of corn, and profits are small, most farmers produce fewer hogs. In time, the number of hogs marketed is reduced enough to cause prices to increase. Farmers then increase the production of hogs to take advantage of the more favorable prices. In two or three years, the number of hogs has increased sufficiently to cause prices to fall again. The complete cycle from low prices through high prices to low prices again for hogs usually takes about four or five years. With beef cattle, changes are made more slowly, and as a result, the cycles are about 15 years in length. These and other cycles are speeded up or retarded by droughts, wars, depressions, government policies, and other factors.

The market reports in the newspapers often indicate what is happening with regard to supply and demand. For example, if market reports in the

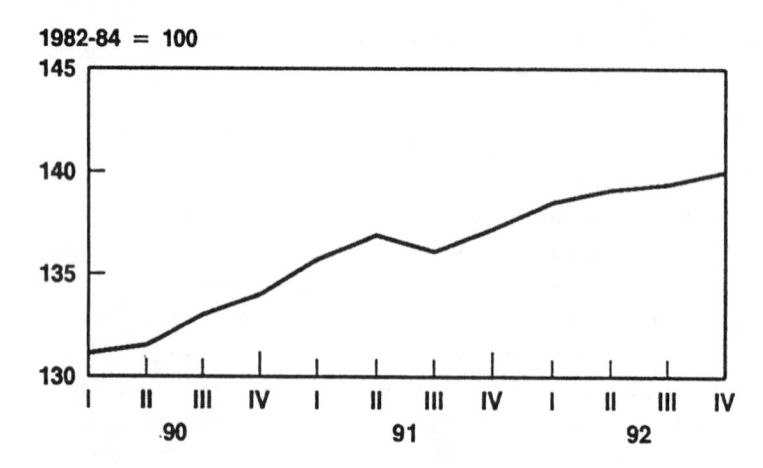

Fig. 21-11. Long-time trends in price fluctuations of farm and nonfarm products are similar. (Courtesy, Bureau of Labor Statistics)

newspaper indicate that hog prices have dropped to a very low level and at the same time that corn prices have increased, a supply-demand imbalance is probable between pork and feed. The price changes may be due to an oversupply of pork and an undersupply of corn. To adjust to this kind of situation, farmers must plan to produce fewer hogs during the following year or take a chance on growing hogs for limited profits.

The supply of farm products is a major factor in causing short-term changes in prices, as demand for these products usually changes more slowly than the supply. As a result, even small increases in production may cause sharp reductions in prices paid to farmers. For example, for crops such as potatoes and wheat, the per capita consumption is fairly constant and is not greatly affected by prices. An increase of 10 per cent in the supply of potatoes may cause prices to fall 40 to 50 per cent. The change in farm prices for eggs from 1971 to 1973, as shown in Fig. 21-12, is a dramatic example of how small changes in production can result in a sharp change in price. While there was only about a 5 per cent reduction in egg production from 1971 to 1973, there was an increase of nearly 90 per cent in the farm price. The downward adjustment in prices was just as rapid as the price rise when production edged upward the following year. The rapid drop in turkey prices in 1974 shows the effect of other factors, such as an increase in broiler production and availability of other competing foods.

Grains and some other farm products are bought and sold on two kinds of markets: the cash market and the futures market. In the cash market, actual grain is transferred from a seller to a buyer. In the futures market, grain is bought or sold through contracts for delivery in specified amounts at a later date.

As an example of a *cash sale*, North Dakota wheat is taken to a country elevator by a farmer who is paid in cash according to the going price for wheat. This price is determined at central grain exchanges where buyers and other sellers meet to trade. The prices paid are determined largely by the estimated supply and the estimated demands. Buyers for flour milling companies and for other commercial concerns are present at the grain exchange where they buy the wheat on the central market.

A futures market is where farmers, elevator operators, millers, and others buy and sell contracts for delivery of commodities at some later date. No grain is actually handled. The *futures markets* are located in the large grain exchanges. In Chicago, for example, the cash and futures markets are located in the Chicago Board of Trade Building.

The prices on the futures market fluctuate as weather and other conditions affect grain crops favorably or unfavorably. The futures market helps to stabilize grain prices throughout the year.

Farmers can use the futures market to establish a price for their products in advance of the time the products will actually be ready to market. Knowing the price to be obtained for their farm products helps farmers in

Eggs: Changes in Production and Farm Prices

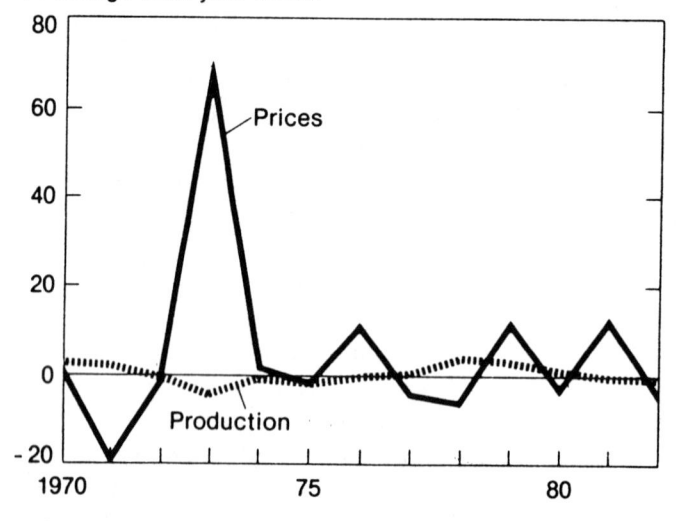

1981 preliminary, 1982 forecast. December 1 previous year through November 30 current year.

Turkeys: Changes in Production and Farm Prices

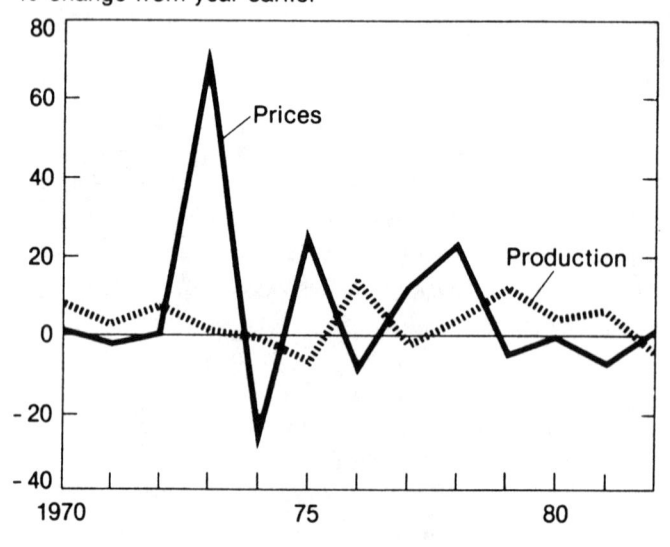

1981 preliminary, 1982 forecast.

Fig. 21-12. Small changes in supply can cause sharp changes in prices. (Courtesy, U.S. Department of Agriculture)

making management decisions. Just as is true for other farm management decisions, before using the futures market, farmers should study the situation carefully.

Quality of product

Another factor that influences prices is quality. In recent years, hogs that have topped the market have usually been of the meat type and have weighed 200 to 225 pounds. These provide pork that is fairly lean and hence preferred by most consumers. Here again, demand is a factor in determining prices.

High-quality eggs, vegetables, and fruits are preferred by most people who are willing to pay more for them than for a lower-quality product. Packaging these quality products for sale to the consumer involves uniformity with regard to size, shape, and color, in addition to freshness and freedom from defects. Premium prices are often paid to farmers who can provide products that meet one or more of the characteristics desired by the consumer.

Trends in eating habits

Although consumer demands for foods are fairly stable, some changes have taken place in recent years. The per capita consumption of poultry and fruits has increased. The per capita consumption of butter, however, has decreased in recent years due to replacement by butter substitutes. Marked decreases have also occurred in the per capita consumption of dairy products, meat products, and potatoes.

Government policies

Various policies of the federal government affect farm prices. One of these policies is the use of *protective tariffs* for many agricultural and industrial products. These tariffs are intended to keep foreign countries from sending goods to the United States and selling them at prices below the cost of producing these goods here. Under usual conditions, tariffs do not increase prices of farm products raised in quantities greater than domestic needs. If such products are sold abroad, the prices must be in line with prices in other countries that compete with the United States for foreign markets.

Cotton and wheat are two important products that have been exported in considerable quantities. In recent years, various price support programs have been developed by the federal government to keep prices of cotton, wheat, corn, rice, and some other products at levels that seem fair to the farmers in the United States who produce them. These programs have

574 AGRISCIENCE IN OUR LIVES

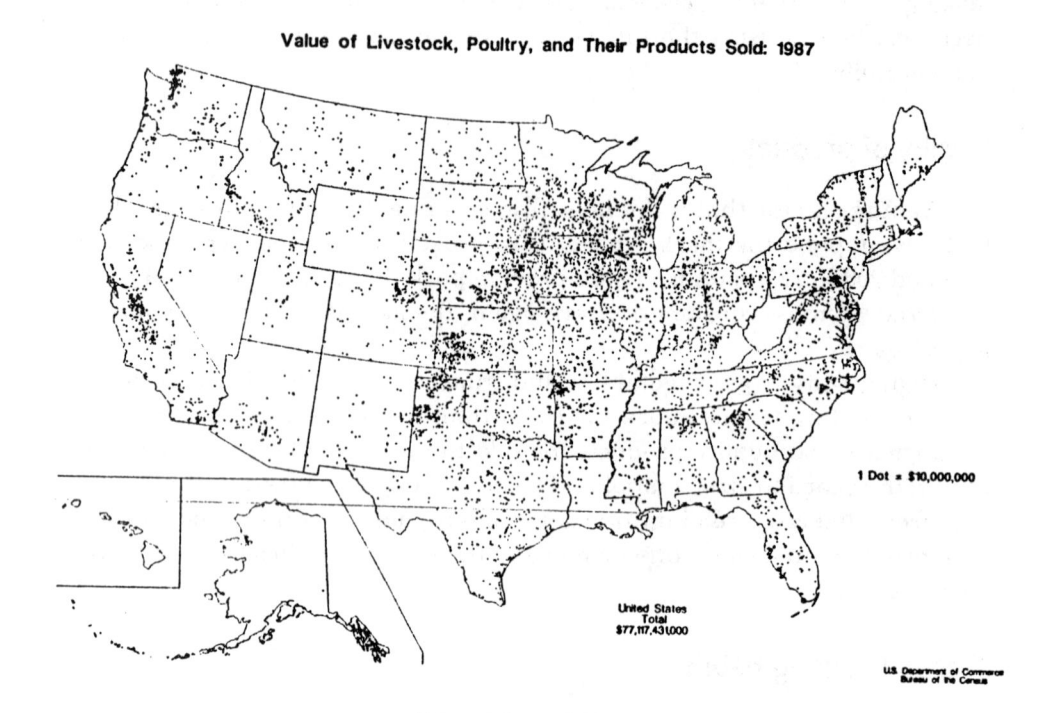

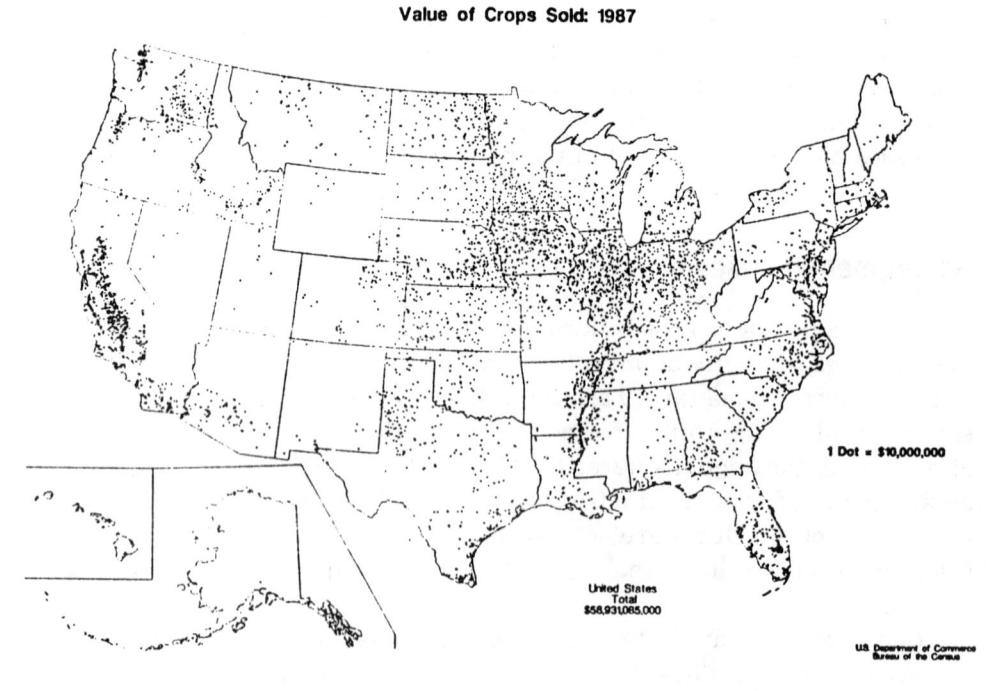

Fig. 21-13. Changes in eating habits are reflected in prices paid the farmer. (Courtesy, U.S. Department of Agriculture)

helped to keep prices of these products from falling to seriously low levels, but they have also tended to restrict sales abroad.

In recent years, the United States has maintained some foreign trade by providing loans to foreign countries that have used the loans to buy U.S. agricultural products. The United States has also donated considerable wheat and other surplus agricultural products to some nations. Furthermore, reciprocal trade agreements have made it possible for the United States to exchange some agricultural products for materials needed here. One of the greatest problems facing this country is to find ways to decrease tariffs so as to increase foreign trade and at the same time to satisfy various groups who feel they might be injured if this is done.

The total amounts of U.S. agricultural products exported in 1989 reached $39.6 billion. These exports are important if we are to strengthen our relations with friendly countries and secure some products from them which we need.

When prices of farm products are discussed, the question of what are "fair" prices to farmers is involved. *Parity* is a term used a great deal today. Parity is a price level for farm products, which is believed to be fair to farmers in relation to the prices for items they buy. It gives a commodity the same purchasing power as it had in a base period. The years 1910 to 1914 at times have been used as a basis for computing parity price levels. Thus, if

Top U.S. Agricultural Markets
Billion Dollars, Fiscal 1992 Forecast

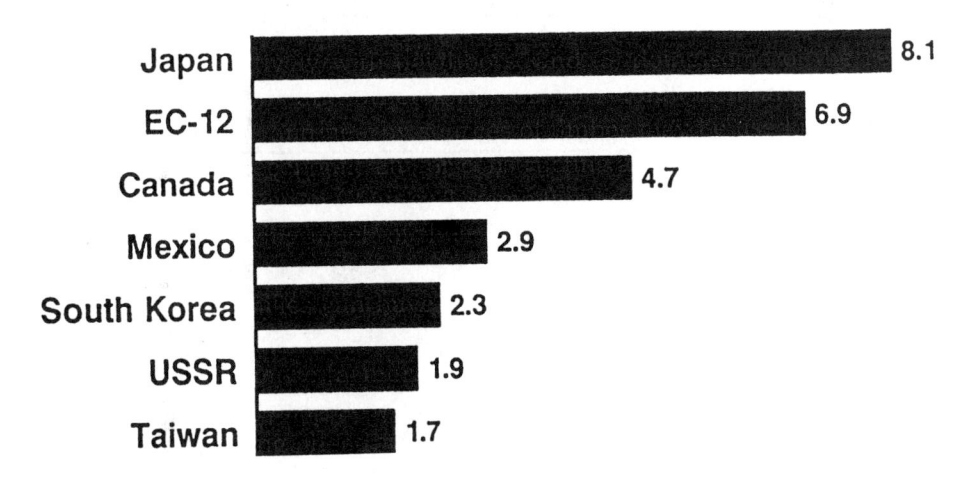

Country	Billion Dollars
Japan	8.1
EC-12	6.9
Canada	4.7
Mexico	2.9
South Korea	2.3
USSR	1.9
Taiwan	1.7

Fig. 21-14. Much of the U.S. agricultural surplus is sold to foreign countries, primarily to developed countries. (Courtesy, U.S. Department of Agriculture)

corn now sold for 100 per cent of parity, a farmer could buy a piece of equipment with the same amount of corn as in 1910 to 1914. Actually, in recent years, corn and some other farm products have sold for considerably less than 100 per cent of parity.

More recent years are now being used as a base period for calculating parity so that a more nearly accurate reflection of the changing price relationships among the different farm products can be obtained. The U.S. Congress ruled in 1948 that an average of the prices for the most recent 10-year period should be used in determining parity.

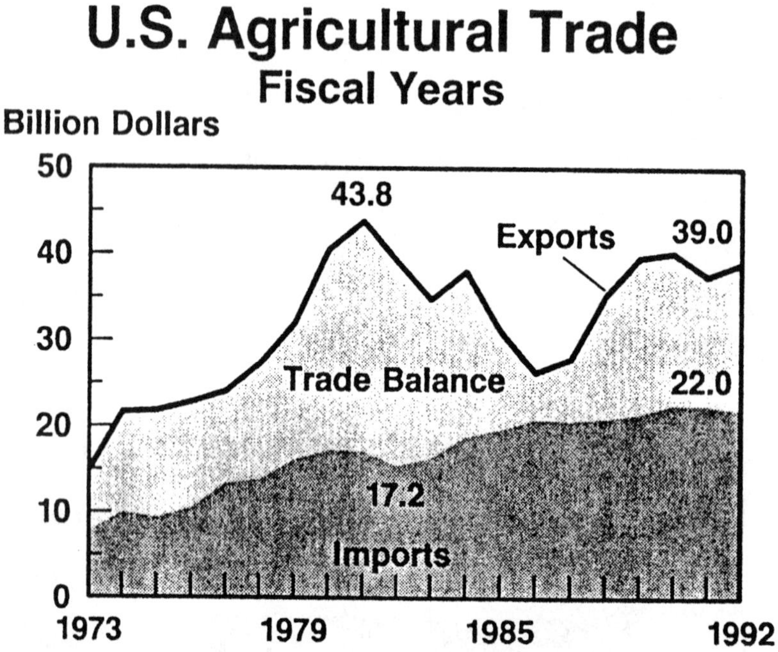

U.S. Agricultural Trade
Fiscal Years

	1973	1975	1977	1979	1981
		Billion dollars			
Exports:					
Agricultural	17.7	21.9	23.6	34.7	43.3
Nonagricultural	52.6	84.7	95.4	143.8	185.6
Imports:					
Agricultural	8.4	9.3	13.4	16.7	16.8
Nonagricultural	60.6	87.6	134.5	189.1	242.2
Trade balance:					
Agricultural	+9.3	+12.6	+10.2	+18.0	+26.5
Nonagricultural	−8.0	−2.9	−39.1	−45.3	−56.6

Fig. 21-15. In 1981, the agricultural trade surplus offset nearly one-half of the nonagricultural trade deficit. (Courtesy, U.S. Department of Agriculture)

Much discussion in Congress and elsewhere has involved questions of parity. Some groups feel that prices of farm products should be supported at a specific level, such as 90 per cent of parity. Others believe that flexible supports should be used; support prices could then be adjusted downward for each product if the supply increased. Most farm organizations and some other groups agree that some methods should be used to prevent farm prices from falling to extremely low levels.

Not all products are priced in a free market. Fluid milk and sugar prices are set by some authority. Natural cheese and cotton are priced in a free market with government price supports providing a "floor." Other products, such as eggs, broilers, and dressed meat, are priced in a free market.

In the free market system, sellers make the best deals they can get or accept the going market value as established by someone else.

Under the administered system—which covers a large share of manufacturing and retail trade—individuals, committees, trade organizations, and announced list prices are important.

Under the authoritarian system—which includes utilities and market orders—prices are usually set by boards, committees, or public agencies.

Prices (and how they are determined) have a fundamental role in both long-term and short-term decisions at all levels of industry.

In the best of all possible worlds, prices should satisfy consumers, help industry put its resources to best use, and facilitate trading.

Most people who have studied the price problem feel that ways to expand the demand for U.S. farm products must be found, both at home and abroad. Unfortunately, production of some farm products is increasing faster than these products can be consumed at prices that would be fair to farmers. Furthermore, some people in the United States and many abroad are not getting enough to eat and wear in order to maintain desirable living standards. The problem is complicated and difficult to solve to the satisfaction of everyone.

How Are Consumers Protected in the Marketing of Farm Products?

Some persons may feel that consumers are the forgotten people in the marketing and distribution of farm products. Actually, they are protected in many ways. One way in which they are protected is that they have a great deal of freedom in the products they buy and the forms in which they purchase these items. If consumers prefer, they may buy foods that are not highly processed and are therefore less expensive than others. For example, if homemakers prefer to do so, they may buy flour and make their own bread more cheaply than they can buy the baked bread. If one kind of food is high in price, consumers frequently may substitute cheaper foods of similar food value. Low-priced cuts of meat, for example, are as nutritious as high-priced steaks.

Consumers in the United States are blessed with an abundance of products raised on farms, and only rarely are there scarcities of any important products. There is every reason to believe that farmers will continue to produce in abundance for many years to come, even though the population is increasing. The various agricultural programs of the federal government are designed to help make sure that high-quality food is available at reasonable prices. These programs include price supports, conservation, credit, school lunch programs, various regulatory activities, agricultural products used in foreign aid programs, and agricultural research and educational programs to farmers. Most of these expenditures have benefited all the people, rather than any one group exclusively.

The U.S. Department of Agriculture and other federal agencies, as well as state and local governments, help in various ways to protect consumers from unsanitary and adulterated food products. Meat animals and meats that enter into interstate trade are inspected by government experts. Most states have laws and provide inspectors for food-handling establishments. Most cities require that milk be produced and handled under sanitary conditions.

Fig. 21-16. Examples of grading labels used to indicate product quality. From top to bottom, left to right, the labels are the following: *in the left row*—(1) butter, (2) nonfat dry milk, and (3) cheese; *in the second row*—(1) eggs and (2) poultry; *in the third row*—(1) canned or frozen fruits and vegetables and (2) fresh fruits and vegetables; and *in the right row*, all the labels are for meat.

Consumers are also protected by various official grades placed on many food products. These grades make it possible for consumers to know the quality of food they buy. Much of the meat handled in grocery stores is stamped with government grade labels. Eggs are graded according to freshness and size and sold in cartons on which the grade is indicated. An Egg Products Inspection Act became effective July 1, 1972. Under this act, all eggs moving in interstate or foreign commerce must meet U.S. standards of quality and weight. The act is designed to assure consumers that only clean, sound shell eggs reach their tables. Milk is retailed in jugs or cartons that carry the grade label. Consumers should become familiar with these grade labels and know what they mean so that they will be able to purchase foods of the desired quality.

Many school children benefit from school lunch programs financed by the federal government in partnership with state and local governments. Each year, millions of youngsters eat lunch at school in all parts of the country at a fraction of what the normal costs would be. Foods high in nutritional value are provided. This program has helped the nation's youth, as well as dispose of some of the surplus agricultural products.

The U.S. Department of Agriculture and state colleges of agriculture provide useful information on foods. This is done through bulletins, radio broadcasts, and other ways. Much research is being conducted on the preparation of foods, improvements of diets, and methods of marketing farm products.

Consumer protection extends to agricultural products other than food. For example, there are quality standards for Christmas trees. U.S. grade standards for Christmas trees, established by the Fruit and Vegetable Division, Consumer and Marketing Service, over a decade ago, can help a person choose a tree that has these characteristics. The grade standards require that a tree be:

1. *Fresh:* with pliable needles that are firmly attached to the branches.
2. *Clean:* at least moderately free of moss, lichen, vines, and other foreign matter.
3. *Healthy:* fresh, natural appearance for particular species.
4. *Well-trimmed:* free of all barren branches below the first whorl and smoothly cut at the butt.

In addition, the specific requirements of each grade are:

1. *U.S. Premium:* not less than medium density; normal taper; and all four sides free from any type of damage.
2. *U.S. No. 1,* or *U.S. Choice:* not less than medium density; normal taper; and three damage-free faces.
3. *U.S. No. 2,* or *U.S. Standard:* light or better density; "candlestick," normal, or flaring taper; and at least two adjacent damage-free faces.

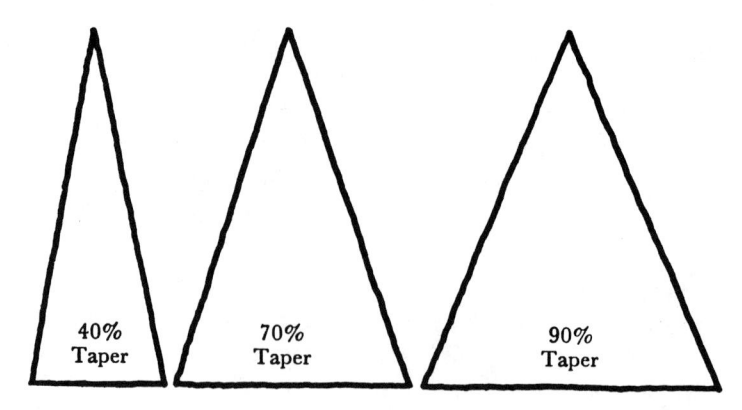

Fig. 21-17. In the U.S. grades, normal taper is 40 to 70 per cent for firs and spruces, 40 to 90 per cent for pines. Tapers greater than these are termed *flaring*. Candlestick taper is less than 40 per cent.

How Can Marketing of Farm Products Be Improved?

Farmers and consumers alike are interested in improving the marketing of farm products.

By using information on price cycles and other information on marketing, as discussed in other portions of this chapter, farmers may improve the marketing of these products. They may consider the kinds of products likely to be in greatest demand and in some cases adjust their farming operations to produce these materials. They may choose methods of marketing that will bring increased returns for the products they sell. In addition, they may produce high-quality products for which consumers are willing to pay higher prices.

Many farmer cooperatives have been formed to provide more bargaining power for farmers in the market place. Many farmers working together can assume some of the risks and provide the capital needed in the complex marketing process. Basically, this kind of activity helps farmers keep control of their products until prices have been set for them.

Various kinds of market information help farmers obtain better prices for their products. Beginning in June, 1963, the farmers of Illinois were provided with a new marketing service known as the Illinois–U.S. Department of Agriculture Grain Market News Service. This service is a cooperative venture between the Illinois Department of Agriculture and the U.S. Department of Agriculture, with each paying half of the costs. Farmers are informed each day of the actual cash prices being paid for grain at grain elevators throughout the state. With this kind of information, farmers are able to determine where to take their grain, and they are able to bargain more effectively at their local elevators.

Marketing Farm Products 581

Fig. 21-18. Selling directly to the consumer can result in an increase in farm income because existing farm labor is used to market farm produce.

New methods of processing, distributing, and advertising farm products may help both producers and consumers. One person has said that good marketing depends on having the right products in the right place, in the right form, at the right price, at the right time. What does this mean to you?

New uses for farm products help in marketing. Scientists are constantly finding new uses for farm products. The term *chemurgy* has been applied to this science. *Chem* is taken from the word *chemistry* and the last part is derived from a Greek word meaning *work*. Thus, *chemurgy* means putting chemistry to work.

Chemurgy is concerned with three major ways to increase the uses of farm products. These are (1) finding new nonfood uses for present farm products, (2) finding new crops for old or new industrial uses, and (3) finding profitable uses for farm waste materials. Many examples could be given. Some milk is used in making high-quality synthetic rubber. Casein from milk is used in making airplane fabric. Some lard and other animal fats are used in making plastic materials for raincoats and other products. Straw, cornstalks, cotton, and other plant materials are used in making rayon. Castor oil is used in making lubricants for jet planes, as this oil is not affected by extremely high or low temperatures. Dried pulp from citrus fruits, from which the juices have been extracted, is used for livestock feed. Large quantities of farm products, including cotton, wool, corn, soybeans, hides from cattle, and many others, are used by the automobile industry. In a recent year, more than a million tons of corn cobs were used in industry. About half of these cobs were used for producing *furfural*, a chemical which is used in making nylon, synthetic rubber, drug products, resins, and refined petroleum.

One of the ways of improving the marketing of farm products is to increase foreign trade, as previously indicated. Wheat and cotton are two important products that are grown in excess of domestic needs. Unless increased amounts of these products are sold abroad or production on farms is restricted, or both, serious surpluses will continue to accumulate.

SUGGESTED ACTIVITIES

1. Describe some of the items formerly provided by pioneer farmers, for their own use, which are now purchased by most farmers. How did the situation then differ from the present with respect to the basic necessities of food, clothing, and housing?

2. Make a list of the foods eaten during the past two days at your family table. In what form was each purchased? Which foods had been serviced most before they were purchased? Which were serviced least?

3. If a terminal or central market for livestock or grain is located in your city or at some nearby city, arrange a class trip or a trip with your parents. What products are assembled there? How were they transported to the market? What happens to them at the market and afterward?

4. Visit an establishment that bottles and distributes milk. Find out how it secures the milk and what is done to the milk before it is distributed.

5. Choose an important farm product raised in your area of the state and trace the stages in processing from the time a farmer sells the raw product until it is purchased by a customer.

6. Keep a daily record of the market prices for some kind of livestock such as hogs. Using the top prices paid each day at some central market, such as Peoria, Illinois, make a graph to show the trends from day to day. Over a period of several weeks, what fluctuations do you notice? How do you explain the fluctuations that take place? (Note: The top prices are reported in most daily papers and over the radio.)

7. Describe the school lunch program in your school. Find out how it is financed and how the foods are secured. What healthful foods are served?

8. Assume that you have a certain quantity of some farm product that you can sell in the next few months. Decide when you can sell it to best advantage and figure the amount it will bring on the day you sell it. During the next few months, check prices to see if you have sold at the best time. (For example, take 1,000 bushels of wheat, or 20 hogs weighing 200 pounds each, or 1,000 bushels of corn, or a specified quantity of some other commodity.)

9. For some vegetables or livestock products that you raise, develop plans for marketing these products to best advantage.

Marketing Farm Products

10. Make a special study of various products made from some important farm products, such as corn, cotton, soybeans, peanuts, and livestock.
11. Concepts for discussion:
    a. Farming profits are determined as much by how well products are marketed as they are by efficiencies of production.
    b. The total marketing process includes all activities related to a product from the time it has been produced to the time it is bought by the consumer.
    c. Farmers sell their produce in order to get needed items that are produced by others.
    d. Farmers receive only a small part of the consumer food dollar.
    e. Prices of farm products fluctuate widely.
    f. Farmers cannot control prices because production cannot be turned on and off easily and because many farm products are perishable.
    g. The difference between what the farmer gets and what the consumer pays is the cost of marketing.
    h. Marketing costs constitute about 60 per cent of the cost of food.
    i. Convenience foods preparation has increased the cost of marketing farm produce.
    j. As the amount of processing done to foods after they leave the farm increases, the farmer's share of the food dollar decreases.
    k. Less than 17 per cent of the disposable income in the United States is spent for food—a very low food cost in terms of income.
    l. The marketing industry for farm products is large and complex.
    m. The agricultural marketing process includes assembling, transporting, grading, processing, storing, buying and selling, financing, and carrying risks.
    n. The development of large and specialized farms has led to changes in marketing practices.
    o. Transportation developments have reduced the importance of nearness to market as a factor in determining where food may be produced.
    p. The prosperity of farmers is tied to the prosperity of the rest of the country.
    q. Prices for farm products rise and fall with supply and demand both seasonally and for longer periods of time.
    r. For products with a fairly stable demand, small changes in production may cause wide changes in price.
    s. The futures market helps to stabilize prices throughout the year.
    t. People generally are willing to pay more for high-quality produce than for low-quality produce.
    u. Trends in eating habits affect the marketing of farm produce.
    v. Parity is a price level for farm products believed to be fair in relation to prices of things purchased.
    w. Some farm products are priced in a free market, while others

are priced by an authority, under an administered system, or in a free market with a government support "floor," or minimum price.
x. Consumers are protected against unsafe agricultural products by a variety of federal, state, and local laws and agencies.
y. Farmer cooperatives provide more bargaining power for farmers.
z. Accurate marketing information helps farmers decide when and where to sell.
aa. New uses for farm products broaden the market possibilities.

REFERENCES

Lee, J. S., J. G. Leising, and D. E. Lawver. *Agricultural Marketing*. Interstate Publishers, Inc., Danville, IL 61832.

U.S. Department of Agriculture. *Farm Index*, a monthly publication. U.S. Government Printing Office, Washington, DC 20402.

U.S. Department of Agriculture. *Food—From Farm to Table*. Yearbook of Agriculture: 1982, U.S. Government Printing Office, Washington, DC 20402.

U.S. Department of Agriculture. *Marketing U.S. Agriculture*. Yearbook of Agriculture: 1988, U.S. Government Printing Office, Washington, DC 20402.

U.S. Department of Agriculture. *New Crops, New Uses, New Markets*. Yearbook of Agriculture: 1992, U.S. Government Printing Office, Washington, DC 20402.

Chapter 19

Managing the Farm Business

At one time, it was frequently said, "Anybody can make a living farming." There may have been some truth to this statement 75 to 100 years ago. Land was cheap. A team or two of horses and a few simple machines represented a small investment. The typical farmer of that time had plenty of labor, and most of the family needs were secured from the farm or obtained by bartering locally. The native fertility of much of the soil was high, and farmers could draw on this fertility with little concern about replacing it.

OBJECTIVES

1. Discuss why some farmers are more prosperous than others.
2. Identify planning necessary to manage a farm business.
3. Describe the purpose of farm business records.
4. Identify ways a small-scale or part-time farm operation can be improved.

Conditions today are quite different. Land suitable for farming is expensive to buy or rent. As discussed in Chapter 8, a high investment in machinery and other equipment is required for most types of farming. In fact, typical farmers of today have more money invested in machinery and other equipment for operating their farms than an entire farm would have cost 50 or more years ago. If they raise livestock, an additional investment is represented. If they own part or all of the land they operate, this represents a further investment.

Most of the products raised on farms are sold on the market and eventually purchased by consumers all over the United States and in some foreign countries. In a recent year, only about 3 percent of the total value of farm products was consumed by the farmers who raised them.

Today, large cash expenditures are needed to operate family-size farm businesses. Money is needed to pay production expenses such as fuel for

Prices Farmers Pay

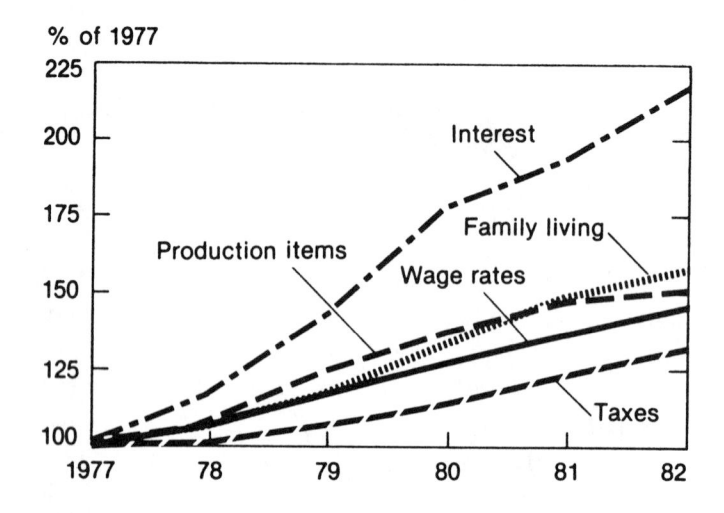

Fig. 22-1. Farmers need cash to pay for operating expenses—production items, wage rates, family living, taxes, interest, etc. (Courtesy, U.S. Department of Agriculture)

tractors, electricity for various uses, fertilizers for crops, repairs and replacements for machinery and other equipment, and feeds to supplement those grown on the farm. Also, farmers need money to pay taxes; and if they have borrowed money, they must have money to pay interest and to pay off those debts. Thus, farmers must manage their farms so as to secure sufficient income to meet these expenses and to pay living and other expenses for their families.

Some people who are not farmers own farms which they rent to others; these farm owners also are interested in efficient management of their farms.

Many people live on small-scale farms. These persons, too, need to manage their farms skillfully in order to secure good returns.

How Much Income Do Farmers Receive?

Many people have the impression that farm owners are making large amounts of money and that they are able to afford expensive luxuries of various kinds. It is true that many farm families are living better than they were formerly, but the same is true of most other families in the United States. Many farm families are having a difficult time meeting the rising costs of operating a farm business.

Average Net Farm Income

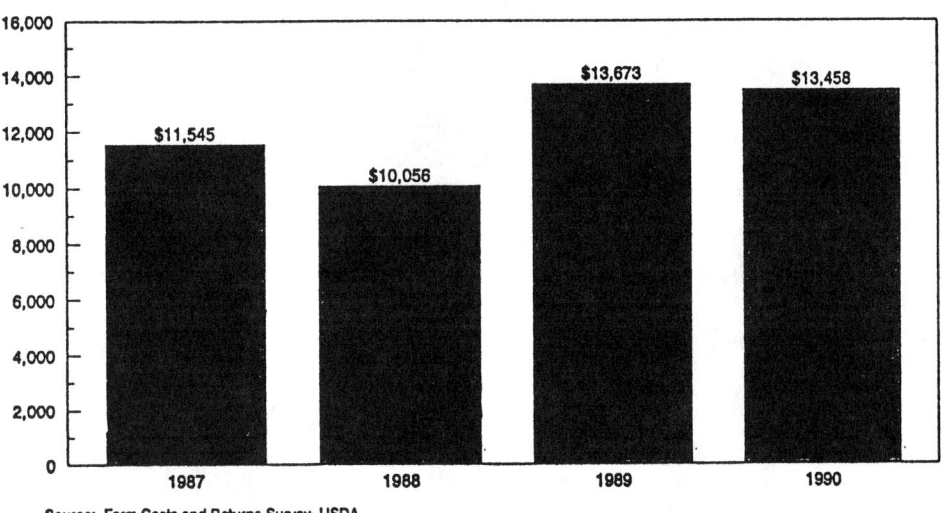

1987	$11,545
1988	$10,056
1989	$13,673
1990	$13,458

Source: Farm Costs and-Returns Survey; USDA.

Fig. 22-2. Average net farm income of operations increased over the 1987-1990 period through higher cash receipts offsetting lower Government payments and higher operating expenses. (Courtesy, U.S. Department of Agriculture)

As pointed out before, the farmer actually receives about 30 cents, on the average, of every dollar spent for U.S. farm-grown food; 49 cents of each dollar spent for meat; 14 cents of each dollar spent for cereal and bakery products; and about 51 cents of each dollar spent for dairy products.

Farm families earned a total of about $90 billion from farm and off-farm sources in 1991. More than half of farm income has been from off-farm sources for the last several years. The average per person disposable income in 1990 was $13,458 for farm people. Thus, farm income continues to be low when the risks and high total investment required are taken into account.

An important factor in the amount of income farmers receive is the portion classified as disposable income. While farmers generally have received more dollars of net income each year, the disposable income per capita has been consistently about a thousand dollars less than that of the nonfarm population. Since the prices paid by farmers for interest, taxes, and wages have continued to rise more rapidly than the prices received for products sold, the financial picture for farmers has continued to look worse each year rather than better.

A factor in maintaining the income per person on farms is the decreasing number of people on farms. Consequently, in recent years, the farm income has been divided among fewer and fewer people.

Fig. 22-3. "The strain is beginning to tell." When costs of items farmers buy rise faster than the prices they receive, they are in trouble. (Courtesy, *The National Grange Monthly*)

In meeting the situation described, farmers can do a great deal to help themselves by improving the management of their farms. It should be recognized that improvements of other kinds are needed also, as discussed in preceding chapters of this book and in the remaining chapters.

Why Are Some Farmers More Prosperous Than Others?

Farmers vary widely in the profits they secure from their farming operations. Even on farms similar in size and quality of land, large differences occur in the net income received by the farmers who operate them. Many factors help to account for these differences in income. Among the factors are the following:

1. A large volume of business.
2. High rates of crop and livestock production.
3. A suitable combination of enterprises.
4. Efficient use of labor and machinery.

Farming has undergone many changes in recent years, and this will continue. One of the major changes is in the amount of money that must be invested to own and operate a farm. Another major change is in the improved management ability required to be successful. For example, in 1940, labor accounted for more than 50 per cent of the resources used in farming. Today, labor accounts for less than 20 per cent of farm resources. Farms are also larger and much more complex. Thus, the farmer today must be a good businessperson and financial manager as well as a good manager of labor. A higher degree than ever before of technical knowledge and skill, as well as the ability to plan wisely for the total farm operation, is also needed.

For financial success in farming, it is important for farmers to manage their businesses so that they are using their land, labor, and capital in a combination that gives the best possible net income. For example, some farmers may have a fairly large gross income, but they may spend money for machinery or buildings beyond the real needs of the farm. In this case, they might be worse off financially than if they had less income and spent it wisely. Some farmers with a small volume of business could increase their net incomes if they bought or rented additional land, raised crops or livestock that utilized more labor and provided greater returns, or increased the yields of crops and the production of livestock. Some farmers struggle along all their lives with low-producing animals, when high-producing animals would require only a little more feed and labor and would give a much higher return. Thus, we see the importance of farmers studying their businesses, finding the factors responsible for keeping their income low, and then improving these factors so that their entire businesses are on a sound basis. By doing this, they are most likely to secure a high income year after year. This is one of the goals of most farmers.

How Can a Farm Business Be Planned for High Earnings?

A farm business does not run itself. Its success cannot be left to chance. As previously indicated, farms that are outwardly similar in many respects vary widely in net income or profits because of differences in the way they are planned and operated.

A farm management specialist at the University of Illinois compared the business records of two 160-acre farms in Illinois. These farms were equal in size and in quality of land, were located in the same community, and were farmed by men of about the same age. During a 22-year period, one farmer earned $300 more net profit *per acre* than the other. The total net profit from this farm during this period was $48,000 more than the other. The difference between the incomes from these two farms was largely due to *management*. In other words, one farmer was better able than the other to organize and operate his business for high returns over a long period of time.

Planning is important if a farm business is to be most successful. In Vermont, 12 farmers planned their businesses carefully. During a six-year period, their results were compared with 12 other farmers on similar farms who did not follow individual farm plans. Even though this latter group kept up-to-date on good farming practices and improved the management of their farms in some ways, they could not keep up with the farmers who followed good plans for their farms. In the sixth year, the farmers with good farm plans averaged about $2,000 more per year in labor income. While this was not enough to make these farmers rich, it was enough to provide many items that could contribute to improved living for their families.

As already emphasized, a good plan for an individual farm is important, and it pays off over a period of years. A farm plan is like a road map. It serves as a guide and a goal for a farmer and farm family. In developing a plan for the farm business, a farmer should take into account many factors. Farmers differ in the amount and kind of land and labor at their disposal. They differ in the capital available in the form of farm buildings and other improvements, machinery and equipment, and amount and kind of livestock. They also differ in their special abilities and interests and in their management ability. All these factors make it desirable to develop a special plan for each farm.

What Are Important Goals for Farmers and Their Families?

How farmers plan their businesses depends on what they consider important in life. Almost all individuals, whether they live on farms or elsewhere, have goals in life. As people grow older, their goals may change. Goals differ among families and among individuals. However, almost everyone will agree that the following are important goals for families:

1. Making money to provide a living for ourselves and our families, an education for our children, some luxuries in life, and security for retirement.
2. Having time for recreation and other leisure-time activities.
3. Being accepted and feeling we are wanted and needed by our families and other people.
4. Carrying out responsibilities as citizens and as community members.
5. Securing personal satisfaction from being successful in a chosen line of work.

Most rural people have goals similar to the ones indicated. By planning and operating their farms efficiently, they can increase their incomes and achieve other goals that they consider important.

Surveying resources

The first step in developing a plan for a farm business is to make a survey of the present conditions on the farm being considered. Farm management specialists refer to this as "taking an inventory of resources." These resources can be classified into four main groups: land, labor, capital, and management.

Land is the resource for farming provided by nature. The land resource inventory should include sketches or maps of the farm with all present field arrangements. Shown on the maps should be fences, ditches, ponds, streams, roads, lanes, and any other physical features that would affect farm planning. Before the inventory is completed, soil tests should be made on all fields to show basic soil fertility needs for at least lime, phosphorus, and potash. A land-use capability map should also be prepared.

Labor refers to the number of people available for performing the physical work on the farm. The total available labor would include that of the operator, the family, and hired labor.

Capital refers to the amount of money invested in the farm business and a listing of the items in which capital is invested, in addition to the land investment. Items such as buildings, machinery, livestock, equipment, feed, fertilizer, and various other materials are included. Some farmers also like to include the amount of money or financial credit available as a capital item, while others prefer to include this as a part of management.

Management consists of the knowledge, experience, and effort of the

Fig. 22-4. *(Left)* Floyd Taylor of Beallsville, Ohio, listens to James Rees, soil scientist, *(right)* explain the soil map of his farm. (Courtesy, U.S. Department of Agriculture)

person who actually makes the decisions about the planning and operation of the farm. As indicated above, financial credit is sometimes included as a part of management. Many farmers own and manage their own farms; some land owners employ professional farm managers to provide this service for them. The likes and dislikes of the farmer and farm family with regard to various kinds of crops and livestock should be considered here also.

Planning a cropping system

After the farm situation as it is at present has been surveyed, a cropping system for the land, which is likely to provide a high income and, at the same time, conserve soil fertility, should be developed. It is also desirable to make a farm map showing the future arrangement of fields. Fields should be arranged in a way that will permit following a systematic rotation on the tillable land and to provide for the efficient use of labor and machinery in working the land. Next, a list of crops adapted to the area and for which markets are available should be developed. Crops that will provide a large amount of feed per acre or a high profit per acre should be given special consideration. Generally speaking, since the land is the basic farm resource, it should be used for the most profitable crops it is capable of producing. Alternative cropping programs that fit the capability of the land should be planned, and estimates of production and income for each calculated before a final decision is made. The final decision will be affected by other factors such as government programs, special disease or weather problems, and the likes or dislikes of the farmer.

Planning a livestock program

After the cropping system has been planned, a livestock program, which will utilize to the best advantage the crops to be grown, should be planned. Again, the various alternatives should be considered and estimates calculated before any decision is made. Other available resources, such as labor, buildings, and equipment, should be considered in a livestock program. It may even be financially more advantageous to market crops through cash sale than through livestock feeding. Often, however, the need for diversification in order to use well all available resources dictates that some livestock enterprises be considered. Then, too, on most farms there is some land suitable only for pasture or other forage crops.

Planning a long-time fertilizer program

After the cropping and livestock systems have been planned, a long-time fertilizer program should be planned. Commercial fertilizers, livestock manure, and crop residues should all be considered. A good land-use plan

Managing the Farm Business

includes provisions for building up and maintaining soil fertility at a level that will make continuously high yields possible.

The amount of fertilizer a farmer can afford to use depends on the cost of the fertilizer and the increase in crop yields that result. By estimating the costs of applying fertilizer and the additional income expected from in-

Fertilizer Used and Prices Paid

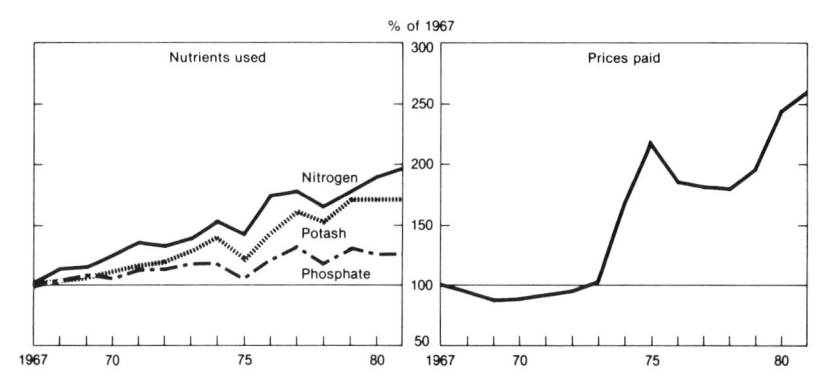

Fig. 22-5. The use of agricultural chemicals has increased over two-thirds since 1967. Prices have risen rapidly with increased usage. (Courtesy, U.S. Department of Agriculture)

Fertilizer Nutrients Used

Fig. 22-6. The rate of increase in pounds of fertilizer used per acre is slowing. (Courtesy, U.S. Department of Agriculture)

creased yields, a farmer can determine the amount of fertilizer that will result in a profit for the work and expense involved. Some plant nutrients, when applied to the fields, help increase crop yields for several years. Thus, the cost of application should be spread over all the crops produced over the same period of years.

Planning a marketing program

The marketing of crops and livestock should also be planned in order to provide for the highest possible returns. In planning the marketing program, a farmer should consider these factors: storage facilities, price cycles, labor efficiency, and the best use of available crops. It is also important to plan the marketing program to provide for cash income throughout the year and especially at times when major bills come due or major purchases must be made. Sometimes the local and state tax assessment regulations dictate the sale of stored crops before the dates for assessment. The need for avoiding extremes of high and low incomes in relation to income tax reporting also should be taken into account. Advance planning of marketing details can avoid hurried or forced sales that could be costly in a market that is fluctuating daily.

Planning for farm buildings and equipment

The need for farm buildings, equipment, and machinery is based mainly on the nature of the cropping program, the livestock program, and the marketing program. Some machinery and power equipment items may be desirable to remove the drudgery from work, even though no additional income is obtained by having them. Building layout and construction, as well as machinery and equipment purchases, should provide for maximum labor efficiency. In general, however, purchases in this category should pay for themselves through increased income. Many farms are low-income farms because their investments in machinery and equipment are too large.

Planning for labor efficiency

If all other farm planning steps have been carefully taken, the best use of labor has already been considered. Sometimes, however, additional workers may be needed to make maximum use of the land and other resources. On some farms there may be some labor unused with the plans as developed. In either case, some replanning needs to be done to make certain that the labor available and the work to be done are in reasonable balance. A study of the total labor needs for the peak and low labor demand periods in the farm operation should also be made. It can be just as costly to have unused labor as it is to be short of labor at critical times.

Planning for use of energy

Fuel prices and fuel availability can affect decisions regarding crops to be grown in the future. While field crops require less energy for producing them, their value is less than other crops, and the cost of the energy used is a greater percentage of their farm value. For example, the farm value of cotton is more than 16 per cent energy; whereas, for corn, it is about 9 per cent.

All fruits and vegetables, however, average only about 6 per cent of farm value for energy cost. Thus, a 10 per cent fuel price increase would have less effect on fruit and vegetable growing than on field crop growing. One exception is citrus fruits, which have an energy value equal to 19 per cent of the total farm value.

Energy Prices Paid by Farmers

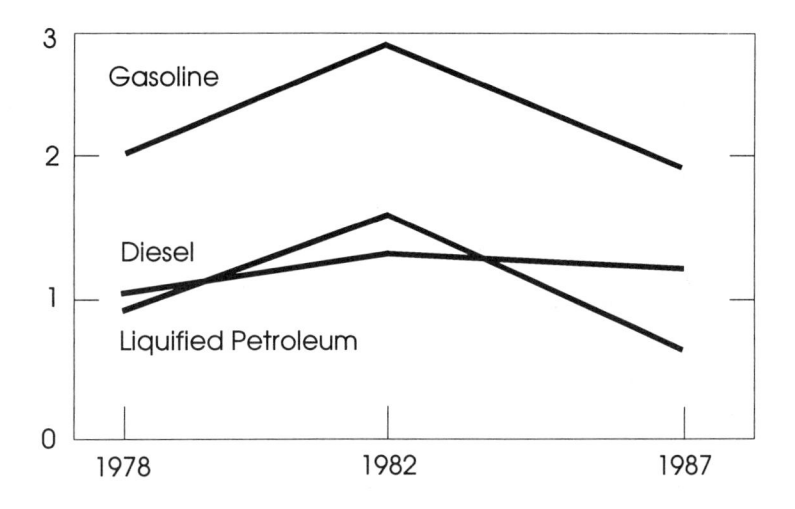

Fig. 22-7. Energy costs continue to rise. (Courtesy, U.S. Department of Agriculture)

Planning the farm purchasing program

This is one of the most neglected phases of farm planning. If the marketing program is to function efficiently and be most effective, it is necessary to have also a sound farm purchasing plan. Many economies can be obtained through practices such as:

1. Buying in quantity whenever feasible.

2. Comparing prices of various sources of farm needs.
3. Buying at the time of the year when prices are most favorable.
4. Giving preference in purchasing to the needs of the most profitable farm enterprises.
5. Consolidating purchasing so that costs for purchasing trips and for delivery can be kept to a minimum.
6. Planning to do purchasing on definite days to reduce labor required.
7. Planning purchases far enough in advance to avoid costly emergency purchases.
8. Analyzing purchases to make sure items were really needed and will pay for themselves by increased farm income.
9. Maintaining a running inventory of purchased items to avoid purchases of items already on hand in sufficient quantities.
10. Coordinating purchasing and marketing programs to eliminate borrowing to the greatest extent possible.

Deciding on leases and agreements

Many farms are operated on partnership or tenant-landlord arrangements. In each case, it is best that there be a written agreement covering points such as:

1. Purpose of the agreement and description of the property involved.
2. Identification and legal relationships of the contracting parties.
3. Legal status of the contract.
4. Period of time covered.
5. Cash rent or other rental arrangement.
6. Contributions to the farm business of each party.
7. Duties of landlord and tenant, or partners.
8. Working days and hours, vacations, living arrangements, sick leave, insurance, and other personal problems and arrangements.
9. Management of the business.
10. Sharing of receipts and expenses.
11. Arrangements for making improvements and repairs.
12. Reservations of special rights to the persons involved.
13. Arrangements for termination of the agreement.
14. Records and accounts to be kept.
15. Arrangements for periodic financial settlements.
16. Arrangements for settling disputes and amending agreements.

While leases and agreements will not cure all problems, they can form the basis for strong and lasting business relationships by providing a written record of what both persons agreed to when they decided to work together. There is also some assurance that adequate consideration has been given to potential problem areas before a commitment to work together is made.

More and more is being heard about setting up farm businesses as corporations. The main advantage of this kind of organization appears to be that ownership of a farm can be in the form of shares that can be bought or sold without selling the farm or dividing it into smaller parts, each of which might be too small to operate profitably. Other factors for farmers to consider in deciding whether to incorporate a farming operation are (1) the ease of continuing operations when a death occurs, (2) the limitation of liability to the assets of the corporation, (3) the added sources of financial credit, and (4) the advantages gained in planning for payment of taxes. Relatively few farms are organized as corporations.

The other major form of farm business organization is the sole proprietorship where one person owns and runs the farm business. Having employees does not change the fact that the one individual, as sole proprietor, has full ownership and control of the farm.

Planning the farm insurance program

When farms were small, with several kinds of crops and livestock produced on each farm, the chances of every crop and livestock enterprise failing in any one year were very small. The great increase in investment in farming and the trend toward a high degree of specialization in crop and livestock production, however, have made the risks of farming more than a single farmer can afford to bear. The failure of a farmer's major enterprise could bankrupt the business. A serious disease, the ravages of winter, or the destruction of a crop by hail could wipe out an entire year's income. As a result, various kinds of insurance, some provided by the federal government, are available as a protection against loss. Insurance against crop failure, fire, storm damage, theft, livestock loss, and many other forms of financial disaster can be purchased. Each farmer must decide how much of the risk to take and how much to guard against in the form of insurance.

However, several kinds of insurance are so important that they should be purchased for every farm business. Among the most important types of insurance are fire insurance on all property, liability and basic insurance on all vehicles on the road, life insurance on the farmer's life, comprehensive public liability insurance, and workers' compensation and unemployment insurance when the farm business qualifies for these. Other types of insurance to consider are employer's liability, life insurance on other members of the family, comprehensive and collision insurance for cars and trucks, crop insurance, and coverage on farm property for certain types of damage.

Implementing the farm plan

Just making farm plans does not put them into effect. It may take four or more years to make the transition from the present farm plan to the new farm plan. Consideration needs to be given to the costs of making the desired changes and to the probable increases in income. If the estimated increased earnings are not sufficient to meet the costs of the changes, some modifications in the plans are probably desirable.

In developing a final plan, the farmer should make goals that supplement securing a high income. All members of the family should share in deciding on these goals. Some families may decide that shorter working hours and an occasional vacation are desirable, rather than expanding the farm business. Some families will want to use part of the income for improving their homes and surroundings.

In making their farm plans, some farmers find it desirable to discuss their problems with county agricultural agents, high school agriculture teachers, extension specialists from colleges of agriculture, soil conservation specialists, and successful farmers in the community. Valuable suggestions are frequently obtained from these persons. Some farmers and urbanites who own farms find it desirable to secure the services of farm management specialists to aid in planning and managing their farms.

How Can a Farm Business Be Operated for Greatest Profits?

After plans have been developed for a farm business, these should be put into operation as rapidly as seems desirable and practical.

The preceding chapters on crops and livestock included many suggestions important in operating a farm efficiently. If a farmer "slips" on one important step, certain phases of the farming program may fail, even though the farmer is following other desirable practices. In producing crops, for example, a farmer may be using certified seed, employing good cultural methods, and performing several other desirable practices. However, if the farmer neglects soil fertility, yields are likely to remain low in spite of the good practices being followed. There is an old saying that "a chain is no stronger than its weakest link." How does this apply to the situation described?

In carrying out plans from year to year and over a period of years, farmers should be flexible enough to take advantage of new opportunities and to adapt to changing conditions. As children get old enough to work, more labor becomes available. As farmers are able to decrease indebtedness, more of their income can be invested in equipment that could not be afforded earlier. Furthermore, they are able to improve the home and add conveniences that make life more enjoyable for the entire family. If they are so inclined, the time saved can be used just to take things easy and to devote more time to the family and to community activities.

The efficient use of labor is important for success in farming. There are many tasks competing for the time of persons on a farm. A farmer must use good judgment in deciding which of many jobs on the farm are most important and which would cause the greatest losses if they were left undone for a period of time. Doing first things first and doing them on time are important. Some jobs, such as repairing buildings and fences, may be safely left for rainy days or other times when important seasonal jobs are less pressing.

Good farmers try to plan combinations of crops and livestock enterprises that will make it possible to distribute their labor as uniformly as possible throughout the year. If a farm business is planned carefully, extreme peaks of labor may be avoided. Machinery kept in good repair prevents costly breakdowns at times when important jobs, such as harvesting crops, should be done as quickly as possible.

If the costs of farming rise faster than the prices received for the products, or if prices fall without corresponding reduction in costs, special attention should be given to costs of production. This does not always mean that a farmer should spend less money. Some expenditures, such as for fertilizer, may return several times their cost. New machines and equipment may save enough in hired labor to justify getting them. Such equipment may also save labor performed by the farmer or the farm family and thereby make it possible to expand the farm business in various ways. Some other practices that will increase production and will usually pay, even if they require additional labor or expense, are (1) feeding good rations to livestock, (2) eliminating crop insects and diseases, and (3) controlling livestock ailments of various kinds.

Some ways to increase production with little or no increases in costs are (1) using good cultural methods for crops, (2) feeding and caring for livestock regularly, (3) doing all work on time, and (4) repairing and operating machinery properly.

Some methods that will reduce costs without cutting production may be followed. These include (1) exchanging labor, (2) owning expensive machines cooperatively with neighbors, and (3) doing some of the repair and construction jobs and other kinds of work for which labor was formerly hired. Woodlots and nearby sources of timber may be used in some cases to supply lumber economically for constructing and repairing farm buildings.

As pointed out in Chapter 6, some families save considerable amounts of money by growing much of their food supply in gardens. By using modern equipment for storing foods in frozen form and by storing foods in other ways, farm families can have a year-around supply of many food products with very little extra labor.

More and more farmers are turning to the computer to perform management tasks such as (1) keeping farm records; (2) analyzing farm record data; (3) planning ahead; (4) preparing income tax returns; (5) making legal searches, such as for water rights statutes; (6) analyzing the labor market; (7) estimating losses from various causes; (8) alternating uses of resources; and (9) even tracing stolen trucks.

Consumer Prices
Percent Changes

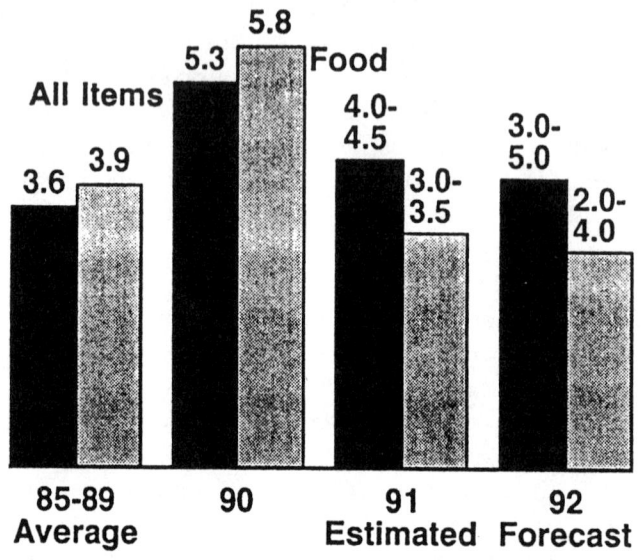

- All Items: 85-89 Average 3.6; 90: 5.3; 91 Estimated 4.0-4.5; 92 Forecast 3.0-5.0
- Food: 85-89 Average 3.9; 90: 5.8; 91 Estimated 3.0-3.5; 92 Forecast 2.0-4.0

Fig. 22-8. The farm value of food is rising more slowly than the retail prices are. (Courtesy, U.S. Department of Agriculture)

Components of Increases in Retail Food Prices

% change in retail price
- 1979: 10.8
- 1980: 8.0
- 1981: 7.3
- 1982: 5.0

Components:
- Higher prices for fish and imported foods
- Higher farm value
- Higher farm-to-retail price spread

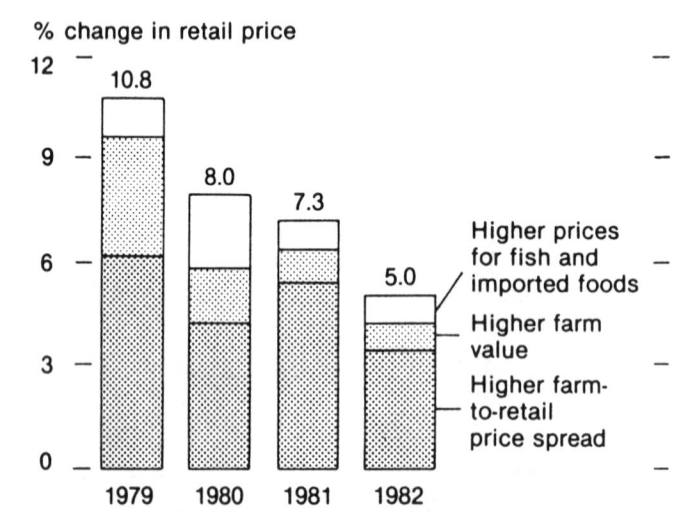

Fig. 22-9. Higher farm value is only a small part of the increase in retail food prices. (Courtesy, U.S. Department of Agriculture)

How Can the Farm Business Be Financed?

Farming today requires the investment of large amounts of money. In addition to the initial investment, large sums of money are needed for operating the farm business. Some farmers may be good producers, but they may be handicapped because they do not handle their finances properly.

Purchasing a farm at a price too high to permit the returns from the business to provide a fair return on the investment is a common mistake. Buying a farm is a long-time investment and 20 to 30 years of earnings are frequently required to pay for it. The problem is likely to be especially serious if these farmers go heavily in debt on the property which they pur-

Fig. 22-10. Three lessons in farm management.

chase. Prices for farm products may have been high enough at the time a farm was purchased to make it seem safe to pay a high price for the farm. However, if farm prices fall, farmers heavily in debt may be in serious trouble. Thus, long-time price trends are important to farmers who borrow money for extended periods.

An expert in farm management has suggested that a farmer should have livestock and equipment debt-free and enough cash to pay upwards of one-third the purchase price of a farm before buying one. Otherwise, it might be best to continue on a rental basis. A farmer should buy land at prices that are not likely to prevent meeting the payments and interest on any indebtedness incurred in the purchase of the farm.

Careful selection of sources of credit for financing farms is important. Farmers may secure credit from several sources, including loans from individuals, private banks, life insurance companies, and federal agencies such as Federal Land Banks. Loans from Federal Land Banks are made through the National Farm Loan Associations, as discussed in Chapter 23. In financing the purchase of a farm, a farmer should secure loans at nominal rates and make plans for paying off the principal. Farmers frequently need money to finance the purchase of equipment and to operate their farms. Short-term loans are needed for such purposes. These may be secured from Production Credit Associations or various other sources, including individuals and private banks. Before borrowing money for any purpose, farmers should make plans for the future with respect to expected income and expenses. If these estimates are used wisely, farmers will borrow only the amounts that can be repaid without undue risks.

Fig. 22-11. Farmers should try to obtain farm loans from lenders with experience in making loans for farming purposes.

One major development to reduce the cost of long-term borrowing is CO-FARM, introduced by the Federal Land Bank of St. Paul.

CO-FARM lets borrowers make payments of real estate loans and allows them to reborrow without negotiating a new loan. This practice reduces the bank cost of providing service to borrowers and cuts borrowers' costs by eliminating additional title searches, recording, and other charges associated with new or refinanced loans.

When seeking a loan, a farmer should go to lenders who have experience in making loans for farming purposes. Many agencies and lending institutions have farm specialists who work with farmers who need to borrow money and who are able to talk with farmers about farm operations as well as about financial matters. This kind of help can be particularly valuable to someone just getting started in farming.

It is desirable for farmers and other persons to study carefully their purchases along all lines. Some farmers purchase pieces of equipment because of high-pressure sales techniques or attractive advertising, when they could get along without these articles. For some pieces of machinery, such as combine harvesters, it may be desirable for two or more farmers to purchase them together. Every farmer should become an expert in buying as in other phases of farming.

What Records Are Needed for a Farm Business?

A group of high school students and their teacher made a field trip to the farm of a successful farmer. While they were discussing various aspects of farming with him, he thought he would have a little fun by asking them to name the most important piece of equipment on the farm. They named everything from pitchforks to tractors, but he told them they were all wrong. He reached down in his pocket and pulled out a pencil with a sharp point and said, "This is it. I use this for keeping various records necessary for running my business successfully."

Records for the farm business are valuable for the following reasons:

1. They show how much the farm business is making or losing and reveal some of the weaknesses that need correcting.
2. They aid in checking on plans for the farm business as these are carried out, and thus provide a basis for revising plans from year to year.
3. They provide information needed in figuring income taxes and social security payments.
4. They provide information helpful in obtaining loans.
5. They provide information for making fair divisions in carrying out partnership and rental agreements.

Most farmers today keep some kind of records for their farm business operations. These may consist only of keeping sales slips for products sold and cancelled checks used for payment of expenses. However, if a farmer wishes to figure net income accurately and to use records in various ways to improve the farm business, a suitable system of records is needed. Most farmers prefer a system of records that is fairly simple and that requires only a small amount of time. The use of computers can reduce the amount of time needed for keeping and using records.

Keeping records for a farm business

In keeping farm business records, a farmer should select a good type of record book to use. Most state colleges of agriculture have developed farm record books that are quite easy to keep. Usually, directions are provided for keeping and summarizing these records. These record books may usually be purchased for a small sum from the state agricultural college or the county agricultural agent; or the high school agriculture teacher will help a farmer secure one. The most suitable record books provide for (1) figuring depreciation on machinery, buildings, and equipment; (2) making opening and closing inventories; (3) recording expenses and income; and (4) other items.

Some farmers obtain help in keeping and using records by attending adult-farmer classes sponsored by high school departments of vocational agriculture. Some join with other farmers in a farm record association and cooperatively employ a field supervisor who helps each farmer-member keep records and use them in improving the farm business.

Keeping records of production

Production records of various kinds are a necessity, especially for the major crop and livestock enterprises. These usually include the amounts of each product raised, and they may include the value of the products. Unless adequate production records are kept, it may be very difficult for a farmer to determine which of the several enterprises are the source of the farm profits.

Crop production records should show (1) all kinds of crops produced, (2) acreage, (3) yields, (4) costs of production, and (5) value of the crop when it was sold or used in some other way. It is not enough, for example, just to know that the corn was fed to the hogs and then to figure the total profit from the two enterprises together. It is entirely possible that a greater profit could have been made by selling the corn rather than by feeding it. It is also possible that the profit was due to an efficient hog enterprise and that the corn enterprise was a losing part of the farm business.

A valuable type of record is an outline map of the farm, which is made each year to show the crops raised on each field. Some farmers record on the map the yield of the crop on each field and the amounts of fertilizer, lime, and manure applied.

Livestock production records should show the births and deaths for each enterprise. The average number of pigs farrowed and raised per litter is helpful in determining the efficiency of the swine enterprise. The total milk or butterfat for the dairy herd may be determined from the milk checks, with additions of the amounts consumed on the farm. From these, the average production per cow per year may be computed. By recording the number of eggs gathered daily or by keeping a record of the dozens of eggs sold and then adding the number consumed on the farm, a farmer can determine the total number of eggs produced. From these, the average production per hen per year may be figured. One basis for computing average egg production in a laying flock is to divide the total number of eggs for the year by the number of hens housed in the fall.

Additional records for livestock enterprises are useful. Besides birth dates, these include breeding dates and names of sires used. These are especially important if purebred livestock are to be registered.

Some dairy farmers belong to Dairy Herd Improvement Associations, which make it possible to secure accurate production records for each cow in the herds of the members. Some hog producers find it valuable to determine weights of litters at 56 days or at a later age. In some cases, farmers belong to Swine Herd Improvement Associations which keep these kinds of records for the farmers who join them.

How Can Records Be Used for Improving a Farm Business?

The primary purpose for keeping farm business records is to improve the farm business. Consequently, it is important for each farmer to summarize and study the farm records carefully.

When a farmer says that the farm made $2,500 last year, this does not mean very much unless we know how this figure was calculated. Does this figure mean only that there was $2,500 more cash on hand at the end of the year? If so, the figure is deceiving because it does not include a possible decrease or increase in inventory, the value of farm-raised products used by the family, the unpaid family labor, and perhaps other items that are important in figuring and interpreting the income from the farm.

Studying farming efficiency

Efficiency in livestock production may be established from the pounds of milk and butterfat produced per dairy cow per year; the number of pigs

Fig. 22-12. An extension service farm management specialist *(right)* is helping this Nebraska farmer look over computer printouts for analyzing cash flow and other aspects of the farm business. (Courtesy, U.S. Department of Agriculture)

raised per litter, the weight per pig in the litter at 56 days (or some other age); and the number of eggs laid per hen per year. Computing the percentage of cows of breeding age that produced calves during the year will help to determine the breeding efficiency of the beef breeding herd. A useful figure for all livestock enterprises is the returns per $100 worth of feed consumed.

By computing the annual cost of machinery and equipment per acre, farmers can determine how efficient their use of machinery has been. By adding items such as the cost of repairs, depreciation, gas and oil, machinery hired, and the farm share of telephone, electricity, and auto costs, farmers can obtain the total annual cost of machinery and equipment.

Improving a farm business

In analyzing the farm business records for their given farms, farmers should consider the important factors that affect farm profits. Some of these important factors, as previously mentioned, are (1) the size or volume of business, (2) the rate of crop and livestock production, (3) the combination of enterprises, and (4) the efficiency in using labor and machinery.

Then, in terms of these factors, farmers can compare their results with those obtained in previous years and can thereby note progress, or lack of progress, in their farm businesses. Furthermore, they should compare their

Managing the Farm Business

results with those obtained by other farmers during the same year, particularly the results of some of the well-established farmers. Next, they should try to determine the strong and weak places in their farm businesses and then consider how to correct the weaknesses.

As measured by net income, gross sales, total acres, number of days of productive labor, and other measures, their businesses may be too small. Thus, to increase the size, farmers may need to farm more acres, add more livestock, increase the yields, etc.

If their crop yields are low, farmers should study their cultural methods. In doing so, they may decide to use more fertilizer, to switch to higher yielding varieties, and/or to use better methods for controlling diseases and insects.

If production per dairy cow is low, farmers should determine ways to increase it. They may decide to change their feed rations, try to improve breeding, provide better housing, and/or establish a better disease and insect control/prevention program.

Farmers may decide to change some of their crop and livestock enterprises. Perhaps changing to some higher profit crops or increasing the acreages of some present crops and reducing those of others would yield greater farm income. In addition, farmers could adjust livestock enterprises

Fig. 22-13 Proper repair and adjustment of farm machinery can be the difference between profit and loss.

to make better use of the crops raised and labor available. Any enterprise that consistently fails to show a reasonable profit in income over expenses should be improved or dropped from the farm business.

Possibly hand labor could be used more effectively if some labor-saving equipment for crops and livestock were purchased or if the chore schedules for livestock were better planned.

Farmers should consider whether or not the methods and times of marketing farm products might be changed to advantage, as was pointed out earlier in this chapter.

The purchasing program for the farm should be analyzed each year to make sure that the best possible economies are obtained in terms of timing of purchases and quantity purchasing. Farmers may discover that some purchased items were actually not needed.

Someone has said, "A person who is an *average* farmer should get better or quit." This is a real challenge to farmers.

Taking care of legal problems

Farmers, like other businesspersons, may face legal problems that need to be handled. In addition, because farmers own animals, use chemicals for production purposes, use chemicals to control diseases and insects, and operate farm equipment, legal challenges are always possible. While legal problems are not as common as other kinds of farming problems, they do occur and when they do, legal services should be obtained.

The kinds of legal matters for which legal assistance would be helpful include buying or selling property, making a will, settling farm boundary disputes, discussing actions of others that affect the farm either above or below ground, determining water rights, allowing trespassing for any purpose, making leases and agreements, making contracts, discussing environmental protection matters, and meeting various state and federal regulations. Many farmers attend adult classes on farm law at the local high school vocational agriculture department to learn about farm law, especially when to seek the services of a lawyer.

How Can the Management of Small-Scale and Part-Time Farms Be Improved?

Some farmers on small farms earn good incomes, but many of these persons have off-farm jobs that provide major portions of the family incomes.

For the low-income farms, there are several ways for farmers to bring about improvements. These include the following:

1. Secure work off the farm to provide additional income for the family. In many public schools, vocational education or some other kind of training should be provided to prepare rural people for well-paying jobs.
2. Increase the size of farm business by increasing the acreage or by improving the land already available. Many farmers need money to do this, and good sources of credit must be provided.
3. Choose kinds of livestock and crops that will provide substantial returns on a small layout, and use good methods for these enterprises.
4. Raise a large amount of food for use by the family.

Many people live in the country and engage in small-scale farming because they like this kind of life. If these people have off-farm jobs to provide a major source of income, they are able to provide a good living for their families. One of the best uses for their small farms is to produce a maximum amount of food for family use. Even though the methods used in small-scale production are, at best, less efficient than in large-scale farming, food products can be raised for considerably less than they would cost in retail stores. Many individuals become very skillful and raise better quality foods than those frequently available in stores. On many small-scale farms, some products may be raised for sale as well as for home use, thus increasing the cash income.

Improving farm living

It is important to emphasize that the real purpose for managing the farm business so as to increase profits is to bring about better living conditions for the farm family. Better living includes better homes with more conveniences, better schools, better churches, better roads, and greater opportunities to travel and enjoy some of the pleasantries of life.

A farmer, in talking to a group of rural people, said, "A successful farmer must establish a good home, make himself a valued part of his community, and operate a 'good' farm, as well as make money. Money is not the greatest measurement of success in farming."

SUGGESTED ACTIVITIES

1. What goals do you and your family consider important? How can you help to attain them?
2. Visit a farm where the business has been carefully planned. Have the farmer describe the main features of the plan and the recent changes made in the farm business operation.
3. Make a list of important jobs on a farm that must be done on time if serious losses are to be avoided. List other jobs that may be delayed for slack periods.

4. For something you expect to raise, such as a garden, chickens, or a sow and litter, draw up a budget for the coming year. What are the anticipated expenses? What income do you expect?

5. For some article you want very much, what are the pros and cons that you should consider before you purchase it? After you have considered all angles, what is your decision?

6. Arrange a trip to a farm where good business records are kept or interview a farm manager who helps farmers to keep records. What are the main features of the business records? Why are these records kept? What uses are made of them?

7. For some livestock or crop you are raising for yourself, plan a system of records to determine the amount of money you make. Provide for an opening and closing inventory and for expenses and receipts. At the end of the year, determine your profit or loss.

8. If you are living on a small-scale farm, what are some of the ways you could help to improve the operation of it?

9. Concepts for discussion:
    a. Large amounts of cash or credit are needed for current farm operation expenses.
    b. The farm of today often represents an investment of nearly $500,000.
    c. Net farm income consists of the difference between expenses and gross income.
    d. Disposable income per capita from farming has been consistently below that of the nonfarm population.
    e. Per capita income from farming has been maintained in part because the total income has been divided among fewer persons because of the decreasing number of farmers.
    f. Prices paid by farmers have risen faster than prices received by them for farm products.
    g. The most prosperous farmers have large-size businesses, high production rates, well-selected enterprises, and efficient use of labor and machinery.
    h. Good management is essential in farming.
    i. Long-time farm planning results in increased income.
    j. The personal goals of the farm family determine the kind of planning needed.
    k. An inventory of resources—land, labor, capital, and management—is essential for farm planning.
    l. Land should be used for the most profitable crops it will produce.
    m. A livestock program should be based on the most profitable way to market crops produced.
    n. Marketing programs that will take advantage of price cycles and provide money when needed should be planned.
    o. Farm building and equipment planning should be guided by the need for labor efficiency.
    p. Soil fertility planning should be directed so that continuously high yields can be obtained.
    q. The farm operation should be planned so that labor and management can be effectively utilized.

r. Farm purchases should be planned so that farmers can secure the most for the money spent and avoid costly borrowing to the greatest extent possible.
s. All joint operations of farms should be covered by written leases or agreements.
t. Insurance that will cover risks farmers cannot handle should be purchased.
u. In farm planning, free expert professional help should be consulted.
v. Proper maintenance of equipment is essential to profitable farming.
w. There are many ways in which operating costs of farming may be reduced.
x. Borrowing money should be carefully planned; money should be obtained from the most economical source.
y. Good records are essential to the management of a farm business.
z. Excess farm labor may often be used in off-farm employment.

REFERENCES

Davis, C. L. *The Farm Corporation.* Cooperative Extension Service, College of Agriculture, University of Georgia, Athens, GA 30602.

Luening, R. A., R. M. Klemme, and W. P. Mortenson. *The Farm Management Handbook.* Interstate Publishers, Inc., Danville, IL 61832.

Newman, M. E., and W. J. Wills. *Agribusiness Management and Entrepreneurship.* Interstate Publishers, Inc., Danville, IL 61832.

Rice, G. D., and R. S. Smith. *Insurance for the Farm Business.* Cooperative Extension Information Bulletin 167, College of Agriculture and Life Sciences, Cornell University, Ithaca, NY 14850.

U.S. Department of Agriculture. *Farm Index,* a monthly publication. U.S. Government Printing Office, Washington, DC 20402.

U.S. Department of Agriculture. *Marketing U.S. Agriculture.* Yearbook of Agriculture: 1988, U.S. Government Printing Office, Washington, DC 20402.

U.S. Department of Agriculture. *New Crops, New Uses, New Markets.* Yearbook of Agriculture: 1992, U.S. Government Printing Office, Washington, DC 20402.

Vocational Agriculture Service. *Law for the Farmer.* VAS 2018d, Vocational Agriculture Service, College of Agriculture, University of Illinois, Urbana, IL 61801.

Chapter 20

Cooperatives in Agriculture

> It ain't the guns nor armament
> Nor the army as a whole
> But the everlastin' teamwork
> Of every bloomin' soul.
> Rudyard Kipling (1865–1936)

The above lines written by a famous English poet express the spirit of cooperation needed for success in many endeavors. This is illustrated further by an old fable about several brothers who spent much of their time quarreling among themselves. Because of their disagreements and the failure to help each other, they did not accomplish much work. Finally, their father called them together. He showed them several sticks that had been gathered and tied in a bundle. He passed the bundle from one son to another and asked each in turn to try to break it. Not one was able to do so. He then untied the bundle and gave a stick to each son. Each son was told to break the stick he held, and this each did easily. By this simple demonstration, the sons were convinced that in group effort there is strength. What are examples from your own experiences that emphasize the value of group effort over individual effort?

OBJECTIVES

1. Define farmer cooperative.
2. Explain how cooperatives differ from other types of businesses.
3. Identify the kinds of activities performed by farmer cooperative organizations.

Why Is Cooperation Important Among Rural People?

Farmers have learned to work together in many ways. In pioneer days, neighbors got together for barn raisings, corn-husking bees, butchering, and other jobs that required many hands. In more recent years, threshing rings provided a way for farmers to exchange labor for threshing grain. In some cases, the threshing outfits were owned cooperatively. Today, this method of threshing grain has been replaced largely by combine harvesters owned by individual farmers or hired by them on a custom basis to do the work. For combines, large crews of workers are not needed. Today, farmers still exchange labor for some kinds of activities. Even barn raisings are not entirely a thing of the past. There are still instances in which farmers get together on the farm of a sick neighbor to do some of the work that needs to be done.

The development of cooperative organizations

One of the interesting developments among rural people is the formation of cooperative organizations to perform some of the business activities of farming. The development of these kinds of organizations has taken place rapidly since about 1900, but some were founded prior to that time. Some kinds of cooperative societies have been developed among urban people too.

A mutual insurance society, started in 1752, was probably the first cooperative organization in the United States. It was formed in Philadelphia, and Benjamin Franklin was one of its founders. It is still operating. Later, mutual insurance societies were formed among farmers. In each of these organizations, farmers who joined agreed to pay a portion of a fire loss to any one of the members. Today, these mutual insurance companies collect a specified sum, called a premium, from each member. These premiums are figured carefully to pay the operating expenses and expected losses.

In 1820, some farmers in Ohio joined together to take their hogs to a central market. About the middle of the nineteenth century, the Mormons in Utah formed mutual irrigation companies in which they pooled their labor and shared the cost to bring water to their crops.

After about 1840, farmers in various parts of the United States started to produce crops and livestock in increased amounts, and they began to meet with difficulties in securing fair prices for these products. Some groups of farmers developed cooperative methods for selling their surplus products. In 1841, a cooperative cheese factory began in a log farmhouse in Wisconsin. The milk from cows owned by six farmers was made into cheese by a farm woman and her son. The cheese was hauled by oxen to the village of Milwaukee where it was sold or traded. A short time later, some farmers in New York set up a cooperative creamery, or butter factory. In 1857, a cooperative grain elevator was established in Wisconsin for marketing the grain produced by the farmers who were members. Other early cooperatives of various kinds were started in other places in the United States. In 1857, the state of New York passed the first farmer cooperative law in the United States to legalize these organizations.

Over a century ago, some people in New England started a buying club,

Cooperatives in Agriculture

in which they supplied members with goods priced as near cost as possible. This was one of the first purchasing or consumer type of cooperative in the United States.

Recent developments in cooperatives

Today, there are many cooperative organizations among farmers. These have developed because of the need for them in the business operations of farming. A family-type farm is an efficient unit for producing raw materials of various kinds. However, because the volume of products raised on any one farm is usually quite small, an individual farmer is frequently handicapped in selling these products from the farm. Similarly, the farmer buys in small quantities and may find it difficult to purchase some things at reasonable prices.

Individually, farmers can do little to improve conditions related to marketing and purchasing. By using teamwork in the form of cooperative organizations, farmers frequently are able to improve their bargaining power in buying and selling.

Cooperative organizations make it possible for farmers to provide themselves with services that would be expensive or impossible to obtain if each farmer tried to secure them alone. The development of rural electrification

Fig. 23-1. Cooperative marketing associations, such as the one indicated by the sign, help to set the price all farmers receive by providing competition to nonfarmer-owned businesses.

is one example. Early in the 1930's, power lines had not been extended into most rural areas, and rates for electricity were high where this had been done. An act of Congress in 1935 provided for the Rural Electrification Administration, which aided farmers to form cooperatives and borrow money for building power lines. Some of these cooperatives built central generating plants, and others purchased electricity from privately owned or publicly owned power plants. By 1954, over 92 per cent of the farms in the United States were provided with electricity. Many of these farms secured electricity from privately owned power lines and privately owned power companies. However, in 1954, the REA cooperative organizations served about half the farms in the United States, and these developments stimulated private companies to expand their services to additional rural areas. By 1980, 99 per cent of all farms had electric service.

Cooperative organizations of various kinds help to make it possible for farmers to continue in the highly valued private enterprise system. By joining together in some kinds of activities, farmers are able to maintain a fair income with a minimum of governmental assistance. Prosperity among farmers helps to keep a strong, prosperous nation. Everyone should become familiar with cooperative organizations among farmers and thereby better understand the importance of these types of group activities to both farmers and the nation.

Agricultural Cooperative Service

As a means of helping farmers to help themselves, the U.S. Congress enacted the Cooperative Marketing Act of 1926. Under this act, the Agricultural Cooperative Service of the U.S. Department of Agriculture provides research, management, and educational assistance to cooperatives to strengthen the economic position of farmers and other rural residents. It works directly with cooperative leaders and federal and state agencies to improve organization, leadership, and operation of cooperatives and to provide guidance for the further development of cooperatives.

The Agricultural Cooperative Service strives to:

1. Help farmers and other rural residents obtain supplies and services at lower cost and get better prices for the products they sell.
2. Advise rural residents on developing existing resources through cooperative action to improve rural living.
3. Help cooperatives improve services and operating efficiency.
4. Inform members, directors, employees, and the public on how cooperatives work and benefit their members and their communities.
5. Encourage international cooperative programs.

In carrying out its responsibilities, the Agricultural Cooperative Service publishes research and educational materials and issues the publication *Farmer Cooperatives*.

What Are Farmer Cooperatives and How Do They Operate?

Farmer cooperatives are misunderstood by many urban people, largely because they know very little about these organizations. Some rural people, too, have mistaken ideas about cooperatives. A study conducted by Iowa State University showed that many urban people do not know that the major purpose of farmer cooperatives is to provide owner-members with goods and services at cost. Many think these organizations are exempt from taxes. Some believe cooperatives are a threat to our way of life. Many persons from farms and elsewhere do not know how these organizations are operated. This study also showed that persons who knew the most about cooperatives were usually in favor of them.

A farmer cooperative is a business organization of agricultural producers that is developed, owned, and operated by the members to perform important services for them at the lowest possible cost. Such organizations among farmers help to improve farm income by providing new, improved, or more economical services than could be provided otherwise.

Cooperatives must be operated efficiently in order to be successful. Some modern methods of operating cooperatives trace back to a small cooperative society founded in 1844 by a few people who lived on Toad Lane in Rochdale, England. There, a few textile workers, whose earnings were meager, banded together to purchase some of the necessities of life at a lower cost than they had been paying. The venture proved successful in spite of many handicaps. The success of this cooperative was due largely to the use of sound rules of operation, which to this day are followed generally by cooperatives in many countries.

The three rules of action most important for the success of cooperative organizations are as follows:

1. *Democratic control by members.* The members of a cooperative are the ones who benefit from it; therefore, they should share equally in controlling it. Democratic control is provided in most cooperatives by limiting each member to one vote. This provision makes it difficult for a few members to gain control of the organization. The laws for cooperatives in most states limit each member to one vote, regardless of the amount of the member's investment in the organization or the extent to which the member uses the organization.

2. *The interest on money invested in a cooperative organization is limited to a fixed percentage.* Capital is needed to operate a cooperative

organization, as is true with any business. Since a cooperative is a nonprofit organization, its earnings represent savings for its members. Hence, these earnings are not used for increasing the interest paid to investors. Federal laws and most state laws specify the maximum interest that may be paid on money invested in a cooperative organization.

3. *Benefits in the form of savings from a cooperative organization are distributed among members in proportion to the amount of business each does with the organization.* These distributed funds are called patronage dividends. For example, one farmer might market $1,000 worth of products through a cooperative during one year. Another farmer might market only $100 worth of products. The first farmer would receive 10 times as much of the savings or earnings (after all expenses of the organization are paid) at the end of the year as the other farmer.

Other important features of most cooperative organizations are:

1. Membership is open to any farmer who is willing to abide by the rules.
2. Prices of products sold by cooperatives are in line with levels in the community.
3. Business is transacted largely on a cash basis.
4. Full information about accounts and audits is made available to members.
5. An educational program is provided so that members and prospective members can become acquainted with the organization's operation.

Organizing and operating a cooperative

Cooperatives vary somewhat in ways of doing business. Nearly every cooperative elects persons to a board of directors that formulates policies and helps make major decisions. These directors often serve without pay, other than expenses. The major responsibility for operating a cooperative is usually placed in the hands of a general manager, who is hired by the board of directors.

Cooperatives are organized according to sound business procedures. In organizing a cooperative, farmers must consider the articles of incorporation, bylaws, marketing agreement, procedure for becoming a member, and area to be served. Cooperatives may be classified as local, regional, federated, or centralized.

The articles of incorporation are a statement of the kind and scope of business the cooperative is designed to do. These articles specify the name of the organization, its purpose, the scope of its authority, its registered agent (who must be a member of the board of directors) and business office address, the number of directors, and the type of capital (stock or nonstock).

The bylaws tell how a cooperative is going to operate. They are more descriptive and detailed than the articles of incorporation. Typically, the bylaws contain provisions for membership eligibility, election of directors, annual meetings, officers' duties, voting rights, dues and assessments, and rate of dividends (if any) to be paid on members' capital invested.

The marketing agreement is used by many marketing cooperatives. This agreement, which may be referred to in the bylaws as a separate document or included as a section of the bylaws, is a contract between the cooperative and a member. It states the duty and intent of the member to deliver a specified amount or per cent of production to the cooperative.

The procedure for becoming a member varies with the type of cooperative. Some associations, particularly marketing associations, require a formal application for membership and issue a membership certificate. In other cooperatives, particularly purchasing cooperatives, membership is identified with patronage. A patron making a purchase becomes eligible for membership if he/she qualifies as a producer.

The classification of a cooperative as local or regional depends upon the geographic area served and the makeup of the membership.

A local cooperative provides service for a community, a county, or several counties. Individuals are the members of a local cooperative.

A regional cooperative is one that usually serves an area making up a number of counties, an entire state, or a number of states. There are two important classifications of regionals based entirely upon the makeup of their membership—federated or centralized (some regionals are combinations of the two).

A federated cooperative is one composed of local cooperatives as its members. These local cooperatives are operated by local managers appointed by and responsible to local boards of directors. The federated cooperative has its own general manager and staff and a board of directors elected by and representing the local associations. The locals are autonomous but depend in varying degrees on the federation for a variety of services.

A centralized cooperative may serve patrons in an area covering at least 5 to 10 counties; it often serves a greater area. A regional centralized association is structurally like a small-scale local cooperative—individuals make up the membership. A centralized cooperative has one central office, one board of directors, and one general manager who supervises the entire operation, which may be conducted through several branch offices.

The actual operation of a cooperative is carried out by the directors, the manager and some assistants, the clerical and operating employees, and the member-patrons.

The directors are elected by the members at the annual meeting. The board of directors is responsible for managing the affairs of the association.

These responsibilities include selecting managerial personnel, delegating authority and assigning duties to managerial personnel, formulating policies, checking to see that policies are being followed, and evaluating the results obtained from the operation of policies adopted.

The board of directors of a cooperative selects the manager, who, in turn, selects any needed assistants. After the board has decided what the cooperative will do, the manager and staff decide how it can best be done. Ideally, the manager's principal tasks are planning, reporting to the board of directors, conferring with members of the management team, maintaining good organizational relations, and overseeing the cooperative's operations.

Clerical and operating employees are selected by the manager. They carry on the various operations in the cooperative business. In many cases, cooperatives have training programs for their employees.

Cooperatives are organized by members and exist only to serve the needs of members. At the annual meeting, members learn how their cooperative is doing and have a chance to make their wishes known. Members are responsible for using the services of their cooperative and encouraging others to do so. As a cooperative grows, members provide a portion of the new captial needed. They are also responsible for serving as directors, working on committees, calling on prospective new members, and representing the cooperative in various community activities.

Well-informed, active members are important in the success of a cooperative. They should think in terms of the greatest good for the greatest number. They should elect knowledgable directors and insist that capable managers be hired. Managers who are experts in their fields and who know how to secure the full confidence and support of the members should be selected.

Every cooperative organization worthy of its name conducts its business democratically. Members are free to express their personal viewpoints at meetings and to help decide how the business is to be conducted in the best interests of the people who belong to it.

How cooperatives differ from other types of businesses

In order to understand cooperatives fully, it is helpful to compare them with other ways of doing business. Several ways of doing business are found in many communities.

One kind of business is the *single-owner type*. In this type, one person is responsible for making the investment and operating the business. This individual gets any profits and takes the losses if they occur. Many retail stores, local service establishments, and other small business firms are of this type. Many farmers operate their farms as single owners. Single-owner businesses are easy to form and to control, since the owner's decisions are final. The operation is flexible, with great individual opportunity.

Cooperatives in Agriculture

A second type of business is known as a *partnership*. In this, two or more persons pool their capital and their abilities in a business. As in the case of a single-owner business, the partners get the profits or take the losses. Some business establishments are of this type. In some cases, fathers and one or more of their children form business partnerships in farming. Partnerships combine the capital and the special abilities of the partners for mutual benefit. However, each partner is liable for all partnership debts, and of course, a partnership is held together only as long as each agrees to it.

A third type of business is a private *corporation* in which the stockholders own the business. In most organizations of this kind, each stockholder votes in proportion to the number of shares of stock owned. The returns on the investment of each stockholder depend on the profits made by the corporation. The primary objective of a private corporation is to make a profit, as is true of the two types of business previously described. Many banks, factories, chain stores, and other establishments are organized as corporations.

Stockholders' liabilities are limited to the number of shares of stock owned. Shares of stock can be bought or sold. The business continues while owners change as a result of the buying and selling of stock. An individual stock owner may have very little influence on company policy. It is legally quite complex to set up a private corporation.

A fourth type of business organization is a *cooperative*. A cooperative is a corporation, but its objective is to *operate at cost* in providing some kind of

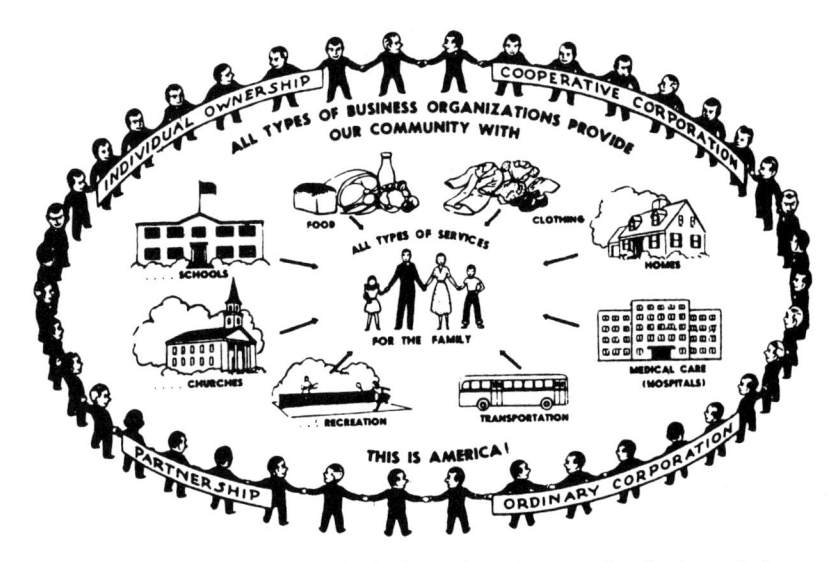

Fig. 23-2. The four ways of doing business shown here are all parts of our private enterprise system, and all have a place in building prosperous communities. (Courtesy, U.S. Department of Agriculture)

service for its members. Any savings resulting from a cooperative's services are returned to the users of its services. Thus, a cooperative is a nonprofit organization that is operated democratically and is in no way in conflict with our democratic form of society. Many types of business firms, in addition to farmers, use various types of cooperatives. Many retail grocers, druggists, bakers, lumber dealers, hardware dealers, and other businesspersons belong to retailer-owned cooperatives. The Associated Press is owned cooperatively by about 2,000 newspapers. The main purpose of these organizations is to enable independent business firms to stay in operation in competition with large establishments. Member-owners have limited liability and control. Complex in organization, cooperatives continue to exist even when ownership changes.

Each of the four types of business organizations has an important place in our economic system, and none of them is in conflict with our way of life.

Cooperatives are sometimes unjustly accused of having certain tax advantages. Cooperatives, like other forms of business, pay many kinds of taxes. Among these are real estate taxes, social security taxes, sales taxes, and many other types. Only in the case of federal income taxes are cooperatives sometimes permitted certain exemptions. Cooperatives have the right to deduct patronage refunds—money collected from the farmer-owners of the cooperatives for services performed, which is found to be in excess of costs of operation—from income before taxes are calculated. (As a matter of clarification, it should be noted that any corporation can make these deductions if it wishes to make these distributions of profits to its customers.) To qualify for such exemptions, however, cooperatives must meet certain rigid requirements. Many cooperatives choose to pay federal income taxes on their savings rather than meet the strict exemption requirements.

What Kinds of Activities Are Performed by Farmer Cooperative Organizations?

Farmer cooperative organizations are numerous and varied in nature. All the major farm organizations (Farm Bureau, Grange, and Farmers' Union) have aided in the development of farmer cooperatives. The American Institute of Cooperation provides educational materials and information about cooperatives. The Cooperative League of America promotes the development of cooperatives to aid consumers. The National Council of Farmer Cooperatives is a national farm organization for farmer-owned marketing and purchasing associations.

Federal and state laws have been passed to legalize cooperative organizations and to help regulate them in desirable ways. In some cases, loans from federal sources have been provided to help in financing cooperative organizations, with provisions for repayment of the amounts borrowed.

Cooperative organizations among farmers may be classified as production, marketing, purchasing, or service cooperatives.

The following are examples of cooperative marketing organizations:

> Livestock shipping associations and commission firms
> Grain elevators for storing and selling grain
> Factories for processing and selling fruits and fruit products
> Plants for grading, packaging, and selling eggs
> Plants for dressing, packaging, and selling poultry
> Cotton gins for separating seeds and baling cotton
> Plants for making butter, cheese, and other dairy products
> Associations for grading and marketing wool
> Associations for marketing milk and other dairy products
> Associations for marketing tobacco

The most important jobs performed by various marketing associations are assembling, grading, packing, processing, financing, storing, transporting, and selling. Some do only one of these jobs in addition to selling, while others perform two or more. These organizations carry farm products one step or more beyond the farm and thereby help farmers to receive more money for their products. Some cooperative marketing organizations handle no products but act as bargaining agencies in selling the products of their farmer-members. In addition to various marketing activities, some of these cooperatives engage in selling selected products to farmers. These organizations are a combined type of marketing and purchasing, or consumer, cooperative.

Consumers of various products usually have a great deal of confidence in brand names used on many high-quality products, and this is especially true of many foods. Some well-known brands of foods sold in retail stores throughout the country have been processed and packaged by cooperatives. Examples are Sunmaid raisins and other dried fruits, Sunsweet dried prunes and apricots, Sunkist oranges and other citrus fruits, Land O' Lakes butter, Eatmor and Ocean Spray cranberries, Flav-R-Pac frozen fruits and vegetables, Tru-Blu-Berries, Texsun grapefruit, Rockingham frozen turkeys, and many others.

The following are examples of activities of *purchasing, or consumer, cooperatives:*

> Sale of feed, seed, fertilizer, and other supplies
> Sale of farm machinery and other equipment
> Sale of gasoline, oil, and other petroleum products

These and other consumer cooperatives provide goods and supplies needed by members. The products are handled at cost, and the savings are distributed to purchasers as patronage dividends. In addition to these kinds of savings, special care is taken to provide products of good quality. Some consumer cooperatives operate as wholesale establishments that sell products to local cooperatives that sell directly to consumers. A few cooperatives

Fig. 23-3. Cooperatives contribute in a major way to the improvement of livestock and crop production.

engage in manufacturing the products they sell, and some engage in producing and refining petroleum products, which are sold by local cooperatives to the users.

The following are examples of *service cooperatives:*

> Rural electrification cooperatives
> Rural telephone cooperatives
> Credit and loan associations
> Insurance societies
> Artificial breeding associations for dairy cattle
> Dairy Herd Improvement Associations
> Irrigation organizations
> Farm management and record keeping associations
> Frozen food locker plants

The service cooperatives provide many kinds of functions, as noted by the above examples. They deal primarily in services other than marketing and purchasing products for the members.

The farmer cooperative organizations in the United States number in the thousands and their members number in the millions. In 1991 according to USDA reports, there were a total of 4,494 marketing, farm supply, and related service cooperatives with a total gross business volume of $90.8 billion. The most recent figures are shown in Table 23-1. Memberships in marketing, farm supply, and related service cooperatives totaled about 4.1 million in 1991, a drop of 4.4 per cent from a year earlier. The drop in membership reflects the decrease in the number of farmers in the United States.

In 1991 four farm commodities accounted for nearly 84 per cent of the net sales of farm products by cooperatives. The four commodities were (1)

Table 23-1. Types of cooperatives, number of associations, and number of members

Type of cooperative	Number of associations	Number of members or participants
Marketing	2,384	1,842,413
Farm supply	1,689	2,024,700
Related services	421	191,457
Federal Land Bank Associations	80	n.a.
Production Credit Associations	72	n.a.
Dairy Herd Improvement Associations	n.a.	31,302
Rural credit unions	751	3,413,000
Rural electric cooperatives	896	11,061,000
Rural telephone cooperative	n.a.	31,302

Source: *Agricultural Statistics, 1992*, U. S. Department of Agriculture, U.S. Government Printing Office, Washington, DC 20402

Farmer Cooperatives in the United States

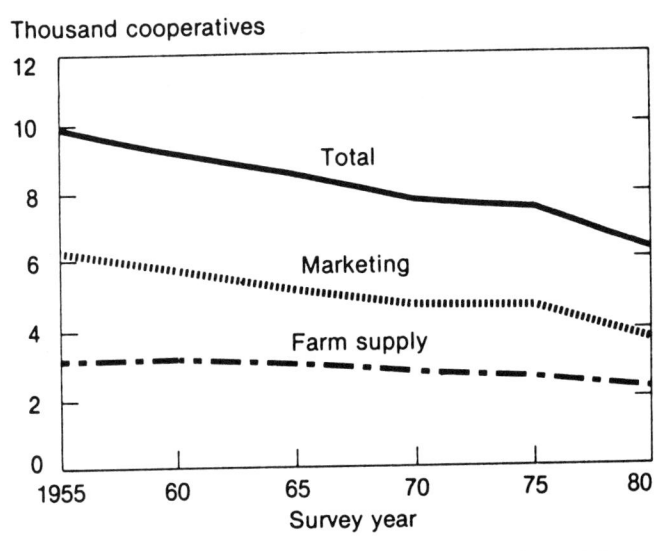

Total includes a small number of cooperatives that provide specialized related services.

Fig. 23-4. The total number of farmer cooperatives continues to decrease as the number of farmers decreases. (Courtesy, U.S. Department of Agriculture)

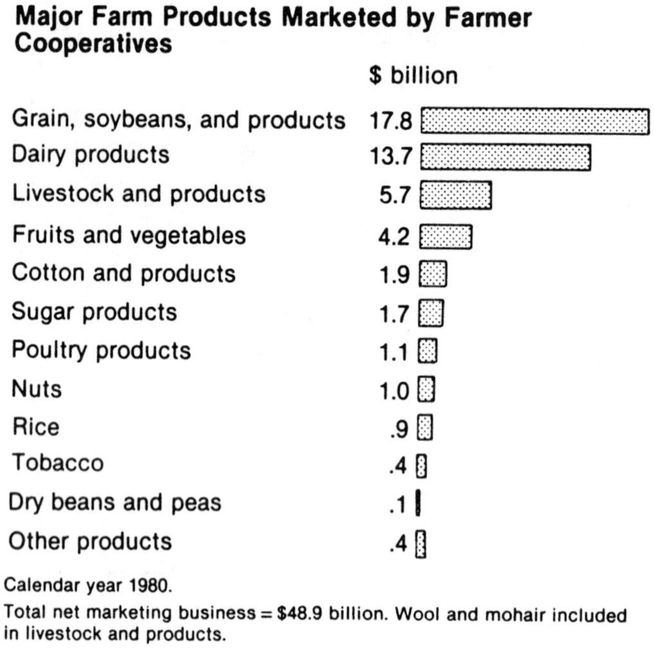

Major Farm Products Marketed by Farmer Cooperatives

$ billion

Grain, soybeans, and products	17.8
Dairy products	13.7
Livestock and products	5.7
Fruits and vegetables	4.2
Cotton and products	1.9
Sugar products	1.7
Poultry products	1.1
Nuts	1.0
Rice	.9
Tobacco	.4
Dry beans and peas	.1
Other products	.4

Calendar year 1980.
Total net marketing business = $48.9 billion. Wool and mohair included in livestock and products.

Fig. 23-5. Grain, soybeans, and their products and dairy products account for a large percentage of the business of farmer marketing cooperatives. (Courtesy, U.S. Department of Agriculture)

Major Farm Supplies Handled by Farmer Cooperatives

$ billion

Petroleum products	5.1
Feed	3.5
Fertilizer	3.5
Farm chemicals	1.0
Building materials	.5
Seed	.5
Machinery and equipment	.4
Meats and groceries	.2
Containers	.1
Other supplies	1.3

Calendar year 1980. Total net farm supply business = $16.1 billion.

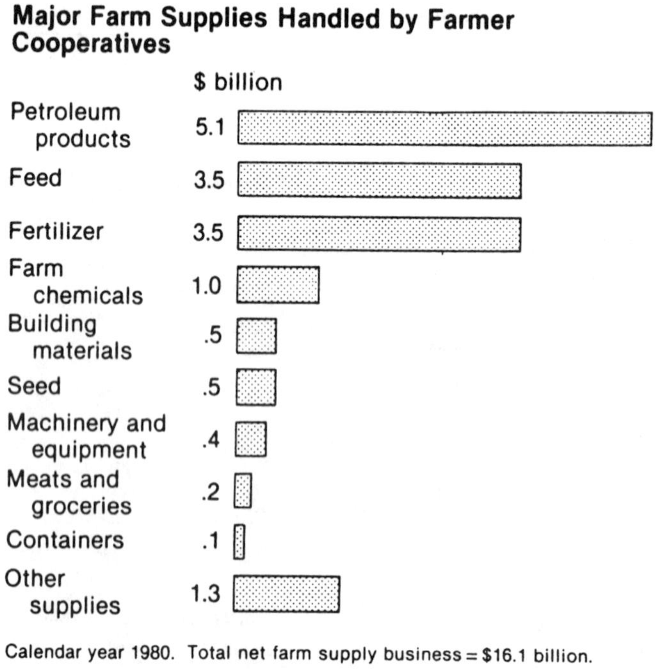

Fig. 23-6. Most of the farm supply business of farmer cooperatives comes from sales of petroleum products, feed, and fertilizer. (Courtesy, U.S. Department of Agriculture)

Share of Products Marketed by Farmer Cooperatives

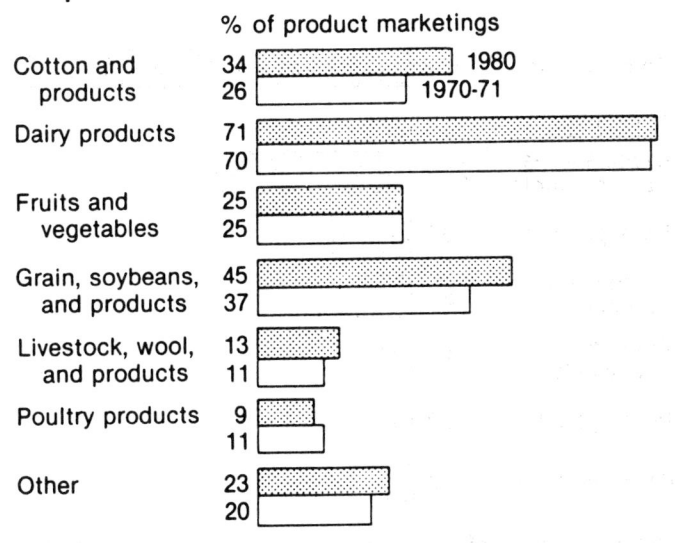

% of product marketings

Product	1980	1970-71
Cotton and products	34	26
Dairy products	71	70
Fruits and vegetables	25	25
Grain, soybeans, and products	45	37
Livestock, wool, and products	13	11
Poultry products	9	11
Other	23	20

Farmers marketed 27 percent of their products through their cooperatives in fiscal year 1971 and 31 percent in 1980.

Fig. 23-7. In 1980, the farmer cooperative share of products marketed ranged from 9 per cent for poultry products to 71 per cent for dairy products. (Courtesy, U.S. Department of Agriculture)

Business Volume of Farmer Cooperatives

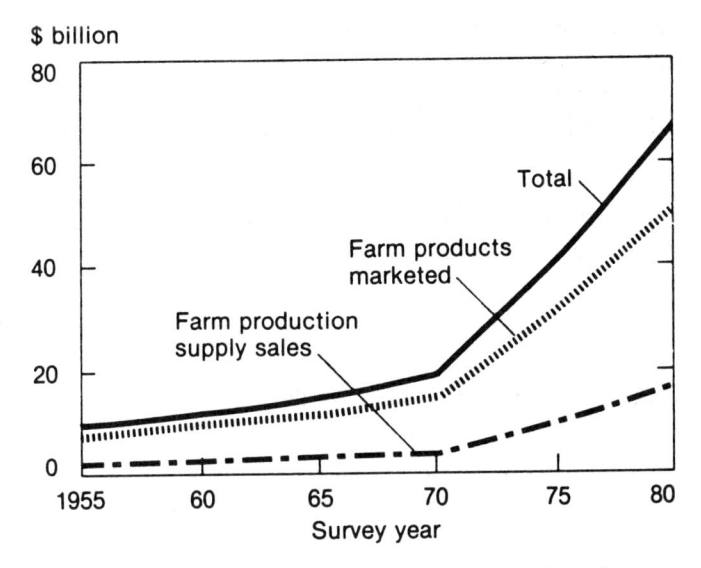

Business volume is on net basis; it excludes intercooperative sales, but includes receipts for specialized services provided to patrons.

Fig. 23-8. The volume of business conducted by farmer cooperatives has increased rapidly since 1970. (Courtesy, U.S. Department of Agriculture)

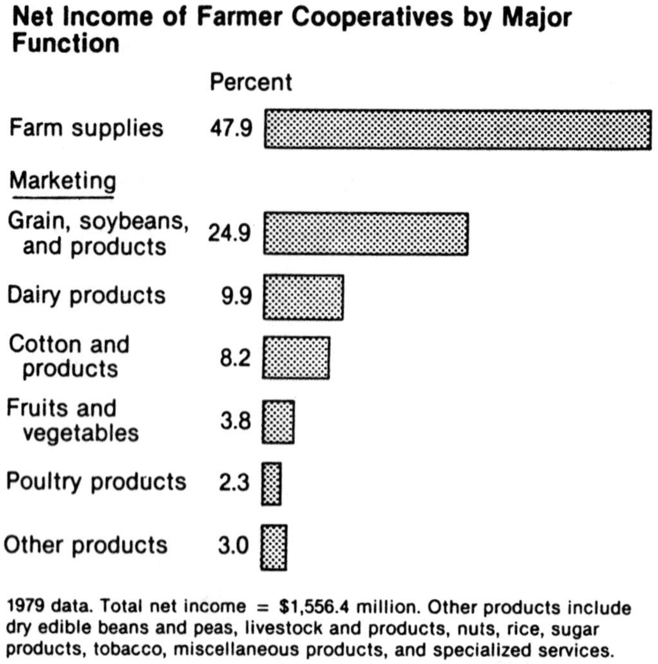

Net Income of Farmer Cooperatives by Major Function

Percent

Category	Percent
Farm supplies	47.9
Marketing	
Grain, soybeans, and products	24.9
Dairy products	9.9
Cotton and products	8.2
Fruits and vegetables	3.8
Poultry products	2.3
Other products	3.0

1979 data. Total net income = $1,556.4 million. Other products include dry edible beans and peas, livestock and products, nuts, rice, sugar products, tobacco, miscellaneous products, and specialized services.

Fig. 23-9. Nearly half of the net income of farmer cooperatives in 1979 was made in the handling of farm supplies. (Courtesy, U.S. Department of Agriculture)

grain, soybeans, and products; (2) dairy products; (3) livestock and products; (4) fruits and vegetables.

Petroleum products, feed, and fertilizer accounted for about 75 per cent of total net supply sales.

Farmer cooperatives handled more than 36 per cent of the farm products marketed in 1991. Cooperatives also handled 40 per cent of all farm supplies in 1991.

Cooperatives average about 900 members each; with farm supply the largest at about 1,200 members each.

What Are Some Examples of Cooperatives?

Studying some examples and noting the ways in which they aid farmers can help us to understand better how cooperatives operate.

Cooperative marketing of grain

Cooperatively owned grain elevators buy, store, and sell grain for farmer-members. By performing these services at cost, they save farmers considerable money. In some cases, they hold grain in storage during seasons of low prices and market it at favorable times throughout the year.

Cooperative marketing of dairy products

Several types of cooperatives have been developed for aiding farmers to get better prices for dairy products. Many of these cooperatives engage in producing cheese and butter. Emphasis is usually placed on producing high-quality products, and farmers are paid premium prices for their high-grade milk and cream.

Land O' Lakes Creamery, Inc., is one of the largest dairy marketing organizations in the world. It is owned and controlled by dairy and poultry farmers in Minnesota, Wisconsin, and nearby states. Land O' Lakes is a central association that includes several hundred cooperative creameries, cheese factories, and dairy plants. It has aided in providing better dairy products for consumers and better prices for producers. In addition to dairy products, it markets poultry and eggs for its members. The central association assembles, grades, packages, and sells products from member organizations. Because of the emphasis on high quality, the brand name has become widely and favorably known. The organization also sells feed, seed, and fertilizers to farmer-members. A herd improvement division provides breeding services for the dairy herds of members and thus helps to increase the production in these herds. In these and other ways, the organization has helped many farmers to become more prosperous.

Various types of cooperative associations are engaged in the marketing of milk that is sold to distributors in cities. Some of these have trucks for hauling the milk from farms to distributors; in addition, they act as bargaining agencies in arriving at prices that distributors pay for the milk. Some of them serve as bargaining agencies only and handle none of the milk. An example of this type is the Pure Milk Association for dairy farmers in the Chicago milk-producing area. This cooperative organization helps dairy farmers to bargain for prices for the milk sold to milk dealers. It also encourages the public to use more milk and other dairy products.

Four large milk-marketing cooperatives in northern Ohio and western Pennsylvania form Milk, Inc. This cooperative is one of the largest in dairying, with membership almost equal to that of Dairymen, Inc.

The Georgia Milk Producers Cooperative Association is a division of Dairymen, Inc., a multi-state cooperative of dairy farmers of the southeastern United States. Along with Georgia producers who were formerly members of Dairymen, Inc., it is the cooperative's largest division in volume of milk produced. Three other divisions of Dairymen, Inc., are the Old Dominion Division at Richmond, Southeast Division at Bristol, and Norfork Division at Chesapeake, which serve dairy farms in Virginia.

Cooperative marketing of livestock and livestock products

Cooperative livestock marketing associations make it possible for farmers to receive more for the animals they send to market. Some of these associations assemble livestock locally and ship them by rail or truck to central markets. In some cases, these animals are sorted according to grades and are sold by commissioners who represent cooperatives at central markets. In other cases, livestock assembled locally by cooperatives are sold directly to buyers for large meatpacking firms.

Many farmers sell wool through cooperative associations. These associations in each state ship the wool in properly labeled sacks to the National Wool Marketing Association in Boston. There it is graded, stored, and sold to manufacturers of woolen products. The sale of wool by this central cooperative is spread over several months in order to secure the best prices possible. Farmers who own the stored wool secure loans representing a portion of the value of the wool stored on their accounts.

The Poultry Producers of Central California is a cooperative organization that markets eggs, produces feeds, and performs other services for poultry raisers in the famous Petaluma area of that state.

Cooperative marketing of fruit

One of the oldest and largest cooperative marketing associations in the United States is Sunkist Growers, Inc., formerly the California Fruit Growers Exchange. It was started over 60 years ago by the growers of citrus fruits. At that time, oranges were a luxury and were used primarily as a special treat at Thanksgiving and Christmas. As early as 1924, the California Fruit Growers Exchange had exported citrus products to most parts of the world. This organization had an important part in making oranges and orange juice a regular part of breakfast for many U.S. families. This was done by good advertising, by improved quality, and by efficient marketing.

The California Prune and Apricot Growers Association is a cooperative that markets these products for many growers. Prior to cooperative marketing, the growers marketed their products individually, and during seasons of large production, the prices fell to low levels. The cooperative organization advertises widely and through its processing plants provides high-quality products. These are placed in attractive packages and sold under the Sunsweet brand to grocers and other distributors. The organization has developed new uses for these products. Today, it markets more prunes than any other organization in the world. Its primary purposes are to market a high-quality product and to obtain for its grower-members the highest possible net profit.

At least 80 per cent of the cranberry crop is marketed through Ocean

Cooperatives in Agriculture

Spray, a New England cooperative. Ocean Spray has its headquarters in Massachusetts near the Cape Cod bogs.

Cooperative marketing of cotton

The Cotton Producers Association, with headquarters at Atlanta, is one of the leading cooperative associations for marketing cotton. By securing prices based on the quality of cotton produced, this organization has added to the income of many farmer-members in Georgia, Alabama, and South Carolina. It provides warehouses for receiving and storing cotton. It also sells high-quality seed, fertilizer, feed, and some other farm supplies. By supplying farmers with seed of improved varieties and by grading, delinting, and treating the cotton seed that farmers plant, the association has helped farmers develop cotton improvement programs.

Cooperative purchasing associations

Some cooperatives handle feed, seed, fertilizer, petroleum products, and many other supplies needed by farmers. Some of these organizations operate on a regional basis and sell to distributing organizations. These associations are large enough to buy products in large quantities and hence can secure them at reduced prices. They distribute these products on a nonprofit basis and thus make further savings possible to farmers. Some of

Fig. 23-10. Grain is being loaded on a truck at an agricultural cooperative.

these cooperatives mix feeds and carry on other processing activities for the products they sell.

The nation's largest farmer-owned purchasing cooperative was formed by the merger of the Cooperative Grange League Federation Exchange and the Eastern States Farmers Exchange in 1964. This cooperative, Agway, Inc., serves the eastern states through many retail outlets. Feed, grain, flour, and cereals represent the chief types of products sold. Other products include fertilizer and lime, petroleum products, seed, seed potatoes, various metal products, building materials, insecticides and other farm chemicals, farm equipment, refrigeration equipment, rope and twine, and tires and auto supplies. It also engages in marketing some kinds of farm products for farmer-members.

The Farm Bureau Cooperative Association in Ohio handles many kinds of commodities needed by farmers. These are sold at wholesale to various member associations that sell to consumers in many parts of Ohio. Feed, petroleum products, fertilizer, seed, farm machinery, steel products, and building materials are among the products sold by this cooperative through its distributors. Similar cooperatives are operated by the Farm Bureau in Michigan, Pennsylvania, Indiana,, and many other states.

The Southern States Cooperative, with headquarters at Richmond, furnishes supplies to retail cooperatives and privately owned service agencies in Virginia, West Virginia, Kentucky, Maryland, Delaware, and Tennessee. Feed, fertilizer, seed, petroleum products, farm equipment, insecticides, and electrical equipment are among the products sold. This organization also provides for the marketing of eggs, grain, and some other products produced by its members.

The Farmers Union Central Exchange, Incorporated, has headquarters at St. Paul. It operates cooperatives in Minnesota and several other northern states. Petroleum products, farm machinery, tires and auto supplies, fertilizer, and building materials are among the products handled.

Cooperative associations for loaning money

Farmers, like other businesspersons, frequently need to borrow money. Up until 1916, this credit was secured entirely from individuals and private agencies, but since then, cooperative organizations have provided a large share of this service. This has made it possible for farmers to borrow money on more favorable terms than formerly. Many of these cooperative agencies are assisted by the federal government through the Farm Credit Administration. These cooperatives are principally of two types: those that provide mortgage loans on farms on a long-term basis and those that provide production loans to farmers on a short-term basis.

Farm mortgage loans to individual farmers are made by Federal Land Banks through Farm Loan Associations in the United States. Short-term loans are secured through local cooperative organizations known as Production Credit Associations. Through them, loans are made to farmers for purposes such as purchasing seed, feed, fertilizer, machinery, and livestock. Most of these loans are repaid in less than a year, although renewals are possible in some cases. Every farmer who borrows from a Production Credit Association becomes a member and purchases shares of stock amounting to about 5 per cent of the amount borrowed, as is also the case in the Farm Loan Association. Both types of associations operate on a nonprofit basis and provide credit to farmers at cost.

Banks for Cooperatives operate in various parts of the United States. These loan money to marketing and purchasing associations of farmers.

In all three types of cooperative credit associations, provisions are made for them to pay back the initial loans from federal sources. In recent years, many associations have paid back all these federal loans and thereby have become completely farmer owned.

Rural electrical cooperatives

Much of the increase in electrification of farms has been made possible by the development of cooperative organizations financed by the Rural Electrification Administration of the federal government. Through cooperative organizations of farmers, power lines have been constructed, and in some cases, central power plants for making electricity have been built. The REA loans to local cooperatives are repaid within 25 years by small amounts added to each consumer's bill. The National Rural Electrical Cooperative Association is a private overall organization for rural electric cooperatives and public power districts in the United States. Of the nearly 2.3 million farms in the United States, 99 per cent receive central station electric service.

From the discussion in this chapter and other information that you may have secured, you should have a better understanding of cooperative organizations among rural people. These organizations help farmers to get better prices for many of the products they raise. Cooperatives of some kinds help farmers to purchase high-quality products at lower costs. Some cooperatives help farmers to provide various services at lower costs than could be secured without such organizations. Thus, in many ways, cooperatives help farmers to help themselves.

SUGGESTED ACTIVITIES

1. Describe some of the activities you have heard about in which several farmers have worked together for the benefit of the group. What benefits were gained?
2. What kinds of business organizations operate in your community? Describe an example of each type.

3. Interview some farmer who is a member of a cooperative. What advantages does the farmer feel are obtained through this organization?

4. With your class, arrange to visit the headquarters of a cooperative organization in your community. Find out from the individuals in charge how it is organized and operated. What products or services are handled? How do individual members participate in its operations? How are patronage dividends distributed? What other features are of special interest?

5. Invite a manager or director of a local cooperative to speak to your class. Ask for an explanation of how the cooperative helps its members and the general public.

6. Organize a panel discussion on a topic such as "the place of cooperatives in rural life" or "how cooperatives serve our community."

7. Study the cooperative movement in some foreign country, such as Sweden or Denmark. Prepare a report to present to your class.

8. Describe some of the activities of an organization to which you belong. Which activities are carried out in a cooperative manner? How could your organization be improved if the spirit of cooperation were applied more extensively than is done at present?

9. Concepts for discussion:
   a. Cooperative action applies to many areas of life.
   b. Farmers have formed many kinds of cooperatives to improve their bargaining power in buying and selling.
   c. Farmer cooperatives are simply one way of doing business.
   d. Cooperatives must use sound business procedures in order to be successful.
   e. Cooperatives are owned and controlled by their members.
   f. Farmer cooperatives are nonprofit organizations, operating at the lowest possible costs to members and with any earnings being credited as farm income to owners in proportion to the volume of business.
   g. Cooperatives serve large areas of the country.
   h. Well-informed, active members are essential to the success of a cooperative.
   i. Cooperatives operate on the principle of one person, one vote.
   j. Cooperatives differ from other ways of doing business in their operation.
   k. There are many nonfarmer kinds of cooperatives.
   l. The federal income tax advantages of cooperatives are available to other businesses that meet the strict exemption requirements for them.
   m. Cooperatives are regulated by federal and state laws.

REFERENCES

Cotterill, R., ed. *Consumer Food Cooperatives*. Interstate Publishers, Inc., Danville, IL 61832.

Newman, M. E., and W. J. Wills. *Agribusiness Management and Entrepreneurship*. Interstate Publishers, Inc., Danville, IL 61832.

Oliver, J. D. *Agricultural Cooperatives*. Curriculum Materials for Agricultural Education, Virginia Polytechnic Institute and State University, Blacksburg 24061, in cooperation with the Virginia State Department of Education, Richmond, VA 23216.

Chapter 21

Keeping Our Environment Clean

It has been only in the past decade that enough people have become concerned to make the pollution of our environment one of the most talked about of our many social and economic problems. Since people are the major source of pollution, and since the population of the world continues to increase daily, it is unlikely that the problem of pollution will ever disappear.

OBJECTIVES

1. Identify some of the ways we pollute our environment.
2. Discuss how pollution can harm wildlife.
3. Discuss air, water, and land pollution.
4. Identify ways everyone can help fight pollution.

How Do People Pollute?

Much of what is called pollution of the environment is simply a result of the activities of people in their struggle to live and to earn a living. The technology of the economy, whether mechanical or chemical, has resulted in actions that have made pollution a by-product of an affluent society. For example, much food is wrapped, placed in a container, then put in a cardboard box, and finally transferred to a plastic bag. Not only does all this packaging add to the price of food, but it also costs extra money to have the waste packaging collected and disposed of by the sanitation department. Recently, the average weight of food packages was reported to have increased over one-third, while the weight of the food in the packages rose only 2.3 percent. Additionally, most of the plastic containers, paper bags, glass bottles, metal cans, and other packaging materials are never reused, and thus, scarce natural resources are wasted. Even the vast outer space surrounding the earth is a part of the environment and can be polluted to an extent that the health and well-being of people will be affected.

Unfortunately, since the time of the Industrial Revolution, people have behaved unwisely with regard to their environment. What many people do not realize is that no material on earth can actually be thrown away, it can only be turned into something else. Thus, the enormous problems of pollution control are to identify more clearly what the pollution problems really are, to determine what can actually be done about these problems, and to decide what should be done first. Basically, the problems of pollution involve, in one way or another, how to use things in ways that will not result in waste products that will destroy the environment and make life impossible.

Pollution and agriculture

Environmental pollution has been a major concern in agriculture over the years. While many of the present problems are a result of technological progress, some of the problems of pollution have existed for a long time and range from the very specific to the very general and complex. The examples are many.

Agriculture accounts for about half the solid wastes produced in the United States. Some of these wastes can be disposed of only at a high cost.

One of the most vivid of the problems of pollution is that of soil in rivers and lakes. In agriculture, this problem has been attacked from the standpoint of erosion control (see Chapter 7). Soil being washed into rivers and streams is now called water pollution as well as soil erosion. How to dispose of animal waste products and the odors from the production and processing of livestock products has long been of concern. These products and odors are now a form of air pollution. The many controls on food production designed to put an abundance of safe food on the table have actually been a form of pollution control.

Feedlots, concentrated animal feeding operations, are especially troublesome. Feedlots have often resulted in the discharge of large amounts of plant and animal waste into streams. Since the decomposition of these materials requires large amounts of oxygen, the use of the oxygen in the water depletes the supply so much that the fish and other forms of aquatic life are unable to live. Bacteria that include waterborne diseases, phosphorus, ammonia-nitrogen, and odors are other pollutants that come from feedlots. In Maryland, an 800-acre oyster and clam bed had to be closed because of bacterial contamination traced to a large feedlot.

The examples of the relationship between pollution and agriculture are almost endless. While pesticides get a major share of attention as pollutants related to agriculture, the others mentioned are also under study and will become more visible as the fight for a clean environment continues.

In addition to sediments and animal waste products, the kinds of pollutants being studied, from the standpoint of agriculture, include plant nutrients, waste from industrial processing of raw agricultural products, forest and crop residues, and inorganic salts and minerals.

Synthetic chemicals are carried to all parts of the earth by air, water, and other means. Some of these chemicals have been found in the bodies of wild animals in every part of the world.

Experience and research are helping people learn how to use chemicals more safely. Biological techniques are being developed to control pests without the use of chemicals. Yet, chemicals are needed to help in food production and to kill disease-bearing insects. About half of the pesticide purchases are used for farming; the other half are used for lawns, golf courses, and other nonfarm purposes.

Persistent pesticides

One of the most dangerous of the agricultural pollutants is the persistent pesticide group.

The persistent pesticide group is made up of the *organochlorines*. These chemicals are persistent because they decompose slowly and, as a consequence, stay active for many years after being used. The main persistent pesticides are aldrin, DDT, and dieldrin. Some others of the same group are benzene hexachloride, heptachlor, lindane, strobane, TDE, and toxaphene. Most farmers, and other agriculturists who use chemicals, will recognize the chemicals named as old friends in the fight against plant and animal pests. The U.S. Department of Agriculture has placed restrictions on the use of these persistent pesticides.

The long life of the persistent pesticides that makes them so dangerous to the environment is also one reason why they are so effective against insects. They are also relatively safe to handle, and they help to control a wide variety of insects.

The long life of the persistent pesticides accounts for the accumulation of their residues in birds and in other animal life in quantities sufficient to cause reproductive failures and death.

Harmful insecticides

Some insecticides are very dangerous to people as well as to the insects they are expected to control. Some insecticides are deadly when used, while others become more dangerous as quantities build up in plant and animal tissue. The residues stay active in soil and water where fish and other wildlife live, and they also accumulate in the fatty tissue of warm-blooded animals, including humans.

The major problem with DDT is that it is long lasting. Because DDT is a stable chemical, it can pass from field to animals and humans and become more dangerous as it accumulates along the food chain.

Often the search for nonpersistent pesticides is not very successful. Parathion is a nonpersistent pesticide, but it is harmful to people in other ways. Parathion is an example of a pesticide that is immediately dangerous. Parathion is a member of the nerve gas chemical family and is absorbed quickly through the skin, causing fainting, nausea, convulsions, and death. Many cases of poisoning and death have been traced to parathion.

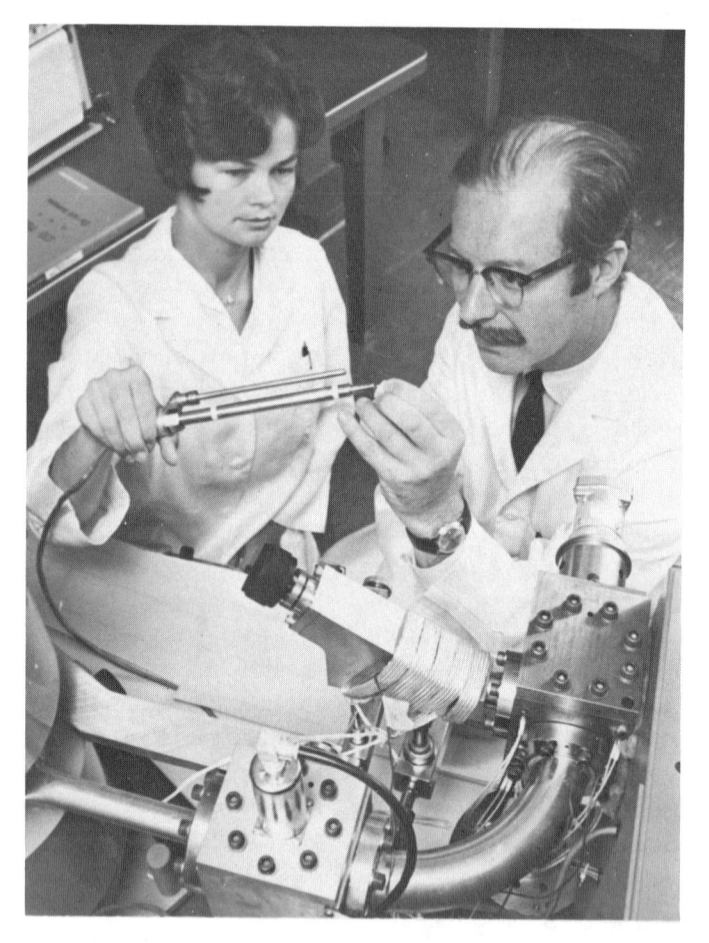

Fig. 24-1. Scientists insert a probe containing a sample into a mass spectrometer to check DDT breakdown. This highly sophisticated equipment permits rapid and accurate identification of the makeup of minute amounts of the organic mixtures.

While parathion, methyl parathion, carbaryl, and similar compounds (organophosphorus and carbamates) can be used to control most pests and do not "persist" in the environment, they are dangerous to handle. They also tend to destroy insect parasites and predators, thus reducing the effectiveness of biological controls and increasing the need for chemical controls.

How Can Pollution Harm Wildlife?

Even when used wisely, insecticides can harm wildlife. Thus, it is important to be very safety conscious in the use of any insecticides.

The greatest exposure of wildlife to insecticides is through the food that wildlife eat. Plants and seeds that have been treated for various purposes with insecticides are the source of much of the danger to the wildlife that use the seeds and plants for food. Wild ducks and wild geese have been killed by the insecticide aldrin when it was used to control the water weevil in rice crops. Young gulls have been killed by the parathion in insects brought to them by their parents. Large losses of animal life have been traced to the use of dieldrin to control the Japanese beetle, the use of heptachlor to control the fire ant, and the use of diazinon to control the chinch bug on lawns and golf courses.

In the Soviet Union, by using chemical pesticides carelessly, two farmers accidently killed more than 50 cranes, 200 rare great bustards, 11 gray geese, and 50 foxes.

Once the insecticides get into the food chain, the effects can be felt by the animals that prey on other animals that have already accumulated insecticides in their bodies. Animals that prey on a variety of other animals may eat several kinds of insecticides in this way, some of which may be more dangerous when they are brought together in the animal's body than when they are eaten separately.

It is even possible for wildlife to be killed months after having eaten pesticides because of the amounts stored in the body fat of the animals. Animals store energy in the form of fat when food is plentiful. Pesticides are also stored in the body fat of animals. When the fat is later used at times of food shortages, the stored pesticides are also released into the animals' bodies. When this occurs, the animals may die.

Meat-eating birds and mammals are most affected by the accumulation of insecticides in other animals. Examples of wildlife affected this way are falcons and hawks, which eat birds; eagles, ospreys, and pelicans, which eat fish; mink, which eat mammals; and aquatic mammals, such as seals, which eat various forms of marine life.

As indicated in the discussion on conservation, the direct killing of wildlife by insecticides is only one part of the problem. The reduction in ability to reproduce and the effect of pesticides on the ability of these animals to find food and to defend themselves against predators and other dangers may have a greater long-term impact on wildlife survival than the direct killing of individual animals. Pollution-caused thin eggshells are especially troublesome to the reproduction of many birds.

Other pollutants also are a serious threat to wildlife. PCB's are known to be a major threat. As the study of pollution continues, the effects of heavy metals, radiation, and air pollution are becoming better known.

Air pollution

With so much air, it does not seem possible there might not be enough good air to breathe. In some places, however, there are already times when just breathing to stay alive is dangerous.

Fumes are certainly near the top of the list as air polluters. Exhaust fumes from various types of vehicles, such as cars, buses, trains, trucks, motorcycles, minibikes, snowmobiles, swamp buggies, planes, ships, and boats, are constantly polluting the air.

Smog is one of the most noticeable evidences of air pollution. Many large U.S. cities, including Atlanta, Baltimore, Los Angeles, New York, and Washington, D.C., experience many times when the air is dangerous to breathe.

Often, smog results from what is called a temperature inversion. This means that the air above a city is just as hot as the air near the ground. When this happens, there is no natural air movement due to temperature differences, and so fumes and other pollutants stay in the air right where they are released into it. People breathing air filled with pollutants, such as sulfur dioxide, often need treatment for sore eyes and sore throats.

Air pollution has been found to make some people ill and to even cause death. If air pollution in major cities were to be reduced by 50 per cent, three to five years could be added to a person's life expectancy.

Agriculture's greatest concern regarding air pollution is the damage it causes to plants and animals. In grazing areas near smoky factories, air pollution has been found to cause abortion in sheep, poor-quality wool, more pasture needed per cow, extra feed and care needed for livestock, stunting of growth in livestock, calcium deficiency in bovine milk, decreased egg production and death in poultry, and illness and death in sheep and cattle. Forests, hundreds of miles from cities, are being damaged by the polluted air from the cities. Many crops are damaged seriously by polluted air.

Water pollution

One of the worst forms of pollution is water pollution. All life can be affected by whatever is carried in water. Some aspects of this are discussed in Chapter 7.

In addition to the pollution from insecticides, waterways can be polluted in other ways. One of the major dangers to streams and rivers is called *eutrofication*. Eutrofication is the excessive growth of plant life in bodies of water as a result of the run-off of nutrients from the land into the waterways, and as a result of the addition of nutrients from other sources, such as phosphorus from detergents. Many nutrients flowing into bodies of water come from the heavy use of fertilizer and the disposal of human and animal

Keeping Our Environment Clean

Fig. 24-2. What looks like snow is actually pollution caused by detergents in Snow Creek, Anniston, Alabama. (Courtesy, U.S. Department of Agriculture)

wastes. As the plant life increases, it fills the waterways and uses up the oxygen supply in the water, thus killing fish and other forms of animal life.

In Lake Michigan, the blanket weed *Cladophora* has become a real nuisance. An algae, *Cladophora* thrives in clear, active water with good light and a supply of nutrients. As the masses of algae grow long, they break off and drift onto the shores to decay and smell like sewage and to plug water intake systems.

Acid run-off from mines and certain industries causes much of the water pollution that has killed so many fish.

Oil spills in the oceans have killed almost all animal life in some of the areas polluted. Many people have given numerous hours and a great amount of money to help clean beaches that have been covered with oil and to help save the lives of birds caught in the oil.

Increasingly intensive use of metals such as mercury, arsenic, and lead, in industry and in agriculture, have raised the amounts of such materials in the environment to the point of endangering life.

The intake of mercury by people through air, food, and water has increased many times in this and other countries in recent years. As early as 1953, there were reports of mercury poisoning in Japan. At one time, over 100 people were killed or disabled in a single community by mercury poisoning. Among the symptoms of mercury poisoning are anxiety, self-consciousness, trouble in concentrating, irritability, headaches, fatigue, and excessive perspiration. Mercury poisoning can lead to senility. Many U.S. fishing areas have been closed because the fish contain unsafe amounts of mercury. Swordfish have been almost entirely removed from the markets.

Infectious disease organisms can be transmitted in water. For example, botulism can be caused in waterfowl when these birds eat contaminated, decaying vegetation containing the causative organism. Organic pesticides can be used in ways that result in pollution of water and the death of wildlife.

Poor practices in farming, forestry, road building, and construction projects that require the movement of soil can result in filling streams and lakes with soil particles. These particles can become so numerous in the water that they can suffocate fish and cut off the sunlight needed by aquatic plant life.

Radioactive wastes can be absorbed by aquatic organisms and cause many kinds of growth and reproductive problems. Even dumping radioactive wastes in the ocean, the most popular dumping site in recent years, is not considered by many scientists to be a safe practice.

Hot water, a by-product of power generation and other industrial activity, can change the temperature in parts or all the bodies of water into which the hot water is discharged. The hot water reduces the capacity of the body of water to hold oxygen, which is needed by marine life.

Land pollution

Most of the concern for the land is with respect to control of erosion and the growing amounts of pesticides and other chemicals accumulating in the soil. There are other forms of land pollution that should be of concern and that are present mostly in the rural areas.

One form of land pollution is the destruction from strip mining. Huge cuts in the earth and huge piles of turned earth and spoils make acres of land useless for any purpose.

The accumulations of junk cars and other old machines are ugly signs of modern civilization and its wastefulness. The litter along the roadside is evidence of the lack of concern of many persons for a healthy and beautiful environment.

The accumulation of salts in the top soil in irrigated lands is a pollution problem that interferes with continued use of the land for production.

Wise use of the land is the key to most pollution control efforts, because it is the use of land for food production, for waste disposal, as a source of minerals, and for other purposes that is the start of most pollution problems.

How Can Pollution Be Combated?

In December, 1970, a federal agency called the Environmental Protec-

Fig. 24-3. This coal seam was uncovered by the Great Page "Walker" Dragline. Scars on the land from strip mining are a major social concern. (Courtesy, U.S. Department of Agriculture)

tion Agency was given the responsibility for the control and regulation of chemical pesticides, as well as many other possible pollutants. This agency could well be the most important agency the federal government has for protecting people from themselves.

The 1972 Federal Water Pollution Control Act (PL 92-500) is likely to have as much effect on our lives as any law ever passed. Under this law, nearly $100 billion in federal spending over the next decade was projected to keep the waters clean. Spending by states, local governments, and private industry probably exceeded this amount. Yet, between 1971 and 1980, nearly 78,000 U.S. citizens became ill after drinking water from community or noncommunity water systems. While the costs seem very high, many individuals believe everything possible must be done to insure clean, safe water. What do you think?

One of the nation's first water pollution control laws, the Refuse Act of 1899, is being used to help control pollution. It bans the discharge of refuse (all foreign substances except liquid sewage) into navigable waters without a permit from the Army Corps of Engineers. Since even small streams of water may be classified as navigable waters, the Refuse Act of 1899 is very useful in the fight against water pollution.

Ecology departments and environmental protection agencies have been established in many states to help combat pollution. Since stopping pollution will mean stopping people from doing many of the things they have been used to doing, these agencies will need support.

Many states have recognized the danger of water pollution and waste and are adopting laws to regulate the use of water.

Permits are now needed in many states to erect or make any structure or deposit in any public waters. Upstream land owners are being held responsible for pollution of water downstream.

Pyrethrum, discovered by the Chinese many years ago, is deadly to insects, but one of the least hazardous of available insecticides. The residues of pyrethrum are chemically unstable and rapidly break down in soil and water into relatively harmless compounds. About half of the world's supply of pyrethrum comes from Kenya. A British firm has developed a synthetic product that is as safe as pyrethrum and is more effective against insects. Thus, its use may be expanded. Most of the present supply of pyrethrum used in the United States goes into household insecticides. Very little is used for farming.

Not only should home owners limit the use of pesticides but they should also use only those that are the safest and that cause the least pollution. In addition to pyrethrum, some of the safest insecticides are sevin (carbaryl), dormant oil, malathion, methoxychlor, and resmethrin. Sevin and malathion can be harmful to bees and are moderately toxic to fish.

Fig. 24-4. Kenya natives harvesting pyrethrum. Pyrethrum residues are chemically unstable and rapidly break down in soil and water into relatively harmless compounds. (Courtesy, U.S. Department of Agriculture)

Keeping Our Environment Clean

There are various ways in which the home gardener can avoid the use of toxic materials. For example, a home cure for the spider mite of ornamentals may be a mixture of 1 cup of buttermilk, 8 cups of wheat flour, and 10 gallons of water.

The composting of crop refuse and fall plowing of the garden can be a big help in insect control.

For some insects, hand picking and disposal may be preferable to the use of chemical controls. Gardeners may well also try using some of the enemies of harmful insects, as reported under biological controls.

Even planting some strong smelling herbs, such as basil, mint, and sage, can help repel garden pests and reduce dependence on chemical insecticides.

Biological controls may be a partial answer to the insecticide problem. With biological controls, however, some insects always escape and cause damage to plants and animals. However, the use of biological controls may help reduce the amounts of chemicals needed.

The best known biological control is the *lady beetle*. It eats aphids, mealybugs, and whiteflies. One active lady beetle may eat several hundred aphids in a day.

The *dragonfly* eats many mosquitoes as well as flies, beetles, moths, and other injurious insects. The dragonfly can travel at speeds of up to 60 miles per hour.

The *fire ant,* a problem in the South, eats housefly larvae, boll weevil grubs, mites, cutworms, and other destructive insects. The fire ant venom also kills some bacteria and molds. Because of its insect-killing characteristics, the fire ant may do more good than harm.

Biological controls of other kinds, such as those for the Japanese beetle and Oriental fruit moth, have been discussed in other parts of this book. These methods will not be as effective as insecticides; therefore, fruit can be expected to be of a lower quality than if insecticides were used.

Rodents may also one day be controlled without the use of chemicals. One of the more exciting possibilities for biological control of the common brown or Norway rat *(Rattus norvegicus)* is through genetics. Dr. Allen J. Stanley, a University of Oklahoma physiology professor, has bred a strain of the brown rat that carries a gene for sterility. A distinctive white spot on the forehead marks the breed.

The male rats are sterile and can produce no offspring. Even though they mate normally, they induce only false pregnancies in the females. The females do produce young on mating with normal rats, but half of the young will carry the gene for sterility.

Controlling rats in this way may make it unnecessary to use chemicals, which may be dangerous to other animals, including people, as well as to rats.

The heavy use of fertilizer has been found to be one cause of water pollution. While not enough is known about the movement of fertilizer in either surface or ground water to indicate how much of the phosphorus and nitrogen in water comes from this source, there are ways to reduce pollution from fertilizer use. Other than limiting its use, the best ways known to reduce pollution from fertilizer are:

1. Develop and use fertilizer with slower release of all nutrients to enable a more complete use of it by plants. Some lawn fertilizers are already prepared in this way.
2. Use cultural practices and timing of application to reduce leaching and loss from erosion.

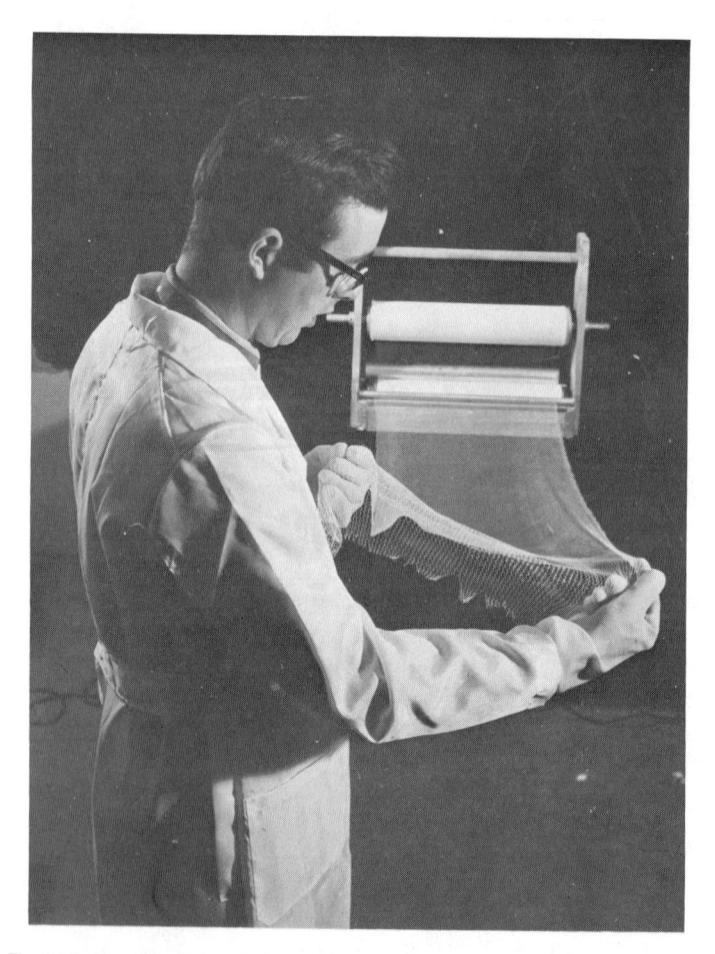

Fig. 24-5. Experiments are being conducted on controlling weeds with loosely woven cloth, treated with weed killer, cut to fit the area to be controlled, and anchored in place. The cloth decomposes before the end of the growing season. This method would eliminate the dangers from spray application of herbicides. (Courtesy, U.S. Department of Agriculture)

3. Place fertilizer in the soil near the plant-root zone to improve use by plants and to reduce loss from leaching and erosion.
4. Apply fertilizer in small amounts more frequently.

Ultra–low-volume sprayers for apple orchards may also make a major contribution to the conservation of fuel and insecticides. A few years ago, orchard sprayers applying up to 400 gallons of spray per acre were common. The low-volume sprayer does the same job, applying only 20 to 40 gallons of a more concentrated spray with 30 per cent less chemical pesticide per acre. Thus, both the energy and the chemical required for spraying are greatly reduced, resulting in both energy conservation and a cleaner environment.

New methods for studying the amounts of pesticides put into the environment are being developed. Two scientists have designed and built laboratory model agroecosystems to simulate the conditions plants grow in naturally so that scientists can measure pesticides left in the soil, water, plants, and air. These new laboratory tools can be adjusted to give the researchers the exact weather conditions they want to test year around. Tools of this kind are needed to help producers test chemicals for controlling pests so that they can grow food without polluting the environment.

Since pollution sources are not limited to the area in which the pollution is found, national laws have been passed to help with pollution control. Some of these laws are (1) the Clean Water and Resource Conservation and Recovery Act; (2) the Safe Drinking Water Act; (3) the Marine Protection, Research, and Sanctuaries Act; (4) the National Ocean Pollution Planning Act; (5) the National Advisory Committee on Oceans and Atmosphere Act; (6) the Toxic Substances Control Act; and (7) the Federal Insecticide, Fungicide, and Rodenticide Act.

How Can Everyone Help Fight Pollution?

There are many ways in which people can help fight pollution and improve the environment. One way to help is to join organizations that concentrate their efforts on how to use the environment wisely. However, before joining, individuals should investigate to make certain these organizations are supporting the same activities that they support. Another way to help is for individuals to avoid polluting in their daily activities. The following are some steps most people could take if they tried.

Avoid air pollution

1. Reduce car operations by walking, riding a bicycle, or riding with someone else.
2. Keep the car tuned up.

3. Buy a compact car.
4. Limit the burning of trash and plant residues. Gases are released when plastics and certain other materials are burned.
5. Use a hand lawn mower on small and level lawns.

Avoid water pollution

1. Use soap rather than detergents when possible.
2. Limit the use of household cleaners, such as ammonia and chlorine, to the bare necessities.
3. Conserve water by running it only when needed.
4. Keep a bottle of water in the refrigerator for cold drinks.
5. Limit the use of water for lawns and plantings.
6. Avoid loading the sewage system with trash.
7. Control soil erosion in the garden and on the home grounds.

Avoid land pollution

1. Use utensils and materials that can be cleaned and used again rather than those that are disposable. Examples of this are shopping bags, lunch boxes, picnic plates, bottles, and clothes.
2. Place litter in containers that prevent spread.
3. Buy products in the largest amounts and largest containers suitable for individual needs. This will reduce disposal problems for wrappings.
4. Buy good-quality items that will last.
5. Return reusable materials to stores. Examples of reusable items are coat hangers, bottles, and jugs.
6. Where possible, save paper, bottles, and cans for recycling.

Use chemicals carefully

1. Select chemicals that persist in the environment for only a short time after their use.
2. When using chemicals, protect animal food and water sources.
3. Avoid the use of insecticides while birds are nesting and migrating.
4. Make sure no chemicals are applied in excess or are left in large amounts exposed on the ground or places where animals eat and drink.
5. Avoid the use of insecticides on or near land that drains directly into ponds.
6. Apply chemicals in ways that will prevent the drifting of the chemicals in the air.

7. Disk or water granular applications of chemicals into the ground so birds cannot pick up the granules.
8. Dispose of, or store carefully, unused portions of chemicals and empty chemical containers.

Promote community action

1. Work with and support clean-up groups.
2. Support pollution control laws.
3. Write and talk to government officials and representatives at all levels about pollution.
4. Report to the appropriate officials any instances of pollution seen in the community.
5. Encourage educational programs about pollution and pollution control.

Controlling Pollution Is a Difficult Task

Controlling pollution is not easy. For example, after laws were passed against the use of phosphate detergents, it was soon discovered that some of the substitutes for the phosphate detergents were more dangerous to people than were the effects of the phosphate detergents.

It has also become clear that halting the use of some of the persistent pesticides may be more dangerous than would be continued use. The World Health Organization, an agency of the United Nations, wants some uses for DDT permitted. The scientists in the World Health Organization have said that efforts to control or to eliminate malaria, sleeping sickness, and elephantiasis would be seriously crippled if DDT use were to be stopped entirely.

DDT is credited with a great increase in the population of robins in Amarillo, Texas. The reason given is that DDT reduced greatly the population of mosquitoes, which carry numerous bird diseases.

The food supply could be seriously affected if certain insecticides were to be banned entirely. However, placing greater emphasis on crop rotations, using sanitation practices, encouraging the development of pest-resistant plant varieties, using existing insects and other organisms that can help control pests, and employing other biological control methods can reduce the use of insecticides significantly.

Thus, people must learn how to use the materials in the environment wisely. The value of using certain materials must be balanced with the danger from using them. The human race cannot survive if it does not bring pollution of the air, land, and water under control. However, people cannot stop using all the materials that have been found to be pollutants. Instead, ways to reduce the problems to a manageable size must be found.

Meanwhile, the problems of pollution will not wait for the results of research. The environment must be preserved with the best-known techniques and kept fit for living while researchers try to discover ways in which that job can be done better.

SUGGESTED ACTIVITIES

1. Develop a checklist of actions that everyone can take to control pollution and then use it to see how well you are doing at home and in school.
2. Survey the local community to see how many examples of pollution you can find. Write an article for the local newspaper to report your survey.
3. Prepare a talk on pollution for presentation to a school assembly and to various community groups.
4. Collect samples of water from local streams and ponds in jars. Let the materials in the water settle to the bottom of the jars and judge which source of water was the cleanest.
5. Invite pollution control officials to speak to your class.
6. Invite local government officials to speak to your class about what they are doing to help control pollution.
7. Use clippings from newspapers and magazines to develop a scrapbook on pollution problems and their elimination.
8. Concepts for discussion:
    a. Pollution is caused by people as they struggle for economic gain.
    b. No material on earth can be thrown away; it can only be turned into something else.
    c. The problems of environmental pollution will never disappear.
    d. Agriculturists have been concerned with pollution in many forms for years.
    e. Agriculture accounts for about half the solid waste products produced in the United States.
    f. Eliminating pollution is costly, and everyone must share in paying for it.
    g. Persistent pesticides are among the most dangerous of agricultural pollutants.
    h. The long life of a pesticide helps make that pesticide both effective against insects and dangerous as a pollutant.
    i. Persistent pesticides become more harmful as they accumulate in the tissues of animals and eventually become concentrated enough to cause harm.
    j. Temperature inversions prevent natural air movement and help bring about smog.
    k. Air pollution may bring about reduced production in plants and animals.
    l. All life on earth may be affected by pollutants in water.

m. Nutrients carried into bodies of water from fertilized fields and human and animal wastes destroy bodies of water through eutrofication.
n. Industrial wastes, such as acids and oil, have killed much plant and animal life in bodies of water of all sizes.
o. Metals, such as mercury, are accumulating in the environment to the point of endangering human life.
p. Combating pollution is everyone's business.
q. Land pollution is ugly and is indicative of a waste of resources.
r. Preventing pollution will mean stopping people from doing things they are used to doing.
s. Although biological controls help prevent pollution from insecticides, they are not as effective as insecticides in controlling insects.
t. Some nonpersistent pesticides are available, and more are being developed.
u. Careful use of agricultural chemicals and fertilizers can do much to reduce pollution.
v. Genetic control of rodents may be possible.
w. Care in daily home activities can do much to reduce pollution.
x. Continuing public support of anti-pollution laws and agencies will be needed to prevent and control pollution.
y. Wise judgments will be needed to balance the value of using certain materials against the danger from using them.

REFERENCES AND ORGANIZATIONS

Defenders of Wildlife, 2000 North Street, N.W., Washington, DC 20036.

Field Museum of Natural History, Roosevelt Road at Lake Shore Drive, Chicago, IL 60605.

Kelley, J. W., et al. *Environmental Awareness* (4-H Leaders' Guide). Cooperative Extension Services of the Northeast States, Cooperative Extension, College of Agriculture and Life Sciences, Cornell University, Ithaca, NY 14853.

National Audubon Society, 950 Third Avenue, New York, NY 10022.

National Wildlife Federation, 1412 16th Street, N.W., Washington, DC 20036.

Natural Resource Booklet Series: Alaska, Arizona, California, Colorado, Idaho, Massachusetts, Montana, Nevada, New Mexico, Ohio, Oregon, Texas, Utah, Washington, and West Virginia. Superintendent of Public Documents, U.S. Government Printing Office, Washington, DC 20402.

Sierra Club, 1050 Mills Tower, San Francisco, CA 94104.

U.S. Department of the Interior, Conservation Yearbook Series:

Yearbook No. 1, Quest for Quality

Yearbook No. 2, The Population Challenge

Yearbook No. 3, The Third Wave

Yearbook No. 4, Man—An Endangered Species

Yearbook No. 5, It's Your World

Yearbook No. 6, River of Life

Yearbook No. 7, Our Living Land

World Wildlife Fund—U.S., 1601 Connecticut Avenue, N.W., Washington, DC 20009.

Chapter 22

Starting a Small-Scale Agricultural Business—Entrepreneurship

A changing agriculture, with the continuing increase in specialization in farming, has resulted in the formation of many kinds of agricultural businesses other than farming. Youth and adults who want to get into an agricultural occupation now have many opportunities, either working for established agricultural businesses or starting (owning) their own agricultural businesses. Many nonfarm agricultural businesses can be started on a small scale just as many farm businesses were started. Of course, the risks and hard work needed to start a small agricultural business should be carefully thought through beforehand. For those persons who are successful, the rewards can be great, including the satisfaction that comes from ownership.

OBJECTIVES

1. Discuss the factors to consider in entrepreneurship.
2. List some of the business fundamentals people need to learn in order to be entrepreneurs.
3. Identify some entrepreneurship opportunities.

Deciding on a Business

Choosing an agricultural business to get started in as an owner is more complex than choosing some other form of occupation. In addition to considering all the factors included in choosing an occupation, the person making a choice of business must take into consideration all the many factors that help to determine the success of a business.

Occupational choice factors to consider

Since operating a business is an occupation, a person can start the selection process with a consideration of the factors used in choosing an occupation, as discussed in Chapter 4. These factors include:

1. The nature of the business/occupation.
2. The non-financial rewards and satisfactions.
3. The personal qualities, interests, and aptitudes that are important for success.
4. The opportunities for getting started.
5. The education and training needed.
6. The financial rewards.
7. The advantages and disadvantages of owning a small business (the occupation).
8. The extent to which a person's own interests and abilities fit the tasks to be performed by the owner of a small business (the occupation).
9. The extent to which an individual's family is interested in and supportive of that person getting started in a particular small business occupation.

Business factors to consider

In addition to the occupational choice factors listed above, anyone trying to decide whether or not to start a small-scale agricultural business should also consider the following factors.

What are the financial resources needed for getting started? When a person works for someone else, that other person (the employer) generally provides the resources needed to do the work. While an automobile, clothing, and personal tools are sometimes required of the employee, these items are usually those the employee would have anyway or could purchase at a reasonable cost. Some business owners even provide loans to employees for the purchase of those items needed so that the employee can start work. In any case, the business owner must provide most of the capital resources needed for a business. The amount of capital required will depend on the kind of business. Farming requires capital for land, machinery, and buildings; whereas, a small nursery can be started with half an acre of land, a storage shed, and some tools. For each kind of small business being considered, the individual should determine all the capital requirements that would be needed. This analysis is especially critical if the individual must borrow money to get started.

What are the labor needs for the business? Many persons start a small-scale business with family labor as a first step toward making the business a full-time operation. If the family is to be the source of labor, the business must be selected to make use of the abilities the family already has or which the family can acquire as the business is started and grows. In addition, the business should be limited to a size that can be handled with the amount of family labor available. In some situations, there may be part-time labor available locally from retired people or persons who want part-time jobs to supplement their incomes. Careful analysis of available capital and the potential cash flow are needed so that the individual can determine the extent to which money would be available to pay for hired labor.

What management requirements must be met? Starting a small-scale business is often a way to learn how to manage. However, there are many management skills and much management knowledge needed even for starting a small business.

A major qualification of the owner-manager of an agricultural business is a knowledge of the technical aspects of the business, for both the agricultural and the business operations. This is especially true for a small-scale business where the management does not have the resources to employ technical expertise. By working for someone else for a period of time, either full time or part time, a person can often acquire the technical knowledge needed. For example, someone who wants to start a small landscaping business may work for a landscaping company on weekends and during vacation. Sometimes a person may go into business on a small scale after working for someone else in the same business full time for a number of years.

Fig. 25-1. Even a small-scale nursery requires the availability of some land.

It is also possible for a person to obtain the technical knowledge and management training needed by attending adult education classes, by taking evening classes at a nearby college or technical school, and by getting help from agencies such as the federal Small Business Administration.

In addition to the technical knowledge required, an owner-manager needs to know about the records to be kept, labor laws, sales taxes, local business ordinances, sources of materials, markets, pricing, inventory management, credit, human relations, and general business procedures.

Is there a market for the product or service? A product or service must be marketable if a profit is to be made. Thus, a study of the market potential for a business is a very important step in helping an individual decide on the business to start.

The market for a business may be local or nationwide. For most small-scale agricultural business ventures, local markets may be anticipated. Part of the reason why owner-managers should base plans on local markets is that in doing so, they will be able to use family labor on a part-time basis to

Fig. 25-2. Small buildings can be constructed for sale to busy home owners.

get started. For some kinds of businesses, however, local markets would not be enough even for a small-scale operation. For each kind of business being considered, the market potential should be carefully studied, including how the product or service will be delivered to the customer. If long-range plans include the expansion of the business into a full-time operation, the study of the market should be expanded accordingly, including consideration of other businesses that provide the same services or products.

Where should the business be located? The location of the business may determine the success of the business. The location is especially important when the business will depend on people passing by for customers. For example, not many homegrown vegetables will be sold to Sunday afternoon travelers if the sales are to be made from a roadside stand located on a little-used country road. Most people want to see flowers before they buy them. Thus, a flower shop must be located where it will be convenient for potential customers.

If the business is to be operated from the home, the kind of business should lend itself to the location of the home, at least for the volume of business needed to get started.

Should a person buy a business or start a new one? Sometimes it may be possible to buy a small business rather than to start one. Even when the business will be operated from the home, it may be advantageous to buy out the equipment and stock from someone who wishes to sell.

Buying a business has its advantages and disadvantages. Buying a business can (1) allow a quicker start, (2) provide ready-made customers, (3) eliminate some competition, (4) reduce the cost of getting established, and (5) yield a base of financial information for estimating costs and profits.

Some of the disadvantages of buying a business are (1) the greater capital resources needed at the very beginning, (2) no time to learn while the business is developing, (3) the possibility of misjudging and buying a loser, (4) the problem of accepting the location or moving the business, (5) the loss of safety that comes from growing into the business as resources permit, (6) the risk of missing some critical and costly aspect of the business in closing the deal, (7) the cost of legal assistance needed for making the purchase, and (8) the cost of going out of business if it is disappointing.

Some business fundamentals

There are many fundamentals or principles of business that people need to learn as they enter the business world as owners. These fundamentals apply regardless of the size of the business. A few of the important fundamentals for getting started in a business are as follows:

1. *Keep the size of the business consistent with the capital resources available and with progress in developing management capability.*

One of the reasons for starting a small-scale business is to give the owner time to grow into the business from both management and capital standpoints.

2. *Select a business that is long on labor needed, typically family labor, and short on finances needed.* The business selected should be one in which family labor, especially the owner's labor, can be used well, and one in which it is possible for available finances to be spent on high-profit items. Labor intensive business activity, rather than capital intensive activity, allows the owner to control the use of available resources and to minimize risks.

3. *Devote adequate time to the development of management capability.* Make use of available learning resources, such as the vocational agriculture teacher, the county extension agent, the Small Business Administration, nearby colleges and technical schools, the bank that carries the business account, and other businesspersons.

4. *Keep available labor fully employed all year.* Have a regular job working for someone else until the workload for the business is such that this is no longer possible. Keep all excess family labor fully employed in the business or elsewhere. Be prepared to make personal and family sacrifices until the business gets off the ground.

5. *Prepare a sound plan of operations and follow it.* The plan should include a month-by-month schedule of cash flow needs, production, labor needs, and marketing for at least a year. Establishing short-term and long-term goals as checkpoints can be very helpful in periodically evaluating and revising the plan.

6. *Study the business continually* to determine for which parts of the business the use of scarce capital will add most to production and profits. Try to keep all aspects of the business growing and improving in a coordinated manner so that one aspect does not lag so much behind other aspects in its development that it causes the business to grow at an unnecessarily slow rate or even to fail.

7. *Maintain an adequate cash operating fund.* Operating funds can be kept earning for the business through an interest-paying checking account. The fund should be sufficient to take care of most of the high cash need times of the business, so borrowing at high interest rates will not be necessary.

8. *Maintain an inventory level that provides for responding to customer requests promptly.* This will require accurate and detailed recordkeeping on a weekly basis as a guide for ordering materials and supplies and for determining and scheduling business activity.

9. *Plan to use as much of the profits as possible to expand the business.* The increase in the value of the business is a form of saving money for the future as well as a way of expanding the business so greater earnings from it can be possible.

10. *Use the physical facilities of the home to the fullest extent possible.* This will help keep capital needs at the lowest possible level and will shift part of the cost of owning and maintaining the home to the business.
11. *Establish the prices charged for products and services at a level that will result in a reasonable profit while still being lower than or equal to the prices charged by competing businesses.* The prices charged must also be at a level customers will be willing to pay. However, no business can continue very long unless the prices charged are enough to pay for the owner's time and capital investment.
12. *Buy good-quality materials and supplies needed at the lowest price possible.* The quality is important whether the material is used for the production of other goods or for resale. Also, the price paid for materials must be low enough to allow for the additional charge to the customer, which is needed to pay for the operation of the business and to provide a margin of profit.
13. *Treat customers and potential customers courteously and fairly.* Often a business succeeds because customers like the way they are treated rather than because of the price they pay.
14. *Treat employees fairly.* Unhappy employees can harm a business by doing poor-quality work, by being discourteous to customers, or by failing to do a full day's work. At the same time, an employer cannot afford to keep someone on the payroll if that person is not contributing enough to the success of the business.

What Are Some Examples of Small-Scale Agricultural Businesses?

There are many examples of small-scale agricultural businesses that can be started on a part-time basis with the home physical facilities as the location. What examples can you add to those described here?

Home gardening

Most large agricultural businesses can also be operated on a small-scale. A good example of small-scale farming is the home vegetable garden. While most families with gardens use the produce themselves (in effect, sell to the family), some home gardeners market the surplus production by selling at roadside stands or by selling to regular customers who pick up the vegetables themselves. A home garden can contribute many hundreds of dollars to the family income as well as provide excellent nutritious food for the family table.

Some of the advantages of a home garden as a small business are that

(1) it probably expands an activity already being performed, (2) the family may already own the equipment and tools, (3) it is labor-intensive, (4) the family already has much of the technical knowledge needed, (5) it requires very little space, (6) the family can use the produce as well as sell it, (7) the entire family can help with the work, and (8) it fits in well with full-time jobs.

If enough land is available, a pick-your-own business in strawberries, raspberries, or other small fruits or vegetables can be developed. It is not wise to use a "pick-your-own" business approach for products such as sweet corn, for which the time for harvesting is not readily apparent.

Small greenhouse

A small greenhouse is another possibility that fits in well with part-time family labor and with very little space. A greenhouse can be used to provide plants needed for the family garden and for the home, as well as plants for sale. Marketing of plants can be through sales to individuals, stores, or larger greenhouse operations. The greenhouse does not need to be located where people will walk by or where traffic is heavy. While there is some expense for constructing, equipping, and operating a greenhouse, the capital required is fairly low. This kind of business can be started on a very small scale and can be expanded fairly easily.

Retail flower shop

Persons who buy flowers are generally in the medium and higher income levels. Thus, a shop should be started only if enough potential customers at that income level live in the community. Growing communities provide the best opportunities for this kind of business. In determining the number of potential customers, it is important to take into account any competing shops that exist. For those individuals who wish to operate retail flower shops on a part-time basis, it may be profitable to start such a business even when the community is too small for a full-time operation.

The availability of materials and supplies is also important, especially if there are no plans to produce some of the flowers in a home greenhouse. While some business can be conducted by telephone, a shop located where people go anyway offers a much better chance for success.

For individuals who are starting on a part-time basis, a home shop with a small greenhouse offers many advantages, the most important being the greater availability of family labor and the lower cost for space. The greatest disadvantages are the inaccessibility to potential customer traffic and the need to learn how to operate two kinds of businesses. It may be necessary to focus on providing flowers for special events rather than focusing on selling flowers to individuals who happen to pass by if a home location with an associated greenhouse is selected as the kind of business.

Fig. 25-3. A roadside stand is used to market the plants grown in the greenhouse *(top)*. A small greenhouse can provide plants for the home garden and for sale *(bottom)*.

Small nursery or Christmas tree farm

If a few acres of land are owned, it is not too difficult or expensive to start a small nursery, to grow Christmas trees, or to do both, although it will require a great deal of technical know-how. Depending on the kinds of plants grown, it will take from one to five or six years before plants are

Fig. 25-4. Plants can be marketed through existing retail outlets.

ready for sale. Christmas trees can be sold on a cut-your-own basis, thus reducing labor needs and avoiding the cutting of trees not sold. This kind of operation lends itself to the use of surplus family labor. In addition, the work can be done when there are no other activities in which the family wishes to engage. Marketing can be to individuals, to landscaping companies, through garden centers, or through other nurseries. Adding land acreage, shifting gradually to more labor-intensive plantings, changing marketing strategies, and expanding into a landscaping business are ways to enlarge nursery businesses.

Beekeeping

People in a great variety of occupations keep bees. Some beekeepers keep bees only as a hobby, while others may invest up to $100,000 in beekeeping as a way to make a living. Whether a person has only one or many hives, success depends on having a good knowledge of the life and behavior of bees. While income from bees is mostly from the honey produced, many thousands of hives are rented for purposes of pollinating fruit and vegetable crops. Even though beekeeping is a skilled occupation, it does lend itself to small-scale operations because of the small number of hives that can be kept and the small amount of space that is required. A small beekeeping operation can make good use of part-time family labor. It can also be expanded easily. The honey can be sold to individuals or through stores. Rental of bee hives for pollination can be done through individual contacts.

Beekeeping can be low cost in that individuals can get out of the business if they do not wish to continue.

Small-engine repair

A small-engine repair business requires little space and only a modest investment for tools and equipment. Part of a garage or a small shed can be converted into a shop for engine work and for temporary storage of customer equipment.

Typically, small-engine work includes repair and maintenance of the equipment for which the engine provides the power. For example, most people expect to have the non-engine parts of a lawnmower repaired and maintained by the same business that repairs and maintains the engine. Because of the increasing number of home gardening and landscaping tools powered by small engines, this business has the potential for rapid expansion in size. Small-engine repair work can be limited, of course, to a part-time operation.

Small-engine repair work is a good example of a labor-intensive business. For the beginner, buying parts and materials only after the need for them has been determined can keep the capital outlay small. In effect, the sales stores maintain the parts inventory. After the business has become established, some common parts, such as spark plugs, should be kept on hand. This kind of business makes good use of spare time, and the schedul-

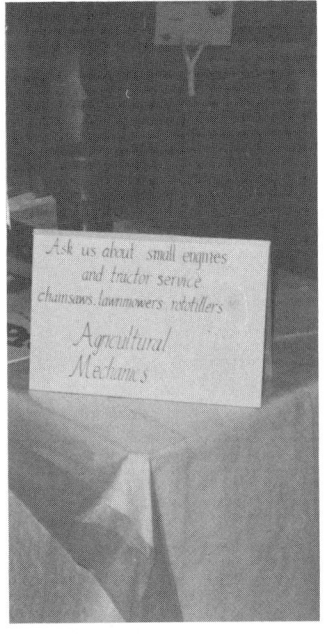

Fig. 25-5. Small-engine repair work is one way to get started in business.

ing of work can fit around the workday for any kind of full-time job. Because of the experience required, small-engine repair does not lend itself to the use of all available family labor, although the bookkeeping and other clerical tasks can be done by persons not expert in engine repair.

Small-engine repair work can be done throughout the entire year, especially if the shop is heated and there is adequate space for temporarily holding customer equipment brought in for repair. Providing a pick-up and delivery service for the equipment is one way to expand a small-engine repair business. A pick-up and delivery service reduces the amount of space needed for temporary storage of customer equipment, and it can increase business by attracting those persons unable to transport their equipment to the repair shop themselves.

What Are the Values of a Small-Scale Business?

Small-scale agricultural businesses can contribute to the family income, to the family food supply, and to the enjoyment that comes from working with growing things. Many small businesses can be pursued as hobbies (see Chapter 3) as well as businesses. Small businesses are especially helpful for utilizing family labor and supplementing the family income. They also serve a useful purpose in helping youth mature into responsible adults. A major value for many families comes from having activities that bring the entire family together for work and enjoyment. For those who wish to go into business full time, a small-scale business is a good way to develop the expertise needed to manage and operate a large business.

SUGGESTED ACTIVITIES

1. By yourself or with another student, prepare a presentation to convince the rest of the class to start a small nonfarm agricultural business of some kind.
2. Organize a panel discussion on the topic of working for an employer versus owning a business.
3. With your class, survey the local community to identify the possibilities that exist for starting small agricultural businesses.
4. Invite two or three owners of small-scale agricultural businesses to tell the class how they got started in business.
5. Concepts for discussion:
   a. Increased specialization in farming has resulted in the formation of many nonfarm agricultural businesses.
   b. It is a good idea for individuals to get some experience as employees in businesses before starting out on their own.
   c. Time and money must be invested for a period of time before income from a business can be obtained.
   d. The owner-manager of a small-scale business must have both management ability and technical knowledge.

e. The success and growth of a small-scale business located in the home depends on the efforts of the entire family.
f. A study of the market potential for a product or service is an important first step in helping an individual decide whether or not to start a business.
g. Sufficient capital should be available to carry a business until income can be obtained from it.
h. Small businesses often succeed because of the hours the owners are willing to work.
i. Even the best managed business will fail if the location selected for the business is not a good one.
j. There is less financial risk in starting a small business for which most of the start-up investment is in the form of labor than for a business for which the start-up investment is mostly money.
k. A business should not be expanded beyond the availability of capital and management.
l. Continued learning by the owner and workers is an important element in the success of a business.
m. A sound plan of operations is important to the success of a business.
n. All parts of a business must grow at the same rate for maximum success.
o. Labor, facilities, and capital must all be fully used for maximum success in a business.
p. The business inventory should be kept as small as possible, but it should still be large enough that customer requests can be met promptly.
q. Some of the profits from a business should be used to expand and improve the business.
r. Money should be borrowed only when the returns from use of the money will be greater than the costs of borrowing.
s. The prices charged for products and services should provide a level of income sufficient to pay for the operation of the business and an adequate return on the capital invested.
t. A business can increase its profit if the materials and supplies needed for operation are carefully purchased.
u. Labor should be employed for a business only when the costs of employment are less than the additional earnings to the business from the work performed.

REFERENCES

Cooperative Extension Services. The Cooperative Extension Service in each state will have a variety of publications on agricultural businesses.

Lee, J. S., and M. E. Newman. *Aquaculture: An Introduction.* Interstate Publishers, Inc., Danville, IL 61832.

Newman, M. E., and W. J. Wills. *Agribusiness Management and Entrepreneurship.* Interstate Publishers, Inc., Danville, IL 61832.

U.S. Department of Agriculture. *American in Agriculture: Portraits of Diversity*. Yearbook of Agriculture: 1990, U.S. Government Printing Office, Washington, DC 20402.

U.S. Department of Agriculture. *Food—From Farm to Table*. Yearbook of Agriculture: 1982, U.S. Government Printing Office, Washington, DC 20402.

U.S. Department of Agriculture. *Living on a Few Acres*. Yearbook of Agriculture: 1978, U.S. Government Printing Office, Washington, DC 20402

Vocational Agriculture Service, College of Agriculture, University of Illinois, Urbana, IL 61801. Many instructional units are available on the subject of agricultural businesses for use in agriculture classes.

Chapter 23

Frontiers of Agriculture

Although many changes have taken place in the past, changes can be expected to occur at an increasing rate in the future. In the book *Alice in Wonderland,* Alice was described as running so fast she was breathless, but she did not seem to be getting any place. When she complained, the Queen said, "Now here, you see, it takes all the running you can do to keep in the same place." In agriculture too, it takes a lot of running to keep pace with new developments.

OBJECTIVES

1. Compare the past to today's developments in agriculture.
2. Discuss the need for improving methods of production and developing new crops.
3. Discuss the impact of leisure time and recreation on rural land.
4. Discuss the importance of natural resource management to the future.
5. Discuss improving rural leadership and relationships.

If we could compare conditions in the pioneer days in the United States with those of the present, we might be able to appreciate how great some of these changes have been. The following description of a pioneer farmer reveals some of the conditions of farm life at that time.

> From his feet to his head the farmer stood in vestment produced on his own farm. The leather of his shoes came from the hides of his own cattle. The linen and woolen that he wore were products that he raised. The farmer's wife or daughter braided and sewed the straw hat for his head. His fur cap was made from the skin of a fox he shot. The feathers of wild fowl in the bed, whereon he rested his weary frame by night, were the results of his own shooting. The pillow cases, sheet and blankets, the comforters, quilts and counterpanes, the towels and tablecloths, were homemade. His harness and lines he cut from hides grown on his farm. Everything about his ox yoke except staple and ring he made. His whip, his ox gad, his flail, axe, hoe, and fork-handle, were his own work. How little he bought, and how much he contrived to supply his wants by home manufacture would astonish this generation.

The mechanization of farming has brought about many spectacular changes. One agricultural specialist has summed up these developments by

Fig. 26-1. The old hitching post, still a useful item in some parts of the country, is a reminder of what was once considered a luxury.

saying that farmers have substituted capital (in the form of machines, motors, and other equipment) for human labor. Today, a typical farmer has thousands of dollars invested in equipment of various kinds. Many changes have been made in crop and livestock production, as described elsewhere in this book. Not too many years ago, an average farm worker produced enough for 8 people. In 1991 an average farm worker produced enough for 115 people; and this efficiency is steadily increasing. Various other developments, such as automobiles, airplanes, improved roads, widespread use of electricity, increased reliance on computers, have changed tremendously the lives of rural people, as well as those of urban people.

Most people probably would agree that these changes have been desirable. However, changes bring problems not easily solved satisfactorily. One farm leader said, "The same era that brought to many farmers the benefits of automobiles, motor trucks, tractors, electricity, radio, and good roads also brought the end of free lands and a declining foreign demand for the produce of new acres."

In developing this great country, pioneer farmers settled new lands, often in the face of severe hardships. The frontier of new land has vanished, and with it, opportunities for increasing numbers of people to engage in farming. In the future, rural people as well as urban people, must face other kinds of frontiers. In meeting the challenge of these frontiers, we should recognize the importance of a satisfactory life for all people in rural and urban communities. Farmers and their families, as well as other people, want the

Frontiers of Agriculture

Fig. 26-2. Increasingly, the laboratory scene symbolizes our changing agriculture. Here, chemical tests of protein content, fat, moisture, and other substances are being made. (Courtesy, U.S. Department of Agriculture)

opportunity to earn enough to live comfortably, to secure a good education, to enjoy the better things of life, to maintain the respect of others. Rural people also want to carry out their responsibilities as citizens.

What Are Some of the Frontiers That Confront Rural People and How Are These Related to All of Us?

From our study and experiences, we have become acquainted with many phases of agriculture and rural life. We have learned that farming is a complicated occupation and that it is a way of life as well as a way to earn a living. Furthermore, we have learned that agriculture is important to everyone and to the maintenance of a strong nation. With a background of these and other features of agriculture, we are in a position to consider some of the frontiers that confront rural people and that await further development. The ways in which many of these are developed are important to all of us, whether we live on farms or elsewhere.

Meeting the challenge of an abundance of farm products

In the United States, fewer and fewer farmers are providing an abundance of farm products for more and more people. In fact, farmers are producing surpluses of some products, and the handling of these surpluses must concern everyone. Fortunately, our problem is overabundance rather than scarcity. Yet, even though surpluses provide a margin of safety, we have not learned how to handle surpluses of farm products without penalizing the farmers who produce them.

In facing the frontier of abundance, many people would agree that the following considerations are important.

1. Farm production should be such that serious surpluses do not accumulate, although ample reserves should be maintained for use in case of adverse weather conditions or other emergencies.
2. Farmers should receive prices for their products that make it possible for them to secure a fair share of the national income.
3. Agricultural products should be available to consumers at prices that are fair and reasonable.
4. Soil and other natural resources should be conserved in such a way that any unneeded capacity can be held in reserve for further production as population increases.
5. Farmers should have as much individual freedom as possible in operating their farms.

It will not be easy to find a satisfactory solution to the problem of agricultural surpluses. The time will probably come when population will catch

Fig. 26-3. A honey-comb box designed to reduce damage to grapefruit during test shipments to Japan. (Courtesy, *Agricultural Research*, U.S. Department of Agriculture)

Frontiers of Agriculture

up with production, but that is not likely for many years. In the meantime, it is probable that only some form of government-guided programs providing for a reduction in the total production of crops in serious oversupply can keep surpluses within manageable proportions. It is also probable, because of the manner in which various farm crops compete with each other and serve as substitutes for each other, that some plan involving all farm crops, rather than a commodity-by-commodity approach, will be needed.

Improving methods of production

Great strides have been made in raising crops and livestock on U.S. farms and ranches. It is true that ample quantities are being produced for a rapidly increasing population; but the time has not yet come when we can become complacent about our food supply. It would take only a minor emergency in the international situation or in certain factors affecting crop production to change surpluses into needed supplies, or even to create a deficit food situation. Furthermore, there are many aspects of production, such as uniformity in quality of product, uniformity of size and shape of product, and lowered costs of production, that are still unconquered frontiers.

Considerable progress in breeding better livestock and plants, through the application of genetics, has been made, but much more improvement is

Farm Fuel Use

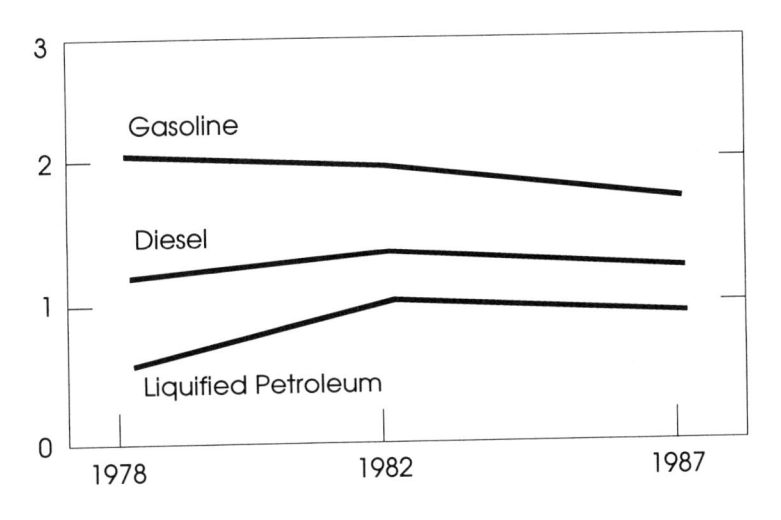

Fig. 26-4. The substitution of diesel fuel for gasoline and the usage of fewer gallons of fuel is a continuing trend. (Courtesy, U.S. Department of Agriculture)

needed. New practices for fertilizing the soil, controlling diseases and harmful insects of plants and animals, feeding livestock, and irrigating the soil are being developed.

Although practical applications of genetic engineering are many years in the future, some of the possibilities are beginning to emerge. It may be possible, for example, to develop crops that fix their own nitrogen just as legumes do and thus reduce fertilizer needs. Fruits and vegetables that are resistant to insects and diseases may soon be available.

Augmentation, adding to the population of beneficial insects already living in an area by releasing more of the same kind grown in "insectaries," is a practice in the United States that shows promise. Other insect control measures being tested by researchers are genetic modification of insects, cultural practices that provide a favorable environment for natural enemies, artificial food and sex attractants, viruses or bacteria, and growth regulators. Since almost 250 insect species now show some resistance to insecticides, new control measures must be found.

Overcoming a shortage of rainfall, by using plastics to cover the soil and by improving water storage in the soil through controlled cracks in the soil and other methods, will make it possible to produce more in the dryer areas. Only a small beginning in research has been made to make better use of the tremendous energy of sunlight in crop production. With these and other practices likely to come in the future, startling opportunities lie ahead for improvement in the production of livestock and crops.

Researchers are studying the possibilities of using the "transfer factor." One lymphocyte, a type of white blood cell, can transfer to another lymphocyte the ability to recognize the danger of a disease organism so that it will start a defense against it. The transfer is done between cells, and the transfer agent is a molecule. Thus far, the cattle transfer factor has been used to treat animals other than cattle, including mice, rabbits, dogs, horses, and monkeys. This new approach to disease control may be useful in treating diseases such as brucellosis of cattle. It may also be a useful approach for controlling human disease. Researchers are also studying the possibility that trees may use a chemical signal to tell other trees nearby when they are under attack. If this is found to be true for all plants, it could signal a whole new approach to protecting plants from insects and disease.

The use of automation in industry, in which machines control their own performance by electronic devices, has become widespread. Some applications of automation are being made in farming, and many others can be expected. Devices for mixing feeds in specified proportions, for watering livestock, and for controlling temperature in farm buildings are a few examples. How atomic power and nuclear energy may be applied to farming in the future is anyone's guess. Certainly, computer technology is a resource that has been applied only in a limited way to the management and operational aspects of food production thus far. Developing new applications of computer

Fig. 26-5. Computers are a very useful tool for the future of agriculture. (Courtesy, M. Thomas)

technology for food production is one of the exciting frontiers awaiting youth today.

Perhaps future scientists may learn the secret of photosynthesis, the process by which plants capture the energy of the sun and store it in the form of starches and sugars. If so, totally new methods of producing food may be developed. Scientists may also find ways to synthesize foods from some materials not now being utilized for this purpose. In many areas in future years, farmers will be challenged by the frontier of new and improved methods of production.

Developing new crops

One of the ways in which agriculture is changing is in the development of new kinds of crops. Extensive research on developing a crop adapted to the United States for producing our own source of rotenone is nearing a break-through. A South American yam, from which the medicine cortisone is derived, may soon become a new U.S. crop.

ARC, a variety of alfalfa that is resistant to anthracnose, the pea aphid, bacterial wilt disease, and the alfalfa weevil, has been developed. An alfalfa plant that is resistant to verticillium, a disease that has the potential for destroying our ability to grow alfalfa, is currently being tested. Thus, these kinds of developments are important in conserving energy as well as in reducing the need for chemicals in insect and disease control. There are thousands of plants over the world that need to be studied so that their possible uses and production in this country can be determined. This is indeed a broad and relatively unlimited frontier of agriculture.

If the costs of producing and securing oil continue to rise, the agricultural industry may have to produce crops that will be used for energy purposes. Experts estimate that about 100 million acres of land would be needed to produce enough crops, such as corn, sugarcane, and sycamore trees, to meet 10 per cent of our national energy needs. Although there are many problems to be solved before this can be done, and although the costs of such energy conversion remain prohibitive, scientists will continue to study this possibility. Researchers have already developed ways in which alcohol can be produced economically from organic waste products, such as timber waste, municipal garbage, and wastepaper, as well as from agricultural surpluses. Experiments conducted in Nebraska have shown that with existing engines, automobiles can run efficiently on a 10 per cent blend of grain alcohol and gasoline. Thus, the day when agriculture will "grow" substitutes for oil is fast approaching.

Using farmland for recreation

Growing new crops is not the only direction in which farmers are turning in order to make profitable use of their land. With the help of government programs, some farmers are turning to recreation as a new cash crop. One 250-acre farm was completely converted into lakes, a golf course, softball fields, riding lanes, and many other kinds of facilities for meetings and games.

This kind of development has been made possible by the increasing number of leisure hours available to the public generally. Turning local farmland into recreational areas will bring many kinds of recreational activities within easy driving distance of many people who would otherwise be unable to enjoy them.

Demand for rural recreation has skyrocketed over the years. Rural recreation provides both opportunities and pitfalls for the farmer who thinks of running a farm recreation enterprise.

Like other farm enterprises, not every attempt to run a recreation business will be successful. The family-sized recreation business is like other part-time operations and must meet the competition or fail.

Most people in the United States (90 per cent according to one study) have simple tastes in outdoor recreation. They prefer sightseeing in a car, picnicking, swimming, fishing, boating, hiking, hunting, camping, and horseback riding. They like to be near water. They prefer rolling woodlands to flat areas, and many of them want and need combinations of natural resources, facilities, and services. The farmer who wants to branch out in the recreation business must provide at least some of these.

Controlling air pollution

According to the Agricultural Research Service of the U.S. Department of Agriculture, air pollution is a growing and serious menace to agriculture.

Air pollutants are costing today's producers of crops, forest and shade trees, and livestock hundreds of millions of dollars a year. As our society becomes more urbanized and mechanized, the impact of these pollutants will multiply in force.

Two of the most damaging pollutants are the smog substances known as ozone and peroxyacetyl nitrate (PAN). Significant crop losses have been caused in recent years along the northeastern seaboard and in the Los Angeles area, although damage has been found near practically every large metropolitan area.

The most obvious damage from air pollutants is to foliage. Other forms of damage may be more serious. Pollutants may reduce photosynthesis, increase respiration, induce early leaf drop, retard growth, and reduce yields. The effects of acid rain on the forests and lakes of the world, as well as on yields of crops, are becoming more serious with each passing year.

The seriousness of the threat of air pollution to our food supply is just now being recognized. A study of this problem is one of the many frontiers in agriculture awaiting the youth of the nation.

Improving marketing and distribution of agricultural products

At present, the United States has surpluses of wheat, cotton, corn, and some other agricultural products, as previously discussed. These surpluses in part are caused by underconsumption by some people in the United States. This is particularly the case with products such as milk, fresh fruits and vegetables, eggs, and meats. Although as a whole, people in this country are eating better than ever before, there are still many people who do not get adequate nutrition.

The problem of fair prices to farmers for the agricultural products they raise continues to be a challenge. In recent years, even when the rest of the economic system has prospered, lowered prices for some important farm products has resulted in lowered returns to farmers. Price supports or stabilization programs in some form help prevent prices to farmers from falling to ruinously low levels. If farmers are to have advantages equal to other groups, effective ways for determining fair prices must be found.

Business and industry are protected to some extent by laws and tariffs, and some industries have tax exemptions to compensate for certain kinds of capital improvements. Steamship companies, airlines, and some other forms of business are subsidized in various ways. Labor groups are protected by minimum wage laws, unemployment compensation, immigration laws, and some forms of guaranteed annual wages. Most of these provisions for non-farm groups are taken for granted. Finding suitable and effective ways for providing farmers with equal advantages is a challenging frontier.

The concept of farm parity came into being during the depression days of the late 1920's and early 1930's. Since the passage of the Agricultural Adjustment Act of 1933, parity has served to provide farmers with a basis for obtaining a fair exchange value for their products. However, the parity concept no longer works well because many changes have taken place in agriculture. Thus, many people are now advocating that a production cost basis be used, with separate calculations being made for different parts of the country. Many problems must be solved before this concept for adjusting farm prices and supports can be used.

We must find ways for improving trade relationships with other coun-

Fig. 26-6. Prompt and accurate reports on prices of farm products will continue to play a major role in marketing. Here, a chart of Florida citrus prices is being prepared. (Courtesy, U.S. Department of Agriculture)

tries so as to exchange some of our surplus agricultural products for products we need. Reciprocal trade agreements and some lowering of tariffs have helped, but developments of these kinds are handicapped by protests from groups in this country who feel they will be injured financially. Some agricultural products, such as coffee, crude rubber, cocoa, bananas, and silk, that cannot be produced in the United States are needed; and so trade is desirable. However, we must find methods for increasing exports of agricultural products from the United States in return for increased imports of non-agricultural materials. Freedom from want in many foreign countries can be obtained in part from increased exports of agricultural products from the United States, but we must be willing to accept more imports in exchange for these products.

Developing new products

Consumers will have more foods like dried fruits, beef sticks, spreadable frostings, and toaster tarts. These foods all have the moisture in them "tied up" with other ingredients so that bacteria do not grow. As a result, these foods do not need to be refrigerated, will keep for a long time, and can be eaten without having water added.

Irradiated foods may be legalized to make it possible for people in the United States to enjoy highly perishable fruits and vegetables from the tropics and other distant places.

Trappers and travelers will turn more to compressed foods, which require less storage space than the traditional packaged foods. Soybean products, such as coagulated soy milk, soyburgers, and meat-like foods, will become more common.

Finding new uses for agricultural products

The science of chemurgy has aided in the consumption of agricultural products by finding new uses for them. Progress in utilizing various agricultural materials once considered as waste also has been made. More than 200 consumer items are derived from corn and about as many from soybeans and cotton. The possibilities are extensive for future development in the uses of agricultural products. Also, uses may be found for additional kinds of plants, which may replace some of the acreages of crops now grown extensively.

Chemurgy is not a complete answer for utilizing surpluses of farm products, but it constitutes a frontier deserving of further development. In the future, new ways probably will be found through chemistry for synthesizing some fibers and foods now produced on farms. These will bring new challenges and the need for further adjustments in farming.

Fig. 26-7. Beef cattle manure is being combined with water in a mixing vat for fermentation and the production of methane gas. (Courtesy, *Agricultural Research*, U.S. Department of Agriculture)

Implementing new developments in irrigation

In addition to supplying moisture, irrigation may help with other problems as well. Some irrigation systems are being used to dispose of liquid wastes from production and processing operations. Fertilizers, herbicides, and insecticides are being spread through irrigation systems.

Plastic pipe is being used to replace open ditches to form a semi-closed irrigation system. By replacing the open main ditches with 2- to 6-inch plastic pipe buried 2 feet deep, a farmer needs almost 40 per cent less water. Also, because so much less water must be transported, the semi-closed system requires fewer pumps, less pumping time, and less energy. Less labor is also required, because it is easier to control the water flow and because the water reaches the plant as soon as the pump is turned on.

Recent undertakings in research to develop food plants resistant to salt indicate that seawater used for irrigation may be possible. At the University of California at Davis, two scientists have made progress toward producing barley plants that can survive with seawater as their source of moisture. Work is also being done to perfect tomato and wheat plants resistant to high concentrations of salt. In the future, it may be possible to grow food for people along the shores of the seas and oceans, with saltwater being used for irrigation and for many of the nutrients needed by plants.

Fig. 26-8. The impressions left by this imprinter-seeder will collect water to give seedlings a better chance in the arid Southwest. (Courtesy, *Agricultural Research*, U.S. Department of Agriculture)

Finding ways to decompose insecticides

Laboratory research may soon find ways to guarantee self-destruction of some currently persistent agricultural chemicals, such as DDT. Such discoveries could make use of very effective chemicals safe for producing food as well as for preserving forests, wild plants, and wild animals.

Making soil surveys from the air

Soil surveys of the future will be made mostly from thermal infrared sensors borne by aircraft or satellites.

Preliminary Agricultural Research Service studies indicate that infrared imagery can detect differences in surface soil temperatures and that these differences may be related to subsurface soil properties influencing the management of agricultural land.

Providing for a rapidly increasing population

In the United States, production of food and fiber has more than kept pace with the increasing population. This will probably continue to be the case for many years.

The main problem at present and for many years to come is to help the developing countries secure greater amounts of food and fiber for their increasing populations. Improved world trade, in which foods and other surplus agricultural products in the United States and some other countries are exchanged for non-agricultural products in countries that need more food and fiber, will help. In addition, people in several developing countries need help in improving their methods of farming.

In the United States, practically all the land suitable for farming is now being used. If ways can be found to provide additional water, some new land areas may be brought into production. However, some acreages are continually being taken out of agricultural production for the expansion of cities, highways, airports and for other non-agricultural uses. In the future, the principal means for increasing the production of agricultural products in the United States will be the improved use of land now being farmed. This has been called a vertical frontier, as compared to the horizontal or geographic frontier of our ancestors who moved on to new land and brought it into production.

More research is needed, as is the increased use of improved methods now known to be effective. Thus, a challenging frontier is the improved use of land now being farmed. This frontier is closely related to the conservation of soil.

Conserving soil and other natural resources

In the United States, considerable progress has been made in the conservation of soil and some other natural resources. Effective soil conservation practices, as discussed in Chapter 7, have been applied to millions of acres. Considerable progress in putting into grass some of the acres subject to water and wind erosion and otherwise poorly suited for cultivated crops has been made. Some attention has been given to reforestation, water control, and wildlife conservation. However, much remains to be done. Much soil is still being lost through wind and water erosion. Water is rapidly becoming a limiting factor in agricultural and industrial development.

We need to develop an increased sense of responsibility for the land, a recognition that we are all affected and that our national security depends on the wise use of our land and other natural resources. We need to use the best methods now known and to find new methods for conserving soil, forests, water, and other natural resources essential for our continued prosperity and security. These comprise a challenging frontier.

Improving rural leadership

There is an increasing need for responsible leadership among rural people. Leaders who are willing to serve the interests and welfare of rural people are needed. At the same time, these leaders must understand the interrelationships of the problems of persons on the land with the welfare of our entire nation. Some leaders who serve rural people are inclined to work for purely political interests. At the grass roots, some desirable leaders are being developed among farmers themselves, and there is need for more of these.

Rural people, through widespread efforts and leadership, must give increasing attention to national policies and programs for agriculture that affect their welfare. This problem will become more difficult as the percentage of people on the land continues to decrease. One of the challenging frontiers for rural youth and other youth is to prepare themselves for roles of leadership so that they will have an understanding of rural and urban groups in our society.

Improving rural-urban relationships

At times, agriculture has had what some people call a poor press. This means that some newspapers and various other publications, radio, and television have at times given urban people distorted ideas about conditions in farming, or they have not given enough attention to the viewpoints of farm people.

This problem of relationships is of special significance to rural people because, from year to year, they are comprising a smaller and smaller percentage of the total population. In 1955, the farm population had fallen to about 13.3 per cent of the total as compared to 30.0 per cent in 1920. This percentage dropped to 1.9 per cent in 1990 The farm population percentage is already less than this in many states.

At no time in history have the rank and file in the United States been better fed than in recent years. They are spending between 16 and 17 per cent of their disposable income for food, and this percentage has remained fairly constant for a long period of years. For each hour of factory labor, more and better food can be bought than in any earlier period.

One government official has stated: "It's a fact that in no other nation today do so few farmers . . . produce so much food and fiber . . . to feed and clothe so many . . . at such a relatively low price."

Thus, it seems important to emphasize that one of the important frontiers in the United States for all people not engaged in farming is to become better acquainted with agriculture and rural life.

Improving rural living

Great strides have been made in equipping farm homes with modern conveniences and other devices of a scientific age. However, many farm homes today are substandard and have few of these comforts.

Improvements have been made in rural schools, but in many places educational facilities are inadequate and better educational opportunities need to be provided. Many rural youth need education to fit them for farming successfully; others need to be educated for nonfarming occupations. Schools and other agencies need to expand the opportunities for adults to continue their education throughout life.

Medical services and hospitals and library facilities need to be improved in many rural communities, although some progress has been made in recent years.

Incomes from the upper one-third of farmers in the United States compare favorably with incomes of urban people. However, many farmers receive incomes that are much too low. Education for improved farming and farm living, improved vocational education for persons who leave farms to secure jobs elsewhere, opportunities to secure loans, and assistance in preparing for and securing off-farm jobs are all needed to help correct the situation.

Farmers and other owners of farmland are carrying more than a fair share of the nation's tax load. Historically, property owners have been considered to be the most wealthy people, with a corresponding ability to pay heavy taxes. With so much of the nation's wealth now being in the form of stocks, bonds, and other similar forms of investments, ownership of property is no longer a fair basis for determining the ability of people to pay taxes.

Increasing numbers of people are establishing homes on small tracts of land on the fringes of cities and along the main highways leading from these cities. These people secure much of their income from nonfarm jobs of various kinds. People who are engaged in small-scale, part-time farming have many opportunities for improved living. However, many of them need help making the best use of land, producing food for home use, and meeting many other problems in this type of farming.

Many rural people have more leisure time than formerly, although in general, the amount of leisure is less than for most urban people. Social activities and various forms of wholesome recreation for youth and adults need to be developed in many rural areas. One of the alarming conditions in rural life is the increase in crimes in some communities.

In many ways, life can be made richer for rural people. The achievement of this kind of life constitutes a challenging frontier for future development.

Making land reforms

Land reform in the United States? Surprisingly, it is something to watch for. The beginnings of support for turning away from the trend toward even larger farms can be seen in actions of some labor unions, environmental protection groups, education associations, farmer organizations, and independent political groups.

The arguments for land reform grow from the problems of an expanding population and the desire for maintaining a healthy and beautiful environment. That owners who work the land and plan to leave it to their children take better care of it is one kind of argument. That large corporations generally are not concerned about soil depletion or land improvement for future generations is another.

There are many people who believe that production would actually be increased by a return to more small-scale farming. Food prices are kept low by the intense competition among many farmers. Increasing the number of small-scale farmers would help assure a continuation of a good supply of low-priced foods.

Bringing back small farms would revive small rural communities and result in a better distribution of the population. This would help eliminate some of the present urban ills.

Returning many people to the land would help preserve our democratic government. Communities in small-farm areas have a larger middle class than other communities, a more stable income pattern, better schools, and more active civic groups. Absentee landlords are generally not interested in raising land taxes for civic purposes. Farming has long been the stronghold of the free enterprise system of small, independent businesspersons who pride themselves on their freedom and work.

Many changes will be needed to make small-scale farming profitable and a good way of life.

1. Tax laws that permit the use of farming for speculation and for tax-loss purposes must be changed.
2. Higher minimum wages set by law for farm laborers would favor the self-employed farmer who uses mostly family labor.
3. Farm subsidy programs need to be changed to favor the small-scale operator by limiting payments to a certain acreage.
4. New zoning laws, other tax changes, moratoriums on land reclamation, and other actions would be needed to effect a return to small-scale farming.

While strong land reform seems still a long way off, the ever-increasing population pressures, with the problems created by these pressures, make future governmental action to achieve land reform more and more likely.

Improving our democratic way of life

In the United States, we are fortunate to be members of a great democracy. Farmers are often called the cornerstone of our democracy, and without doubt, they are a major stabilizing force for our democratic way of life. The real strength of our nation and of our democratic way of life lies close to the soil and in no small degree depends on an abundance of products raised by farmers.

U.S. farmers are the most efficient producers of food and fiber in the world, and our future welfare depends on helping them to secure a satisfying living. The family farm prevails in the United States, and there is good reason to believe that this accounts for much of the success in U.S. agriculture.

Farmers can prosper only if the entire nation prospers. Therefore, it is important to maintain full employment and full production in all phases of U.S. industry and business. In the long run, the success of our democracy depends on everyone, farmers and others, working for the best interests of all. This is one of our most important frontiers. Elmer Davis, a noted journalist, once said, "Our future will be what we are strong enough and resolute enough and intelligent enough to make it."

Young people are asking, "Where do we fit into this changing world?" In closing, it is well to re-emphasize that young people have a place in rural life and in the nation at large. As Eleanor Roosevelt said:

> Young people are going to be valuable to their community in proportion to the time and trouble they take to study the questions which face the community. . . . Let us not hang back from new things. Cling to the old things that are good, but remember that we live in a world which must go forward. We cannot stand still, so let young people take an interest and study every new thing that comes their way. Let us encourage them not to shy away from anything because it is new.

New frontiers await the on-coming generation. Young people should move forward with confidence in building a richer and better life for the rural United States and for the entire nation. This is the greatest challenge and the most important frontier of all.

SUGGESTED ACTIVITY

1. Concepts for discussion:
   a. Farmers are substituting capital for labor to increase efficiency of production.
   b. We have not yet learned how to benefit most from agricultural food surplus production.

Frontiers of Agriculture

c. It would take only a minor emergency to change our surplus food situation into a food deficit situation.
d. The search for ways to increase food production is a continuing one.
e. Farming will become even more mechanized in the future than it is today.
f. There are thousands of plants yet to be studied so that their agricultural uses can be determined.
g. Improved productivity in agriculture has made possible the use of land for recreational purposes.
h. There is still much to be discovered in all aspects of the agricultural industry.
i. Food can be produced synthetically.
j. There is always a need for good leadership in agriculture.
k. Rural areas provide an escape from crowded urban living.
l. Land is a limited resource in which the general public has a large stake.

REFERENCES

Lee, J. S., and D. L. Turner. *Introduction to World AgriScience and Technology.* Interstate Publishers, Inc., Danville, IL 61832.

Osborne, E. W. *Biological Science Applications in Agriculture.* Interstate Publishers, Inc., Danville, IL 61832.

U.S. Department of Agriculture. *Agricultural Research, Farm Index,* and *Soil Conservation,* monthly publications. U.S. Government Printing Office, Washington, DC 20402.

U.S. Department of Agriculture. *American in Agriculture: Portraits of Diversity.* Yearbook of Agriculture: 1990, U.S. Government Printing Office, Washington, DC 20402.

U.S. Department of Agriculture. *Food—From Farm to Table.* Yearbook of Agriculture: 1982, U.S. Government Printing Office, Washington, DC 20402.

U.S. Department of Agriculture. *New Crops, New Uses, New Markets.* Yearbook of Agriculture: 1992, U.S. Government Printing Office, Washington, DC 20402.

U.S. Department of Agriculture. *Our American Land.* Yearbook of Agriculture: 1987, U.S. Government Printing Office, Washington, DC 20402.

Index

A

Aberdeen Angus, 294
Accidents, farm, 50
Acid rain, 202
Acres of land, in 50 states, 13
Acres of land owned, federal government, 13
Activities, group, 87
Agricultural Cooperative Service, 616-617
Agricultural industry
    assets, 1-2
    changes, 3-11
    number of workers, 2
Agricultural Marketing Service, 58
Agricultural occupations, 101-137
    advantages and disadvantages of, 124
    chart, 112-113
    choosing, 114-115
    computer technology in, 114
    education and training required in, 120-124
    finding a job in, 124-126
    flexibility in, 121, 123
    getting established in, 126-128
    in agricultural services, 107
    in business and industry, 105
    in ecology, 110
    in farming, 102
    in production agriculture, 104-105
    in professions in agriculture, 109-110
    job questionnaires, 117-119
    major groups, 103
    opportunities, 120
    qualifications, 119-120
    rewards
        financial, 124
        nonfinancial, 116
    successful careers in, 128-136
Agricultural Research Service, 60
Agriculture
    census, 11
    definition of, 1
    frontiers, 669-687
    state departments of, 60
    study of, 25-27, 30
Air pollution, 642
Air pollution control, 649-650
Alfalfa
    origin of, 467
    plant features of, 451
Anemia, 422, 425
American Saddlebred Horse, 303, 304
Animal and Plant Health Inspection Service, 59, 404
Animals, farm (See Farm animals)
Annuals, 447
Anther, 444
Anthrax, 402
Appaloosa, 304
Apples, origin of, 472
Artificial insemination, 335
Ayrshire, 288

B

Bacteria, 406
Bacterial wilt, 536
Bailey, Liberty Hyde, 131
Barley, origin of, 470
Beef cattle
    associations, 314
    branding, 345
    breeds, 291-297
    caring for, 393-397
    feeding, 367-370
    improving, 344-346
    selecting, 345
Beef, retail cuts of, 557
Belgian, 302, 303
Berkshire, 299
Bermuda grass, 142, 144
Biennials, 447
Biological control(s), 647
Blackberries, 189
Blackleg, 416
Bloat, 411
Bot, 425
Boy Scouts, 93
Brahman, 294
Branding, 395
Breeding, of plants, 472-474
Breed registry associations, 314-317
Brown Swiss, 288, 291
Brucellosis, 413, 421
Building arrangement, 154
Building mechanization, 253-254
Buildings and equipment
    for hogs, 390-391
    for livestock, 376-381
    planning for, 594
Burbank, Luther, 475-476
Businesses, types of, 620-622

C

Camp Fire, Inc., 94
Carbohydrates, 354
Carpet grass, 142
Carver, George Washington, 132-133
Census, agriculture, 11
Centipede grass, 142
Charolais, 294
Chemical pollution, control of, 650-651
Chemurgy, 8, 581
Chester White, 299
Cheviot, 305
Chickens
    breeds of, 310-312
    caring for, 386-390
        chicks, 386
        laying flock, 387
    classes of, 310
    controlling diseases and parasites in, 416-419
    culling, 342
    egg quality of, 389-390
    feeding of, 362-365
    improving, 341-342
Chilling, 424
Chlorophyll, 443
Cholera, 419
Christmas tree grades, 579

Chromosomes, 326-332
Churches, 64, 95
Citrus fruits, 196
Classification, of farm animals, 275
Classification, of farm crops, 447-449
Clovers, origin of, 471
Clovers, plant features of, 451
Clydesdale, 302, 303
Coccidiosis, 416
Columbia, 305
Commodity Credit Corporation (CCC), 58
Communications, 4, 65
Community improvement, 61-67
Computer programs, 114, 566, 567
Concentrates, 358
Conservation
    and land use, 228-230
    and soil loss, 205
    definition of, 203
    of forests and woodlots, 218-221
    of soil, 210-213
    of water, 213-218
    of wetlands, 221-222
    of wildlife, 222-228
    organizations, 232-233
    tillage, 212
    value of, 203-205
Contouring, 211
Cooperatives
    activities of, 622-628
    and other types of businesses, 620-622
    definition of, 617
    development of, 614-616
    examples of, 628-633
        cotton, 631
        dairy, 629
        electrical, 633
        financial, 632
        fruit, 630
        grain, 628
        livestock, 630
    marketing of, 623
    operation of, 617-618
    organizing, 618-620
    purchasing (consumer), 623, 631-632
    service of, 624
    types and number of, 625
    volume of business of, 624-628
Cooperative State Research Service, 60
Corn
    cost of production, 14
    cultivating, 529
    harvesting and storing, 530
    hybrid seed, 526
    importance of, 433
    irrigation of, 530
    origin of, 468-470
    plant features of, 449
    planting, 528
    seedbed, 526
    soil for, 525
Corriedale, 305
Cotswold, 308
Cotton, origin of, 472
Cover crops, 212
Crops, farm (See Farm crops)
Cropland, changes in, 10
Cropping program, 507
Cropping system, 592
Crossbred, 287
Crossbreeding, 334, 476-482
Currants, 191

D

Dairy cattle
    associations, 314
    barns, 383
    breeds, 287-291
    calves, 385
    caring for, 381-386
    feeding, 360-362
    improving, 338-341
    labor, 385
    linear classification of, 341
    production, 339
    product quality, 384-385
    selecting, 338
    sire, 339
    type, 338
Dairy cow, parts of, 276
Dallisgrass, 483
Darwin, Charles, 461
Desert landscaping, 156
Diet, 167
Digestible nutrients, 358
Digestion of feeds, 282-283, 357-358

Disease(s)
    causes of, 405-407
    control of, 59, 410-425
    definition of, 405
    economic losses from, 401-402
    in gardens, 184
    in relation to human health, 401-403
    of fruit trees, 195-196
    prevention of, 408-410
Domestication, of animals, 319-323
Dormancy, 446
Dorset, 307
Dragonfly, 647
Drug problem, 81
Duroc, 299

E

Earning a living, 76
Earning money, 79
Economic conditions, changes in, 8
Economic Research Service, 60
Education, 60
Egg Products Inspection Act, 404
Electrification, 251-253
Electronic marketing, 564-568
Employment, in agriculture, 101
Employment, in United States, 101
Energy use, 595
Enterotoxemia, 416
Environmental protection, 637-654
Equipment, farm
    operation, 256
    photo quiz, 261-269
    selection, 255
    service and repair, 256
Erosion
    control, 213, 517
    prevention, 210-213
    signs of, 207
    water, 207
    wind, 209-210
Erysipelas, 402, 421
Extension Service, 60

Index

F

Fall color, of trees, 431
Farm animals
    as beasts of burden, 272
    as clothing source, 272
    as food source, 273-274
    breeds of, 286-312
    by-products of, 274
    caring for, 375-400
    classification of, 275
    domestication of, 319-323
    feeding of, 351-373
    health of, 401-426
    importance of, 279
    improvement of, 319-350
    intelligence of, 281-282
    stomach capacity of, 282
    traits of, 281-286
    use of feed for, 352-357
Farm Bureau, 53
Farm Credit Administration, 58
Farm crops
    and climatic adaptation, 465
    and disease resistance, 464
    and insect resistance, 464
    breeding of, 472-474
    characteristics of, 427, 459
    classification of, 447-449
    factors affecting, 440-441
    forage, cultural practices for, 538-544
    hunger signs in, 503
    importance of improvement, 462-464
    improving, 461-487
    origin of, 466-472, 474-475
    planting and cultivating, 523-549
    principal, 432-440
        fiber, 438-439
        forage, 435-436
        grain, 432-435
        oil, 440
        roots and tubers, 437-438
        sugar, 439-440
        wood, 440
    production of, 5
    quality of, 464
    rotation adaptation of, 466
    seed selection for, 484-486
    strains of, 475
Farmers Home Administration, 58
Farmers' Union, 53-54
Farm homes, 4

Farming
    animal-power era, 236
    changes in, 3-11
    definition of, 1
    electrification, 251, 252
    factors in success of, 22-24
    hand-power age, 236
    kinds of, 15-17
    mechanical-power era, 238
    mechanization of, 235-269
    nature of, 11
    occupation of, 14
    operating equipment for, 255-257
    operating expenses in, 3
    production assets in, 13
    regions, 19-20
    small-scale, 54
Farming efficiency, 605-606
Farm plan, 597
Farm population, 37-39
Farms
    definition of, 20
    investment in, 13-14
    kinds of, 15-20
    number of, 6
    size of, 6, 20, 21
Farm shop, 256-257
Fats, 354
Feeding
    beef cattle, 367-370
    chickens, 362-365
    dairy cattle, 360-362
    hogs, 365-367
    horses, 371-372
    livestock, 351-373
    sheep, 370-371
Feeds
    digestion of, 357-358
    nutrients in, 353-357
    selecting, 358
Fertilizer(s)
    analysis, 512
    applying, 513-514
    buying, 512-513
    commercial, 510
    program, 592-593
    ratio, 512
Fescues, 142
FFA, 30, 90, 92-93
Fiber crops, 438-439
Fife, David, 475
Financing the farm business, 601-603
Fire ant, 647

Fires, 51
Flower clock, 154
Flowers, 153, 154
Foliar feeding, 502
Food(s)
    and proper nutrition, 45
    assistance programs, 59
    Basic Four, 46-47
    Basic Seven, 167
    consumption, 556
    costs, 556
    preservation, 196-198
Food Safety and Inspection Service, 403
Footrot, 416
Forage crops
    fertilizing, 540
    harvesting and storing, 540-542
    in pastures, 542, 544
    kinds of, 435-437
    preparing seedbed for 540
Foreign Agricultural Service, 59
Forests and woodlots
    conserving, 219-220
    establishing, 220-221
Forest Service, 58
4-H Clubs, 90
Fowl cholera, 418
Fowl pox, 418
Frontiers of agriculture
    challenge of abundance, 671-673
    conserving soil, 682
    controlling air pollution, 677
    farmland for recreation, 676-677
    insecticide decomposition, 681
    irrigation, 680
    land reform, 685
    marketing, 677-679
    methods of production, 673-675
    new crops, 675-676
    new products, 679
    rural leadership, 683
    rural living, 684
    rural-urban relations, 683
    soil surveys, 681
    world food needs, 682
Fruit production, 454
Fruits, storing, 196-197
Fruit trees

controlling insects and
  diseases, 195-196
cultivating, 195
planting, 194
pruning, 195
selecting kinds and
  varieties, 194
Future Homemakers of America
  (FHA), 93
Futures market, 571
Futures trading, 59

G

Gardening
  amount to grow in, 180
  citrus trees in, 196
  controlling insects and
    diseases in, 184-185
  cultivating, 183
  equipment, 173
  fertilizing, 179-181
  harvesting, 185,186, 187
  location of, 171
  planning, 173, 177, 179
  planting, 182
  preparing seedbed for,
    179-181
  raising currants and
    gooseberries, 191-192
  raising grapes, 192-193
  raising raspberries and
    blackberries, 189-191
  raising strawberries,
    187-189
  raising tree fruits, 193, 196
  storing vegetables and
    fruits, 196-198
  varieties of vegetables,
    175-177
  vegetables to grow, 172,
    175
Gastroenteritis, 421
Genes, 330-332
  dominant, 330
  recessive, 330
Germination, 442
Girl Scouts, 94
Goals, management, 590
Goats, associations of, 317
Goitre, 405
Gooseberries, 191
Government, federal, 67
Government in agriculture, 56-60

Government policies, marketing,
  573-577
Governments, local, 66-67
Grade, 286
Grading farm products, 58-59
Grading labels, 578
Grains, small
  cultural practices for,
    531-535
  plant features of, 450-451
Grange, 52
Grapes, 192-193
Grasses
  origin of, 471
  plant features of, 451-452
Grass waterways, 211
Grimm, Wendelin, 475
Ground covers
  kinds of, 146
  planting, 146
Growth stimulants, 357
Guard dogs, 397
Guernsey, 288

H

Hackney Pony, 303
Hampshire, 299, 307
Hardware disease, 414
Hay, standards, 359
Health, animal, 59, 401-426
  Animal and Plant Health In-
    spection Service, 59, 404
  importance of, 401-403
  maintaining, 408-425
Health facilities, 64
Heaves, 425
Heiser, Leslie, 133-134
Hemp, origin of, 472
Heredity, livestock, 325-332
Hereford, 293-294
Hobbies, 162-163
Hog cholera, 419
Hogs (See Swine)
Holstein-Friesian, 287
Home life, 88
Horse Protection Act, 404
Horses
  associations of, 315-316
  breeds of, 302-304
  caring for, 397-399

  controlling diseases and
    parasites in, 424-425
  draft, 302-303
  feeding, 371-372
  improving, 349
  light, 303-304
  parts of, 278
  selecting, 348-349
  training, 398-399
Horticultural specialty crops, 457
Humus, 492
Hunger signs, 503
Hybridization, 476-482
Hydrologic cycle, 215-216
Hydroponics, 493-494

I

Immunization, 409
Inbreeding, 334
Income, wise use of, 47
Influenza, 419, 424
Insecticides, 639
Insects
  injury to lawns, 147
Insurance, crop, 59-60
Insurance program, 597
International Farm Youth
  Exchange Program, 97
Intestinal worms, 419
Investment
  per farm, 13
  per farm worker, 13
Iron deficiency, 503
Irrigation, 162, 515-517

J

Jersey, 287-288

K

Kentucky bluegrass, 142, 452
Kidneyworms, swine, 422
Kinyon, Mary 134
Kline, Allan B., 134

L

Labor
  efficiency, 594

Index

hand, 239
hours per unit, 241, 242
migrant, 11
Land-grant colleges, 57
Land pollution, 644
Land pollution control, 650
Land, reclaiming, 213
Landscaping
  building arrangements, 154
  desert, 156
  drives and walkways, 148
  flowers, 153
  ground covers, 146
  guidelines, 156-157
  hobbies, 162-163
  irrigation, 162
  lawns (See Lawns)
  pest control, 162
  planning, 139, 141
  plant arrangements in, 148-150
  pruning and caring for trees and shrubs in, 159-161
  purposes for plants in, 140
  selecting trees and shrubs in, 151-152
  transplanting trees and shrubs in, 157-159
  windbreaks, 155
Land use, conservation, 228-230
Land-use planning, 228-230
Lawns
  controlling insects and diseases of, 147-148
  fertilizing, 144-146
  improvement of, 141-146
  kinds of grasses in, 141-143
  mowing, 146
  planting, 143
  rolling, 146
  seeding, 143-144
  shady, 146
  sprigging or plugging, 144
  watering, 145, 146
  weed control of, 145, 148
Layering, 190, 193
Leases and agreements, 596-597
Legal problems, 608
Leghorn(s), 311, 312
Legumes, 431
Leicester, 308
Library service, 62-63
Lice, 419, 421
Life cycle, plants, 442

Lime, 509, 511
Lincoln, 308
Little Country Theatre, 50
Livestock
  appearance (type), 336
  artificial insemination in, 335
  breed registry associations of, 314-317
  buildings and equipment, 376-381
  caring for, 375-400
  characteristics, 286-312
  crossbred, 287
  domestication, 319-323
  feeding, 351-373
  grade, 286
  growth stimulants, 357
  identification, 395
  improving, 319-350
  leading states, 279-281
  nutrients in feed, 353-357
    carbohydrates, 354
    fats, 354
    minerals, 353-355
    proteins, 354
    vitamins, 356-357
    water, 356
  pedigree, 287
  production, 5, 279-281
  production levels, 332-333
  program, 592
  purebreds, 286
  registration certificate, 287
  salt, 355
  scrub, 287
  selection, 336-338
  use of feed for, 352-357
Lohman, Andrew G., 131-132
Lumber production, 457

M

Magnesium deficiency, 503
Maintenance ration, 353
Management, farm, 585-611
  buildings and equipment, 594
  cropping system, 592
  efficiency, 605-606
  energy, 595
  fertilizer program, 592-593
  financing, 601-603
  goals, 590
  income, 586, 588
  insurance, 597
  labor, 594

  leases and agreements, 596-597
  legal problems, 608
  livestock program, 592
  marketing program, 594
  operations, 598-599
  planning, 589-590
  purchasing, 595-596
  records, 603-606
  small-scale farming, 608-609
  surveying resources, 591-592
Mange, 421
Marketing
  consumer protection, 577-579
  cooperatives, 628-631
  cost components of, 553
  definition of, 552
  eating trends, 573
  electronic, 564
  exports, 575-576
  farm value, 554
  government policies, 573, 577
  grades, 578, 579
  importance, 552-557
  improvement in, 580-582
  laws and orders, 58-59
  plan implementation, 597
  program, 594
  quality, 573
  steps in
    assembling, 559
    buying and selling, 563
    financing, 568
    grading, 561
    packaging, 563
    processing, 562
    storing, 563
    transporting, 560
  supply and demand, 569-573
Mastitis, 414
McLean County System of Swine Sanitation, 421-422
Mechanization of farming, 235-269
  buildings, 253-254
  effects of, 238-243
  electrification, 251-253
  equipment, general, 244
  farm shop, 256-257
  future of, 257-259
  harvesting equipment, 248-251
  operating equipment, 256

planting equipment, 247-248
safety, 257
selecting equipment, 255
servicing and repairing equipment, 256
tillage equipment, 244-246
Melon production, 454
Mendel, Gregor, 474
Merino, 308, 310
American, 308, 310
Delaine, 308, 310
Microorganisms, 492
Migrant labor, 11
Milk fever, 412
Milk, how manufactured, 283-284
Milking, 382-383
Minerals, 354-355
Mites, 419
Montadale, 307
Morrow Plots, 490
Muck, 491
Mustangs (broncos), 304
Mutations, 329

N

Nail, 236
National Agricultural Library, 60
National Farmers Organization (NFO), 54
National Herb Garden, 29
Nature Conservancy, The, 221
Nematodes, 633
Newcastle disease, 417
New Hampshires, 311
Nitrogen, 511, 513
Nitrogen deficiency, 503
Nut production, 456-457
Nutrients, in feeds, 353-357

O

Oats, origin of, 471
Occupations (See Agricultural occupations)
Oil crops, 440
Orange, navel, origin of, 467
Organic matter, 507-508, 511
Organizations
establishing, 98
for older youth, 95
for rural youth, 89
rural, 52-54
Ornamental plants, toxic, 405
Ornithosis, 402
Ovary, 444
Ovules, 444
Ox warbles, 414, 415

P

Parasites
as cause of disease, 406-407
in cattle, 414
in chickens, 419
in hogs, 421-422
in horses, 425
in sheep, 423-424
Parity, 575-577
Part-time farming, 608-609
Pasture, 359, 542
Peanuts, origin of, 472
Pedigree, livestock, 287, 336
Percheron, 302, 303
Perennials, 447
Performance (production), 337
Personal problems, 82
Pesticides, 639
Petals, 444
Pfister, Lester, 129
Phosphorus, 511, 512, 513
Phosphorus deficiency, 503
Photosensitization, 416
Photosynthesis, 428, 443
Pinkeye, 412
Pistil, 444
Plant Kingdom, 447
Plant nutrients, 501-502
Plants
as food source, 429-430
as shelter and clothing source, 430
breeding of, 472-482
classification of, 447-449
cooling effect of, 431-432
elements for growth of, 443
fall color of, 431
growth of, 442-444
life cycle of, 442
reproduction of, 444-446
root systems of, 444
soil fertility of, 431
species of, 441
toxic, 405
uses of, 429-432
water requirements of, 444
Plant Variety Protection Act, 486
Plymouth Rocks, 311
Pneumonia, 416
Poisoning, 424
Poland China, 299
Pollen, 444
Pollination, 445
Pollution
agricultural, 638-640
air, 642
and wildlife, 641
combating, 644-652
Federal Water Pollution Control Act, 645
land, 644
Refuse Act, 645
water, 217-218, 642-644
Ponds, farm, 212
Population
farm, 6
of United States, 7
rural, 37-39
Potash, 511, 513
Potassium deficiency, 503
Potatoes
cultural practices for, 544-546
origin of, 467
plant features of, 453-454
Poultry (See Chickens)
Price supports, 59
Production, of farm enterprises, leading states in, 17-19
Progeny, 337
Proteins, 354
Proved sire, 337
Prunes, origin of, 467-468
Pruning
berries, 191
currants and gooseberries, 191
fruit trees, 195
grapes, 193
trees and shrubs, 159-161
Pseudorabies, 422
Psittacosis, 402
Pullorum disease, 417
Purchasing program, 595-596

Index

Purebreds, 286-287
Pyrethrum, 646

Q

Quarter Horse, 303, 304

R

Rabies, 402
Rambouillet, 310
Raspberries, 189
Rations, 353
Records, farm business, 603-606
Recreation, 49, 84
Red Angus, 294
Red Poll, 294
Registration certificate, 287
Relationships
    national, 70
    rural, 8
    rural-urban, 69
Reproduction, plants, 444-446
Research, on plant
    improvement, 482-484
Resources, farm, 591
Retirement, security for, 48
Rhinitis, 421
Rhode Island Reds, 311
Rice, origin of, 471
Rickets, 405
Roads, 65
Rocky Mountain Spotted Fever, 403
Rodents, 647
Romney, 308
Roots and tubers, 437-438
Root systems, 444
Roughages, 357
Roundworms, intestinal, 421-422
Ruminants, 281
Rural Development Act, 68
Rural Electrification
    Administration (REA), 58, 616
Rural life, in United States, 37-74
    churches, 64-65
    communication, 65
    family life, 41-44
    government, federal, 56-60, 67
    governments, local, 66-67
    health facilities, 64
    homes, 41-43
    library service, 62-64
    nutrition, 45-47
    organizations, 52-54
    population, 37-39
    preventing accidents, 50-51
    preventing fires, 51-52
    recreation and hobbies, 49-50
    schools, 61-62
    security for retirement, 48
    taxes, 70
Rural living, and youth, 75
Ruritan, 54, 95
Rye, origin of, 471

S

Safety, equipment operation, 255-257
St. Augustine grass, 142
Salt, for livestock, 355
Santa Gertrudis, 294
Savell, James M., 130
Sayre, Mrs. Raymond, 130
Schools, improving, 61-62
School, staying in, 76
Scrub, 287
Seed, 442
Sepals, 444
Septic sore throat, 402
Sex determination, livestock, 327
Sheep
    associations of, 316
    breeds of, 304-310
    caring for, 393-397
    controlling diseases and parasites in, 423-424
    feeding, 370-371
    improving, 346-348
    parts of, 277
    providing protection from predators of, 397
    selecting, 346-347
Shetland Pony, 303, 304
Shire, 302, 303
Shorthorn, 291
Shropshire, 308
Small-scale business
    choosing, 655-659
    examples of, 661-666
        beekeeping, 664
        Christmas tree farm, 663
        greenhouse, 662
        home gardening, 661
        nursery, 663
        retail flower shop, 662
        small-engine repair, 665
    values of, 666
Small-scale farming, 608-609
Social Security, 48
Soil
    applying lime to, 509
    cropping program, 507
    deficiencies, 503
    description of, 491-493
    erosion, 205
    fertility, 489-491, 501-502
    fertility loss, 503-504
    fertility maintenance, 504-519
    formation, 491-493, 494-497
    land capability classification, 500-501
    needs, 505-506
    providing organic matter for, 507-508
    testing, 506
    tilling, 518-519
    types, 497-501
    water, 216-217, 515
Soil conservation districts, 230
Soil Conservation Service, 59, 68
Sorghum(s)
    cultural practices for, 547-548
    kinds of, 435
Southdown, 308
Soybeans
    cultural practices for, 535-538
    harvesting and storing of, 538
    origin of, 471
    planting of, 536
    production of, 434
    seed for, 536
Spotted, 299
Stamens, 444
Standardbred (American Trotter), 303
Statistical Reporting Service, 60
Stomach worms, 423
Strawberries, 187-189
Streams and rivers, 217
Strip cropping, 211

Suffolk, 308
Sugar crops, 439-440
Supply and demand, 569-573
Swine
    associations, 315
    breeds, 298-302
    caring for, 390-393
    controlling diseases and
        parasites of, 419-422
    feeding, 365-367
    handling, 391
    improving, 342-344
    parts of, 277
    selecting, 342-344

T

Tamworth, 300
Tapeworm, 403
Tariffs, 573
Taxes, 70
Tennessee Valley Authority, 230
Tennessee Walking Horse, 303, 304
Terraces, 212
Teton Dam, 205
Texas cattle fever, 407
Thoroughbred, 303
Thrush, 424-425
Tick, Gulf Coast, 414
Ticks, 424
Tillage, 518-519
Tillage, conservation, 212
Timothy, 451
Tobacco, origin of, 471, 472
Tomatoes, origin of, 472
Transplanting, of vegetables, 182-183
Transportation, 4

Trees and shrubs
    selecting, 151-152
    transplanting, 157-159
Trichinosis, 403
Tuberculosis, 403, 413, 417-418
Tularemia, 402
Turkey, wild, 320

U

Undulant fever, 402
U S. Department of Agriculture, 57, 59, 403-405

V

Vegetables
    production of, 454-456
    storing, 196-197
    varieties of, 175, 176-177
Venezuelan equine
    encephalomyelitis (VEE), 424
Vesicular stomatitis, 416
Veterinary Services, 404-405
Vibriosis, 416
Virus-Serum-Toxin Act, 404
Vitamins, 356-357

W

Water
    conservation, 213-218
    controlling pollution of, 217-218, 650
    controlling streams and rivers of, 217
    hydrologic cycle, 215-216
    pollution, 642-644
    soil, 216-217, 515
    use, 215

Watermelon, origin of, 472
Weeds
    poisonous, 405
Welsh Pony, 303
Wetlands, conservation of, 221-222
Wheat, 435
    origin of, 470
Wildlife
    and pollution, 641
    conservation, 222-228
    endangered species, 222, 225
    extinct species, 222
    harmful species, 226-227
Windbreak, 155
Wood crops, 440
Worms, 425

Y

Yorkshire, 300
Youth
    and drugs, 81
    and earning a living, 76
    and group activities, 87
    and home life, 88-89
    and recreation, 84-87
    and solving personal
        problems, 82-84
    and staying in school, 76-79
    centers, 88
    exchange programs, 97
    needs and problems, 75
    organizations, 89-99

Z

Zoning, 60
Zoysia grass, 142, 144